POLLUTION PREVENTION

The Waste Management Approach for the 21st Century

POLLUTION PREVENTION

The Waste Management Approach for the 21st Century

by

R. Ryan Dupont
Louis Theodore
Kumar Ganesan

LEWIS PUBLISHERS
Boca Raton London New York Washington, D.C.

Library of Congress Cataloging-in-Publication Data

Dupont, R. Ryan.
  Pollution prevention : the waste management approach for the 21st century / R. Ryan Dupont, Louis Theodore, Kumar Ganesan.
    p. cm.
  Includes bibliographical references and index.
  ISBN 1-56670-495-2 (alk. paper)
   1. Pollution prevention—Case studies. 2. Factory and trade waste—Management—Case studies. 3. Industries—Energy conservation. 4. Industrial safety. I. Theodore, Louis. II. Ganesan, Kumar, 1945– III. Title.
  TD897.D87 1999
  363.73'7—dc21   99-047203
                                                                                    CIP

This book contains information obtained from authentic and highly regarded sources. Reprinted material is quoted with permission, and sources are indicated. A wide variety of references are listed. Reasonable efforts have been made to publish reliable data and information, but the authors and the publisher cannot assume responsibility for the validity of all materials or for the consequences of their use.

Neither this book nor any part may be reproduced or transmitted in any form or by any means, electronic or mechanical, including photocopying, microfilming, and recording, or by any information storage or retrieval system, without prior permission in writing from the publisher.

All rights reserved. Authorization to photocopy items for internal or personal use, or the personal or internal use of specific clients, may be granted by CRC Press LLC, provided that $.50 per page photocopied is paid directly to Copyright Clearance Center, 222 Rosewood Drive, Danvers, MA 01923 USA. The fee code for users of the Transactional Reporting Service is ISBN 1-56670-495-2/00/$0.00+$.50. The fee is subject to change without notice. For organizations that have been granted a photocopy license by the CCC, a separate system of payment has been arranged.

The consent of CRC Press LLC does not extend to copying for general distribution, for promotion, for creating new works, or for resale. Specific permission must be obtained in writing from CRC Press LLC for such copying.

Direct all inquiries to CRC Press LLC, 2000 N.W. Corporate Blvd., Boca Raton, Florida 33431.

Trademark Notice: Product or corporate names may be trademarks or registered trademarks, and are used only for identification and explanation, without intent to infringe.

Visit the CRC Press Web site at www.crcpress.com

© 2000 by CRC Press LLC
Lewis Publishers is an imprint of CRC Press LLC

No claim to original U.S. Government works
International Standard Book Number 1-56670-495-2
Library of Congress Card Number 99-047203
Printed in the United States of America   4 5 6 7 8 9 0
Printed on acid-free paper

PREFACE

> Beginning is the most important part of the work.
> Plato (427-327 B.C.), *The Republic*, Book I

The technical community, particularly the engineering profession, has expanded its responsibilities to society to include the management of wastes, with particular emphasis on control by pollution prevention. The term pollution prevention, in this text, is defined as that process or operation that attempts to reduce or eliminate the generation of wastes and/or pollutants that are emitted into the environment. Unfortunately, the term pollution prevention has come to mean different things to different people. Other terms - including waste minimization (a term that the EPA has discouraged the use of because it focuses on hazardous wastes only), waste reduction, source reduction, and pollution minimization (a common term in industry) - have come, in a very general sense, to be used interchangeably with pollution prevention. However, irrespective of the term employed, the main focus of environmental management today and in the future will be to reduce or eliminate waste streams entering the environment.

The reader should also note that there are two other areas of pollution prevention that need to be included in any environmental management analysis. The first of these areas is energy conservation (EC). EC programs have resulted in cost-saving measures that have directly reduced waste production. Thus, energy conservation is directly related to pollution prevention since a reduction in energy consumption corresponds to less energy demand and, consequently, less pollutant output. The second additional area of concern is that of health, safety, and accident prevention (HS&AP). Accidents like Chernobyl and Bhopal have increased both public and industry awareness of the potential dangers of the mismanagement of hazardous chemicals, and have helped stimulate regulatory policies concerning emergency planning, emergency response, etc. HS&AP issues have directly contributed to the development of OSHA and EPA regulations regarding the handling, storage, and use of hazardous chemicals, have led to reductions in the release of hazardous chemicals to the environment, and are therefore directly related to pollution prevention. Further, these areas are also inter-related. A pictorial representation of these relationships is provided below. Although both the EC and HS&AP issues are addressed in this book, the main focus of this text is on waste reduction.

Increasing numbers of engineers, technicians, and maintenance personnel are being confronted with problems in this most important area. Since a preventative approach to environmental management is a relatively new concept, the environmental engineers of today and tomorrow must develop a proficiency and an improved understanding of not only waste treatment but also pollution prevention techniques in order to cope with these challenges and changes ahead. Although this is not the first professional book to treat this

Preface

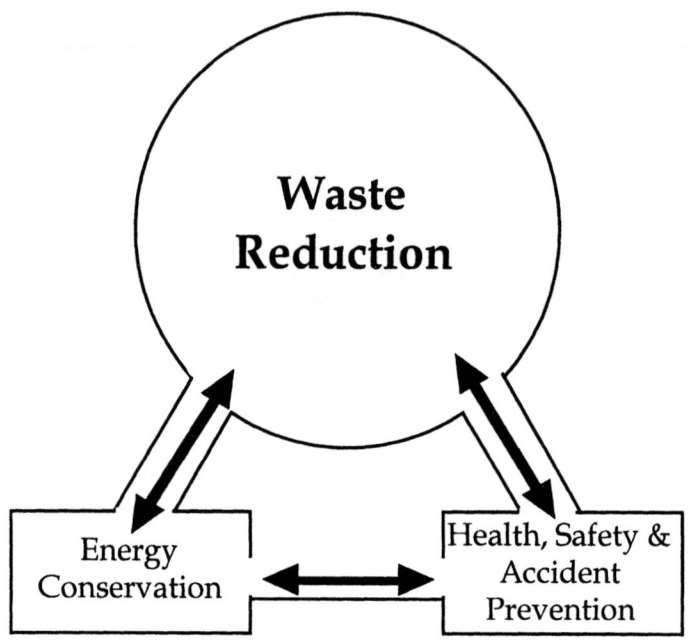

Idealized Pollution Prevention Relationship

particular subject, it is one of the few books dealing with the technical and engineering aspects of pollution prevention that may be used as a textbook.

This text-reference book is intended primarily for regulatory personnel, practicing engineers and engineering/science students, and contains engineering methods for source reduction and the general technical aspects of pollution prevention. It is assumed that the reader has already taken basic courses in physics and chemistry, and has a minimum background in mathematics through calculus. The authors' aim is to offer the reader the fundamentals of this subject with appropriate practical applications to pollution prevention, and to provide an introduction to the specialized literature in this and related areas. Readers are encouraged, through the references, to continue their own development beyond the scope of this book.

As is usually the case in preparing a book, the problem of what to include and what to omit has been particularly difficult. However, every attempt has been made to offer engineering and science course material to individuals with a technical background at a level that should enable them to better cope with some of the complex problems encountered in pollution prevention today.

The book is divided into three parts: Process and Plant Fundamentals, Pollution Prevention Options, and Pollution Prevention Applications with case studies. The general subjects of process and plant fundamentals, equipment, and calculations are examined in Part 1. A separate chapter is devoted to process diagrams and economic considerations. Part 2 covers the broad subject of pollution prevention options. This section includes chapters on pollution prevention assessments, source reduction, and recycling methods. Separate sections on ISO 14000, multimedia approaches, life-cycle analysis, energy conservation, accident management, and environmental justice are also contained in Part 2. Part 3 is highlighted by individual chapters devoted to specific industrial applications involving pollution prevention with illustrative case study examples.

The authors cannot claim sole authorship for the material in this text. The present book has evolved from a host of sources including: notes, homework problems and lecture handouts prepared by L. Theodore for a one-semester, three-credit Pollution Prevention graduate course at Manhattan College; L. Theodore, J. Reynolds, and F. Taylor's Wiley-Interscience text, *Accident and Emergency Management*; a host of Theodore Tutorial Problem Workbooks, including *Pollution Prevention*, *Mass Transfer Operations*, and *Hazardous Waste Incineration*, L. Theodore; L. Theodore and Y.C. McGuinn's *Pollution Prevention* text; and U.S. EPA/625/7-88/003 *Waste Minimization Opportunity Assessment Manual* and various U.S. EPA pollution prevention documents. Although the bulk of the material is original and/or taken from sources that the authors have been directly involved with, every effort has been made to acknowledge material drawn from other sources. The authors trust that their apology will be accepted for any error(s) or omission(s), and changes will be included in a later printing or edition.

For some, particularly for novices, the book may serve as a starting point that will allow them to become acquainted with the pollution prevention field. For others, who would classify themselves as experts, the book could serve as a reference text. It may also be useful as a tool for training in industry, in government, or in academia. The book should be valuable to engineers in regulatory agencies and industry, to technicians, and to maintenance personnel. The aim of the authors is to provide, in a thorough and clear manner, a book covering both the fundamentals of pollution prevention and their application to real-world problems. It is hoped that this text will serve both industry and government in attempting to reduce and/or eliminate waste releases that can result in negative impacts to both human health and the environment.

The preparation of the first edition of this book was funded, in part, by the USEPA's Air Pollution Training Institute, located in Research Triangle Park, North Carolina. The authors gratefully acknowledge this support. Young C. McGuinn of Merck served as a co-author on an earlier version of this work. Her contribution to this text is also appreciated and acknowledged.

Ryan Dupont
Louis Theodore
Kumar Ganesan

Smithfield, Utah

INTRODUCTORY COMMENTS REGARDING POLLUTION PREVENTION

Applying pollution prevention strategies – the environmental management option of the future – will not eliminate all wastes from all production processes. Rather, pollution prevention strategies offer a cost-effective means of minimizing the generation of waste and complying with local, state and federal environmental and health and safety regulations. Source reduction is the first step in the traditional hierarchy of integrated waste management options and represents the most preferable of the available waste generation and pollution prevention steps. The next step in such a hierarchy is the responsible recycling of any wastes that cannot be reduced at the source. When recycling is conducted in an environmentally sound manner, it shares many of the same advantages of source reduction, such as conserving energy and other resources, reducing the reliance on raw materials, and reducing the need for end-of-pipe treatment or containment of wastes. Wastes that cannot be "feasibly" recycled should be treated in accordance with environmental standards that are designed to reduce both the hazard and volume of waste streams. Finally, any residues remaining from the treatment of waste should be disposed of safely in order to minimize their potential for release into and impact on the environment. While this integrated waste management approach includes treatment and ultimate disposal, the concepts of pollution prevention that are highlighted in this text are focused on the first two waste management options, namely source reduction and closed-loop recycling. Some additional aspects of pollution prevention that are relevant to current and future practices and that are covered in the text include related areas of: energy conservation; health, safety, and accident management; ISO 14000; state regulations concerning pollution prevention; and environmental accounting. This latter item is of particular interest as industries can be expected to balance costs and benefits when evaluating pollution prevention opportunities, considering such factors as costs to carry out process changes versus benefits in terms of savings in raw material and operating expenditures, avoided pollution control costs, avoided liability, and improved relations with local communities.

About the Authors

R. RYAN DUPONT is Professor of Civil and Environmental Engineering and Head of the Environmental Engineering Division at Utah State University, Logan. Dr. Dupont received a Ph.D. in Environmental Health Engineering from the University of Kansas in 1982, and has been conducting applied environmental research at the Utah Water Research Laboratory and teaching undergraduate and graduate environmental engineering students at Utah State University since 1982. In 1998, he was named by the American Society of Engineering Educators an Outstanding Young Engineering Educator. Throughout his career, Dr. Dupont has been involved in the development and preparation of educational resource material to support undergraduate environmental engineering education. Dr. Dupont has written in the areas of Accident and Emergency Management, VCH Publishers; Air Toxics Management, Gordon and Breach; Environmental Management, Lewis Publishers; Hazardous Waste Incineration, John Wiley; and Pollution Prevention, Gordon and Breach.

LOUIS THEODORE is Professor of Chemical Engineering at Manhattan College, Riverdale, New York. He received his M.CH-E and Eng.Sc.D. from New York University. For the past 37 years, Dr. Theodore has been a successful educator, researcher, professional innovator, and communicator. Included in Dr. Theodore's 76 text/reference books are *Pollution Prevention*, Van Nostrand Reinhold; *Air Pollution Control Equipment*, Prentice-Hall; *Introduction to Hazardous Waste Incineration*, Wiley-Interscience; and section author/editor in Perry's *Handbook of Chemical Engineering*, McGraw-Hill. He is the co-founder of Theodore Tutorials, a company specializing in providing training needs to industry, government and academia. Dr. Theodore is the recipient of the International Air and Waste Management Association's prestigious Ripperton award that is 'presented to an outstanding educator who through example, dedication and innovation has so inspired students; to achieve excellence in their professional endeavors;' and the recipient of the American Society for Engineering Education AT&T Foundation award for 'excellence in the instruction of engineering students.'

KUMAR GANESAN is a Professor and Chair of the Environmental Engineering Department at Montana Tech of the University of Montana, Butte. As a scientist for the National Environmental Engineering Research Institute in India, he completed several major air quality engineering projects. He earned his Ph.D. in Engineering from Washington State University, Pullman in 1981. He is the managing editor of *Air Toxics, Problems and Solutions*, by Gordon and Breach Publishers, and is actively involved in developing web-based environmental engineering courses for delivery at the undergraduate level.

Contributors

Chapter 1
Thomas McKee, Project Engineer
New York Power Authority
Buchanan, NY

Chapter 2
Lindsy Varghese, Chemical
 Engineering Graduate Student
Manhattan College
Bronx, NY

Chapter 3
Sean Guinan, Electronic Document
 Supervisor
Louis Frey, Inc.
New York, NY

Chapter 4
Bella De Vito, Consultant
Bronx, NY

Amy Leahy, Chemistry Instructor
Yonkers, NY

Chapter 5
Brett Elias, Process Engineer
Komline Sanderson
Peapack, NJ

Chapter 6
Ronald Carpino, Process Engineer
New York Power Authority
Indian Point, NY

Chapter 7
Mathew Reagan, Environmental
 Engineer
Metcalf & Eddy
New York, NY

Sangeeta Gokhale, Consultant
Lake Grove, NY

Chapter 8
Mary Wrieden, Development
 Engineer
Komline Sanderson
Peapack, NJ

Chapter 9
Keith Colacioppo, Process Engineer
Clairol
Stamford, CT

Chapter 13
Mary Wrieden, Development
 Engineer
Komline Sanderson
Peapack, NJ

Chapter 14
Joseph Lanzillotti, Environmental
 Engineer
Hazen and Sawyer
New York, NY

Appendix A
Kerwin Somoza, Environmental
 Engineer
New York State Department of
 Environmental Conservation
Tarrytown, NY

Appendix B
Roberto Diaz, Laboratory Instructor
Manhattan College
Bronx, NY

Appendices C and D
Robert Molter, Project Engineer
Compac Corp.
Netcong, NJ

Appendix E
Karen Counes
New York, NY

Table of Contents

Part 1. Process and Plant Fundamentals..1

Chapter 1 Introduction..3
    1. Definition of Pollution Prevention..3
    2. Units...3
        2.1. Conversion of Units...4
        2.2. Dimensional Analysis..5
    3. Problem Solving Methodology..6
    4. Illustrative Examples..8
    5. Summary..9
    6. Problems..9
    7. References...10

Chapter 2 Definitions..11
    1. Fundamental Definitions...11
        1.1. Temperature..11
        1.2. Pressure...12
        1.3. Moles and Molecular Weights...13
    2. Physical Properties..13
        2.1. Density...13
        2.2. Viscosity...14
        2.3. Heat Capacity...14
        2.4. Thermal Conductivity..14
        2.5. Diffusivity..15
        2.6. Vapor Pressure...15
        2.7. Boiling Point..15
        2.8. Freezing Point..15
    3. Chemical Properties..16
        3.1. Flammability (or Explosion) Limit..16
        3.2. Flash Point..17
        3.3. Autoignition Temperature (AIT)...17
        3.4. Heat (Enthalpy) of Reaction..17
        3.5. Heat (Enthalpy) of Combustion...18
        3.6. Gross Heating Value..19
        3.7. Net Heating Value..19
        3.8. Theoretical Adiabatic Flame
             Temperature (TAFT)..19
    4. Biological Properties...20
        4.1. Biochemical Oxygen Demand (BOD)...20
        4.2. Chemical Oxygen Demand (COD)..20
        4.3. Ultimate Oxygen Demand (UOD)...20
    5. Properties of Mixtures...21
        5.1. pH...21
        5.2. Partial Pressure...21
        5.3. Humidity...21

Table of Contents

    6. Illustrative Examples..22
    7. Summary...23
    8. Problems...25
    9. References..26

Chapter 3 Conservation Laws and Basic Principles....................................27
    1. Conservation of Mass...27
    2. Conservation of Energy...28
    3. Ideal Gas Law..29
    4. Phase Equilibrium...30
    5. Stoichiometry..32
    6. Thermochemistry..33
        6.1. Chemical Reaction Equilibrium.......................... 33
        6.2. Chemical Kinetics...35
    7. Illustrative Examples...36
    8. Summary..38
    9. Problems..39
    10. References... 40

Chapter 4 Unit Operations..43
    1. Quantitative Approaches... 43
    2. Unit Operation Classification..44
    3. Mass Transfer..46
        3.1. Contact of Immiscible Phases..............................47
            3.1.1. Gas-Gas..47
            3.1.2. Gas-Liquid... 47
            3.1.3. Gas-Solid..48
            3.1.4. Liquid-Liquid...49
            3.1.5. Liquid-Solid...49
            3.1.6. Solid-Solid... 50
        3.2. Miscible Phases Separated by a Membrane......................50
            3.2.1. Gas-Gas.. 50
            3.2.2. Liquid-Liquid... 50
            3.2.3. Solid-Solid... 51
        3.3. Direct Contact of Miscible Phases......................51
    4. Illustrative Examples...52
    5. Summary..53
    6. Problems..54
    7. References...54

Chapter 5 Plant Equipment.. 55
    1. Chemical Reactors..55
        1.1. Reactor Definition..55
        1.2. Reactor Types..56
    2. Heat Exchangers...58
    3. Mass Transfer Equipment...60
        3.1. Distillation..60
        3.2. Adsorption... 61

        3.3. Absorption..62
        3.4. Evaporation..62
        3.5. Extraction..63
        3.6. Drying..63
    4. Illustrative Examples...63
    5. Summary..65
    6. Problems..66
    7. References...67

Chapter 6 Ancillary Processes and Equipment..............................69
    1. Conveyance Systems..69
        1.1. Pipes..69
        1.2. Ducts...70
        1.3. Fittings..71
        1.4. Valves..72
        1.5. Fans...74
        1.6. Pumps..74
        1.7. Compressors..75
        1.8. Stacks..75
    2. Utilities..75
        2.1. Electricity..76
        2.2. Steam..76
        2.3. Water..76
            2.3.1. Cooling Water...77
            2.3.2. Potable and General Use Water..............................77
            2.3.3. Demineralized Water...77
        2.4. Refrigeration..77
        2.5. Compressed Air...78
        2.6. Inert Gas Supplies...78
    3. Material Transportation and Storage Equipment.....................78
        3.1. Gases...78
        3.2. Liquids..79
        3.3. Solids..79
    4. Instruments and Controls...80
        4.1. Feedback Loop I&C Systems...80
        4.2. Automatic Trip Systems and Interlocks.........................82
    5. Illustrative Examples...83
    6. Summary..85
    7. Problems..85
    8. References...86

Chapter 7 Waste Treatment Processes and Equipment..................87
    1. Overall Considerations..87
    2. Air Pollution Control Equipment and Processes.....................89
        2.1. Particulate Control Devices...89
            2.1.1. Cyclone Separators..90
            2.1.2. Electrostatic Precipitators..90
            2.1.3. Baghouses and Fabric Filters..................................91

Table of Contents

    2.1.4. Venturi Scrubbers.. 91
   2.2. Gaseous Control Devices.. 92
    2.2.1. Absorbers.. 92
    2.2.2. Adsorbers.. 93
    2.2.3. Vapor Incinerators and Flares................................ 95
  3. Water Pollution Control Equipment and Processes.................... 95
   3.1. Physical Treatment.. 96
    3.1.1. Flocculation and Sedimentation........................... 97
    3.1.2. Dissolved Air Flotation....................................... 97
    3.1.3. Filtration.. 98
    3.1.4. Centrifugation.. 98
    3.1.5. Air Stripping... 98
    3.1.6. Steam Stripping.. 99
    3.1.7. Resin Adsorption... 100
    3.1.8. Electrodialysis.. 100
   3.2. Chemical Treatment.. 101
    3.2.1. Calcination... 101
    3.2.2. Electrolysis.. 102
    3.2.3. Hydrolysis... 102
    3.2.4. Neutralization... 103
    3.2.5. Oxidation... 104
    3.2.6. Photolysis.. 105
    3.2.7. Reduction.. 105
   3.3. Biological Treatment... 105
    3.3.1. Activated Sludge... 106
    3.3.2. Waste Stabilization Ponds and Lagoons............... 107
    3.3.3. Trickling Filters.. 107
    3.3.4. Anaerobic Digestion.. 108
    3.3.5. Composting.. 108
    3.3.6. Enzyme Treatment... 109
  4. Solid Waste Control Equipment and Processes....................... 109
   4.1. Physical Treatment.. 110
    4.1.1. Shredding.. 110
    4.1.2. Air Classification.. 110
    4.1.3. Compaction.. 111
   4.2. Thermal Treatment... 111
    4.2.1. Incinerator Principles.. 112
    4.2.2. Rotary Kiln Incinerators.................................... 114
    4.2.3. Fluidized Bed Incinerators................................. 116
    4.2.4. Multiple Hearth Incinerators............................... 117
  5. Illustrative Examples.. 118
  6. Summary... 123
  7. Problems... 125
  8. References... 127

Chapter 8 Process Diagrams... 129
  1. Flow Sheets... 129
   1.1. Block Diagrams... 130

  1.2. Graphic Flow Diagrams...131
  1.3. Process Flow Diagrams...131
  1.4. Process Piping and Instrumentation Flow Diagrams......132
  1.5. Tree Diagrams...139
    1.5.1. Fault Tree Analysis..139
    1.5.2. Event Tree Analysis...140
 2. Process Simulations..140
 3. Pollution Prevention and Flow Sheets....................................141
 4. Illustrative Examples..141
 5. Summary..144
 6. Problems..146
 7. References..148

Chapter 9 Economic Considerations...149
 1. Economic and Costing Procedures...150
 2. Environmental Accounting...151
  2.1. What Is Environmental Accounting?.............................151
  2.2. What Is an Environmental Cost?..................................152
  2.3. Identifying Environmental Costs..................................153
  2.4. Applying Environmental Accounting
    to Cost Allocation..156
  2.5. Applying Environmental Accounting
    to Capital Budgeting..159
  2.6. Applying Environmental Accounting
    to Process/Product Design..160
 3. Capital Equipment Costs..160
 4. Operating Costs..162
 5. Project Evaluation..163
 6. Perturbation Studies in Optimization.....................................166
 7. Plant Siting and Layout..167
  7.1. Raw Materials..167
  7.2. Transportation...168
  7.3. Process Water..168
  7.4. Waste Disposal..169
  7.5. Fuel and Power..169
  7.6. Labor..169
  7.7. Weather..169
 8. Factors in Planning Layouts...170
  8.1. Methods of Layout Planning..170
  8.2. Modeling System Layouts..170
  8.3. Principles of Plant Layout..171
 9. Plant and Process Design..173
  9.1. Research Considerations..173
  9.2. Process Development Considerations..........................173
  9.3. Process Design...173
  9.4. Design for the Environment (DfE) Initiative................174
 10. Illustrative Examples..174
 11. Summary..177

Table of Contents

    12. Problems...179
    13. References..180

Part 2. Pollution Prevention Principles..183

Chapter 10 From Pollution Control to Pollution Prevention......................185
    1. Overview..185
    2. Federal Regulations...188
        2.1. The Pollution Prevention Act of 1990..........................191
        2.2. EPA's Pollution Prevention Strategy............................192
    3. Progress Toward Pollution Prevention..................................195
        3.1. Promoting Pollution Prevention at EPA.......................195
        3.2. Promoting Pollution Prevention at
            Other Federal Agencies...197
        3.3. Promoting Pollution Prevention in Industry.................198
            3.3.1. Waste Reduction Always
                  Pays (WRAP)..199
            3.3.2. Chemical Manufacturers
                  Association (CMA) Activities.........................201
        3.4. Promoting Pollution Prevention at the State
            and Tribal Level..204
        3.5. Promoting Pollution Prevention at
            Educational Institutions...207
        3.6. Promoting Pollution Prevention through
            Community and Non-Profit Organizations....................208
    4. The Future of Pollution Prevention..209
    5. Illustrative Examples..210
    6. Summary...212
    7. Problems...214
    8. References..215

Chapter 11 Additional Pollution Prevention Components.........................217
    1. Introduction..217
    2. Energy Conservation..217
        2.1. Heat Exchangers..219
        2.2. Ovens and Furnaces..220
        2.3. Insulation...221
        2.4. Boilers..222
        2.5. Chillers and Cooling Towers...223
        2.6. Compressed Air Systems...223
        2.7. Lighting...224
        2.8. Motors and Drives...225
        2.9. Power Factor Issues...226
    3. Accident and Emergency Management..................................227
    4. Health Risk Assessment...229
    5. Multimedia Analysis..231
    6. Life Cycle Analysis..234

      6.1. The LCA Process..236
      6.2. The Life Cycle Checklist......................................237
      6.3. The Life Cycle Assessment Worksheet........................237
      6.4. The Life Cycle Inventory Analysis..............................238
   7. Sustainable Development...240
   8. ISO 14000..242
      8.1. Company Concerns..242
      8.2. The Certification Process......................................241
      8.3. Benefits and Pitfalls of ISO 14000..............................244
   9. Miscellaneous Pollution Prevention Considerations...............245
      9.1. Small Quantity Generators..................................... 245
      9.2. Domestic Activities.. 247
      9.3. Ethical Considerations..248
      9.4. Environmental Justice..249
   10. Illustrative Examples..251
   11. Summary..257
   12. Problems..258
   13. References...258

Chapter 12 Pollution Prevention Opportunity Assessment........................261
   1. Planning and Organization.. 263
   2. Assessment Phase..264
   3. Feasibility Analysis Phase...267
   4. Implementation..268
   5. Pollution Prevention Incentives...270
      5.1. Economic Benefits..270
      5.2. Regulatory Compliance... 270
      5.3. Reduction in Liability...271
      5.4. Enhanced Public Image... 272
   6. Impediments to Pollution Prevention...................................272
   7. Illustrative Examples...274
   8. Summary...277
   9. Problems..278
   10. References... 279

Chapter 13 Source Reduction...281
   1. Source Reduction Options...281
      1.1. Procedural Changes..281
      1.2. Technology Changes.. 284
      1.3. Input Material Changes..285
      1.4. Product Changes..286
   2. Impediments to Achieving Source Reduction.........................286
   3. Illustrative Examples..287
   4. Summary...291
   5. Problems..291
   6. References.. 294

Table of Contents

Chapter 14 Recycling...295
    1. Recycling Options...295
        1.1. Direct Reuse On-Site...297
        1.2. Additional On-Site Recycling..297
        1.3. Recovery Off-Site..297
        1.4. Sale for Reuse Off-Site..297
        1.5. Energy Recovery..298
        1.6. Regulatory Considerations...298
    2. Recycling Technologies..300
        2.1. Vapor-Liquid Separation..300
            2.1.1. Distillation..300
            2.1.2. Evaporation..301
            2.1.3. Gas Absorption..302
        2.2. Solid-Liquid Separation...302
            2.2.1. Filtration...302
            2.2.2. Centrifugation..302
            2.2.3. Sedimentation..303
        2.3. Liquid-Liquid Separation...303
            2.3.1. Liquid-Liquid Extraction..303
            2.3.2. Decantation..304
        2.4. Solute Recovery...304
            2.4.1. Membrane Separation..304
            2.4.2. Ion Exchange...305
            2.4.3. Precipitation...306
    3. Material and Waste Exchanges...306
    4. Illustrative Examples...307
    5. Summary..310
    6. Problems..311
    7. References...312

Part 3. Pollution Prevention Applications...313

Chapter 15 Commercial Printing...315
    1. Process Description...315
        1.1. Image Processing..319
        1.2. Plate Processing..320
        1.3. Printing...322
    2. Waste Description...322
    3. Pollution Prevention Options..324
        3.1. Process Modifications..325
            3.1.1. Image Processing...325
            3.1.2. Plate Processing...326
            3.1.3. Makeready...327
            3.1.4. Printing and Finishing...327
        3.2. Recycling and Resource Recovery...328
        3.3. Good Housekeeping and Operating Practices...................................329
    4. Case Study...330
        4.1. Process and Waste Description..330

        4.2. Pollution Prevention via Process
            Modification 331
        4.3. Pollution Prevention via Resource Recovery
            and Recycling 332
        4.4. Pollution Prevention via Better
            Housekeeping and Operating Practices 333
    5. Summary 334
    6. References 336

Chapter 16 Metal Finishing Industries 337
    1. Process Description 337
    2. Waste Description 337
    3. Pollution Prevention Options 337
        3.1. Process Modifications 340
            3.1.1. Process Baths 340
            3.1.2. Rinse Systems 342
        3.2. Recycling and Resource Recovery 343
        3.3. Good Housekeeping and Operating Practices 345
    4. Case Study 347
        4.1. Problem Description 347
        4.2. Pollution Prevention Techniques Used 347
        4.3. Benefits 347
    5. Summary 347
    6. References 349

Chapter 17 Electronics Industry 351
    1. Process Description 351
        1.1. Ingot Growth 351
        1.2. Ingot Sandblasting and Cleaning 353
        1.3. Ingot Cropping 353
        1.4. Wafer Slicing 353
        1.5. Wafer Washing 353
        1.6. Wafer Lapping 353
        1.7. Wafer Etching 354
        1.8. Polishing 354
        1.9. Epitaxial Growth 354
    2. Waste Description 354
    3. Pollution Prevention Options 355
        3.1. Process Modifications 355
        3.2. Recycling and Resource Recovery 356
        3.3. Good Housekeeping and Operating Practices 356
    4. Case Studies 357
        4.1. Case Study 1 - Electronic Circuit
            Manufacturing 357
            4.1.1. Problem Description 357
            4.1.2. Pollution Prevention Techniques
                  Used 357
            4.1.3. Benefits 357

Table of Contents

        4.2. Case Study 2 - Manufacture of Printed
            Circuit Boards..357
            4.2.1. Problem Description..357
            4.2.2. Pollution Prevention Techniques
                Used...357
            4.2.3. Benefits..358
    5. Summary..358
    6. References...359

Chapter 18 Drug Manufacturing and Processing Industry..........................361
    1. Process Description..361
        1.1. Fermentation...361
        1.2. Chemical Synthesis..361
        1.3. Natural Extraction..363
        1.4. Formulation..363
    2. Waste Description...364
        2.1. Fermentation...364
        2.2. Chemical Synthesis..364
        2.3. Natural Extraction..365
        2.4. Formulation..365
    3. Pollution Prevention Options...365
        3.1. Process Modifications..365
        3.2. Recycling and Resource Recovery...................................368
        3.3. Good Housekeeping and Operating Practices.................369
    4. Case Study..370
        4.1. Problem Description..370
        4.2. Pollution Prevention Techniques Used............................370
        4.3. Benefits..370
    5. Summary..370
    6. References...372

Chapter 19 Paint Manufacturing Industry..373
    1. Process Description...373
    2. Waste Description...375
    3. Pollution Prevention Options...375
        3.1. Process Modifications..375
        3.2. Recycling and Resource Recovery...................................377
        3.3. Good Housekeeping and Operating Practices.................377
    4. Case Study..377
        4.1. Process and Waste Description.......................................378
            4.1.1. Solvent-Based Paints..378
            4.1.2. Water-Based Paints..378
        4.2. Pollution Prevention via Process Modifications..............381
        4.3. Pollution Prevention via Resource Recovery
            and Recycling...382
        4.4. Pollution Prevention via Better Housekeeping
            and Operating Practices...383

  5. Summary 384
  6. References 386

Chapter 20 Pesticide Formulating Industry 387
  1. Process Description 387
  2. Waste Description 390
  3. Pollution Prevention Options 391
    3.1. Process Modifications 391
    3.2. Recycling and Resource Recovery 392
    3.3. Good Housekeeping and Operating Practices 392
  4. Case Study 392
    4.1. Process and Waste Description 393
    4.2. Audit Findings 393
    4.3. Pollution Prevention via Better Housekeeping
      and Operating Practices 393
  5. Summary 394
  6. References 395

Chapter 21 Pulp and Paper Industry 397
  1. Process Description 397
  2. Waste Description 400
  3. Pollution Prevention Options 401
    3.1. Process Modifications 401
      3.1.1. Pulping Process Modifications 401
      3.1.2. Bleaching Process Modifications 402
    3.2. Recycling and Resource Recovery 403
    3.3. Good Housekeeping and Operating Practices 404
  4. Case Study 405
    4.1. Process and Waste Description 405
    4.2. Pollution Prevention via Better Housekeeping
      and Operating Practices 405
    4.3. Benefits 405
  5. Summary 405
  6. References 406

Appendices
  Appendix A: Units, Conversion Factors and Mathematical Symbols. 409
  Appendix B: State Pollution Prevention Programs 421
  Appendix C: Environmental Organizations 451
  Appendix D: Pollution Prevention Software 455
  Appendix E: Pollution Prevention Contacts in Education 457

Index 493

List of Tables

Table

1.1	Common Systems of Units	4
1.2	English Engineering Units	5
1.3	SI Units	6
1.4	Prefixes for SI Units	7
3.1	Values of R in Various Units	31
5.1	Solution for Example Problem 5.1	64
5.2	Input Data for Example Problem 5.2	65
8.1	Typical Equipment Designations	138
9.1	Developments in Economic Analysis Approaches	151
9.2	Types of Management Decisions Benefiting from Environmental Cost Information	153
9.3	Fabricated Equipment Cost Index	161
10.1	Evolution of Waste Management Approaches in the United States	185
10.2	The 33/50 Program Target Chemicals	193
10.3	Guiding Principles for Responsible Care^{SM} A Public Commitment	205
11.1	Pollution Prevention Options and Their Effect on Heat Exchanger Performance and Energy Efficiency	221
11.2.	Primary SIC Groups Likely to Contain SQGs	246
11.3	Major Waste Types Generated by SQGs	246
14.1	Selected Waste Exchanges Operating in the United States	308
15.1	Wastes from Commercial Printing Processes	323
15.2	Economic Comparison of On-Site versus Off-Site Ink Recycling	333
16.1	Wastes Generated by the Metal Finishing Industry	340
16.2	Applications of Recovery Technologies	344
18.1.	Commonly Recycled Solvents in the Pharmaceutical Industry	368
19.1.	Raw Materials Used in Plant B	380
19.2.	Economics of On-Site Distillation	384
20.1.	Pesticide Production in the United States	388
21.1.	Raw Wastewater Loadings for Different Pulping Processes	400

List of Figures

Figure

4.1	Hypothetical Flow System	44
5.1	Common Distillation Column Configurations	61
7.1	Common Process/Waste Treatment System Configuration with Product/Waste Separation within the Production Process Showing Conventional and Pollution Prevention Operating Mode Options	88
7.2	Common Process/Waste Treatment System Configuration with Product/Waste Separation Subsequent to the Production Process Showing Conventional and Pollution Prevention Operating Mode Options	88
8.1	Fundamental Operations Described by Block Diagrams	131
8.2	Commonly Used Valve Symbols	133
8.3	Typical Instrumentation Symbols for Temperature	134
8.4	Typical Instrumentation Symbols for Pressure	135
8.5	Typical Instrumentation Symbols for Flow	136
8.6	Typical Instrumentation Symbols for Liquid Level	137
8.7	Vessel with Level Control on Outlet. FIC, Flow Indicator Controller; LG, Level Gauge; LIC, Level Indicator Controller; LAWL, Level Alarm for Water Level; M, Motor	138
9.1	Examples of Environmental Costs Incurred by Firms	154
9.2	Example of Misallocation of Environmental Costs Under Traditional Cost System	158
9.3	Example of Revised Cost Accounting System	158
9.4	Typical Master Plot Plan	171
10.1	U.S. EPA's Integrated Waste Management and Pollution Prevention Hierarchy	187
10.2	Pollution Prevention and Waste Management Options	189
11.1	Distribution of U.S. EPA Climate Wise Project Types Reported in 1998	219
11.2	Action Plan Checklist from EPA's Climate Wise Program	220
11.3	Hazard Risk Assessment Flowchart	229
11.4	Hazard Risk Assessment Flowchart	231
11.5	Multimedia Approach for a Hypothetical Chemical Plant	233
11.6	Overall Multimedia Process Flow Diagram for a Hypothetical Product	234
11.7	Life Cycle Stages for a Hypothetical Product	235
11.8	An Example Life Cycle Checklist	238
11.9	An Example Life Cycle Assessment Worksheet	239
12.1	Pollution Prevention Opportunity Assessment Procedure	262
13.1	Typical Process Schematic	282

13.2	Typical Process Schematic with Technology or Process Change(s)	282
13.3	Typical Process Schematic with Procedural Change(s)	282
13.4	Typical Process Schematic with Process Change(s)	282
13.5	Source Reduction Options and Alternatives	282
14.1	Typical Process Schematic	296
14.2	Basic Process Schematic with Recycle Back to Feed	296
14.3	Basic Process Schematic with Recycle Back to Process	296
14.4	Basic Process Schematic with Additional Treatment for By-Product Recovery	296
14.5	Basic Process Schematic with Additional Treatment and Energy Recovery	296
15.1	The Gravure Printing Process	317
15.2	The Offset Lithography Printing Process	317
15.3	Commercial Offset Lithographic Printing Operation	318
16.1	Typical Zinc Plating Process	338
16.2	Typical Metal Anodizing Operation	339
16.3	Closed-Loop Water Recovery System	346
16.4	Open-Loop Rinse Water Recovery System	346
17.1	Gallium Arsenide Ingot Growth	352
17.2	Liquid Encapsulated Czochralski Ingot Growth System	352
18.1	Typical Fermentation Process Flowsheet	362
18.2	Example Chemical Synthesis Process Flowsheet	363
18.3	Hollow Fiber Contained Liquid Membranes	367
19.1	Block Diagram of Typical Paint Manufacturing Steps	374
19.2	Block Flow Diagram of the Solvent-Based Paint Process in Plant B	379
19.3	Block Flow Diagram of the Water-Based Paint Process in Plant B	379
20.1	Block Diagram for Liquid Pesticide Formulation	389
20.2	Block Diagram for Solid Pesticide Formulation	390
21.1	Representative Bleached Kraft Mill Processes and Pollutant Loads	398
21.2	Stone Groundwood Pulp Mill	399
21.3	Process Flow Schematic of Closed-Cycle Mill	402

PART 1
PROCESS AND PLANT FUNDAMENTALS

Waste disposal, particularly of hazardous wastes, is becoming increasingly difficult and costly. For example, landfilling, traditionally the favored disposal option for a majority of hazardous wastes, is subject to strict controls (Land Ban Regulations) in the United States. The threat of future legal liabilities has further reduced the attractiveness of this option. In addition, Congressional legislation on pollution prevention (through the Pollution Prevention Act of 1990), on the safe handling and storage of hazardous chemicals (through the Superfund Amendment and Reauthorization Act, SARA, Title III Community Right-to-Know legislation of 1986; and the Clean Air Act Amendments of 1990 requiring Risk Management Plans and Off-Site Consequence Analysis) has, and will continue to have, a significant impact on industrial practice.

The key regulation regarding pollution prevention, the Pollution Prevention Act of 1990, reinforced the U.S. EPA's environmental management hierarchy. This regulation established as national policy and highest priority the prevention or reduction of pollution at the source wherever feasible. In this environmental management hierarchy, the next priority for pollution that cannot be prevented is recycling and reuse. In the absence of feasible prevention or recycling opportunities, the last resort is the treatment and disposal of pollution in an environmentally safe manner. The Act, and its implications, are treated in greater detail in Part 2 of this book.

In the working definition currently used by U.S. EPA, pollution prevention consists of source reduction and closed-loop recycling. Of the two approaches, source reduction is preferable from an environmental perspective as it is often less costly, and presents much less risk to workers, the community, and the environment than collection, treatment, and recycling of a waste stream once it is produced. Source reduction and closed-loop recycling each compromise a number of practices and approaches which are discussed briefly in Chapter 1 and in more detail later in the text.

In the evaluation of pollution prevention options it is critical that all pollutant emissions into air, water, and land be carefully considered. This requires what has come to be defined as a "multimedia" approach to environmental assessments (see Chapter 10 in Part 2 for more details), since the transfer of pollutants from one medium to another is not pollution prevention. The removal of trichloroethylene from wastewater using activated carbon, for example, is not pollution prevention, as the pollutants are merely transferred from one medium (wastewater) to another (carbon, as solid waste) during this treatment process.

Ideally, of course, source reduction is the best alternative for companies to use so they produce less waste in the first place. Numerous companies have established formal pollution prevention programs and have reported significant successes in cutting the amount of wastes they produce. Much of the activity is economically driven, and faced with public pressure and rapidly rising treatment and disposal costs, there is little doubt that industry will continue this effort to reduce waste generation rates.

Since many of the pollution prevention efforts are plant and/or process related, Part 1 of the book is devoted to process and plant fundamentals. The material is presented in a traditional engineering format, but with some specific references to pollution prevention. The remaining two parts of the book are focused on plant-related applications and case studies. The chapters that follow in Part 1 contain material on: basic definitions, systems of units, dimensional analysis, and problem solving methodologies (Chapter 1); definitions of key physical and chemical properties (Chapter 2); conservation laws and basic principles underlying the proper design and operation of a chemical process (Chapter 3); unit operations (Chapter 4); plant process equipment including reactors, distillation columns, and heat exchangers (Chapter 5); ancillary processes and equipment including utilities (Chapter 6); waste treatment processes and equipment (Chapter 7); process diagrams illustrating various arrangements of plant equipment, valves, piping, and control systems (Chapter 8); and economic considerations necessary to properly evaluate pollution prevention alternatives (Chapter 9).

From a scientist's or engineer's perspective, Part 3 could be considered the heart of this text. Although Part 1 presents much valuable information of a general engineering nature, and Part 2 introduces and reviews the pollution prevention problem, it is Part 3 that deals directly with specific process industry pollution prevention applications and case study discussions.

Those readers with an engineering background and/or process experience may choose to skip the material presented in Part 1 and proceed directly to Part 2.

CHAPTER 1

Introduction
Contributing Author: Thomas McKee

1. DEFINITION OF POLLUTION PREVENTION

Pollution prevention in the EPA's definition involves the reduction, to the extent feasible, of generated waste. It includes any source reduction or closed-loop recycling activity undertaken by a generator that results in either: (1) the reduction of the total volume or quantity of waste, or (2) the reduction of toxicity of the waste, or both, so long as such reduction is consistent with the goal of minimizing present and future threats to human health and the environment.[1] Source reduction is defined as any activity that reduces or eliminates the generation of waste at the source, usually within a process. A material is "recycled" if it is used, reused, or reclaimed (40 CFR 261.1[c][7]). A material is "used or reused" if it is either: (1) employed as an ingredient to make a product, including its use as an intermediate (however, a material will not satisfy this condition if distinct components of the material are recovered as separate end products, as when metals are recovered from metal containing secondary materials), or (2) employed in a particular function as an effective substitute for a commercial product (40 CFR 261.1 [c][5]). A material is "reclaimed" if it is processed to recover a useful product or if it is regenerated. Examples include the recovery of lead from spent batteries and the regeneration of spent solvents (40 CFR 261.1 [c][4]).

The reader should note that the working definition of pollution prevention in this text also requires energy conservation and health, safety and accident management considerations. All these interrelated effects are considered in more detail in Part 2 of this book.

2. UNITS

A significant amount of a practicing engineer's time is spent in converting data and equations from one set of units to another. As long as scientists provide data and information in grams and centimeters while engineers calculate in terms of pounds and feet, this confusion in terminology will continue to exist.

The units in this text are consistent with those adopted by the engineering profession in the United States. For engineering work, SI (System International) and English units are most often employed; in the United States, English engineering units are generally used, although efforts continue to obtain universal adoption of SI units for all engineering and science applications. SI units have the advantage of being based on the decimal system. There are other systems of units; some of the more common are shown in Table 1.1.[2]

Table 1.1. Common Systems of Units.

System	Length	Time	Mass	Force	Energy	Temperature
SI	meter	second	kilogram	newton	joule	Kelvin, degree Celsius
cgs	centimeter	second	gram	dyne	erg, joule, or calorie	Kelvin, degree Celsius
fps	foot	second	pound	poundal	foot poundal	degree Rankine, degree Fahrenheit
American Engineering	foot	second	pound	pound (force)	British thermal unit, horsepower hour	degree Rankine, degree Fahrenheit
British Engineering	foot	second	slug	pound (force)	British thermal unit, foot pound (force)	degree Rankine, degree Fahrenheit

English engineering units will primarily be used through out this book. Tables 1.2 and 1.3 present units for the English and SI systems, respectively. Some of the more common prefixes for SI units are given in Table 1.4. The reader should refer to Appendix 1 for further details.

2.1. Conversion of Units

Equations used in engineering as well as in the physical sciences are dimensional equations. The choice of the fundamental system of units determines the magnitude of the quantity described. Since the choice is optional, it is usually made to suit the convenience of a certain segment of the scientific community. For example, astronomers measure distance in light-years, nuclear physicists employ angstroms, engineers use feet, etc. Engineers in the United States generally employ English units, but technical individuals need to be able to convert from one system of units to another.

Converting a measurement from one unit to another can conveniently be accomplished by using unit conversion factors. These factors are obtained from a simple equation that relates the two units numerically. For example,[3] from

$$1 \text{ foot (ft)} = 12 \text{ inches (in)} \tag{1.1}$$

the following conversion factor can be obtained:

$$12 \text{ in}/1 \text{ ft} = 1 \tag{1.2}$$

Since this factor is equal to unity, multiplying some quantity, e.g., 18 ft, by this factor cannot alter its value. Hence:

Table 1.2. English Engineering Units.

Physical Quantity	Name of Unit	Symbol for Unit
Length	foot	ft
Time	second	s
Mass	pound (mass)	lb
Temperature	degree Rankine	°R
Temperature (alternate)	degree Fahrenheit	°F
Moles	pound-mole	lbmol
Energy	British thermal unit	Btu
Energy (alternate)	horsepower-hour	hp-hr
Force	pound (force)	lbf
Acceleration	foot per second squared	ft/s^2
Velocity	foot per second	ft/s
Volume	cubic foot	ft^3
Area	square foot	ft^2
Frequency	cycles per second, Hertz	cycles/s, Hz
Power	horsepower, Btu per second	hp, Btu/s
Specific heat capacity	British thermal unit per (pound mass-degree Rankine)	Btu/lb-°R
Density	pound (mass) per cubic foot	lb/ft^3
Pressure	pound (force) per square inch	psi
	pound (force) per square foot	psf
	atm, bar	atm, bar

$$18 \text{ ft } (12 \text{ in}/1 \text{ ft}) = 216 \text{ in} \qquad (1.3)$$

Note that in Equation 1.3 the units of feet on the left-hand side cancel out, leaving only the desired units of inches.

2.2. Dimensional Analysis

One of the properties of equations that has a rational basis and is deduced from general relations is that they must be dimensionally homogeneous, or consistent. This can be demonstrated theoretically. If two sides of an equation should have different dimensions the equation is in error. Thus, one test for the consistency of an equation is whether all the terms in the equation contain the same units. If this condition is satisfied, the equation is said to be dimensionally consistent (correct); if not, it is incorrect. Note however, that this aforementioned property of dimensional homogeneity does not apply to empirical (not based on theoretical or physical properties) equations.

In summary, physical equations must be dimensionally consistent and, for the equality to hold, each term in the equation must have the same dimensions. This condition can be checked when solving engineering type

Table 1.3. SI Units.

Physical Quantity	Name of Unit	Symbol for Unit
Length	meter	m
Time	second	s
Mass	kilogram, gram	kg, g
Temperature	Kelvin	kg, g
Temperature (alternate)	degree Celsius	°C
Moles	gram-mole	gmol
Energy	joule	J, kg-m^2/s^2
Force	newton	N, kg-m/s^2, J/m
Acceleration	meters per second squared	m/s^2
Velocity	meters per second	m/s
Volume	cubic meter, liters	m^3, L
Area	square meter	m^2
Frequency	Hertz	Hz, cycles/s
Power	watt	W, kg-m^2/s^2, J/s
Specific heat capacity	Joules per (kilogram-Kelvin)	J/kg-K
Density	kilograms per cubic meter	kg/m^3
Pressure	pascal	Pa
	newton per square meter	N/m^2
Pressure (alternate)	bar	bar

problems. Throughout the text, and in particular in the pollution prevention applications, care should be exercised in maintaining the dimensional formulas of all terms and the dimensional homogeneity of each equation. This approach should help the reader to more easily attach physical significance to the equations presented in later chapters.

3. PROBLEM SOLVING METHODOLOGY

In addition to working with units, an engineer or scientist must also be comfortable solving problems. Therefore, solved examples and chapter problems can be found throughout the book. There are three illustrative examples and six problems provided with each chapter in Parts 1 and 2. A systematic approach to problem solving can help one avoid solving problems haphazardly. A recommended ten-step procedure for solving problems is presented below:

Table 1.4. Prefixes for SI Units.

Multiplication	Factors	Prefix	Symbol
1,000,000,000,000,000,000	$=10^{18}$	exa	E
1,000,000,000,000,000	$=10^{15}$	peta	P
1,000,000,000,000	$=10^{12}$	tera	T
1,000,000,000	$=10^{9}$	giga	G
1,000,000	$=10^{6}$	mega	m
1000	$=10^{3}$	kilo	k
100	$=10^{2}$	hecto	h
10	$=10^{1}$	deka	da
0.1	$=10^{-1}$	deci	d
0.01	$=10^{-2}$	centi	c
0.001	$=10^{-3}$	milli	m
0.000001	$=10^{-6}$	micro	μ
0.000000001	$=10^{-9}$	nano	n
0.000000000001	$=10^{-12}$	pico	p
0.000000000000001	$=10^{-15}$	femto	f
0.000000000000000001	$=10^{-18}$	atto	a

1. State what is known.
2. State what must be found.
3. Provide a simple sketch, e.g., a line diagram, labeling relevant information.
4. State all assumptions.
5. Select the appropriate and correct equation to describe the system.
6. Ensure that a consistent and appropriate set of units is utilized.
7. Substitute in numerical values to solve the equation.
8. Perform any additional calculational analyses.
9. Analyze the results.
10. State any relevant comments.

The above orderly approach to solving problems will reduce errors, save time, and will result in a clearer understanding of the limitations of a particular solution.

4. ILLUSTRATIVE EXAMPLES

Example 1.1. The heat capacity of a substance is defined as the quantity of heat required to raise the temperature of that substance by 1 degree. If the heat capacity of methanol is 0.61 cal/g-°C @ 60°F convert this value to English Engineering units.

SOLUTION. Calculate the heat capacity of methanol in Btu/lb-°F by multiplying the heat capacity of methanol by the corresponding conversion factors to obtain the desired units.

(0.61 cal/g-°C) (454 g/lb) (Btu/252 cal) (°C/1.8 °F) = 0.61 Btu/lb-°F

NOTE: 1.0 Btu/lb-°F is equivalent to 1.0 cal/g-°C.

Example 1.2. Specific gravity is defined as the ratio of two densities, that of the subject of interest to that of a reference substance. The reference substance is normally water. If the specific gravity of a liquid (methanol) is 0.92 @ 60°F, determine its density. Assume the density of water is 62.4 lb/ft^3 at 60°F.

SOLUTION. Write the equation for the definition of specific gravity.

Specific Gravity = (density of methanol)/(density of water)

Substitute the known quantities into the equation and solve for the unknown quantity (density of methanol).

Density of methanol = (0.92) (62.4 lb/ft^3) = 57.4 lb/ft^3

Example 1.3. A regulatory agency stipulates that the maximum concentration of benzo(a)pyrene in drinking water should not exceed 100 ng/L. Express this concentration in lb/1000 U.S. gal.

SOLUTION. Calculate the unit conversion by multiplying the maximum concentration of benzo(a)pyrene by the corresponding conversion factors to obtain the desired units.

(100 ng/L) (1 g/10^9 ng) (1 lb/454 g)(3,785 L/1000 U.S. gal)
= 8.34 x 10^{-7} lb/1000 U.S. gal

The same solution may also be obtained by using the conversion factor for water of 1 mg/L = 1 ppm$_w$ = 8.34 x 10^{-6} lb/gal.

(100 ng/L) (1 mg/10^6 ng) (8.34 x 10^{-6} lb/gal) (mg/L)
= 8.34 x 10^{-10} lb/gal (1000 gal/1000 U.S. gal)
= 8.34 x 10^{-7} lb/1000 U.S. gal

5. SUMMARY

1. A significant amount of a practicing engineer's time is spent in converting data and equations from one set of units to another. As long as scientists provide data and information in grams and centimeters while engineers calculate in terms of pounds and feet, this confusion in terminology will continue to exist. The units in this text are consistent with those adopted by the engineering profession in the United States. For engineering work, SI (System International) and English units are most often employed; in the United States, English engineering units are generally used, although efforts are still under way to obtain universal adoption of SI units for all engineering and science applications.

2. Equations used in engineering as well as in the physical sciences are dimensional equations. The choice of the fundamental system of units determines the magnitude of the quantity described. Since the choice is optional, it is usually made to suit the convenience of a certain segment of the scientific community.

3. Converting a measurement from one unit to another can conveniently be accomplished by using unit conversion factors. One of the properties of equations that has a rational basis and is deduced from general relationships is that they must be dimensionally homogeneous, or consistent.

4. Engineers and scientists should employ appropriate problem solving methodologies when dealing with technical issues. This will reduce errors, save time, and result in a clearer understanding of the limitations of a particular solution.

6. PROBLEMS

1. If a 250-cm^3 volume of liquid weighs 550 g, determine what volume of liquid will weigh 3.2 g.

2. If a manufacturing process yields 54 g of solid pollutant each day, determine the amount of pollutant produced in 1 year considering the process is performed continuously 5 days per week.

3. If the density of carbon tetrachloride is 99.2 lb/ft^3 @ 68°F and atmospheric pressure, determine the density in g/cm^3 at the same conditions.

4. Thermal conductivity provides a measure of how fast (or how easily) heat flows through a substance and is defined as the amount of heat that flows per unit time through a unit area of unit thickness as a result of a unit difference in temperature. If the thermal conductivity of methanol is 0.0512 cal/m-s-°C @ 60°F, convert this value to English engineering units (Btu/ft-hr-°F).

5. The viscosity of a fluid provides a measure of the fluid's resistance to flow. If the viscosity of methanol is 0.64 cP @ 60°F, convert this value to English engineering units (lb, ft, and s).

6. Express the concentration 72 g of HCl in 128 cm^3 of water into terms of fraction and percentage by weight, ppm, and molarity.

7. REFERENCES

1. U.S. EPA, *Report to Congress*, Office of Solid Waste, Washington, D.C., U.S. EPA/530-SW-86-033, 1986.
2. Weast, R. C., Ed., *CRC Handbook of Chemistry and Physics*, 51st edition, CRC Press, Boca Raton, Florida, 1971.
3. Theodore, L., Reynolds, J., and Taylor, F., *Accident and Emergency Management*, Wiley-Interscience, New York, 1989.

CHAPTER 2

Definitions

Contributing Author: Lindsy Varghese

Physical and chemical properties are important considerations in any study of pollution prevention. A substance may exhibit certain characteristics under one particular set of reaction or process conditions of temperature, pressure, and composition. However, these conditions are often changed when an attempt is made to implement a pollution prevention strategy. To promote a better understanding of such effects, many of which are covered in Parts 2, and 3, some definitions and certain key physical and chemical properties are briefly reviewed in this chapter.

1. FUNDAMENTAL DEFINITIONS

1.1. Temperature

Whether in the gaseous, liquid, or solid state, all molecules possess some degree of kinetic energy; that is, they are in constant motion–vibrating, rotating, or translating. The kinetic energies of individual molecules cannot be measured, but the combined effect of these energies in a very large number of molecules can be. This measurable quantity is known as *temperature* and is a macroscopic concept only (i.e., it does not exist on the molecular level).

Temperature can be measured in many ways; the most common method of which makes use of the expansion of mercury with increasing temperature (on process equipment, however, thermocouples or thermistors are more commonly employed). The two most commonly used temperature scales are the Celsius and Fahrenheit scales. The Celsius scale is based on the boiling and freezing points of water at 1 atm pressure; to the former, a value of 100°C is assigned, while the latter is assigned a value of 0°C. On the older Fahrenheit scale, the corresponding temperatures are 212°F and 32°F, respectively. Equations 2.1 and 2.2 show the conversion from one temperature scale to the other.

$$°F = 1.8(°C) + 32 \qquad (2.1)$$
$$°C = (°F - 32)/1.8 \qquad (2.2)$$

The volume of a gas would theoretically be zero at a temperature of approximately - 273°C or - 460°F. This temperature, which is known as absolute zero, is the basis for the definition of two absolute temperature scales, the Kelvin (K) scale and Rankine (°R) scale. The former scale is defined by shifting the Celsius scale by 273°C so that 0 K is equal to - 273°C. Equation (2.3) shows this relation.

$$K = °C + 273 \qquad (2.3)$$

The Rankine scale is defined by shifting the Fahrenheit scale 460°F, so that

$$°R = °F + 460 \tag{2.4}$$

As one might expect, temperature is an important parameter in pollution prevention studies as it has a significant impact on chemical reaction rates and many physical properties (vapor pressure, solubility, density, etc.) of chemicals.

1.2. Pressure

In the gaseous state, molecules possess a high degree of translational kinetic energy, which means that they are able to move quite freely throughout the body of a gas. If the gas is in a container, the molecules are constantly bombarding the container walls. The macroscopic effect of this bombardment by the large number of molecules making up the gas is called *pressure*. The pressure exerted by a gas is described by quantifying the force these molecules exert per unit area.

A number of units are used to express a pressure measurement. Some are based on an explicit unit of force per unit area, for example, pound (force) per square inch (psi) or dyne per square centimeter (dyne/cm^3). Others are based on a fluid height, such as inches of water (in H$_2$O) or millimeters of mercury (mm Hg), which produce an equivalent pressure at the bottom of the column. Units such as these are convenient when the pressure is indicated by a difference between levels of a liquid, as in a manometer or barometer. Barometric pressure is a measure of the ambient air pressure. Standard barometric pressure is 1 atm and is equivalent to 14.7 psi or 29.92 mm Hg.

Measurements of pressure by most gauges indicate the difference in pressure either above or below that of the surrounding atmosphere. If the system pressure is greater than atmospheric, the gauge (reading) is positive. If the gauge (reading) is negative, it is less than atmospheric pressure; the term vacuum designates a negative gauge pressure. Gauge pressures are usually identified by the letter g after the pressure unit (e.g., psig). Since gauge pressure is the pressure relative to the prevailing atmospheric pressure, the sum of the two gives the absolute pressure and is often indicated by the letter a after the unit (e.g., psia):

$$Pa = P + Pg \tag{2.5}$$

A pressure of 0 psia is the lowest possible pressure theoretically achievable, i.e., a perfect vacuum.

Pressure, like temperature, can have a significant impact on any process change due to its potential effect on process chemistry and reaction rates, on the potential release of reaction chemicals resulting from pressure differentials across a process, and on the intrinsic safety of a process.

1.3. Moles and Molecular Weights

An atom consists of protons and neutrons in a nucleus surrounded by electrons. An electron has such a small mass relative to that of the proton and neutron that the weight of the atom, the atomic weight, is approximately equal to the sum of the weights of the particles in its nucleus. Atomic weight may be expressed in atomic mass units per atom or in grams per gram-atom. The number of atoms in one gram-atom, called Avogadro's number, is 6.02×10^{23}.

The *molecular weight* (MW) of a compound is the sum of the atomic weights of the atoms that make up one molecule of the material. Units of grams per gram-mole (g/gmol) are used for molecular weight. One gram-mole contains 6.02×10^{23} molecules. In the English system, one pound-mole (lbmol) contains $454 \times 6.02 \times 10^{23}$ molecules.

A process stream seldom consists of a single component. It may also contain two or more phases, or a mixture of one or more solutes in a liquid solvent. For mixtures of substances, it is convenient to express compositions in mole fractions or mass fractions. The following definitions are often used to represent the composition of component A in a mixture of components.

$$w_A = \text{mass of A/total mass of stream} = \text{mass fraction of A} \quad (2.6)$$
$$y_A = \text{moles of A/total moles of stream} = \text{mole fraction of A} \quad (2.7)$$

Trace quantities of substances in solutions are often expressed in parts per million (ppm) or as parts per billion (ppb). These concentrations are presented on a mass per volume basis for liquids, and on a mass per mass basis for solids. Gas concentrations are usually represented on a volume basis (i.e., ppm_v), and the following equations apply.

$$ppm_v = 10^6 \, y_A \quad (2.8)$$
$$ppm_v = 10^3 \, ppb_v \quad (2.9)$$

2. PHYSICAL PROPERTIES

2.1. Density

The *density* of a substance is the ratio of its mass to its volume and may be expressed in units of pounds per cubic foot (lb/ft^3) or kilograms per cubic meter (kg/m^3). For solids, density can be determined easily by placing a known mass of the substance in a liquid and measuring the displaced volume. The density of a liquid can be measured by weighing a known volume of the liquid in a graduated cylinder. For gases, the ideal gas law (to be discussed later) can be used to calculate the density from the temperature, pressure, and molecular weight of the gas.

The specific gravity (SG) is the ratio of the density of a substance (ρ) to the density of a reference substance at a specific condition.

$$SG = \rho/\rho_{ref} \quad (2.10)$$

The reference most commonly used for solids and liquids is water. At its maximum density, which occurs at 4°C, this reference density is 1.000 g/cm^3, 1000 kg/m^3, or 62.4 lb/ft^3.

2.2. Viscosity

Viscosity is a property associated with a fluid's resistance to flow; more precisely, this property accounts for the energy losses resulting from shear stresses that occur between different portions of the fluid, moving at different velocities. The absolute viscosity μ has units of mass per length-time; the fundamental unit is the poise, which is defined as 1 g/cm-s. Viscosities are frequently given in centipoises (0.01 poise), the abbreviation for which is cP. In English units, the absolute viscosity is expressed as pounds (mass) per foot-second (lb/ft-s). The absolute viscosity depends primarily on temperature and to a lesser degree on pressure. The kinematic viscosity, υ, the absolute viscosity divided by the density of the fluid, is useful in certain fluid flow applications. The units for this quantity are length squared per time (e.g., ft^2/s). Because fluid viscosity changes rapidly with temperature, a numerical value of viscosity has no significance unless the temperature is specified.

2.3. Heat Capacity

The *heat capacity* of a substance is defined as the quantity of heat required to raise the temperature of a unit mass by 1°. The term *specific heat* is frequently used in place of *specific heat capacity*. This is not strictly correct because, traditionally, specific heat has been defined as the ratio of the heat capacity of a substance to the heat capacity of water. However, since the specific heat of water is approximately 1 cal/g-°C or 1 Btu/lb-°F, the term specific heat has come to imply heat capacity. For gases, the addition of heat to cause a 1° temperature rise may be accomplished either at constant pressure or at constant volume. Since the amounts of heat necessary are different for the two cases, subscripts are used to identify which heat capacity is being used – Cp for constant pressure, C_v, for constant volume. This distinction does not have to be made for liquids and solids, since there is little difference between the two. Values of heat capacity are available in the literature.[1]

2.4. Thermal Conductivity

Experience has shown that when a temperature difference exists across a solid body, heat energy will flow from the region of high temperature to that of low temperature until thermal equilibrium is reached. This mode of heat transfer, in which vibrating molecules pass kinetic energy through a solid, is called *conduction*. Liquids and gases also transport heat in this fashion. *Thermal conductivity* provides a measure of how fast heat flows through a substance, and is defined as the amount of heat that flows in a unit of time through a unit surface area of unit thickness as a result of a unit difference in temperature. Typical units for thermal conductivity are Btu/h-ft-°F.

2.5. Diffusivity

The *diffusion coefficient*, or *diffusivity* D, is a measure of the rate of transfer of one substance through another by molecular diffusion. The diffusivity for component A migrating through a solute B, for example, is given by the ratio of the flux J_A (mass transferred per unit time through a unit area) to its concentration gradient dC_A/dz:

$$D = -J_A/(dC_A/dz) \quad (2.11)$$

The negative sign indicates that diffusion occurs in the direction of concentration decrease. Diffusivity is a function of temperature, pressure, concentration, phase, and the number and nature of the other components within the mixture.

2.6. Vapor Pressure

Vapor pressure is an important property of liquids and, to a much lesser extent, of solids. If a liquid is allowed to evaporate in a confined space, the pressure of the vapor phase increases as the amount of vapor within the space increases. If there is sufficient liquid present, the pressure in the vapor space eventually comes to equal exactly the pressure exerted by the liquid at its own surface. At this point, a dynamic equilibrium exists in which vaporization and condensation take place at equal rates and the pressure in the vapor space remains constant. This gas pressure exerted at equilibrium is called the vapor pressure of the liquid. Solids, like liquids, also exert a vapor pressure. Evaporation of solids (sublimation) is noticeable only for the few solids characterized by appreciable vapor pressures.

2.7. Boiling Point

The *boiling point* is the temperature at which the vapor pressure of a liquid is equal to its surrounding pressure. Since the vapor pressure remains constant during boiling, so does the liquid temperature. The higher the system pressure, the higher the temperature must be to induce boiling. The boiling point is specific to each individual substance and is a strong function of pressure.

2.8. Freezing Point

The *freezing point* is the temperature at which the liquid and solid state can co-exist in equilibrium at a given pressure. At this point the rate at which the substance leaves the solid state equals the rate at which it leaves the liquid state.

3. CHEMICAL PROPERTIES

3.1. Flammability (or Explosion) Limit

Flammability limits for a flammable gas define the concentration range of a gas-air mixture within which an ignition source can start a self-propagating reaction. The minimum and maximum fuel concentrations in air that will produce a self-sustaining reaction under given conditions are called the lower flammability limit (LFL) and the upper flammability limit (UFL), respectively. The abbreviations LEL and UEL, for lower and upper explosivity limits, are sometimes used. The range of flammability limits becomes greater when the ignition energy is higher, moving the UFL to a higher concentration. The flammability limits also increase when the initial pressure and /or initial temperature at which the ignition source is activated increases. In addition, the increase in pressure and temperature will increase the rate at which a flame propagates through a gas. Changing the oxygen content or adding an inert gas to the gas-air mixture can significantly alter flammability limits. The heat capacity of the diluent, or inert gas, plays a role in flammability because the diluent will act as a heat sink. Thus, carbon dioxide is a better diluent than nitrogen because it has a higher heat capacity. The flammability limits also increase for drier mixtures. Tables of flammability limits of flammable gases are available in the literature.[2]

The flammability of a gas mixture can be calculated by using Le Chatelier's law given the flammability of the gas components:

$$LFL_{mix} = \frac{1.0}{(f_1/LFL_1) + (f_2/LFL_2) + \ldots + (f_n/LFL_n)} \quad (2.12)$$

$$UFL_{mix} = \frac{1.0}{(f_1/UFL_1) + (f_2/UFL_2) + \ldots + (f_n/UFL_n)} \quad (2.13)$$

where $f_1, f_2, \ldots, f_n$ = volume or mole fraction of each of the *n* components; LFL_{mix}, UFL_{mix} = mixture lower and upper flammability limits respectively, in volume or mole fraction $LFL_1, LFL_2, \ldots, LFL_n$ = component lower flammability limits in volume or mole fraction; $UFL_1, UFL_2, \ldots, UFL_n$ = component upper flammability limits in volume or mole fraction.

If data are not available for a particular gas mixture, it is possible to estimate the flammability limit by taking data for a similar material and applying Equation 2.14:

$$LFL_A = (M_B/M_A) LFL_B \quad (2.14)$$

where M_A and M_B are the relevant molecular weights of components A and B, respectively.

3.2. Flash Point

The *flash point* of a flammable liquid is defined as the temperature at which the vapor of the liquid is the same as the vapor pressure corresponding to the lower flammability limit concentration. The three major methods of measuring the flash point are

1. The Cleveland open cup method,
2. The Penskey-Martens closed cup tester, and
3. The tag closed cup method

These are experimental tests that measure the lowest temperature at which application of a test flame causes the vapor overlying the sample to ignite. Tables of flash points for selected substances are available in the literature.[3]

Since liquids are condensed and fuel vapor (not the liquid fuel) is the ignition participant, the ignition concepts developed for gases often apply to liquids as well. In general, liquid reactions are inclined to be more sluggish and to have longer time lags than gaseous reactions. Time is required for the liquid to vaporize so that it can mix with air and become ignited.

The physical properties of a flammable solid, such as hardness, texture, waxiness, particle size, melting point, plastic flow, thermal conductivity, and heat capacity, impart a wide range of characteristics to the flammability of solids. A solid ignites by first melting and then producing sufficient vapor, which in turn combines with air to form a flammable mixture.

3.3. Autoignition Temperature (AIT)

The *autoignition temperature* (AIT) or the *maximum spontaneous ignition temperature* is defined as the minimum temperature at which combustion occurs in a bulk gas mixture when the temperature of a flammable gas-air mixture is raised in a uniformly heated apparatus. The AIT represents a threshold below which chemicals and combustibles can be handled safely. The AITs of selected substances are available in the literature.[4] The AIT is strongly dependent on the nature of hot surfaces. The AIT may be reduced by as much as 100-200°C when the surfaces are contaminated by dust. When the temperature of a flammable mixture is raised to or above the autoignition temperature, ignition is not instantaneous. Most notably in liquids, there is a finite delay before ignition takes place, that is, a lapse between the time a flammable mixture reaches its flame temperature and the first appearance of a flame. An equation that correlates with the ignition temperature is also available in the literature.[4]

3.4. Heat (Enthalpy) of Reaction

Many chemical reactions evolve or absorb heat. When applying energy balances (see Chapter 3) in technical calculations the heat (enthalpy) of reaction is often indicated in mole units so that they can be directly applied to show its chemical change.

To simplify the presentation that follows, examine the equation:

$$aA + bB \rightarrow cC + dD \quad (2.15)$$

The heat (enthalpy) of reaction, ΔH, is given by:

$$\Delta H = c\,(\Delta H_f)_C + d\,(\Delta H_f)_D - a\,(\Delta H_f)_A - b\,(\Delta H_f)_B \quad (2.16)$$

Thus, the heat (enthalpy) of a reaction is obtained by taking the difference between the heat (enthalpy) of formation of products and reactants. If the heat (enthalpy) of reaction is negative (exothermic), as is the case of most combustion reactions, then energy is liberated due to chemical reaction. Energy is absorbed if ΔH is positive (endothermic).

The standard heat of reaction, $\Delta H°$, is given by:

$$\Delta H° = c\,(\Delta H_f°)_C + d\,(\Delta H_f°)_D - a\,(\Delta H_f°)_A - b\,(\Delta H_f°)_B \quad (2.17)$$

Thus, the standard heat (enthalpy) of a reaction is obtained by taking the difference between the standard heat (enthalpy) of formation of products and reactants. Once again if the standard heat (enthalpy) of reaction of formation is negative (exothermic), as is the case of most combustion reactions, then energy is liberated due to chemical reaction. Energy is absorbed if $\Delta H°$ is positive (endothermic).

Tables of standard enthalpies of formation combustion and reaction are available in the literature[5] for a wide variety of compounds. It is important to note that these are valueless unless the stoichiometric equation and the state of reactants and products are included.

Enthalpy of reaction and standard enthalpy of reaction are not always employed in engineering reaction/combustion calculations. The two other terms that have been used are the gross (or higher) heating value and the net (or lower) heating value. These are discussed later in this section.

3.5. Heat (Enthalpy) of Combustion

The *heat (enthalpy) of combustion* at temperature T is defined as the enthalpy change during the chemical reaction where 1 mole of material is burned in oxygen, where all reactants and products are at temperature T. This quantity finds extensive application in calculating enthalpy changes for incineration reactions, and is often given in the literature for a standard state temperature of 60°F (16°C). Although much of the literature data on standard heats of reaction are given for 25°C, there is little sensible enthalpy change between these two temperatures and the two sets of data may be considered compatible. Thus, the standard heat enthalpy of combustion is obtained by taking the difference between the standard heat of formation $\Delta H°_f$ of the products and that of the reactants. If the standard heat of combustion is negative (exothermic), as is the case with most combustion /incineration reactions, energy is liberated as a result of the reaction. Energy is absorbed if enthalpy of combustion is positive (endothermic). Extensive standard heat of formation and standard enthalpy of combustion data at 298 K are provided in the literature.[6]

3.6. Gross Heating Value

The heating value of a waste and/or fuel is one of its most important chemical properties. This value represents the amount of heat evolved in the complete reaction (usually combustion) of a given quantity of the waste and/or fuel.

The *gross heating value* or *higher heating value* (HHV) represents the enthalpy change or heat released when a compound is stoichiometrically combusted at 60°F, with the final (flue) products at 60°F and any water present in the liquid state. Stoichiometric combustion requires that no oxygen be present in the flue gas following combustion of the starting fuel.

3.7. Net Heating Value

Due to the varying amounts of water that may be produced in a combustion reaction, a second method for expressing heating value is also in use, the *net heating value*. The net heating value is similar to the gross heating value except that it is calculated with the produced water in the vapor state. The net heating value is also known as the *lower heating value* (LHV).

3.8. Theoretical Adiabatic Flame Temperature (TAFT)

Most combustion reactions do not operate at stoichiometric or zero percent excess air. Incomplete combustion and high carbon monoxide levels would result under this condition. If all the heat liberated by the reaction goes into heating up the products of combustion (the flue gas) the temperature achieved is defined as the *flame temperature*. If the reaction process is conducted adiabatically, with no heat transfer loss to the surroundings, the final temperature achieved in the flue gas is defined as the *adiabatic flame temperature*. If the combustion process is conducted with theoretical or stoichiometric air (0% excess air), the resulting temperature is defined as the *theoretical flame temperature*. Theoretical or stoichiometric air is defined as that exact amount of air required to completely react with a compound to produce oxidized end products. Any air in excess of this stoichiometric amount is defined as excess air.

If the reaction is conducted both adiabatically and with stoichiometric air, the resulting temperature is defined as the *theoretical adiabatic flame temperature (TAFT)*. It represents the maximum temperature the products of combustion (flue) can achieve if the reaction is conducted both stoichiometrically and adiabatically. For this condition, all the energy liberated from combustion at or near standard conditions (ΔH_c° and/or ΔH°_{298}) appears as sensible heat in raising the temperature of the flue products, ΔH_p, that is:

$$\Delta H_C^\circ + \Delta H_P = 0 \qquad (2.18)$$

where ΔH_C° = standard heat of combustion at 25°C; and ΔH_P = enthalpy change of the products as the temperature increases from 25°C to the theoretical adiabatic flame temperature.

4. BIOLOGICAL PROPERTIES

4.1. Biochemical Oxygen Demand (BOD)

Over the years, several different processes have been developed for the treatment of wastewater. Many, including most biological processes, are based on technology originally developed for municipal wastewater treatment plants. One of the key parameters of a wastewater is its oxygen demand: the amount of oxygen needed to convert the organic materials in the wastewater to oxidized end products. *Biochemical oxygen demand (BOD)* is the amount of oxygen used by microbes to oxidize waste materials, usually measured for a standardized 5 day period. BOD_5 is the designation used for this biochemical oxygen demand of waste expressed over a 5 day incubation period at 20°C. The end products of this biochemical oxidation reaction are carbon dioxide, water, partially oxidized materials, and more microbes.

4.2. Chemical Oxygen Demand (COD)

Chemical oxygen demand (COD) is the amount of oxygen, expressed in parts per million (ppm) or milligrams per liter (mg/L), consumed under high temperature, and low pH conditions in the chemical reaction of organic material in a waste stream with a strong oxidant, potassium permanganate. This number is particularly relevant to industrial wastes where toxicity to biological systems could be suspected. If a waste stream has a high COD and no BOD, waste toxicity should be investigated. The end products of this chemical oxidation reaction are carbon dioxide, water, and other inorganic species.

4.3. Ultimate Oxygen Demand (UOD)

Ultimate oxygen demand (UOD) is a calculated oxygen demand that assumes oxidation of all species to their most highly oxidized, stable form, e.g., CO_2, H_2O, etc. In general the chemical oxygen demand (COD) will be less than the ultimate oxygen demand (UOD). This results because the COD test does not completely oxidize all species. The BOD of a waste is always less than the COD, and generally only 50% to 70% of the UOD. This is due both to the inability of the microbes to completely oxidize all materials and to the incorporation of some of the partially oxidized materials into the microbial cell mass.

5. PROPERTIES OF MIXTURES

5.1. pH

An important chemical property of an aqueous solution is its pH value, which indicates the acidity or basicity of the substance. In a neutral solution such as pure water, the hydrogen (H^+) and hydroxyl (OH^-) ion concentrations are equal. At ordinary temperatures, this concentration is:

$$[H^+] = [OH^-] = 10^{-7} \text{ g-ion/L} \tag{2.19}$$

The notation g-ion is the unit for gram-ion, which represents Avogadro's number (6.02×10^{23}) of ions. The pH is a direct measure of the hydrogen ion concentration and is defined by:

$$pH = -\log[H^+] \tag{2.20}$$

Thus, an acidic solution is characterized by a pH below 7 (the lower the pH, the higher the acidity), a basic solution by a pH above 7, and a neutral solution by a pH of 7.

5.2. Partial Pressure

In engineering practice, mixtures of gases are more often encountered than single or pure gases. The ideal gas law (see Chapter 3) is based on the number of molecules present in the gas volume; the nature of molecules is not of significance. This law applies equally well to mixtures and to pure gases. Since pressure is caused by gas molecules colliding with the walls of a container, it seems reasonable that the total pressure of a gas mixture is made up of pressure contributions from each of the component gases. These pressure contributions are called partial pressures. Dalton defined the partial pressure of a component as the pressure that would be exerted if the same mass of the component gas occupied the same total volume alone at the same temperature as the mixture. The sum of these partial pressures equals the total pressure:

$$P = P_1 + P_2 + P_3 + \ldots + P_n = \sum_{i=1}^{n} P_i \tag{2.21}$$

For ideal gases, partial pressure is defined by:

$$P_i = y_i P \tag{2.22}$$

where y_i is the mole fraction of component i.

5.3. Humidity

Curves showing the relative humidity (ratio of the mass of the water vapor in the air to the maximum water vapor the air can hold at that temperature, i.e, if the air were saturated) of humid air appear on psychrometric charts.[7] The curve for 100% relative humidity is also referred to as the saturation curve. The abscissa of the humidity chart is air temperature, also known as the dry-bulb temperature (TDB). The wet-bulb temperature (TWB) is another measure of humidity. It is the temperature at which a thermometer with a wet wick wrapped around the bulb stabilizes. As water evaporates from the wick to the ambient air the bulb is cooled; the rate of cooling depends on how humid the air is. No evaporation occurs if the air is saturated with water, hence at that point the TWB and TDB are the same.

The lower the humidity, the greater the difference between these wet and dry bulb temperatures. On a psychometric chart, constant wet-bulb temperature lines are straight with negative slopes. Finally, the value of TWB corresponds to the value of the abscissa at the point of intersection of this line with the saturation curve.

6. ILLUSTRATIVE EXAMPLES

Example 2.1. The temperature in an air pollution control device is 2000°F. Convert this to °R, °C and K.

SOLUTION: Calculate the temperature in °R.

$$°R = °F + 460 = 2000 + 460 = 2460°R$$

Calculate the temperature in °C.

$$°C = (5/9) (°F - 32) = (5/9) (2000-32) = 1093°C$$

Calculate the temperature in K.

$$K = °C + 273 = 1093 + 273 = 1366 \text{ K}$$

Example 2.2. A mixture contains 20 lb of O_2, 2 lb of SO_2, and 3 lb of SO_3. Determine the weight and mole fractions of each component.

SOLUTION: Determine the molecular weight of each component.

Molecular weight of O_2 = 32 g/gmol
Molecular weight of SO_2 = 64 g/gmol
Molecular weight of SO_3 = 80 g/gmol

Calculate the weight fractions of each component.

Compound	Weight	Weight Fraction
O_2	20	20/25 = 0.8
SO_2	2	0.08
SO_3	3	0.12
	25	1.0

Calculate the mole fraction of each component.

Compound	Weight	Molecular Weight	Moles	Mole fraction
O_2	20	32	20/32 = 0.625	0.625/0.725 = 0.8621
SO_2	2	64	0.0625	0.0862
SO_3	3	80	0.0375	0.05172
			0.725	1.0

Example 2.3. A wastewater contains 250 mg/L aminobenzoic acid, $H_2NC_6H_4CO_2H$. What is the ultimate oxygen demand of this water? The molecular weight of aminobenzoic acid is 137.13 g/gmol.

SOLUTION: Determine the number of moles of aminobenzoic acid in 1 L of solution.

Number of moles = (250 mg/L)/(137.13 g/gmol) = 1.82 millimol/L

Also determine the end products from oxidizing 1 mole of acid and the atoms of oxygen (O) required.

Formula: $H_2NC_6H_4CO_2H = C_7H_7NO_2$. The end products are:

CO_2: 7 gmol, O_2: 14 atoms of O
H_2O: 3.5 gmol, O_2: 3.5 atoms of O
NO_3: 1 gmol, O_2: 3 atoms of O

The total oxygen needed is the atoms of O = 14 + 3.5 + 3 = 20.5. Since the amount of oxygen already present is 2 atoms, the net or ultimate oxygen demand is:

UOD = 20.5 - 2 = 18.5 atoms of oxygen = 9.25 moles O_2

7. SUMMARY

1. Physical and chemical properties are important considerations in any study of pollution prevention. A substance may exhibit certain characteristics under one set of conditions of temperature, pressure, and composition. However, conditions are often changed when an attempt is made to implement a pollution prevention program. To promote a better understanding of these effects, some definitions and certain key physical and chemical properties were summarized in the chapter.

2. Temperature can be measured in many ways; the most common method makes use of the expansion of mercury with increasing temperature. (On process equipment, however, thermocouples or thermistors are more commonly employed.). The most commonly used temperature scales are the Celsius and Fahrenheit scales. The two absolute temperature scales, the Kelvin and Rankine scales, are also commonly used in science and engineering applications.

3. A number of units are used to express a pressure measurement. Some are based on a force per unit area: for example, pound (force) per square inch (psi) or dyne per square centimeter (dyne/cm^2). Others are based on a fluid height, such as inches of water (in H$_2$O) or millimeters of mercury (mmHg). Barometric pressure is a measure of the ambient air pressure. Standard barometric pressure is 1 atm and is equivalent to 14.7 psi and 29.92 in Hg.

4. Trace quantities of substances are very often expressed in parts per million (ppm) or, if the amount is even smaller, as parts per billion (ppb). These are expressed on a volume basis for gases, and the following equations apply.

$$ppm_v = 10^6 \, y_A$$
$$ppm_v = 10^3 \, ppb_v$$

5. The density of a substance is the ratio of its mass to its volume and may be expressed in units of pounds per cubic foot (lb/ft^3) or kilograms per cubic meter (kg/m^3).

6. Viscosity is a property associated with a fluid's resistance to flow. The absolute viscosity, μ, has units of mass per length-time; the fundamental unit is the poise, which is defined as 1 g/cm-s.

7. The heat capacity of a substance is defined as the quantity of heat required to raise the temperature of that substance by 1°; the specific heat capacity is the heat capacity on a unit mass basis.

8. Flammability limits for a flammable gas define the concentration range of a gas-air mixture within which an ignition source can start a self-propagating reaction. The minimum and maximum fuel concentrations in air that will produce a self-sustaining reaction under given conditions are called the lower flammability limit and upper flammability limit, respectively.

9. The flash point of a flammable liquid is a temperature at which the vapor pressure of the liquid is the same as the vapor pressure corresponding to the lower flammability limit concentration.

10. The autoignition temperature, or the maximum spontaneous ignition temperature, is defined as the minimum temperature at which combustion occurs in a bulk gas mixture when the temperature of a flammable gas-air mixture is raised in a uniformly heated apparatus.

11. The heat (enthalpy) of a reaction is obtained by taking the difference between the heat (enthalpy) of formation of products and reactants. If the heat (enthalpy) of reaction is negative (exothermic), as is the case for most combustion reactions, then energy is liberated due to chemical reaction. Energy is absorbed if the heat enthalpy of reaction is positive (endothermic). The standard heat (enthalpy) of a reaction is obtained by taking the difference between the standard heat (enthalpy) of formation of products and reactants.

12. The heat (enthalpy) of combustion at a given temperature is defined as the enthalpy change during a chemical reaction where 1 mole of material is burned in oxygen, where all reactants and products are at that temperature.

13. The gross heating value of a waste and/or fuel is one of its most important chemical properties. This value represents the amount of heat

evolved in the complete combustion reaction of a given quantity of the waste and/or fuel.

14. Due to varying amounts of water that may be involved in the products formed as a result of combustion of a chemical, a second heating value known as net heating value is also in use. The net heating value is similar to the gross heating value except the product water is in the vapor state.

15. If a combustion process is conducted both adiabatically and with stoichiometric air, the resulting temperature of the flue gas is defined as the theoretical adiabatic flame temperature.

16. An important chemical property of an aqueous solution is its pH value which indicates the acidity or basicity of the solution. In a neutral solution, such as pure water, the hydrogen (H^+) and hydroxyl (OH^-) ion concentrations are equal.

17. In engineering practice, mixtures of gases are more often encountered than single or pure gases. Dalton defined the partial pressure of a component as the pressure that would be exerted if the same mass of the component gas occupied the same total volume alone at the same temperature as the mixture. The sum of these partial pressures equals the total pressure.

18. Absolute humidity is the ratio of the mass of water vapor to the mass of dry gas in a system. Relative humidity is a measure of how much vapor the gas contains relative to the maximum amount it can hold.

19. The biochemical oxygen demand is the amount of oxygen used by microbes to oxidize waste materials, usually measured for 5 days at 20°C in the dark.

20. Chemical oxygen demand is the equivalent amount of oxygen required to chemically oxidize waste materials to end products of carbon dioxide, water and other oxidized inorganic species.

21. Ultimate oxygen demand is a calculated oxygen demand that assumes oxidation of all species to their most highly oxidized, stable form.

8. PROBLEMS

1. The height of a liquid column of mercury (density = 848.7 lb/ft^3) open to the atmosphere is 2.493 ft. Determine the pressure at the base of the column in psia, psig, and in inches of water.

2. An analysis of each gas and their mole percent from the analysis of a flue gas is given below:

$$N_2 - 79\%$$
$$O_2 - 5\%$$
$$CO_2 - 10\%$$
$$CO - 6\%$$

Determine the average molecular weight of the flue gas.

3. What is the pH of a solution whose hydrogen ion concentration is 10^{-5} g-ion/L? State whether the solution is acidic or basic.

4. A flue gas (MW = 30 g/gmol, dry basis) is being discharged from a scrubber at 180°F (dry bulb) and 125°F (wet bulb). The gas flow rate on a dry basis is 10,000 lb/h. The humidity at the dry bulb temperature of 180°F is 0.0805 lb H_2O/lb dry air. Determine the mass and molar flow rate of the wet gas.

5. What are the structural and molecular formula for the following toxic compounds?
   a. Tetrachloroethane
   b. Formaldehyde
   c. Carbon tetrachloride

9. REFERENCES

1. Theodore, L., Reynolds, J. R,, and Taylor, F., *Accident and Emergency Management*, Wiley-Interscience, New York, New York, 1989.
2. The World Bank, *Manual of Industrial Hazard Assessment Techniques*, Office of Environmental and Scientific Affairs, London, 1985.
3. Lees, F. P., *Loss Prevention in the Process Industries*, Vol.1, Butterworth, Boston, 1980.
4. Semenov, N. N., *Some Problems in Chemical Kinetics And Reactivity*, Pergamon Press, London, 1959.
5. Theodore, L., and Reynolds, J. R., *Introduction to Hazardous Waste Incineration*, Wiley-Interscience, New York, New York, 1988.
6. Theodore, L., and Moy, J. E., *A Theodore Tutorial, Introduction to Hazardous Waste Incineration*, Theodore Tutorials, East Williston, New York, 1994.
7. Green, D., and Perry, J., *Perry's Chemical Engineer's Handbook*, 7th ed., McGraw-Hill, New York, New York, 1998.

CHAPTER 3

Conservation Laws and Basic Principles
Contributing Author: Sean Guinan

This chapter examines and reviews some of the fundamental principles that engineers and scientists employ in performing design calculations and predicting the performance of plant equipment. Topics include the conservation laws for both mass and energy, the ideal gas law, phase equilibrium, stoichiometry, thermochemistry, chemical reaction equilibrium, and chemical kinetics. These basic principles will assist the reader in acquiring a better understanding of some of the material that appears later in this book.

1. CONSERVATION OF MASS[1]

The conservation law for mass can be applied to any process or system. The general form of this law is given in Equation 3.1:

$$\text{mass in - mass out + mass generated = mass accumulated} \qquad (3.1)$$

Equation 3.1 may be applied to the total mass involved or to a particular species, on either a mole or mass basis. The conservation law for mass can be applied to steady-state or unsteady-state processes and to batch or continuous systems. To isolate a system for study, it is separated from its surroundings by constructing a control boundary or envelope around it. This control boundary may be real (e.g., the walls of a vessel) or imaginary. Mass crossing the boundary and entering the system is part of the mass-in term in Equation 3.1, whereas that leaving the system is part of the mass-out term. Equation 3.1 may be written for any compound whose quantity is not changed by chemical reaction, or for any chemical element, regardless of whether it has participated in a chemical reaction. It may be written for one piece of equipment, around several pieces of equipment, or around an entire process. It may be used to calculate an unknown quantity directly, to check the validity of experimental data, or to express one or more of the independent relationships among the unknown quantities in a particular problem.

A steady-state process is one in which there is no change in conditions (temperature, pressure, etc.) or rates of flow with time at any given point in the system. Under steady-state conditions, the accumulation term in Equation 3.1 is zero. If there is no chemical reaction, the generation term is also zero. All other processes are classified as unsteady-state. In a batch process, the container holds the product or products. In a continuous process, reactants are continuously fed to a piece of equipment or to several pieces in series, and products are continuously removed from one or more points. A continuous process may or may not be operating at steady-state. As indicated previously,

Equation 3.1 may be applied to the total mass of each stream (referred to as an overall or total material balance) or to the individual components of the streams (referred to as a componential or component material balance). Often the primary task in preparing a material balance is to develop the quantitative relationships among the inflow and outflow streams.

The conservation law for mass finds its major application in performing resource utilization and pollution prevention audits. A *waste minimization opportunity assessment* is a systematic, planned procedure with the objective of identifying ways to reduce or eliminate waste. The procedure consists first of a careful review of a plant's operation and waste streams, and the selection of specific areas to assess. After a specific waste stream or area is established as the focus of the resource utilization and pollution prevention audit, a number of options that can potentially minimize materials use and waste production are developed and screened. Third, the selected options are evaluated for technical and economic feasibility. Finally, the most promising options are selected for implementation.[2] Additional details on the assessment process, providing a direct application of the conservation law for mass, are provided in Chapter 11.

2. CONSERVATION OF ENERGY

A presentation of the conservation law for energy would be incomplete without a brief review of some introductory thermodynamic principles. Thermodynamics is defined as the science that deals with the relationships among the various forms of energy.

A system may possess energy as a result of its temperature, velocity, position, molecular structure, surface, etc. The energies corresponding to these conditions are internal, kinetic, potential, chemical, surface, etc. Engineering thermodynamics is founded on three basic laws. Energy, like mass and momentum, is conserved. Application of the conservation law for energy gives rise to the first law of thermodynamics. This law, in steady-state form for batch processes, is presented here (potential, kinetic, and other energy effects are not considered).

$$\Delta E = Q - W \tag{3.2}$$

For flow process,

$$\Delta H = Q - W \tag{3.3}$$

where Q = heat energy transferred across the system boundary; w = work energy transferred across the system boundary; W = mechanical work energy transferred across the system boundary; E = internal energy of the system; H = enthalpy of the system, and ΔE, ΔH = changes in internal energy and enthalpy during the process, respectively.

The internal energy and enthalpy in Equations 3.2 and 3.3, as well as in the other equations in this discussion, may be on a mass or a mole basis, or they may represent the total internal energy and enthalpy of the entire system. Most industrial facilities operate in a steady-state flow mode. If no significant mechanical or shaft work is added or withdrawn from the system, Equation 3.3 reduces to:

$$Q = \Delta H \qquad (3.4)$$

If a unit or system is operated adiabatically, where $Q = 0$, Equation 3.4 then becomes:

$$\Delta H = 0 \qquad (3.5)$$

Although the topics of material and energy balances are covered separately in this section, it should be emphasized that this segregation does not occur in reality; one must work with both energy and material balances simultaneously.

Perhaps the most important thermodynamic function the engineer works with is the enthalpy. Enthalpy is defined by:

$$H = E + PV \qquad (3.6)$$

where P and V are system pressure and volume, respectively.

The terms E and H are state or point functions. By fixing a certain number of variables on which the function depends, the numerical value of the function is automatically fixed; that is, it is single-valued. For example, fixing the temperature and pressure of a one-component, single-phase system immediately specifies the enthalpy and internal energy of the system. The change in enthalpy as the system undergoes a change in state from (T_1, P_1) to (T_2, P_2) is given by:

$$\Delta H = H_2 - H_1 \qquad (3.7)$$

The correlations needed to calculate the values of enthalpy are not presented here; rather, the reader is directed to the literature for this information.[3]

3. IDEAL GAS LAW

The ideal gas law was derived from experiments in which the effects of pressure and temperature on gaseous volumes were measured over a moderate range of temperatures and pressures. As a general rule, this law works best when the molecules of the gas are far apart, that is, when the pressure is low and the temperature is high. Under these conditions, the gas is said to behave ideally. For engineering calculations the ideal gas law is almost always assumed to be valid since it generally works well for the temperature and pressure ranges used in most engineering applications.

The two precursors of the ideal gas law are Boyle's law and Charles's law. Boyle found that the volume of a given mass of gas is inversely proportional to the absolute pressure if the temperature is kept constant; that is:

$$P_1V_1 = P_2V_2 \qquad (3.8)$$

where V_1 and V_2 are the volumes of gas at absolute pressures P_1 and P_2, respectively. Charles found that the volume of a given mass of gas varies directly with the absolute temperature at constant pressure:

$$V_1/T_1 = V_2/T_2 \qquad (3.9)$$

where V_1 and V_2 are the volumes of gas at absolute temperatures T_1 and T_2, respectively. Boyle's and Charles's laws may be combined into a single equation in which neither temperature nor pressure need be held constant:

$$P_1V_1/T_1 = P_2V_2/T_2 \qquad (3.10)$$

This equation indicates that for a given mass of a specific gas, PV/T has a constant value. Since, at the same temperature and pressure, volume and mass must be directly proportional, this statement may be extended to:

$$PV/nT = R \qquad (3.11)$$

where n is the number of moles and R is the universal gas constant. Equation 3.11 is called the ideal gas law. Numerically, the value of R depends on the units of P, V, T, and n. Tabulated values of R are provided in Table 3.1.

4. PHASE EQUILIBRIUM

The term *phase*, for a pure substance, indicates a state of matter, that is, solid, liquid, or gas. For mixtures, however, a more stringent connotation must be used, since a totally liquid or solid system may contain more than one phase. A phase is characterized by uniformity or homogeneity, i.e., the same composition and properties must exist throughout the phase region. At most temperatures and pressures, a pure substance normally exists as a single phase. At certain temperatures and pressures, two or perhaps even three phases can coexist in equilibrium.

The most important equilibrium phase relationship is that between liquid and vapor. Raoult's and Henry's laws theoretically describe liquid-vapor behavior and, under certain conditions, are applicable in practice. Raoult's law, sometimes useful for mixtures of components of similar structure, states that the partial pressure of any component in the vapor is equal to the product of the vapor pressure of the pure component and the mole fraction of that component in the liquid, that is:

$$P_i = P'_i x_i \qquad (3.12)$$

where P_i = partial pressure of component i in the vapor; P'_i = vapor pressure of pure component i; and x_i = mole fraction of component i in the liquid. If the gas phase is ideal, Equation 3.12 becomes:

Table 3.1. Values of R in Various Units.

R	Temperature Scale	Units of V	Units of n	Units of P
10.73	°R	ft^3	lbmol	psia
0.7302	°R	ft^3	lbmol	atm
21.85	°R	ft^3	lbmol	in Hg
555.0	°R	ft^3	lbmot	mmHg
297.0	°R	ft^3	lbmol	in H$_2$O
0.7398	°R	ft^3	lbmol	bar
1545.0	°R	ft^3	lbmol	psfa
24.75	°R	ft^3	lbmol	ft H$_2$O
1.9872	°R	—	lbmol	—
0.0007805	°R	—	lbmol	—
0.0005819	°R	—	lbmol	—
500.7	°R	—	lbmol	—
1.314	K	ft^3	lbmol	atm
998.9	K	ft^3	lbmol	mmHg
19.32	K	ft^3	lbmol	psia
62.361	K	L	gmol	mmHg
0.08205	K	L	gmol	atm
0.08314	K	L	gmol	ba
8314	K	L	gmol	Pa
8.314	K	m^3	gmol	Pa
82.057	K	cm^3	gmol	atm
1.9872	K	—	gmol	—
8.3144	K	—	gmol	—

$$y_i = (P'_i/P) x_i \qquad (3.13)$$

where y_i = mole fraction of component i in the vapor; and P = total system pressure.

Unfortunately, highly concentrated mixtures follow Raoult's law. A more empirical relationship used for representing data for many dilute ystems is Henry's law, valid for low solution concentrations or low values of x_i:

$$P_i = H_i x_i \qquad (3.14)$$

where H_i = Henry's law constant for component i (with units of pressure). Values for Henry's law constant can be found in the literature.[3]

Many air-water equilibrium calculations are accomplished using an air-water phase equilibrium constant K_i. This constant has been referred to in industry as a componental split factor since it provides the ratio of the mole fractions of a component in two equilibrium phases, air and water. In the environmental field it is typically designated as an equilibrium distribution coefficient. The defining equation for this constant is:

$$K_i = y_i/x_i \tag{3.15}$$

This equilibrium constant is a function of the system temperature and pressure, and the nature and mole fraction of these components in the system. However, as a first approximation, K_i is generally treated as a function of temperature and pressure only. For this condition, K_i may be approximated by:

$$K_i = P'_i/P \tag{3.16}$$

5. STOICHIOMETRY

When chemicals react, they do so according to a strict proportion. When oxygen and hydrogen combine to form water, the ratio of the amount of oxygen to the amount of hydrogen consumed is always 7.94 by mass and 0.500 by moles. The term *stoichiometry* refers to this phenomenon, which is sometimes called the *chemical law of combining weights*. The reaction equation for the combining of hydrogen and oxygen is

$$2\ H_2 + O_2 \rightarrow 2\ H_2O \tag{3.17}$$

In chemical reactions, atoms are neither generated nor consumed, merely rearranged with different bonding partners. The manipulation of the coefficients of a reaction equation so that the number of atoms of each element on the left is equal to that on the right is referred to as *balancing* the equation. Once the equation is balanced, the small whole number molar ratio that must exist between any two components of the reaction can be determined simply by observation: these are known as *stoichiometric ratios*. There are three such ratios (not counting the reciprocals) in the above reaction. These are:

2 mol H_2 consumed/mol O_2 consumed
1 mol H_2O generated/mol H_2 consumed
2 mol H_2O generated/mol O_2 consumed

The unit *mol* represents either the *gmol* or the *lbmol*. Using molecular weights, these stoichiometric ratios (which are molar ratios) may easily be converted to mass ratios. For example, the first ratio above may be converted to a mass ratio by using the molecular weights of H_2 (2.016) and O_2 (31.999) as follows:

(2 gmol H_2 consumed) (2.016 g/gmol) = 4.032 g H_2 consumed
(1 gmol O_2 consumed) (31.999 g/gmol) = 31.999 g O_2 consumed

The mass ratio between the hydrogen and oxygen consumed is therefore:

$$4.032/31.999 = 0.126 \text{ g } H_2 \text{ consumed/g } O_2 \text{ consumed}$$

Molar and mass ratios are used in material balances to find the amounts or flow rates of components involved in chemical reactions.

Multiplying a balanced reaction equation through by a constant does nothing to alter its meaning. The reaction shown in Equation 3.17 above is often written:

$$H_2 + 0.5 \, O_2 \rightarrow H_2O \qquad (3.18)$$

There are times, however, when care must be exercised because the solution to the problem depends on the way the reaction equation is written. This is the case with chemical equilibrium problems and problems involving thermochemical reaction equations.

6. THERMOCHEMISTRY

All chemical reactions either absorb or produce energy. Heat is the form of energy most commonly released or absorbed in any chemical reaction. The heat generated from a chemical reaction is transferred from the system to its surroundings. Any process that gives off heat is called an *exothermic process*. Conversely, any process in which heat has to be supplied to the system by the surroundings is called an *endothermic process*.[4]

Consider now the energy effects associated with a chemical reaction. To introduce this subject, the reader is reminded that engineers and applied scientists are rarely concerned with the magnitude or amount of energy in a system; their primary concern is with the changes that occur in the amount of energy in the system. It has been found in measuring energy changes for systems that enthalpy is the most convenient term to work with. There are many different types of enthalpy effects; these include sensible heat, latent heat, and heat of reaction.

As described in Chapter 2, the heat of reaction is defined as the enthalpy change of a system undergoing chemical reaction. If the reactants and products are at the same temperature and in their standard states, the heat of reaction is termed the standard heat of reaction. For engineering purposes, the standard state of a chemical may be taken as the pure chemical at 1 atm pressure. Heat of reaction data for many reactions are available in the literature.[3] Additional details on this topic are provided in Chapter 2; the reader is referred to the sections in that chapter dealing with: Heat of Reaction, Standard Heat of Reaction, Heat of Combustion, Net Heating Value, Gross Heating Value, and Theoretical Adiabatic Flame Temperature.

6.1. Chemical Reaction Equilibrium

With regard to chemical reactions, two important questions are of concern to the engineer:

1. How far will the reaction go?
2. How fast will the reaction go?

Chemical thermodynamics provides the answer to the first question; however, it tells nothing about the second. Reaction rates fall within the domain of chemical kinetics and are treated later in this section. Both equilibrium and kinetic effects must be considered in an overall engineering analysis of a chemical reaction.

Chemical reaction equilibrium calculations are structured around another thermodynamic term called the free energy. This so-called free energy (G) is a property that cannot be defined easily without some further basic grounding in thermodynamics. No such attempt is made here, and the reader is directed to the literature[2] as necessary to review basic concepts regarding free energy principles. Free energy has the same units as enthalpy and internal energy, and may be on a mole or total mass basis.

Consider the equilibrium reaction:

$$aA + bB \leftrightarrow cC + dD \qquad (3.19)$$

where A, B, C, D = chemical formulas of the reactants and products; and a, b, c, d = stoichiometric coefficients for the reaction. The double arrowed symbol is a reminder that the reacting system is at equilibrium.

For this reaction, the change in free energy, ΔG, is given by

$$\Delta G = cG_C + dG_D - aG_A - bG_B \qquad (3.20)$$

where G_i = free energy of species i.

The following equation is used to calculate the chemical reaction equilibrium constant K at a temperature T:

$$\Delta G_T = RT (\ln K) \qquad (3.21)$$

The problem that remains is to relate K to understandable physical quantities. For gas phase reactions the term K in Equation 3.21 may be approximately represented in terms of the partial pressures of the components involved. This relationship is given by Equations 3.22 and 3.23 as:

$$K = K_P \qquad (3.22)$$

where K is an equilibrium constant based on partial pressures and defined by:

$$K_P = P^c_C P^d_D / (P^a_A P^b_B) \qquad (3.23)$$

where P_A = partial pressure of component A, and so forth.

The definition of K_P obviously applies to the reaction of Equation 3.19. Assuming that a K value is available or calculable, this equation may be used to determine the partial pressures of the participating components at equilibrium. For liquid phase reactions, K is given approximately by:

$$K = K_C \tag{3.24}$$

where $K_C = C_C^c \, C_D^d / (C_A^a \, C_B^b)$; C_C = concentration of component C (e.g., gmol/L), and so forth.

6.2. Chemical Kinetics

Chemical kinetics involves the study of reaction rates and the variables that affect these rates. It is a topic that is critical for the analysis of reacting systems. The objective here is to develop a working understanding of this subject that will permit the application of chemical kinetics principles to the subject of pollution prevention. The topic here is treated from an engineering point of view, that is, in terms of physically measurable quantities.

The rate of a chemical reaction can be described in any of several different ways. The most commonly used definition involves the time rate of change in the amount of one of the components participating in the reaction. This rate is usually based on some arbitrary factor related to the reacting system size or geometry such as volume, mass, or interfacial area. The definition shown in Equation 3.25, which applies to homogeneous reactions, is a convenient one from an engineering point of view.

$$R_A = (1/V)(dn_A/dt) \tag{3.25}$$

where R_A = reaction rate based on component A; V = volume of reacting system; and n_A = number of moles of A at time, t. If the volume term is constant, one may write Equation 3.25 as:

$$R_A = dC_A/dt \tag{3.26}$$

where C_A is the molar concentration of A at time t. Based on Equation 3.26, the reaction rate is positive if species A is being formed (since C_A increases with time), and negative if A is reacting (since C_A decreases with time). The rate is zero if the system is at chemical equilibrium.

An equation expressing the rate in terms of measurable and/or desirable quantities may now be developed. Based on experimental evidence, the rate of reaction is a function of the concentration of the components present in the reaction mixture, temperature, pressure, and catalyst variables. In equation form:

$$R_A = R_A(C_i, P, T, \text{catalyst variables}) \tag{3.27}$$

Equation 3.27 may be condensed into:

$$R_A = (+/-)k_A f(C_i) \tag{3.28}$$

where k_A incorporates all the variables in Equation 3.27 other than the concentration variable. The (+/-) notation is included to indicate whether component A is being consumed (-) or produced (+). The k_A may be regarded as a constant of proportionality; it is termed the reaction velocity constant.

Although this "constant" is independent of concentration, it is a function of other variables. This term is very definitely influenced by temperature and catalyst activity. For most applications, it is assumed that k is solely a function of temperature. Thus,

$$R_A = (+/-)k_A(T) f(C_i) \tag{3.29}$$

The effect of temperature on k is generally represented by the *Arrhenius equation* and the values of k are obtained from experimental data. The Arrhenius equation is written as follows:

$$K = A\, e^{-(E_a/RT)} \tag{3.30}$$

where E_a = activation energy of the reaction, kJ/mol; R is the universal gas constant = 8.314 J/K-mol; T = absolute temperature, K; and e represents the base of the natural logarithm scale. The quantity A represents the *collision frequency factor*. It can be treated as a constant for a given reacting system over a wide temperature range, suggesting that the rate constant is directly proportional to A. Moreover, because of the minus sign associated with the exponent E_a/RT, the rate constant decreases with increasing activation energy and increases with temperature. This equation is also expressed in a more useful form by taking the natural logarithm of both sides:

$$\ln k = \ln A - E_a/RT \tag{3.31}$$

7. ILLUSTRATIVE EXAMPLES

Example 3.1. The following information is provided for an ideal gas: Pressure = 1.0 atm, temperature = 60°F, molecular weight of gas = 29 lb/lbmol.

Determine the density of the gas in lb/ft³.

SOLUTION. Rewrite the ideal gas law in terms of the density, ρ.

$$PV = nRT$$
$$n = m/MW$$

where m = mass of gas; and MW = molecular weight.

Thus,

$$PV = (m/MW)\, RT$$

and

$$m/V = \rho = P\,(MW)/RT$$

Substituting in the above equation gives:

$$\rho = P(MW)/RT$$
$$= (1 \text{ atm})(29 \text{ lb/lbmol})/[(0.73 \text{ atm-ft}^3/\text{lbmol-}°R)(60 + 460) °R]$$
$$= 0.0764 \text{ lb/ft}^3$$

Note: The choice of R is arbitrary, provided consistent units are employed. Since the molecular weight of the given gas is 29 lb/lbmol, the calculated density can be assumed to apply to air. The effect of pressure, temperature, and molecular weight on density can also be obtained directly from the ideal gas law. Increasing the pressure and molecular weight increases the density; while increasing the temperature decreases the density.

Example 3.2. An external gas stream is fed into an air pollution control device at a rate of 10,000 lb/h in the presence of 20,000 lb/h of air. Due to environmental constraints, 1,250 lb/h of a conditioning agent is added to assist the treatment of the gas stream. Determine the rate of product gases exiting the unit in lb/h.

SOLUTION. First, write the conservation law for mass to the control device on a rate basis.

$$\text{(rate of mass in)} - \text{(rate of mass out)} + \text{(rate of mass generated)}$$
$$= \text{(rate of mass accumulated)}$$

Apply this equation subject to the conditions in the problem statement. Remember that mass is not generated and steady-state conditions apply, therefore:

$$\text{(rate of mass in)} = \text{(rate of mass out)}$$

or simply

$$m_{in} = m_{out}$$

Substituting yields:

$$m_{in} = 10,000 + 20,000 + 1,250 = 31,250 \text{ lb/h}$$

Since $m_{in} = m_{out}$, the product gas flow rate is then:

$$m_{out} = 31,250 \text{ lb/h}$$

Example 3.3. The reaction equation for the combustion of propane is given below:

$$C_3H_8 + O_2 \rightarrow CO_2 + H_2O$$

1. Balance the above equation and determine the ratio of reactants to products.

2. Using the balanced equation, determine the scf of air required for stoichiometric combustion of 1.0 scf propane (C_3H_8) and calculate both the stoichiometric air requirements and flue gas produced from the complete combustion of propane.

SOLUTION. The balanced stoichiometric equation is:

$$C_3H_8 + 5\ O_2 \rightarrow 3\ CO_2 + 4\ H_2O$$

Although the number of moles on both sides of the equation do not balance, the masses of reactants and products (in accordance with the conservation law for mass) must balance.

The scf of O_2 required for the complete combustion of 1 scf of propane can now be determined:

$$\text{scf of } O_2 = 5 \text{ scf}$$

Thus, there are 5 moles of O_2 required per mole of propane since moles are directly proportional to the volume (scf). From a mass balance, 44 lb of propane have reacted with 160 lb of O_2 to form 132 lb of CO_2 and 72 lb of H_2O with an initial and final total mass of 204 lb.

The amount of N_2 in a quantity of air that contains 5 scf of O_2 is given by the following knowing that the nitrogen to oxygen volume (or mole) ratio in air is 79/21:

$$\text{scf of } N_2 = (79/21)\,(\text{scf of } O_2) = (79/21)\,(5) = 18.81 \text{ scf}$$

The stoichiometric amount of air is then:

$$\text{scf of air} = \text{scf of } N_2 + \text{scf of } O_2 = 18.81 + 5 = 23.81 \text{ scf}$$

In order to calculate the amount of flue gas produced, one must include the nitrogen from the air in addition to the products of combustion, i.e., carbon dioxide and water vapor.

$$\text{scf of flue gas} = \text{scf of } N_2 + \text{scf of } CO_2 + \text{scf of } H_2O$$
$$= 18.81 + 3.0 + 4.0 = 25.81 \text{ scf}$$

8. SUMMARY

1. The conservation law for mass can be applied to any process or system. The general form of this law is given by:

mass in - mass out + mass generated = mass accumulated

2. A system may possess energy as a result of its temperature, velocity, position, molecular structure, surface, etc. The energies corresponding to

these conditions are internal, kinetic, potential, chemical, surface, etc. Engineering thermodynamics is founded on three basic laws. Energy, like mass and momentum, is conserved. Application of the conservation law for energy gives rise to the first law of thermodynamics.

3. For engineering calculations the ideal gas law is almost always assumed to be valid, since it generally works well for the temperature and pressure ranges used in most engineering applications.

4. The most important equilibrium phase relationship in pollution prevention applications is that between liquid and vapor. Raoult's and Henry's laws theoretically describe liquid-vapor behavior and are applicable in practice under concentrated and dilute solution concentrations conditions, respectively.

5. In chemical reactions, atoms are neither generated nor consumed, merely rearranged with different bonding partners. The manipulation of the coefficients of a reaction equation so the number of atoms of each element on the left is equal to that on the right is referred to as *balancing* the equation. Once the equation is balanced, the small whole number ratio that must exist between any two components of the reaction can be determined simply by observation. These are known as *stoichiometric ratios*.

6. Another important consideration is the energy effects associated with a chemical reaction. The heat of reaction is defined as the enthalpy change of a system undergoing chemical reaction.

7. With regard to chemical reactions, two important questions are of concern to the engineer: How far will the reaction go? and How fast will the reaction go?

8. Chemical kinetics involves the study of reaction rates and the variables that affect these rates. It is a topic that is critical for the analysis of reacting systems. The rate of a chemical reaction can be described in any of several different ways. The most commonly used definition involves the time rate of change in the amount of one of the components participating in the reaction. This rate is usually based on some arbitrary factor related to the reacting system size or geometry, such as volume, mass, or interfacial area.

9. PROBLEMS

1. The following information is provided for an ideal gas:

   Standard volumetric flow rate of a gas stream = 12,000 scfm
   Standard conditions = 600°F, 1 atm
   Actual operating conditions = 700°F, 1 atm

   Determine the actual volumetric flow rate in acfm.

2. Atmospheric concentrations of toxic pollutants are usually reported using two types of units: mass of pollutant per volume of air, i.e., mg/m^3, mg/m^3, ng/m^3, etc.; or parts of pollutant per parts of air (by volume), i.e., ppm_v, ppb_v, ppt_v, etc. In order to compare data collected at different conditions, actual concentrations are often converted to standard temperature

and pressure (STP). If the concentration of chlorine vapor is measured to be 5 mg/m³ at a pressure of 600 mmHg and a temperature of 10°C, determine the concentration in units of ppm$_v$ and in units of mg/m³ (STP = 25°C, 1 atm).

3. Calculate the time required to achieve a 99.99% conversion of benzene at a temperature of 980°C. The Arrhenius constants are:

Frequency factor (A) = 3.3 x 10¹⁰ s
Activation energy (E) = 35,900 cal/gmol.

4. The following reaction velocity constant data were obtained for the reaction between two inorganic chemicals.

T, °C	k, L/gmol-h
0	5.20
20	12.0
40	21.0
60	39.0
80	60.0
100	83.0

Using the values of k at 40°C and 100°C, calculate the constants of the Arrhenius equation. Use these to obtain k at 75°C.

5. A total of 18.7 x 10⁶ Btu/h of heat are transferred from the flue gas of an incinerator. Calculate the outlet temperature of the gas stream using the following information.

Average heat capacity, C_P, of gas = 0.26 Btu/lb-°F
Gas mass flow rate, m = 72,000 lb/h
Gas inlet temperature, T_i = 1200°F.

6. A flue gas (MW = 30 g/gmol, dry basis) is being discharged from a scrubber at 180°F (dry bulb) and 125°F (wet bulb). The gas flow rate on a dry basis is 20,000 lb/h.

    a. What is the dew point of the wet gas?
    b. What is the mass flow rate of the wet gas in lb/h?
    c. What is the actual volumetric flow rate of the wet gas in acfh and acfm?

10. REFERENCES

1. Theodore, L., and Reynolds, J. R., *Introduction to Hazardous Waste Incineration*, Wiley-Interscience, New York, New York, 1988.
2. U.S. EPA, *Waste Minimization Opportunity Assessment Manual*, Hazardous Waste Engineering Research Laboratory, Office of Research and Development, Cincinnati, Ohio, U.S. EPA/625/7-88/003, 1988.

3. Weast, R. C., *CRC Handbook of Chemistry and Physics*, 51st ed., CRC Press, Boca Raton, Florida, 1971.
4. Chang, R., *Chemistry*, 4th ed., McGraw-Hill, Inc., New York, New York, 1991.

CHAPTER 4

Unit Operations

Contributing Authors: Bella De Vito and Amy Leahy

In many industrial sectors, including the chemical, food and biological processing industries, there are many similarities in the manner in which the entering feed materials are modified or processed into final products. These seemingly different chemical, physical, or biological processes can be broken down into a series of separate and distinct steps called unit operations. These unit operations are common to all types of diverse, process industries. For example, the unit operation *distillation* is used to purify or separate alcohol in the beverage industry and hydrocarbons in the petroleum industry. *Drying* of grain and other foods is similar to drying of lumber, filtered precipitates, and rayon yarn. The unit operation *absorption* occurs in the absorption of oxygen from air in a fermentation process or in a sewage treatment plant, and in absorption of hydrogen gas in a process for liquid hydrogenation of oil. *Evaporation* of salt solutions in the chemical industry is similar to evaporation of sugar solutions in the food industry. *Settling and sedimentation* of suspended solids are similar in the sewage and the mining industries. *Flow* of liquid hydrocarbons in the petroleum refinery and flow of milk in a dairy plant are carried out in a similar fashion. Thus, the most efficient method of organizing the subject matter of unit operations is based on two facts: (1) although the number of individual processes is great, each one may be separated into a series of steps, called operations, each of which in turn appears in process after process; and (2) the individual operations have common techniques and are based on the same scientific principles. The unit operation concept, therefore, is this: By systematically studying these operations themselves—operations that clearly cross industry and process lines—the treatment of all processes is unified and simplified.

1. QUANTITATIVE APPROACHES

It is necessary to understand the basic function of each unit operation in a process in order to properly apply mass and energy balances to the process. The conservation laws discussed in the previous chapter may be applied at the macroscopic, microscopic, or molecular level. One can best illustrate the differences in these methods with an example. Consider a system (Figure 4.1) in which a fluid is flowing through a cylindrical tube. The system is defined as the fluid contained within the tube between the Points 1 and 2 at any time, t. If one is interested in determining changes occurring at the inlet and outlet of the system, the conservation law is applied on a macroscopic level to the entire system. The resultant equation (usually algebraic) describes the overall changes occurring to the system without regard for internal variations to the system.

Figure 4.1. Hypothetical flow system.

This macroscopic approach is the conventional approach for the analysis of unit operations, has received the bulk of attention in the literature, and is the approach employed in this text. The microscopic approach is employed when detailed information concerning the behavior within the system is required. In the microscopic approach, the conservation laws are then applied to a differential element within the system which is large compared to an individual molecule, but small compared to the entire system. The resultant equations (usually differential) are then expanded, via an integration, to describe the behavior of the entire system. This is defined as the transport phenomena approach. Details of transport phenomenon principles and applications are also available in the literature.[1]

The molecular approach involves the application of the conservation laws to individual molecules. This leads to a study of statistical and quantum mechanics, both of which are beyond the scope of this text. In any case, the description of individual elements at the molecular level is of little value to the engineer. However, the statistical averaging of molecular quantities in either a differential or finite element within a system (i.e., the microscopic approach) leads to a more meaningful description of the behavior of a system. Using the macroscopic approach, the transfer process, whether it be mass, energy or momentum, can be simply described as shown below:

$$\text{Rate of Transfer} = \frac{(\text{Driving Force})(\text{Area Available for Transfer})}{(\text{Resistance to Transfer})} \quad (4.1)$$

For mass transfer, the equation becomes:

$$\text{Mass Transfer Rate} = \frac{(\Delta \text{ Concentration})(\text{Mass Transfer Area})}{(\text{Resistance to Mass Transfer})} \quad (4.2)$$

2. UNIT OPERATION CLASSIFICATION

This chapter examines a variety of classes of unit operations, particularly those employed in pollution prevention activities. The basic principles underlying these operations are reviewed. Application to specific unit operations (and description of the accompanying equipment) is accomplished in Chapter 13 in Part 2 of this book. Unit operations also receive some treatment in Chapter 5, "Plant Equipment," and Chapter 6, "Ancillary Equipment."

Unit operations deal mainly with the transfer and change of energy and materials by physical means. Some, but not all, of the important unit operations that can be combined in various sequences in a process are classified and described below. It should be noted that categories 3 through 10 are classified as mass transfer operations, and are elaborated on later in this chapter.

1. Fluid flow—the principles that govern the flow or transportation of any fluid from one point to another.

   The flow of fluids is an important factor in many unit operations. Fluids are understood to include liquids, gases, and vapors. The handling of fluids is usually simpler, cheaper, and less troublesome than the handling of solids. Consequently, attempts are made to handle and transport materials in the form of fluids, solutions, or suspensions whenever possible. When materials can not be handled in this form one resorts to the handling of solids.

   In most chemical and physical processes, materials must be moved through the process equipment. When fluids are moved, energy must be supplied by a pump, blower or other means. A quantitative evaluation of the energy balance in flowing fluids requires a knowledge of the fundamental thermodynamics and mechanics of fluids. This information is available in the literature.[2]

   Details regarding pipes, ducts, fittings, valves, fans, pumps, blowers, compressors, etc., are available in Chapter 5, "Plant Equipment" and in Chapter 6, "Ancillary Equipment". More extensive information is available in the literature.[3]

2. Heat transfer—a unit operation that deals with the principles that govern accumulation and transfer of heat and energy from one position to another. Many chemical reactions progress more rapidly or are more efficient if the temperature is elevated above ambient. Furthermore, chemical reactions usually release or absorb heat. Therefore, it is often necessary to heat or cool the reactants and products making heat transfer an extremely important unit operation in many industrial processes. Heat transfer is also involved in the vaporization or condensation of a process stream. It is often possible to heat one process stream while cooling another in a piece of equipment called a heat exchanger. In this device, the fluids flow past each other and are usually separated by a metal wall through which heat is transferred from the hotter stream to the colder. Recovery of heat from product streams is often economically favorable. More discussion of heat transfer equipment and basic heat transfer concepts is provided in Chapter 5, "Plant Equipment."

3. Evaporation—a special case of heat transfer, which deals with the thermal removal of a volatile solvent, such as water, from a nonvolatile solute, such as salt or any other material in solution through the application of thermal energy.
4. Drying—an operation in which volatile liquids (usually water) are removed from solid materials not in solution through the application of thermal energy.
5. Distillation—an operation whereby components of a liquid mixture are separated by boiling because of their differences in vapor pressure.
6. Absorption—a process whereby a component is removed from a gas stream by dissolution of it in a liquid.
7. Liquid-liquid extraction—process in which a solute in a liquid solution is removed by dissolution into another liquid solvent that is relatively immiscible upon contact with the solution.
8. Liquid-solid leaching—involves treating a finely divided solid with a liquid that dissolves and removes a solute contained in the solid.
9. Crystallization—the removal of a solute, such as a salt, from a solution by precipitating the solute from the solution.
10. Mechanical-physical separations—involve separation of solids, liquids, or gases by mechanical means, such as filtration, settling, and size reduction, which are often classified as separate unit operations.

3. MASS TRANSFER

Unit operations are applicable to processes that are essentially physical in nature. A substantial number of the unit operations are concerned with the problems of changing the composition of solutions and mixtures through methods that do not involve chemical reactions. Most frequently these composition changes are analytical; that is, it is often desirable to separate the original substance into its component parts. Such separations may be entirely mechanical, such as the separation of solid from liquid during filtration or the classification of a granular solid into fractions of different particle size by screening. On the other hand, if the operations involve changes in composition of solutions, they are known as diffusional, or mass-transfer, operations, and it is these with which we are most often concerned. The substance undergoing the change in composition is most frequently itself a solution, but it may be a physical mixture or one that exhibits the properties of both.

Diffusional operations are characterized by a transfer of one substance through another, usually on a molecular scale. For example, when water evaporates from a pool into an air stream flowing over the water surface, molecules of water vapor diffuse through those of the gas at the surface into the main portion of the air stream, where they are carried away. On other occasions, one molecular species may diffuse through another which is itself diffusing in the opposite direction. In every case, mass transfer is a result of a concentration difference, the diffusing substance moving from a location of high concentration to one of low concentration. This concentration difference is the driving force for the transfer of mass in the same way that a temperature difference is the driving force for the transfer of heat between bodies.

It is useful to classify diffusional operations and to cite examples of each. Diffusional separations may be brought about by:

1. Contact of two immiscible phases, with mass transfer or diffusion through the surface between the phases.
2. Contact of miscible phases separated by a permeable or semipermeable membrane, with diffusion through the membrane.
3. Direct contact of miscible phases.

3.1. Contact of Immiscible Phases

The three states of aggregation permit six possible combinations of states: gas-gas, gas-liquid, gas-solid, liquid-liquid, liquid-solid, and solid-solid, each of which are discussed below.

3.1.1. Gas-Gas

Since all gases are completely soluble in each other, this category of operation cannot generally be practically realized.

3.1.2. Gas-Liquid

If all components of the system are present in appreciable amounts in both gas and liquid phases, the operation, known as distillation, may be employed. In this instance the gas phase is created from the liquid phase by application of heat. For example, if a liquid solution of acetic acid and water is partially vaporized by heating, it is found that the newly created vapor phase and the residual liquid both contain acetic acid and water, but in proportions that are different for the two phases and different from those in the original solution. If the vapor and residual liquid are separated physically from one another and the vapor is condensed, two solutions, one richer and the other poorer in acetic acid, are obtained. In this way a certain degree of separation of the original components is accomplished. Conversely, should a vapor mixture of the two substances be partially condensed, the newly formed liquid phase and the residual vapor will differ in composition. In both instances an interdiffusion of both components between the phases eventually establishes their final composition.

All the components of the solutions involved may not be present in appreciable amounts in both gas and liquid phases, however. If the liquid phase is a pure liquid containing but one component whereas the gas contains two or more, the operations are known as humidification or dehumidification, depending upon the direction of the diffusion. For example, contact of dry air with liquid water results in evaporation of some of the water into the air (humidification of the air). Conversely, contact of very moist air with pure liquid water may result in condensation of part of the moisture in the air (dehumidification). In both cases relatively little air dissolves in the water, and for most practical purposes it is considered that only water vapor diffuses from one phase to the other.

Both phases may be solutions, each containing only one common component that distributes between phases. For example, if a mixture of ammonia and air is brought into contact with liquid water, a large portion of the ammonia, but relatively little air, will dissolve in the liquid, and in this way the air-ammonia mixture may be separated. The operation is known as gas absorption. On the other hand, if air is brought into contact with an ammonia-water solution, some of the ammonia leaves the liquid and enters the gas phase. This operation is known as desorption, or stripping, a common wastewater treatment method.

To complete the classification, consider the case where the gas phase contains but one component and the liquid several, as in evaporation of a saltwater solution by boiling. Here the gas phase contains only water vapor, since the salt is essentially nonvolatile. Such operations do not depend on concentration gradients, but rather on the rate of heat transfer to the boiling solution, and are consequently not considered diffusional separations. Should the salt solution be separated by diffusion of the water into an air stream, however, the operation then becomes one of desorption, or stripping, and is considered a diffusional separation process.

3.1.3. Gas-Solid

It is convenient to classify mass transfer operations in the gas-solid category according to the number of components that appear in two phases. If a solid solution were to be partially vaporized without the appearance of a liquid phase, the newly formed vapor phase and the residual solid would each contain all the original components but in different proportions. The operation is then *fractional sublimation*. As in distillation, the final compositions are established by interdiffusion of the components between the phases. Although such an operation is theoretically possible, it is usually not practical because of the inconvenience of dealing with solid phases in this manner. However, all components may be present in both phases.

If a solid that is moistened with a volatile liquid is exposed to a relatively dry gas, the liquid leaves the solid and diffuses into the gas, an operation generally known as drying, or sometimes as desorption. An example of this process is the drying of laundry by exposure to air. There are many industrial counterparts such as the drying of lumber or the removal of moisture from a wet filter cake by exposure to dry gas. In these cases, the diffusion is, of course, from the solid to the gas phase. If the diffusion takes place in the opposite direction, the operation is known as adsorption. For example, if a mixture of water vapor and air is brought into contact with activated silica gel, the water vapor diffuses to the solid, which retains it strongly, and the air is thus dried.

In other instances a gas mixture may contain several components, each of which is adsorbed on a solid but to different extents (fractional adsorption). For example, if a mixture of propane and propylene gases is brought into contact with a molecular sieve, the two hydrocarbons are both adsorbed, but to different extents, thus leading to a separation of the gas mixture.

In a case in which the gas phase is a pure vapor, such as in the sublimation of a volatile solid from a mixture with one that is nonvolatile, the operation is dependent more on the rate of application of heat than on the concentration difference. This process is essentially a nondiffusional separation. The same is true of the condensation of a vapor to a pure solid, where the rate of condensation depends on the rate of heat removal, not on mass transfer limiting diffusional processes.

3.1.4. Liquid-Liquid

Separations involving the contact of two insoluble liquid phases are known as *liquid-extraction* operations. A simple example is a familiar laboratory procedure: if an acetone-water solution is shaken in a separatory funnel with carbon tetrachloride and the liquids are allowed to settle, a large portion of the acetone will be found in the carbon-tetrachloride-rich phase and will thus have been separated from the water. A small amount of the water will also have been dissolved by the carbon tetrachloride, and a small amount of the latter will have entered the water layer, but these effects are relatively minor. As another possibility, a solution of acetic acid and acetone may be separated by adding it to the insoluble mixture of water and carbon tetrachloride. After shaking and settling, both acetone and acetic acid will be found in both liquid phases, but in different proportions. Such an operation is known as *fractional extraction*.

3.1.5. Liquid-Solid

Fractional solidification of a liquid, where the solid and liquid phases are both of variable composition containing all components but in different proportions, is theoretically possible but is not ordinarily carried out because of practical difficulties in handling the solid phase and because of the very slow diffusion rates in the solid.

Cases involving distribution of a substance between the solid and liquid phases are common, however. Dissolution of a component from a solid mixture by a liquid solvent is known as *leaching* (sometimes called solvent extraction). The leaching of gold from ore by cyanide solutions and the leaching of cottonseed oil from cottonseeds by hexane are two examples of this leaching process. Diffusion is, of course, from the solid to the liquid phase. If the concentration gradient driving diffusion is in the opposite direction, the operation is known as *adsorption*. Thus, colored material that contaminates impure cane sugar solutions may be removed by contacting the liquid solutions with activated carbon, whereupon the colored substances are retained on the surface of the solid carbon and removed from solution.

When the solid phase is a pure substance and the liquid solution is being separated, the operation is called *crystallization*, but as ordinarily carried out crystallization rates are more dependent on heat-transfer rates than on solution concentrations. The reverse operation is *dissolution*. No known operation is included in the category involving a pure liquid phase.

3.1.6. Solid-Solid

Because of the extraordinarily slow rates of diffusion within solid phases, there is no industrial separation operation in this solid-solid category.

3.2. Miscible Phases Separated by a Membrane

In operations involving miscible phases separated by a membrane, the membrane is necessary to prevent intermingling of the phases. It must be differently permeable to the components of the solutions, however, if diffusional separations are to be possible.

3.2.1. Gas-Gas

The operation in the gas-gas category is known as *gaseous diffusion*, or *effusion*. If a gas mixture whose components are of different molecular weights is brought into contact with a porous diaphragm, the various components of the gas will diffuse through the pores at different rates. This leads to different compositions on opposite sides of the diaphragm and, consequently, to separation of the gas mixture. In this manner, large-scale separation of the isotopes of uranium, in the form of uranium hexafluoride, is carried out.

3.2.2. Liquid-Liquid

The separation of a crystalline substance from a colloid, by contact of their solution with a liquid solvent with an intervening membrane permeable only to the solvent and the dissolved crystalline substance, is known as *dialysis*. For example, aqueous beet sugar solutions containing undesired colloidal material are freed of the latter by contact with water with an intervening semipermeable membrane. Sugar and water diffuse through the membrane, but the larger colloidal particles cannot. *Fractional dialysis* for separating two crystalline substances in solution makes use of the difference in membrane permeability of the substances. If an electromotive force is applied across the membrane to assist in the diffusion of charged particles, the operation is *electrodialysis*. If a solution is separated from the pure solvent by a membrane that is permeable only to the solvent, the solvent diffuses into the solution–an operation known as *osmosis*. This is not a separation operation, of course, but if the flow of solvent is reversed by superimposing a pressure to oppose the osmotic pressure, the process is labeled *reverse osmosis*.

3.2.3. Solid-Solid

The operation in the solid-solid category has found little, if any, practical application in the chemical process industry.

3.3. Direct Contact of Miscible Phases

Operations involving direct contact of miscible phases, because of the difficulty in maintaining concentration gradients without mixing of the fluid, are not generally considered practical industrially except in unusual circumstances.

Thermal diffusion involves the formation of a concentration difference within a single liquid or gaseous phase by imposition of a temperature gradient upon the fluid, thus making possible a separation of the components of the solution. This process has been used, for example, in the separation of uranium isotopes in the form of uranium hexafluoride.

If a condensable vapor, such as steam, is allowed to diffuse through a gas mixture, it will preferentially carry one of the components along with it, thus making a separation by the operation known as *sweep diffusion*. If the two zones within the gas phase where the concentrations are different are separated by a screen containing relatively large-size openings, the operation is called *atmolysis*.

If a gas mixture is subjected to a very rapid centrifugation, the components will be separated because of the slightly different forces acting on the various molecules, owing to their different masses. The heavier molecules thus tend to accumulate at the periphery of the centrifuge.

The engineer faced with the problem of separating the components of a solution ordinarily must choose among several possible methods. Although the choice is usually limited because of the peculiar physical characteristics of the materials to be handled, the necessity for making a decision nevertheless almost always exists. Until the fundamentals of the various operations are clearly understood, of course, no basis for such a decision is available, but it is good at least to establish the nature of the alternatives at the beginning.

The engineer may sometimes choose between using a diffusional operation of the sort discussed in this book or a purely mechanical separation method. For example, in the separation of a desired mineral from its ore, it may be possible to use either the diffusional operation of leaching with a solvent or the purely mechanical methods of flotation. Vegetable oils may be separated from the seeds in which they occur by extraction or by leaching with a solvent. A vapor may be removed from a mixture with a permanent gas by the mechanical operation of compression or by the diffusional operations of gas absorption or adsorption. Sometimes both mechanical and diffusional operations are used, especially where the former are incomplete, as in processes for recovering vegetable oils wherein extraction is followed by leaching. A more commonplace example is the wringing of water from wet laundry followed by air drying. It is characteristic that at the end of a solely mechanical operation the substance removed is pure, whereas if removed by diffusional separation methods it is associated with another substance.

Frequently a choice may be made between a diffusional operation and a chemical reaction to bring about a separation. For example, water may be removed from an ethanol-water solution either by causing it to react with unslaked lime or by special methods of distillation. Hydrogen sulfide may be separated from other gases either by absorption in a liquid solvent or by chemical reaction with ferric oxide. Chemical methods ordinarily destroy the substance removed, whereas diffusional methods usually permit its eventual recovery in an unaltered form without great difficulty.

There are also choices to be made within diffusional operations. For example, a gaseous mixture of oxygen and nitrogen may be separated by preferential adsorption of the oxygen on activated carbon, by absorption, by distillation, or by gaseous effusion. A liquid solution of acetic acid may be separated by distillation, by liquid extraction with a suitable solvent, by absorption with a suitable solvent, or by adsorption with a suitable adsorbent.

The principal basis for choice in most cases is economics: the method that costs least is usually the one to be used. Other factors, including legal and/or environmental constraints, also influence the decision, however. The simplest operation, although it may not be the least costly, is sometimes desired because it will be trouble-free. Sometimes a method will be discarded because imperfect knowledge of design methods or unavailability of data for design will not permit results to be guaranteed. Favorable previous experience with a particular method is often given strong consideration. Cost, however, remains one of the prime factors, and is considered in detail in Chapter 8.

Additional details on separation processes can be found later in this text. In Part 2, separation for recycling purposes is presented in Chapter 14, "Recycling."

4. ILLUSTRATIVE EXAMPLES

Example 4.1. Discuss the difference between the macroscopic, microscopic, and molecular transport approaches.

SOLUTION. The macroscopic approach is concerned with the examination of the overall changes between the inlet and the outlet of a system without regard to internal variations within the system. The microscopic approach is concerned with the examination of changes within and across a system which ultimately leads to a determination of the net changes across the entire system. Finally, the molecular approach is concerned with the examination of the behavior of individual molecules within a system as a means to determine the change across the entire system.

Example 4.2. As part of his investigation, Detective Theodore paid a visit to the office of a missing bookmaker, the location where the bookmaker was last seen. In examining the premises, he noticed a half-empty, 7-cm tall coffee mug on the missing bookmaker's desk. Based on his experience and the stain left in the mug, Detective Theodore estimated that the bookmaker had been missing for approximately 2 weeks. Qualitatively explain how this brilliant sleuth reached this conclusion. It may be assumed that the coffee mug was initially full to 0.5 cm from the brim at the time of the bookmaker's disappearance.

SOLUTION. Detective Theodore instinctively realized that the time it took for the coffee in the mug to evaporate would reflect how long the bookmaker had been missing. In mass transfer terms, it is a case of evaporation of a single component, that is, essentially pure water into a stagnant gas (air) at room temperature and 1atm of pressure. Combining Stefan's Law with the evaporation rate results in an equation[4] which may be solved for the length of time required for the 0.5 cm of water/coffee to evaporate from the mug.

5. SUMMARY

1. In many industrial sectors, including the food and biological processing industries, there are many similarities in the manner in which the entering feed materials are modified or processed into final products. The seemingly different chemical, physical, or biological processes can be isolated into a series of separate and distinct steps called unit operations. Unit operations deal with the transfer and change of energy and materials by physical means. Some, but not all, of the important unit operations that can be combined in various sequences in a process are classified and described in the chapter.
2. Fluid flow concerns the principles that determine the flow or advective movement of any fluid from one point to another.
3. Heat transfer is the unit operation that deals with the principles that govern accumulation and transfer of heat and energy from one place to another.
4. Evaporation is a special case of heat transfer, which deals with the evaporation of a volatile solvent, such as water, from a nonvolatile solute, such as salt, in solution.
5. Drying is an operation in which volatile liquids (usually water) are removed from solid materials through the application of thermal energy.
6. Distillation is an operation whereby components of a liquid mixture are separated by boiling, because of their differences in vapor pressure.
7. Absorption is a process by which a component is removed from a gas stream by dissolution of it in a liquid.
8. Liquid-liquid extraction is a process in which a solute in a liquid solution is removed by dissolution into another liquid solvent that is relatively immiscible upon contact with the solution.
9. Mechanical-physical separations involve separation of solids, liquids, or gases by mechanical means, such as filtration, settling, and size reduction, which are often classified as separate unit operations.

6. PROBLEMS

1. Explain why the practical engineer/scientist invariably employs the macroscopic approach in the solution of real world problems.
2. Discuss the primary mass transfer problem/difficulty associated with the direct contact of miscible phases versus the direct contact of immiscible phases.
3. Discuss what mass transfer principle was behind the gas masks employed by the "doughboys" in World War I to prevent problems with poisonous gas releases.
4. Explain why a large open bottle of vinegar ultimately fills the room with the odor of vinegar.
5. One of the basic principles of an artificial kidney is to pass blood through a membrane tube. On the other side of the membrane tube wall is a dialyzing saline solution into which unwanted substances in the blood pass by diffusion. Using Equation 4.2, provide at least three ways of improving the efficiency (the mass transfer rate) in an artificial kidney.
6. Quantitatively explain how the brilliant sleuth in Illustrative Example 4.2 reached his conclusion.

7. REFERENCES

1. Bird, R. B., Stewart, W. E., and Lightfoot, E. N., *Transport Phenomena*, John Wiley & Sons, Inc., New York, New York, 1960.
2. McCabe, W. L., Smith, J. C., and Harriot, P., *Unit Operations of Chemical Engineering*, 5th ed., McGraw-Hill, New York, New York, 1993.
3. Perry, R. H. and Chilton, C. H., *Chemical Engineers' Handbook*, 5th ed., McGraw-Hill, New York, New York, 1973.
4. Coulson, J. M. and Richardson, J. F., *Chemical Engineering.*, Vol. 1, 4th ed., Pergamon Press, New York, New York, 1990.

CHAPTER 5

Plant Equipment
Contributing Author: Brett R. Elias

This chapter provides details on a number of commonly used process units: reactors, heat exchangers, columns of various types (distillation, absorption, adsorption, evaporation, extraction), dryers, and grinders. Details of other process unit operations specific to pollution prevention are available in the literature, and in Chapter 6. The reader should note that the general subjects of health risk assessments, and accident and emergency management are not addressed in this chapter. Although these two concerns are noted in some detail in Chapter 10 of Part 2 of this book, chemical processes and their accompanying equipment should always be designed to permit safe operation.

1. CHEMICAL REACTORS

1.1. Reactor Definition

The reactor is often the heart of a chemical process. It is the place in the process where raw materials are usually converted into products, and reactor design is therefore a vital step in the overall design of the process.

The treatment of reactors in this section is restricted to a discussion of the appropriate reactor types for a process. The design of an industrial chemical reactor must satisfy requirements in four main areas.

1. *Chemical Factors.* These involve mainly the kinetics of the reaction. The design must provide sufficient residence time for the desired reaction to proceed to the required degree of conversion.

2. *Mass Transfer.* The reaction rate of heterogeneous reactions may be controlled by the rates of diffusion of the reacting species, rather than the chemical kinetics.

3. *Heat Transfer Factors.* These involve the removal, or addition, of the heat of reaction.

4. *Safety Factors.* These involve the confinement of any hazardous reactants and products, as well as the control of the reaction and the process conditions.

The need to satisfy these interrelated, and often contradictory, factors makes reactor design a complex and difficult task. However, in many instances one of the factors predominates, hence determining the choice of reactor type and the design method for the overall process or system being evaluated.

1.2. Reactor Types

The characteristics normally used to classify reactor designs are

1. Mode of operation, i.e., batch or continuous.
2. Phases present, i.e., homogeneous or heterogeneous.
3. Reactor geometry, i.e., flow pattern and manner of contacting the phases.

The five major classes of reactors are

1. Batch reactor.
2. Stirred tank reactor.
3. Tubular reactor.
4. Packed bed (fixed) reactor.
5. Fluidized bed reactor.

In a batch process, all the reagents are added at the beginning of the reaction, the reaction proceeds, and the compositions change with time. The reaction is stopped and the product is withdrawn when the required conversion has been reached. Batch processes are suitable for small-scale production and for processes that use the same equipment to make a range of different products or grades. Examples include pigments, dyestuffs, pharmaceuticals, and polymers.

In continuous processes, the reactants are fed to the reactor and the products are withdrawn continuously, and the reactor is usually operated under steady-state conditions. Continuous production normally entails lower production costs than batch production, but lacks the flexibility of batch production. Continuous stirred tank reactors (CSTRs) are usually selected for large-scale production. Processes that do not fit the definition of batch or continuous are often referred to as semi-continuous and semi-batch processes. In a semi-batch reactor, some of the reactants may be added to, or some of the products withdrawn from, the batch as the reaction proceeds. A semi-continuous process is basically a continuous process that is interrupted periodically for the regeneration of catalyst, for instance.[1]

Homogeneous reactions are those in which the reactants, products, and any catalysts used form one continuous phase, gaseous or liquid. Homogeneous gas phase reactors are almost always operated continuously, whereas liquid phase reactors may be batch or continuous. Tubular (pipeline) reactors are normally used for homogeneous gas phase reactions. Both tubular and stirred tank reactors are used for homogeneous liquid phase reactions.

In a heterogeneous reaction two or more phases exist, and the overriding problem in reactor design is to promote mass transfer between/among the phases. The possible combinations of phases are

1. Liquid-liquid, i.e., with immiscible liquid phases.
2. Liquid-solid, i.e., with one or more liquid phases in contact with a solid; the solid may be a reactant or catalyst.
3. Liquid-solid-gas, i.e., where the solid is normally a catalyst.
4. Gas-solid, i.e., where the solid may take part in the reaction or act as a catalyst.

The reactors used for established processes are usually unique and complex designs that have been developed over a period of years to suit the requirements of a particular process. However, it is convenient to classify flow reactors into the broad categories discussed below.[1]

A stirred tank (agitated) reactor consists of a tank fitted with a mechanical agitator and (usually) a cooling jacket or coils. These are operated as batch or continuous reactors. Several reactors may be used in series. The stirred tank reactor can be considered the basic chemical reactor, modeling on a large scale the conventional laboratory flask. Tank sizes range from a few liters to several thousand liters. This equipment is used for homogeneous and heterogeneous liquid-liquid and liquid-gas reactions, and for reactions that involve finely suspended soils and solids that are held in suspension by agitation. Since the degree of agitation is under the designer's control, stirred tank reactors are particularly suitable for reactions that require good mass transfer or heat transfer efficiencies.

When operated as a continuous process, the composition in the reactor is constant and the same as the product stream. Except for very rapid reactions, this limits the conversion that can be obtained in a one stage stirred tank reactor.

Tubular reactors are generally used for gaseous reactions, but are also suitable for some liquid phase reactions. If high heat transfer rates are required, small-diameter tubes are used to increase the ratio of surface area to volume within the reactor. Several tubes may be arranged in parallel, connected to a manifold, or fitted into a tube sheet in an arrangement similar to a shell and tube heat exchanger (see the following section, "Heat Exchangers"). For high temperature reactions, the tubes may be placed in a furnace.[1]

There are two basic types of packed-bed reactors: those in which the solid is a reactant and those in which the solid is a catalyst. Many examples of the first type can be found in the extractive metallurgical industries. In the chemical process industry, the designer normally employs the second type, i.e., catalytic reactors. Industrial packed-bed catalytic reactors range in size from units with small tubes, a few centimeters in diameter, to large-diameter packed beds. Packed-bed reactors are used for gas and gas-liquid reactions. Heat transfer rates in large-diameter packed beds are poor, and where high transfer rates are required, fluidized beds should be considered.[1]

The essential feature of a fluidized-bed reactor is that the solids are held in suspension by the upward flow of the reacting fluid. This promotes high mass and heat transfer rates and good mixing. Heat transfer coefficients in the order of 200 W/m-°C for jackets and internal coils are typically obtained. The solids may be a catalyst, a reactant (in some fluidized combustion processes), or an inert powder added to promote heat transfer.

Although the principal advantage of a fluidized bed over a fixed bed is the higher heat transfer rate, fluidized beds are also useful when large quantities of solids must be transported as part of the reaction processes.

Operational factors that contribute to waste and emissions in chemical reactors include incomplete conversion resulting from inadequate temperature control, by-product formation resulting from inadequate mixing, and catalyst deactivation resulting from poor feed control or purity control. Wastes are often generated in a chemical reactor if the reactor is improperly designed or the catalyst selection is not correct.

2. HEAT EXCHANGERS

Heat transfer is often only one of several unit operations involved in a process, and the interrelations among the different unit operations involved must be understood. Equation 4.1, as applied to heat transfer, requires that there be a driving force present for heat to flow. This driving force is the temperature difference between a warmer and colder body.

Practically all of the operations that are carried out in industry involve the production or absorption of energy in the form of heat. The laws governing the transfer of heat and the types of equipment that consider heat flow as one of their main objectives are therefore of great importance. Heat may flow by one or more of the following three basic mechanisms:

1. *Conduction*: When heat flows through a body by the transference of the momentum of individual atoms or molecules without mixing, it is said to flow by conduction. For example, the flow of heat through the wall of a furnace or the metal shell of a boiler takes place by conduction.
2. *Convection*: When heat flows by actual mixing of warmer portions with cooler portions of the same material, the heat transfer mechanism is known as convection. Convection is restricted to the flow of heat in fluids. Rarely does heat flow through fluids by pure conduction without some convection resulting from the eddies caused by the changes of density of the fluid with changes in temperature. For that reason the terms "conduction" and "convection" are often used together, although in many cases heat transfer is primarily by convection. For example, the heating of water flowing by a hot surface is an example of heat transfer due primarily to convection. Heat transfer by convection due to density differences is defined as *natural convection*.
3. *Radiation*: A body emits radiant energy in all directions. If this energy strikes a receiver, a portion of it may be transmitted, a portion may be reflected, and a portion may be absorbed. It is this absorbed portion which represents heat transfer in the form

of radiation, i.e., the transfer of energy through space by means of electromagnetic waves. Radiation passing through empty space is not transformed to heat or any other form of energy. If however matter appears in its path, the radiation will be transmitted, reflected, or absorbed by that matter. For example, fused quartz transmits practically all the radiation that strikes it. A polished opaque surface or mirror will reflect most of the radiation impinging on it, while a black, dull, or matted surface will absorb most of the radiation it receives The absorbed energy will be transformed quantitatively into heat. Note that it is only the absorbed energy that appears as heat.

The transfer of heat to and from process fluids is an essential part of most chemical processes. The chemical process industries use four principal types of heat exchangers.

1. *Double-pipe exchanger* - the simplest type, used for cooling and heating.
2. *Shell and tube exchangers* - used for all applications.
3. *Plate and frame exchangers (plate heat exchangers)* - used for heating and cooling.
4. *Direct or contact exchangers* - used for cooling and quenching.

The word *exchanger* applies to all types of equipment in which heat is exchanged, but is often used specifically to denote equipment in which heat is transferred between two process streams. An exchanger in which a process fluid is heated or cooled by a plant service stream is referred to as a heater or cooler.

One of the simplest and cheapest types of heat exchangers is the concentric pipe arrangement known as the *double-pipe heat exchanger*. Such equipment can be made from standard fittings and is useful where only a small heat transfer area is required. Several units can be connected in series to extend their capacity.

The *shell and tube exchanger* is by far the most commonly used type of heat transfer equipment in the chemical and allied products industries. The advantages of this type of heat transfer device include:

- Large surface area in a small volume.
- Good mechanical layout, i.e., good shape for pressure operations.
- Reliance on well-established fabrication techniques.
- Wide range of construction materials available.
- Easily cleaned equipment.
- Well-established design procedures.

Essentially, shell and tube exchangers consist of a bundle of tubes enclosed in a cylindrical shell. The ends of the tubes are fitted into tube sheets, which separate the shell-side and tube-side fluids. Baffles are provided in the shell to direct the fluid flow and increase heat transfer.

In *direct contact heat exchange*, there is no wall to separate hot and cold streams, and high rates of heat transfer are achieved. Applications include reactor off-gas quenching, vacuum condensers, desuperheating, and humidification. Water-cooling towers are particularly common examples of direct contact heat exchangers. In direct contact cooler condensers, the condensed liquid is frequently used as the coolant.

Use of direct contact heat exchangers should be considered whenever the process stream and coolant are compatible. The equipment is simple and cheap and is suitable for use with heavily fouling fluids and with liquids containing solids. Spray chambers, spray columns, and plate and packed columns are used for direct contact heat exchanger reactors.

Heat exchangers contribute to waste generation by the presence of cling (process side) or scale (cleaning side). This can be corrected by designing for lower film temperature and high turbulence.

3. MASS TRANSFER EQUIPMENT

There are numerous types of mass transfer equipment employed in the manufacturing industry. Some of the more common pieces of equipment are presented below. The reader is referred to the literature[2,3] for additional details on not only those listed below but also those not covered in this section.

3.1. Distillation

Distillation is probably the most widely used separation process in the chemical industry. Its applications range from the rectification of alcohol, which has been practiced since antiquity, to the fractionation of crude oil. The separation of liquid mixtures by distillation is based on differences in volatility among components. The greater the relative volatilities, the easier the separation. Vapor flows up a column and liquid flows countercurrently down the column. The vapor and liquid are brought into contact on plates, or inert packing material. A portion of the condensate from the condenser is returned to the top of the column to provide liquid flow above the feed point (reflux), and a portion of the liquid from the base of the column is vaporized in the reboiler and returned to provide the vapor flow.

In the stripping section, which lies below the feed, the more volatile components are stripped from the liquid. Above the feed, in the enrichment or rectifying section, the concentration of the more volatile components is increased. Figure 5.1 shows a *distillation column* producing two product streams, referred to as tops and bottoms, from a single feed. These columns are occasionally used with more than one feed, and with side streams withdrawn at points throughout the column (Figure 5.1(b)). This does not alter the basic operation, but it does complicate the analysis of the process to some extent. If the process requirement is to strip a volatile component from a relatively nonvolatile solvent, the rectifying section may be omitted, and the column is then called a *stripping column*.

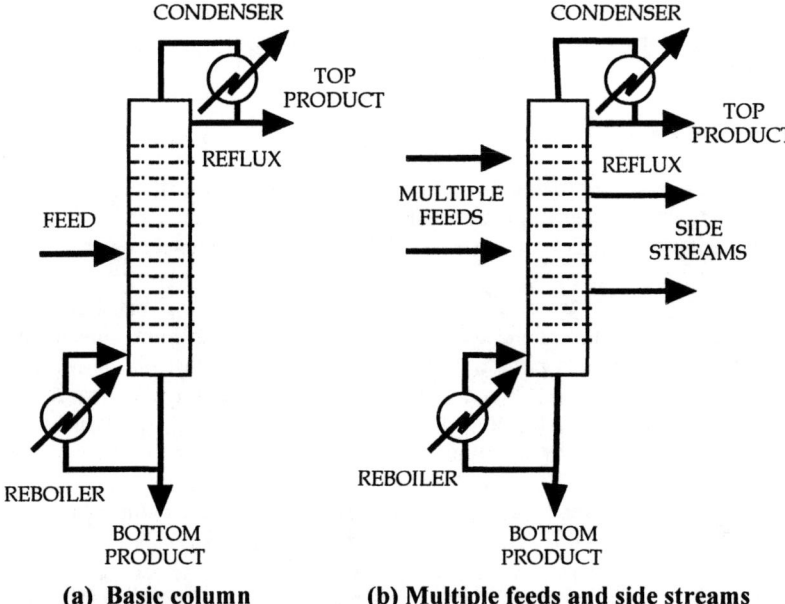

Figure 5.1. Common distillation column configurations.

In some operations where the top product is required as a vapor, the liquid condensed is sufficient only to provide the reflux flow to the column, and the condenser is referred to as a *partial condenser*. When the liquid is totally condensed, the liquid returned to the column will have the same composition as the top product. In a partial condenser, the reflux will be in equilibrium with the vapor leaving the condenser. Virtually pure top and bottom products can be achieved by using multiple distillation stages or, sometimes, additional columns.

3.2. Adsorption

In the *adsorption* process, one or more components in a mixture are preferentially removed from the mixture by a solid (referred to as the adsorbent). Adsorption is influenced by the surface area of the adsorbent, the nature of the compound being adsorbed, the pressure of the operating system (liquid application), and the temperature of operation. These are important parameters to be aware of when designing or evaluating an adsorption process since the possibility for an explosion or fire exists in the adsorption column.

The adsorption process is normally performed in a column. The column is run as either a packed- or fluidized-bed reactor. The adsorbent, after it has reached the end of its useful life, can either be discarded or regenerated. This operation can be applied to either a gas mixture or a liquid mixture. The reader is directed to the literature for further information on the adsorption process,[4] and to Chapter 6 of this text where additional information is provided.

3.3. Absorption

The process of *absorption* conventionally refers to the intimate contacting of a mixture of gases with a liquid so that part of one or more of the constituents of the gas will dissolve in the liquid. The contact usually takes place in some type of packed column.

Packed columns are used for the continuous contact between liquid and gas. The countercurrent packed column is the most common type of unit encountered in gaseous pollutant control for the removal of the undesirable gas, vapor, or odor. This type of column has found widespread application in the chemical industry. The gas stream moves upward through the packed bed against an absorbing or reacting liquid that is injected at the top of the packing. This results in the highest possible efficiency. Since the concentration in the gas stream decreases as it rises through the column, there is constantly fresher liquid available for contact. This provides a maximum average driving force for the diffusion process throughout the bed. The reader is directed to the literature for further information on the adsorption process,[4] and to Chapter 6 of this text where additional information is provided.

3.4. Evaporation

The processing industry has given the operations involving heat transfer to a boiling liquid the general name *evaporation*. The most common application of evaporation is the removal of water from a processing stream. Evaporation is used in the food, chemical, and petrochemical industries, and it usually results in an increase in the concentration of a solution until it forms a thickened slurry or syrup. Applications of evaporation for material processing include sugar slurry thickening, the concentration of dispersed kaolin clay, and bentonite clay dewatering. The factors that affect the evaporation process are concentration in the liquid, compound solubility, system pressure and temperature, scaling, and materials of construction.

An *evaporator* is a type of heat-transfer device designed to induce boiling and evaporation of a liquid. The major types of evaporators include:

- Open kettle or pan evaporators.
- Horizontal-tube natural convection evaporators.
- Vertical-tube natural convection evaporators.
- Forced-convection evaporators.

The efficiency of an evaporator can be increased by operating the equipment in a single- or multieffect mode. Triple-effect evaporators are implemented in kaolin clay processing. In this process the steam utilized for evaporating the solution liquid is passed by the slurry three times before exiting the system. This approach decreases the volume or pressure of steam required to meet the target evaporation rates, thus reducing energy and materials used in the process.

3.5. Extraction

Extraction (sometimes called *leaching*) encompasses liquid-liquid as well as liquid-solid systems. *Liquid-liquid extraction* involves the transfer of solutes from one liquid phase into another liquid solvent. It is normally conducted in mixer-settlers, plate and agitated-tower contracting equipment, or packed or spray towers. Liquid-solid extraction, in which a liquid solvent is passed over a solid phase to remove some solute, is carried out in fixed-bed, moving-bed, or agitated-solid columns.

3.6. Drying

Drying involves the removal of relatively small amounts of water or organic liquids from a solid phase. This can be contrasted to evaporation processes that removes large amounts of water from starting liquid solutions. In many applications, such as in corn processing, drying equipment follows an evaporation step to provide an ultra high solids content final product stream.

Drying, in either a batch or continuous process, removes liquid as a vapor by passing warm gas (usually air) over, or indirectly heating the solid phase. The drying process is carried out in one of four basic dryer types. The first type is a *continuous tunnel dryer*. In a continuous dryer, trays with wet solids are moved through an enclosed system and warm air is blown over the trays. Similar in concept to the continuous tunnel dryer, *rotary dryers* consist of an inclined rotating hollow cylinder. The wet solids are fed in one side and hot air is usually passed countercurrently over the wet solids. The dried solids then pass out the opposite side of the dryer unit. *Indirect drying* equipment heats either a drum or paddle surface which is in contact with the wet solids. The solids are fed across the outside of the hot transfer surface, dried, and discharged continuously. The final type of dryer is a *spray dryer*. In spray dryers a liquid or slurry is sprayed through a nozzle, and fine droplets are dried by a hot gas passed either cocurrently, countercurrently, or co-/countercurrently past the falling droplets. Despite the differing methods of heat transfer, the continuous tunnel, indirect, rotary, and spray dryers can all reduce the moisture content of solids to less than 0.01% when designed and operated properly.

4. ILLUSTRATIVE EXAMPLES

Example 5.1. List the advantages and disadvantages of the three major classes of reactors – batch, CSTR, and fluidized bed – employed in industry.

SOLUTION: Details are provided below in tabular form.

Table 5.1. Solution for Example Problem 5.1.

Type of Reactor	Use in Industry	Advantages	Disadvantages
Batch	1. Small scale productions.	1. High conversion per unit volume for one pass.	1. High operating cost (labor).
	2. Intermediate or one-shot production.	2. Flexibility of operation - same reactor can be used to produce one product one time and different product the next.	2. Product quality more variable than with continuous operation.
	3. Pharmaceutical.	3. Easy to clean.	
	4. Fermentation.		
CSTR	1. Liquid phase reactions.	1. Continuous operation.	1. Lowest conversation/unit volume.
	2. Gas-liquid reactions.	2. Easily adaptable to two phase (gas-liquid) reactions.	2. By-passing and channeling possible with poor agitation.
	3. Solid-liquid reactions.	3. Good temperature control.	
		4. Good control.	
		5. Simplicity of construction.	
		6. Low operating (labor) cost.	
		7. Easy to clean.	
Fluidized Bed	1. Gas-solid reactions.	1. Good mixing.	1. Complex bed fluid mechanics.
	2. Gas-solid catalyzed reactions.	2. Catalyst can be continuously regenerated with use of an auxiliary loop.	2. Severe agitation can result in catalyst destruction, dust formation and carry over in product streams.
		3. Good uniformity of temperature.	3. Uncertain scale-up.

Example 5.2. Describe the main differences between gaseous adsorption and gaseous absorption.

SOLUTION. Gaseous adsorption refers to a process where one or more constituents from a mixture are removed by a solid substance in a column. The process is influenced by several variables including the surface area of the adsorbent, the nature of the compound being adsorbed, and system temperature and pressure.

Gaseous absorption refers to the intimate contacting of a mixture of gases with a liquid so that part of one or more of the constituents of the gas will dissolve in the liquid. The process usually occurs in a packed column operated in a countercurrent mode.

Example 5.3. Calculate the upper and lower flammability limits (UFL, LFL) of a gas mixture entering an adsorber which consists of 30% methane, 50% ethane, and 20% pentane by volume (y_i). Flammability limits for these compounds are given below. Utilize the following equations in your solution:

$$\text{Upper Flammability Limit} = \frac{1}{y_1/UFL_1 + y_2/UFL_2 + \ldots + y_n/UFL_n}$$

$$\text{Lower Flammability Limit} = \frac{1}{y_1/LFL_1 + y_2/LFL_2 + \ldots + y_n/LFL_n}$$

where UFL_i, LFL_i = upper and lower flammability index for component i in the mixture, respectively.

Table 5.2. Input Data for Example Problem 5.2.

Component	LFL, (vol%)	UFL, (vol%)
Methane	4.6	14.2
Ethane	3.5	15.1
Pentane	1.4	7.8

SOLUTION. Using the equations given above, calculate the lower flammability limit of the mixture in percent.

$$LFL = \frac{1}{30/4.6 + 50/3.5 + 20/1.4} = 0.0285 = 2.85\%$$

Calculate the upper flammability limit of the mixture in percent:

$$UFL = \frac{1}{30/14.2 + 50/15.1 + 20/7.8} = 0.125 = 12.5\%$$

5. SUMMARY

1. This chapter provides an overview of a number of commonly used process units: reactors, heat exchangers, columns of various types (distillation, absorption, adsorption, evaporation, extraction), and dryers.

2. The *reactor* is often the heart of a chemical process. It is the location within the process where raw materials are usually converted into products. Reactor design is therefore a vital step in the overall design of the process. The five major classes of reactors are:

- Batch reactor.
- Stirred tank reactor.
- Tubular reactor.
- Packed bed (fixed) reactor.
- Fluidized bed reactor.

3. The transfer of heat to and from process fluids is an essential part of most chemical processes. The most commonly used type of heat transfer equipment is the shell and tube heat exchanger. The chemical process industries use four principal types of *heat exchanger*:

66 Chapter 5

- Double-pipe exchanger
- Shell and tube exchanges
- Plate and frame exchangers (plate heat exchangers)
- Direct or contact exchanges

4. *Distillation* is probably the most widely used separation process in the chemical industry. Its applications range from the rectification of alcohol to the fractionation of crude oil. The separation of liquid mixtures by distillation is based on differences in volatility among components.

5. The *adsorption* process is normally performed in a column. The column is run in either a packed- or fluidized-bed mode. The adsorbent, after it has reached the end of its useful life, can be either discarded or regenerated.

6. The process of *absorption* conventionally refers to the intimate contacting of a mixture of gases with a liquid so that a portion of one or more of the constituents of the gas will dissolve in the liquid. The contact usually takes place in some type of packed column reactor.

7. The processing industry has given operations involving heat transfer to a boiling liquid the general name *evaporation*. The most common application is the removal of water from a processing stream. Evaporation is used in the food, chemical, and petrochemical industries, and it usually results in an increase in the concentration of a certain species to form a thickened slurry or syrup.

8. *Extraction* (sometimes called *leaching*) encompasses liquid-liquid as well as liquid-solid systems. Liquid-liquid extraction involves the transfer of solutes from one liquid phase into another liquid solvent. It is normally conducted in mixer-settlers, plate and agitated-tower contracting equipment, or packed or spray towers. Liquid-solid extraction, in which a liquid solvent is passed over a solid phase to remove some solute, is carried out in fixed-bed, moving-bed, or agitated-solid columns.

9. *Drying* involves the batch or continuous removal from solids of relatively small amounts of water or organic liquids to produce ultra low water content (less than 0.01wt% water) product. Drying removes the liquid as a vapor using warm gas (usually air) which is passed over the solid surface, or by indirectly heating the wet solid phase.

6. PROBLEMS

1. Discuss the different types of heat exchangers that are employed in the chemical process industry.

2. Discuss the major differences between gaseous adsorption and liquid adsorption.

3. Provide qualitative differences between gaseous absorption and gaseous stripping.

4. Given the mass flow rate of a liquid stream of 400 pound per minute (pounds/min), with an average heat capacity of 0.78 Btu/lb-°F, determine the required heat rate in Btu/min to change the liquid stream from 200°F to 1200°F.

5. A gas mixture entering an adsorber has an upper flammability limit of 12.5% and a lower flammability limit of 2.85% consisting of methane, ethane, and pentane. Given that the concentration of methane is 30vol%, calculate the concentrations of the other two components of the gas mixture. Flammability limits for these compounds are given in Table 5.2. Utilize the equations provided in Example 5.3 for calculating the upper and lower flammability limits.

6. A relatively large laboratory with a volume of 1,100 m^3 at 22°C and 1 atm contains a reactor which may emit as much as 0.75 gmol of a hydrocarbon (HC) into the room if a safety relief valve ruptures. If the hydrocarbon concentration in the room air becomes greater than 425 parts per billion (ppb) it constitutes a health and safety hazard.

   a. Suppose the reactor valve ruptures and the maximum amount of HC is emitted almost instantaneously. Assume that the air flow in the room is sufficient to make the room behave like a continuous stirred tank reactor (CSTR), i.e., the air composition is spatially uniform. Calculate the concentration of hydrocarbon in the room in units of ppb. Is there a health risk that exists under this release scenario?
   b. From a health and safety and risk management point of view, what can be done to either decrease the environmentally hazardous nature of this reactor or improve its safety?
   c. From a pollution prevention perspective, what might be done to implement source reduction measures for this reactor?

7. REFERENCES

1. Coulson, J. M., Richardson, J. F., and Skinnott, R. K., *An Introduction to Chemical Engineering*, Pergamon Press, Elmsford, New York, 1983.
2. McCabe, W. L., Smith, J. C., and Harriot, P., *Unit Operations of Chemical Engineering*, 5th ed., McGraw-Hill, New York, New York, 1993.
3. Perry, R. H. and Chilton, C. H., *Chemical Engineers' Handbook*, 5th ed., McGraw-Hill, New York, New York, 1973.
4. Theodore, J. and Buonicore, A. J., *Air Pollution Control Equipment, Volume II: Gases*, CRC Press, Boca Raton, Florida, 1988.

CHAPTER 6

Ancillary Processes and Equipment
Contributing Author: Ronald J. Carpino

The discussion in this chapter begins with a consideration of devices for conveying gases and liquids to, from, or between units of process equipment. This equipment can be a significant source of materials loss in an industrial facility, and analysis of such equipment often leads to the discovery of significant pollution prevention opportunities. Some of the devices discussed are simply conduits for the movement of material (e.g., pipes, ducts, fittings, stacks); others control the flow of material (e.g., dampers, valves); still others provide the mechanical driving force for flow (e.g., fans, pumps, and compressors). This chapter also covers storage facilities, holding tanks, materials-handling devices and techniques, and utilities (e.g., gas, steam, water). The last section of this chapter discusses the role different types of instrument schemes play in controlling all of the aforementioned process equipment.

As indicated in Chapter 5, the general subject areas of health risk assessments and accident and emergency management are treated in Chapter 10 of Part 2 of this text. These are considerations that need to be applied to not only traditional process equipment but also ancillary equipment. This means that the entire process needs to operate in a manner that is inherently safe for personnel, equipment, and the general public.

1. CONVEYANCE SYSTEMS

Pipes, ducts, fittings, pump selection, valves, fans, compressors, and so on may all contribute to waste generation. Pollution prevention and loss prevention can be implemented by the use of seal-less pumps, bellows-seal valves, and other specified equipment. Selection of proper equipment in the construction and design phase of a conveyance system is important. Some of the more important conveyance-related equipment is discussed below.

1.1. Pipes

The most common conduits for fluids are *pipes* and *tubing*. Both usually have circular cross sections, but pipes tend to have larger diameters and thicker walls than does tubing. Because of their heavier walls, pipes can be threaded, whereas tubing cannot. Process systems usually handle large flow rates that require the larger diameters associated with pipes.

Tubing and pipes are manufactured from a large variety of materials. The initial selection of the material for piping primarily depends on the compatibility of the fluid with the piping material in terms of corrosion, and the normal system operating pressure. If special piping is required to accommodate corrosive liquids or high standards of purity, stainless steel, nickel alloys, or materials of high resistance to heat and mechanical damage are used. Steel pipe can be lined with tin, plastic, rubber, lead, cement, or other coatings for special purposes. If corrosion problems or contamination are controlling factors, the use of a nonmetallic pipe such as glass, porcelain, thermosetting plastic, or hard rubber is often acceptable.

There are several techniques used to join pipe sections. For small pipes, threaded connectors are the most common; for larger pipes (typically above 2-1/2 in nominal), flanged fittings, or welded connections, are normally employed.

1.2. Ducts

While pipes and tubing are used as conduits for the conveyance of liquids or gases, *ducts* are used only for gases. Pipes, with their thicker walls, can be used for flows at higher pressures whereas ducts are relatively thin walled (1/16 in max.) and are generally employed for gas flow pressures below 15 psig. Pipes are usually circular in cross section. Ducts come in many shapes (circular, oval, rectangular, etc.). In general, ducts are much larger in cross section than pipes because the gases typically transported have low densities and require high volumetric flow rates. Ducts are most often constructed of field-fabricated galvanized sheet steel, although other materials such as fibrous glass board, factory-fabricated round fiberglass, spiral sheet metal, and flexible duct materials are becoming increasingly popular. Other duct construction materials include black steel, aluminum, stainless steel, plastic and plastic-coated steel, cement, asbestos, and copper. All duct work can be fitted with jacketed insulation or sprayed with a fibrous insulation to minimize any heat transfer occurring between the process gas flow and the environment. The duct can also be fabricated with acoustical insulation to reduce the noise that may be generated from high gas velocities in the duct.

Duct fittings are similar to pipe fittings in that they allow connections to be made between duct sections and enable flow to be diverted where needed. Typical duct fittings include long and short radius elbows of all degrees, transitions, reducers, "T"s, and "Y"s. Joining duct work of relatively small cross-sectional area, typically below 500 in^2 (e.g., 32 in by 16 in nominal), only requires an inert sealing compound and external restraining guides and clamps (commonly referred to as "slips" and "slides"). The preferred sealing compound for low temperature (below 120°F) and/or non-corrosive gas transport applications is RTV silicon. Those gas transport systems that require higher temperatures and/or corrosive resistance should use an elastomer for joint sealer. Joining ducts of larger sizes requires the use of thicker sheet metal so as to fabricate a "flanged" surface as part of the duct. The flanged surfaces of ducts that are joined are typically sealed with an inert gasket material appropriate for the intended service (e.g., red rubber below 120°F or "Viton" below 350°F) and securely fastened to each other.

In order to properly support a system of ducts and prevent seam leaks, all four sides of the duct work should be creased and at least one support should be located at or near every connection (typically every 6 to 8 ft).

Dampers, *economizers* (unlike those used in HVAC applications), and *variable air volume (VAV) boxes* are employed to increase the efficiency of duct systems. Dampers act as flow control valves for gas flow. They are typically actuated by air piston positioners controlled by another system variable (see Instruments and Controls below) or a system "mode" selector switch. Economizers act as regenerative heat exchangers for a duct system. Typically a hot process flow line or a waste gas (e.g., furnace flue discharge) is placed on the "shell" side and a cooler gas which requires preheating (e.g., furnace combustion air intake) is passed through the "tube" side of the economizer to take advantage of available low-grade heat. Finally VAV boxes allow the system to intake or discharge a constant mass of gas, regardless of its volume. This is very useful if intake gas temperatures vary more than a few degrees and/or if the process that utilizes the gas to produce a product changes its production level due to demand.

In many pollution prevention applications, the corrosion resistance of duct materials deserves special consideration. Since material costs generally increase as a function of corrosion resistance, the selection of material must be made based on the desired operating life of the process being designed. For maximum resistance to moisture or corrosive gases, stainless steel and copper are used where their cost can be justified. Aluminum sheet is used where lighter weight and superior resistance to moisture are required.

1.3. Fittings

A fitting is a piece of equipment that has one or more of the following functions:

- The joining of two pieces of straight pipe (e.g., couplings and unions).
- The changing of pipeline diameter (e.g., reducers and bushings).
- The changing of pipeline direction (e.g., elbows).
- The changing of pipeline direction and diameter (e.g., reducer elbows and street elbows).
- The splitting of a stream into multiple streams (e.g., "T"s, "Y"s, and manifolds).
- The joining of multiple streams (e.g., "T"s, "Y"s, and manifolds).
- The mixing of multiple streams (e.g., blender).

A *coupling* is a short piece of pipe whose ends are threaded on the inside (some plastics are not) used to connect straight sections of pipe. A *union* is also used to connect two straight sections but differs from a coupling in that it can be opened conveniently without disturbing the rest of the pipeline, whereas when a coupling is opened, a considerable amount of piping must usually be dismantled. A *reducer* is a coupling for two pipe sections of different diameter. A *bushing* also connects pipes, and/or fittings, of different diameters, but, unlike the reducer, it is threaded on the outside of one end and on the inside of the other. This allows a larger pipe to screw onto the outside and the smaller pipe to screw into the inside of the bushing. An *elbow* is an angled coupling or bushing used to change flow direction, usually by 30°, 45°, or 90°. An elbow that is an angled bushing is referred to as a "street" elbow.

A *"T"* is also used to change flow direction and/or allow future system add-ons, but is more often used to split one stream (supply) into two, and later recombine two streams into one (return). A *"Y"* is similar to the "T" in that it is used to split a stream and/or later combine two streams. A *blender* is a fitting that introduces one stream of a liquid into another in order to achieve a homogenous mixture of the two constituents. Multiple "T" and "Y" sections can be combined to form manifolds of various configurations to split and/or combine multiple process streams as necessary for a given process.

1.4. Valves

Valves have one main function in a pipeline, that is, to control and/or redirect the amount of flow passing through various sections of system piping. There are many different types of valves, but the two most commonly used are the *gate valve* and the *globe valve*. The gate valve contains a disk that slides at right angles to the flow direction. This type of valve is used primarily for on-off control of liquid flows. Typically, 70% of system flow will occur in the first 30% of valve stem travel because small adjustments in disk travel cause extreme changes in the flow cross-sectional area. Therefore, this type of valve is not suitable for adjusting flow rates.

Unlike the gate valve, the globe valve is designed for flow control. Liquid passes through a globe valve via a somewhat torturous route. In one form, the globe valve seal is a horizontal ring into which a plug with a beveled edge is inserted when the valve is shut. Better control of flow is achieved with this type of valve due to the fact that stem movement results in relatively small changes in cross–sectional flow area. However, the pressure drop across a fully open globe valve is greater than a comparable gate valve.

Other available valve types include: *check valves*, which permit flow in one direction only; *butterfly valves*, which operate in a damperlike fashion by rotating a flat plate to either a parallel or a perpendicular position relative to the flow; *plug valves*, in which a rotating tapered plug provides on-off service; *needle valves*, a variation of the globe valve, which give improved flow control; *diaphragm valves*, a valve specially designed to handle very viscous liquids, slurries, or corrosive liquids that might clog the moving parts of other valves; and *ball valves*, valves that provide easy on–off service in addition to good flow control with only small pressure drops when the valves are fully open.

All of the above valves may be obtained with a packing gland, bellows seal, or packingless sealing system. Valves which pass a relatively non–toxic fluid, such as water/steam or nitrogen, should have a packing gland. Valves which are used to control the flow of highly toxic materials, such as any radioactive effluent, should be purchased with a bellows seal so that any packing leak that may occur will be contained and diverted away from plant personnel into a collection area. Systems using fluids that can easily leak through conventional packing systems and cause a safety concern, such as fine oils or hydrogen gas, should be provided with packingless valves.

Depending on their location, size, and process application, valves in a given system may require a variety of different *actuators*. The most common actuator type is the manual *handwheel*. This simple actuator allows an operator to locally isolate or throttle a given process fluid. However, this type of actuator should only be specified if the valve does not require an excessive opening force (i.e., large pressure drop across its seat), does not require significant stem travel, and is accessible and/or is rarely operated (i.e., startup or shutdown).

The next most popular actuator used throughout industry is the air powered actuator. The *air operated valve (AOV)* is commonly used in applications which require a valve to open and shut quickly (e.g., flow or pressure control) and/or that must be placed in a "fail-safe" position. A fail-safe position refers to the state (i.e., open, throttled, shut, energized, or de-energized) that a given component would revert to, in order to minimize plant equipment damage and personnel endangerment, if that component were to become damaged or faulted. AOVs are typically designed to stroke using an incoming air pressure between 40 and 80 psig, are relatively inexpensive to purchase, and are easy to install and maintain.

In addition to AOVs, *motor operated valves (MOVs)* and *hydraulic operated valves (HOVs)* are also available. An MOV is typically used for very large valves (greater than 12-in nominal) whose handwheels would require many turns to stroke fully open and close, but whose pressure drop is too large for an AOV to overcome. Although MOVs have a stronger opening and closing force (due to the use of a high torque motor) and can be placed where an air supply is not accessible, they are generally slow to stroke and expensive to purchase and maintain. The placement of a MOV in a system must also be considered carefully for the following reasons: (1) if the motor were to lose electrical power (e.g., a blackout) or become faulted (e.g., a short), an operator would be unable to remotely stroke the valve, (all MOVs are provided with a handwheel for such an occasion); (2) since the MOV cannot stroke with an unpowered or faulted motor, the valve could remain in a dangerous position for quite some time (e.g., an open acid tank drain valve), and since this type of valve actuator cannot stroke to a safe position when required to do so, it is not considered fail safe.

HOVs are by far the most expensive valve actuators due to their initial cost, installation, and specialized maintenance that is required to keep them operating properly. An HOV consists of one or two motors, significant high pressure tubing, other valves, and hydraulic fluid. However, HOVs can be designed to be fail-safe, can open a valve against extremely high pressure drops, and are capable of very fine process control adjustments (e.g., flow or pressure control).

1.5. Fans

The terms *fans* and *blowers* are often used interchangeably, and only one major distinction is made between them in the discussion that follows. Generally speaking, fans are used for low pressure drop operation, generally below 2 psig. Fans are usually classified as either the centrifugal or the axial-flow type. In centrifugal fans, the gas is introduced into the center of the revolving wheel (the eye) and is discharged at angles from the rotating blades. In axial-flow fans, the gas moves directly through the axis of rotation of the fan blades. Both types of fans are used widely throughout industry.

Blowers are generally employed when pressure heads in the range of 2 to 15 psig are required. Blowers are only offered in axial–flow configurations and may require more than one stage to boost system pressures. However, if operation at higher pressures is required, large volume centrifugal or reciprocating compressors may be required.

1.6. Pumps

Pumps may be classified as *reciprocating, rotary,* or *centrifugal*. The first two are referred to as *positive-displacement pumps* because, unlike the centrifugal type, the liquid or semi-liquid flow is divided into small portions as it passes through these pumps.

Reciprocating pumps operate by the direct action of a piston on the liquid contained in a cylinder within the pump. As the liquid is compressed by the piston, the higher pressure forces it through discharge valves to the pump outlet. As the piston retracts, the next batch of low-pressure liquid is drawn into the cylinder and the cycle is repeated.

The rotary pump combines rotation of the liquid with positive displacement. The rotating elements mesh with elements of the stationary casing in much the same way that two gears mesh. As the rotation elements come together, a pocket is created that first enlarges, drawing in liquid from the inlet or suction line. As rotation continues, the pocket of liquid is trapped, reduced in volume, and then forced into the discharge line at a higher pressure. The flow rate of liquid from a rotary pump is a function of the pump size and speed of rotation, and is slightly dependent on the discharge pressure. Unlike reciprocating pumps, rotary pumps deliver nearly constant flow rates. Rotary pumps are used on liquids of almost any viscosity as long as the liquids do not contain abrasive solids.

Centrifugal pumps are widely used in the process industry because of their simplicity of design, low initial cost, low maintenance, and flexibility of application. Centrifugal pumps have been built to move as little flow as a few gallons per minute against a pressure of several hundred pounds per square inch. In its simplest form, this type of pump consists of an impeller rotating within a casing. Fluid enters the pump near the center of the rotating impeller and is thrown outward by centrifugal force. The kinetic energy of the fluid increases from the center of the impeller to the tips of the impeller vanes. This high velocity is converted to a high pressure as the fast-moving fluid leaves the impeller and is driven into slower moving fluid on the discharge side of the pump.

1.7. Compressors

Compressors operate in a manner similar to that of pumps and have the same classification, i.e., rotary, reciprocating, and centrifugal. Compressors are used for gas conveyance and processing, and an obvious difference between compressors and pumps is the large decrease in volume resulting from the compression of a gaseous stream as compared with the negligible change in volume caused by the pumping of a liquid stream.

Centrifugal compressors are employed when large volumes of gases are to be handled at low-to-moderate pressure increases (0.5 to 50 psig). *Rotary compressors* have smaller capacities than centrifugal compressors, but can achieve discharge pressures up to 100 psig. *Reciprocating compressors* are the most common type of compressors used in industry and are capable of compressing small gas flows to as much as 3,500 psig. With specially designed compressors, discharge pressures as high as 25,000 psig can be reached, but these devices are capable of handling only very small gas volumes and do not work well for all gases.

1.8. Stacks

Gases are discharged into the ambient atmosphere by means of *stacks* (referred to as chimneys by some in industries) of several types. *Stub or short stacks* are usually fabricated of steel and extend a minimum distance up from the discharge of an induced draft fan. These are constructed of steel plate, either unlined or refractory lined, or entirely of refractory and structural brick. Tall stacks, which are constructed of the same materials as short stacks, provide a driving force (draft) of greater pressure difference than that resulting from the shorter stacks. In addition, tall stacks ensure more effective dispersion of the gaseous effluent into the atmosphere. Some chemical and utility applications use metal stacks that are made of a double wall with an air space between the metal sheets. The insulating air pocket created by the double wall prevents condensation on the inside of the stack, thus avoiding corrosion of the metal.

2. UTILITIES

Today the word *utilities* generally designates the ancillary services needed in the operation of any production process. These services are normally supplied from a central location on-site or from off-site utility providers and usually include:

1. Electricity
2. Steam for process heating
3. Water for cooling, potable use, and steam production

4. Refrigeration
5. Compressed air
6. Inert gas supplies

The production and use of these utility services can have a significant impact on the energy demand of a given process or facility, and the audit of energy demands is an important part of a complete pollution prevention assessment. Energy efficiency and conservation issues are discussed in more detail in Chapter 10, while a general discussion of the characteristics of these ancillary services is provided below.

2.1. Electricity

The power required for processes (motor drives, lighting, and general use) may be generated on-site, but more often it is purchased from a local utility. Evaluation of the efficiency of motors, lighting, etc., is an important part of a pollution prevention opportunity assessment, as less power demand can be equated to less overall pollutant generation by a facility. In addition, reduced power costs by reducing both average and peak power demand provide direct monetary benefits to the facility in reduced utility bills, providing quantitative incentives for implementing pollution prevention options.

2.2. Steam

The steam for process heating is generated in either fire- or water-tube boilers, using the most economical fuel available. The process temperatures required can usually be obtained with low pressure steam (typically 25 psig), with higher steam pressures needed only for high process temperature requirements. A significant pollution prevention, energy and cost reduction opportunity may exist in a facility's steam generation system, as many facilities operate at higher than necessary steam pressures, increasing their energy demands for steam production and the cost of producing this steam. In addition, repair and replacement of leaking steam and condenser lines will prevent the wasting of steam, and the energy associated with it.

2.3. Water

A range of water needs typically exist within a facility, all of which may have different water quality demands for specific uses. Significant pollution prevention and waste reduction opportunities exist in the use of water in a facility as many liquid streams can be reused for multiple purposes, i.e., using cooling water blowdown for part rinsing or boiler blowdown for cooling water, etc. Water quality needs for specific unit operations within a facility should be clearly defined so that these reuse opportunities can be effectively evaluted

2.3.1. Cooling Water

Natural and forced-draft cooling towers are generally used to provide the cooling water required on a site, unless water can be drawn from a convenient river or lake in sufficient quantity. Seawater to brackish water can be used at coastal sites, but if used directly will necessitate more expensive materials of construction for heat exchangers because of potential corrosion problems resulting from the high dissolved solids content of this water source. Often this cooling water does not have to be of high purity and can be taken from blowdown streams from boilers or process lines that require much more stringent water quality conditions.

2.3.2. Potable and General Use Water

The water required for general purposes on a site usually is taken from the local supply, unless a cheaper source of suitable quality water (e.g., a river, lake, or well) is available. If the cost of this supply is low, incentives for water use reduction will also be low. It should be remembered, however, that the cost of the supply may be a small portion of the overall cost of managing this water, particularly if it is used in process rinsing, general facility cleaning and wash-down, etc. When the water becomes contaminated, costs for its handling increase exponentially. In addition, waste treatment becomes progressively less efficient and more costly as a waste stream is diluted, and load penalties may be incurred based on the volume of waste discharged to publicly owned treatment works (POTW). It should be evident, then, that numerous benefits can be associated with increased water use efficiency within a plant.

2.3.3. Demineralized Water

Water from which all the minerals have been removed by ion exchange must be used where ultrapure water is needed to meet process demands and strict boiler feedwater requirements. Mixed and multiple-bed ion exchange units are used for this purpose, with resins exchanging multi-valent cations for hydrogen. Boiler water as condensed steam and process water must be removed on a routine basis (blow down) to prevent the build-up of unwanted constituents within these systems, and this ultrapure water may be effectively reused within a plant for other unit operations that demand much less stringent water quality characteristics, i.e., cooling water, process rinse water, etc.

4.4. Refrigeration

Refrigeration is needed for processes that require temperatures below those that can be economically obtained with cooling water. Chilled water can be used to lower process temperatures down to approximately 10°C. For lower temperatures, to -30°C, salt brines (NaCl and $CaCl_2$) are used to distribute the "refrigeration" around the site from a central refrigeration unit.

Vapor compression machines are normally used for this purpose. As with boilers, evaluation of the operating conditions of a chiller can lead to significant pollution prevention opportunities. Consider, for example, that a 1% improvement in chiller efficiency can be expected for each 1°F increase in the chiller site setpoint.[1]

2.5. Compressed Air

Compressed air is needed for general use and for pneumatic controllers that usually serve as chemical process plant controllers. Air is often distributed at a pressure of 100 psig. Rotary and reciprocating single-stage or two-stage compressors are normally used to generate compressed air within a facility. Instrument air must be dry (-20°F dew point) and clean (free from oil). Compressed air also represents potentially signficant cost and energy saving and pollution prevention opportunities as the production of compressed air is highly inefficient ($\approx$90% of the energy used is converted into heat while only 10% is actually used to compress the air).[1] Significant improvements in the production of compressed air can result from modifications of distribution pressure, repair of air line leaks, use of outside air at the intake, etc. These options are discussed in more detail in Chapter 10.

2.6. Inert Gas Supplies

Large quantities of inert gas are required for the inert blanketing of tanks and for reactor, tank, and line purging. This gas is usually supplied from a central facility. Nitrogen is normally used and can be manufactured on-site in an air liquefaction plant or purchased as liquid in tankers. General pollution prevention considerations for the management of gaseous materials, i.e., pressure reduction to lowest practical level to minimize fugitive emissions via leaks; insulation of tanks, lines, etc., to lower vapor pressure and potential fugitive emissions; inspection for and repair of line leaks; etc., should be considered for these inert gas supplies to minimize the amount of these materials that must be used per unit of product generated.

3. MATERIAL TRANSPORTATION AND STORAGE EQUIPMENT

This section deals with equipment needed for the handling, storage, and transportation of solids, liquids, and gases. Each type is considered below individually.

3.1. Gases

The type of equipment best suited for the transportation of gases depends on the differential and operating pressures, and flow rates required. In general, *fans* are used where the pressure drop in the transportation system is small and operating pressures are low. *Axial-flow compressors* are employed for high flow rates and moderate differential pressures, while *centrifugal compressors* are chosen for applications requiring high flow rates and, by staging, high differential pressures. *Reciprocating compressors* can be used over a wide range of pressures and capacities but are normally specified in preference to centrifugal compressors only where high pressures are required at relatively low flow rates.

Gases are stored at high pressures to meet a process requirement and to reduce the required storage volumes. The volume of some gases can be further reduced by liquefying them by pressure or refrigeration. Cylindrical and spherical vessels (Horton spheres) are normally used.

3.2. Liquids

The transportation of liquids is usually accomplished with *pumps*. The type of pump used, i.e., centrifugal, reciprocating, diaphragm, or rotary (gear or sliding vane), depends on the operating pressures and capacity range needed in a given application. Liquids are usually stored in bulk in vertical cylindrical steel tanks, although horizontal cylindrical tanks and rectangular tanks can also be used for storing relatively small quantities liquids.

Fixed and floating-roof tanks can be used for liquid storage. In fixed-roof tanks, toxic or volatile liquids will tend to emit hazardous vapors through a vent line when they are filled. Various configurations can be used to prevent, or at least limit, these releases to the environment. Such designs include routing the vapors to a pollution control system, using a flare to thermally destroy the emissions, or using a compressor to recover vapors as a condensed liquid. In a floating-roof tank a movable cover floats on the surface of the liquid and is sealed to the tank walls using a variety of configurations.

Floating-roof tanks are used to eliminate evaporation losses and, for flammable liquids, to obviate the need for inert gas blanketing to prevent the formation of an explosive mixture above the liquid, as can occur with a fixed-roof tank.

3.3. Solids

Solids are usually more expensive to move and store than liquids or gases. The best equipment to use depends on a number of factors including material throughput, length of travel, change in elevation, and nature of the solids (size, bulk density, angle of repose, abrasiveness, corrosiveness, wet or dry, etc.).

Belt conveyors are the most commonly used type of equipment for the continuous transport of solids. They can carry a wide range of materials economically over both short and long distances, either horizontally or at an appreciable angle, depending on the angle of repose of the solids. *Screw conveyors*, also called *worm conveyors*, are used for free-flowing materials. The modern conveyor consists of a helical screw rotating in a U-shaped trough. This type of device can be used horizontally or, with some loss of capacity, at an incline to lift materials.

Where a vertical lift is required, the most widely used equipment is the *bucket elevator*, consisting of buckets fitted to a chain or belt, which passes over a driven roller or sprocket at the top extent of travel. Bucket elevators can handle a wide range of solids, from heavy "lumps" to fine powders, and are suitable for use with wet solids and slurries.

The simplest way to store solids is to pile them on the ground in the open air. This is satisfactory for the long-term storage of materials that do not deteriorate on exposure to the elements, i.e., for example, coal, which is seasonally stockpiled at utilities, and for which fugitive emissions from the pile are not an air quality concern. For large stockpiles, permanent facilities are usually used for distributing and reclaiming the material. At permanent facilities, traveling cranes, grabs, and drag scrapers are used to feed belt conveyors. Where the cost of recovery from the stockpile is large compared with the value of the stock held, storage in silos or bunkers should be considered.

Overhead bunkers, also called bins or hoppers, are normally used for the short-term storage of materials that must be readily available within the process. The units are arranged so that the material can be withdrawn at a steady rate from the base of the bunker onto a suitable conveyor for addition where needed within the process.

4. INSTRUMENTS AND CONTROLS[2,3]

In chemical processes it is essential that the operator and/or production engineer verify that any given process is functioning properly. In order to perform this job function properly the individual must have access to numerous types of process data. In typical chemical processes many instruments are used to measure, indicate, and record process conditions indicated by process data, such as flow rates, compositions, pressures, and temperatures.

To best aid an operator in controlling a given chemical process, it is often desirable to employ automatically controlled systems. This allows the instruments to not only measure, indicate, and record a variable, but maintain a system variable at a predetermined value as well. Automatic controllers establish a given system parameter at a set value through the use of either "open" or "closed" feedback loops.

4.1. Feedback Loop I&C Systems

Open feedback loop instrumentation and control (I&C) systems are typically less expensive to purchase and much easier to install and maintain when compared to *closed loop I&C systems*. However, open loop systems cannot be relied upon to correct a given out of range process condition; it can only maintain the status quo of a given variable. This is due to the fact that an open loop system only measures one variable and controls it based on a provided input, not its effect on any other system parameters. Consider the following example: It has been calculated that a heat exchanger requires 100 gpm of 75°F river water to cool an important process liquid to 120°F. An open loop system design could be used to modulate the heat exchanger cooling water outlet based on the output of a cooling water flow transmitter. Therefore, if the flow transmitter measured 90 gpm of cooling water, the cooling water valve would automatically stroke more open until the flow transmitter increased to 100 gpm, and vice-versa if the flow transmitter measured 110 gpm. However, problems with this open loop system design arise whenever the river water temperature changes. This results in process liquid temperatures being higher than 120°F when the river warms up and cooler than 120°F when the river gets colder. Since the automatic valve positioner is only concerned with the amount of flow that passes through the valve regardless of the important process liquid final temperature, the feedback system is deemed "open".

Closed feedback loop I&C systems are typically more expensive to purchase and more difficult to install and maintain. However, closed feedback I&C systems are simply much more reliable for ensuring a process is operating at optimum conditions. These systems control a variable by directly measuring its effects on another system variable. Consider the above example: It is critical that an important process liquid be cooled to 120°F to allow it to safely react with another constituent. A closed loop system design could automatically modulate the heat exchanger cooling water outlet valve based on the outlet temperature of the process liquid, not the flow rate of cooling water as stated above. Therefore, as process liquid temperature increases above 120°F the cooling water valve would stroke more open, and vice-versa if the process liquid temperature decreased below 120°F. Since the automatic valve positioner is only concerned with the final temperature of the important process liquid regardless of what amount of cooling water flow it requires to remain at 120°F, the feedback system is deemed "closed".

Automatic systems have been coupled together, through the use of computers, and make it possible to run entire processes with out any direct human control. A process engineer determines the best operating parameters for a given set of conditions and the computer controls all of the "open" and "closed" loop systems to produce a product at optimum process operating efficiency. If the systems are not operating at optimum, as compared to the standards inputted by the process engineer, the computer initiates corrective actions, thereby effectively operating the process as one "closed" loop system. However, each process should still be analyzed for all normal and emergency conditions of operations, and the suitable I&C provided accordingly.

Once an instrument operating range, duty, and environment have been calculated and determined, the instrument should be selected with the following design specifications in mind. First, all I&C must be of the "fail-safe" type. Second, the I&C must be constructed of a corrosion-resistant material and/or encased in an "environmentally" qualified enclosure, when appropriate. Third, I&C must be relatively easy to access for calibration and maintenance. Fourth, separate indicators should be used for each critical point of operation. Fifth, I&C needs to provide both audible and visual alarms to permit operators to identify any type of process deviation occurring. Lastly, all I&C must be completely tested prior to initial service and periodically during plant life to ensure all the automatic functions (e.g., process equipment sequencing, alarms, and automatic system trips) operate as designed.

Following these specifications will allow I&C schemes to maintain process variables within the known operating limits and to detect any hazards if they develop. Instruments must be installed to maintain the key process variables within their design and operating ranges. Critical process variables, such as temperature, pressure, and flow, should be coupled with automatic alarms. Training programs should stress the following concerning alarms: alarms should not be ignored; appropriate operator action(s) should be known and should be immediately taken to correct the problem; and management should be immediately notified of the problem whether it was corrected or not. All critical processes should also be equipped with a system "trip" signal if the operator cannot respond quickly enough to correct the critical system alarm condition.

4.2. Automatic Trip Systems and Interlocks

There are three basic components to an *automatic trip system*. First, a sensor monitors the process control variable of interest and sends an output signal when a preset value for that variable is exceeded (e.g., pump discharge pressure goes above design pressure). Second, pneumatic and/or electric relays transfer the signal to an actuator and/or contactor (e.g., the pump discharge pressure switch actuates a relay in the pump motor breaker). Lastly, the receiving device carries out the appropriate response (e.g., the motor breaker goes open de-energizing a pump motor).

The automatic safety trip may be designed as a separate system or it may be part of a process control loop. However, since adding a trip system to a control loop may present a greater potential for failure (due to having more components), it is safer to design for a separate system. In either case, these safety systems should be inspected and/or tested on a routine, periodic basis.

System designers should also integrate the use of *interlocks* into a given safety scheme. Interlocks ensure that an operator follows the required sequence of actions in a simple, complex, or rarely performed process, from simple batching up to and including plant start-up/shutdown. Examples of interlocks are as follows: ensuring a centrifugal pump is started with its discharge valve shut to prevent a dangerous pump runout condition; to not allow a pump to start without the suction valve being full open; or to not allow a positive displacement pump to start until all the valves in its flow path are opened.

Through the use of system process alarms, automatic trips, and interlocks, both design and process engineers can be reasonably assured that a well employed I&C system will ensure system integrity, personnel safety, and optimum product yield. Additional details are provided in the "Process Piping and Instrumentation Flow Diagrams" section of Chapter 8.

5. ILLUSTRATIVE EXAMPLES

Example 5.1. A 250 psig boiler requires a maximum of 750 gpm of preheated feedwater during peak production. The temperature of the preheated water is kept relatively constant at 120°F due to a heat exchanger that uses steam from the boiler. The feedwater originates from a 26 ft high vented tank that collects the condensate. Note: The tank is kept at a constant 23 ft level due to a makeup water system, and at 90°F due to condensate return and ambient losses. The total pressure loss across the piping, heat exchanger, and fittings is 75 psi. Assuming that the pump efficiency is 75% and the motor efficiency is 92%, determine the pump motor horsepower and available net positive suction head (NPSHA) if the pump is located at the base of the tank and the boiler inlet is 46 ft above the pump.

SOLUTION. The required pump pressure increase, ΔP, is given by:

$$\Delta P = d\,(h_2 - h_1) + (P_2 - P_1) + P_L = 334.9 \text{ lb/in}^2$$

where d, density = 61.71 lbs/ft^3 = 0.03571 lbs/in^3; h_2, boiler height = 46 ft = 552 in; h_1, initial water height = 23 ft = 276 in; P_2, boiler steam pressure = 264.7 lb/in^2; P_1, initial water pressure = 14.7 lb/in^2; and P_L, system pressure losses = 75 lb/in^2.

The required pump hydraulic horsepower, HHP, is given by:

$$\text{HHP} = (Q)\,(\Delta P) = 150 \text{ hp}$$

where Q, volumetric flowrate = 750 gpm = 1.673 ft^3/sec; and ΔP, required pump pressure increase = 334.9 lb/in^2 = 48225 lb/ft^2.

Note: 550 (ft)(lb)/sec = 1 HP.

The required pump brake horsepower, BHP, is given by:

$$\text{BHP} = \text{HHP}/(P_{EFF}) = 200 \text{ hp}$$

where P_{EFF}, pump efficiency = 75%.

Therefore the required pump motor horsepower, MHP, is given by:

$$\text{MHP} = (\text{BHP})(\text{SF})/(M_{EFF}) = 250 \text{ hp}$$

where SF, service factor = 115%; and M_{EFF}, motor efficiency = 92%.

Note: Motor service factor ensures the motor will not burn up under heavily loaded conditions or excessive starting duties, i.e., a factor of safety.

The NPSHA is given by:

$$\text{NPSHA} = H_S + H_A - H_V = 55.3 \text{ ft}$$

where H_S, static water pressure head available to the pump = 23 ft; H_A, ambient pressure head exerted on tank water surface = (14.7 psi)(2.31 ft/psi); and H_V = water vapor pressure head at 90°F = (0.7 psi)(2.31 ft/psi).

Example 5.2. A 2,000 barrel (bbl) atmospheric pressure tank is frequently filled with cyclohexane. To avoid fire and explosion hazards and as a pollution prevention measure, the vapor space is inerted with a nitrogen blanket at ambient temperature and pressure.
 a. Determine the amount of cyclohexane vapor vented if there is no vent recovery system installed on the tank.
 b. Determine how much cyclohexane can be returned to the tank if the vented gas were compressed to 100 psig and cooled to 85°F as an alternative pollution prevention measure.

GIVEN: the tank is at 80°F, cyclohexane molecular weight (MW) = 84.16 g/gmol, cyclohexane vapor pressure at 80°F = 2.051 psia, and at 85°F = 2.320 psia. Note: assume the vapor in the tank is saturated with cyclohexane.

SOLUTION:
 a. The total number of moles of cyclohexane, N_T, is given by:

$$N_T = PV/RT = 28.48 \text{ lbmol}$$

where V, volume = (2000 bbl) (5.614 ft³/bbl); R, gas constant = 10.73 psia-ft³/lbmol-°R; T, temperature = 540°R; and P, pressure = 14.7 psia.

Since the fraction of the volume occupied by cyclohexane is P_{C1}/P_{CT}, the moles of cyclohexane released, N_{C1}, are given by:

$$N_{C1} = (N_T)(P_{C1})/(P_T) = 3.97 \text{ lbmol}$$

where P_{C1}, cyclohexane vapor pressure = 2.051 psia; and P_T, total tank pressure = 14.7 psia.

The moles of nitrogen released, N_N, are then given by:

$$N_N = (N_T)(P_{N1})/(P_T) = 24.51 \text{ lbmol}$$

where P_{N1}, nitrogen partial pressure = 14.7 psia – 2.051 psia.

 b. In a similar manner, the moles of cyclohexane that exist as vapor at 85°F and 100 psig (114.7 psia), N_{C2}, are given by:

$$N_{C2} = (N_N)(P_{C2})/(P_{C2}) = 0.506 \text{ lbmol}$$

where P_{C2}, cyclohexane vapor pressure = 2.320 psia; and P_{N2}, nitrogen partial pressure = 114.7 psia – 2.320 psia.

Therefore, the mass "lost" in the first and second case, M_{C1} and M_{C2}, is given by:

$$M_{C1} = (N_{C1})(MW) = (3.97 \text{ lbmol}) (84.16 \text{ lb/lbmol}) = 334 \text{ lbs}$$
$$M_{C2} = (N_{C2})(MW) \; (0.506 \text{ lbmol}) (84.16 \text{ lb/lbmol}) = 43 \text{ lbs}$$

The second pollution prevention measure has resulted in a decrease of 291 lb (87% reduction) of cyclohexane emissions to the atmosphere.

6. SUMMARY

1. Equipment/devices employed for transporting liquids, gases, and solids include: *pipes, ducts, fittings, stacks, valves, fans, pumps, compressors, belt and screw conveyors,* and *bucket elevators*.
2. Utilities are required in the operation of any production process. These services are normally supplied from a central site facility and usually include: electricity; steam for process heating, water for cooling, potable use, and steam production; refrigeration; compressed air; and inert gas supplies.
3. General pollution prevention considerations for transport and storage systems include such things as: detection and repair of leaks; reductions in operating pressures to minimum required levels to reduce fugitive emissions; collection and recovery of vapors from storage tank operations; inspection and repair of pump seals, connectors, etc.
4. Instruments and controls are at the heart of plant operations. Through use of *open and/or closed feedback loops, computers, alarms, system trips, and interlocks*, a process engineer can ensure that a given process is running both safely and efficiently so that material and energy utilization is carried out at an optimum rate.

7. PROBLEMS

1. Every time an automobile gas tank is filled, the vapor space in the tank is displaced to the environment. Since gasoline is a hydrocarbon, it will eventually be involved in the production of low-level ozone upon solar irradiation. Assume the tank vapor space, the air, and the gasoline are all at 20°C.
GIVEN: The vapor space of the tank is saturated with gasoline and under these conditions the gasoline has a vapor phase mole fraction = 0.4. The molecular weight of the vapor = 70 g/gmol and the gasoline has a specific gravity = 0.62. A typical automobile averages 20 mpg, travels 12,500 mi/yr, and has a 15 gal capacity tank.
   a. Determine the volume of gasoline lost to the air when filling the gas tank (assume 2 gallon reserve left in the tank).
   b. Determine how much gasoline is annually lost from 50 million cars.
   c. What is the value of this loss if gasoline averages $1.19 per gallon?

2. A cooling tower system supplies water to the services identified below at 90°F with a return temperature of 115°F.
GIVEN: Cooling water heat capacity is 1.00 Btu/lb-°F, heat of vaporization at cooling tower operating conditions is 1030 Btu/lb, and the density of water at cooling tower operating conditions is 62.0 lb/ft^3.
   a. How much fresh water make-up is required if 5% of the return water is sent to "blow-down"?

b. Determine the total flow rate of cooling water required for the services listed below.

Process Unit	Unit Heat Duty (Btu/hr)
1	12,000,000
2	6,000,000
3	23,500,000
4	17,000,000
5	31,500,000

3. Determine how many pounds per hour of steam would be required for the following two cases if a plant has the heating requirements shown in the table below.

GIVEN: The Properties of saturated steam are as follows:

Pressure Provided, psig	75	500
Saturation Temp, °F	320	470
Enthalpy of Vaporization, Btu/lb	894	751

a. If steam were provided at 500 psig.
b. If steam were to be provided at both 500 psig and 75 psig pressures.

Process Unit	Unit Heat Duty (Btu/hr)	Required Temperature (°F)
1	10,000,000	250
2	8,000,000	450
3	12,000,000	400
4	20,000,000	300

8. REFERENCES

1. Sprague, B., *Manufacturing Assessment Planner (MAP) Toolkit*, Michigan Manufacturing Technology Center (MMTC) and CAMP, Inc., Ann Arbor, Michigan, 1999.
2. Carpino, R., Department of Chemical Engineering, Manhattan College, personal notes, 1993 and 1996.
3. Liptak, B. G., *Instrument Engineers' Handbook*, Chilton Book Company, Wayne, Pennsylvania, 1985.

CHAPTER 7

Waste Treatment Processes and Equipment
Contributing Authors: Matthew Reagan and Sangeeta Gokhale

The basis for design and performance of a variety of air, water, and solid waste control equipment and processes is briefly summarized in this chapter. This information is provided in the context of a pollution prevention text to indicate control methods that must be installed if contaminant releases exceed acceptable emission levels required by local, state, or federal regulatory agencies. It is also important to understand the nature and potential costs of treatment requirements of various waste streams generated in manufacturing and chemical production processes so that the true benefits of pollution prevention can be incorporated into cost and feasibility studies for options identified as part of a facility pollution prevention opportunity assessment (Chapter 11).

Waste treatment involves the processing of a waste stream in a separate unit operation in order to lessen or remove the impact of the substance once it is released into the environment. Figures 7.1 and 7.2 show possible configurations for waste treatment after a process. In Figure 7.1 the waste comes off the process separately from the process. From there it is sent to the waste treatment unit for further processing. More common is the situation in Figure 7.2, where the product and the waste leave the process in the same stream, and a separation process is utilized to separate the waste so it can be further treated. It should be noted that the latter configuration provides the best opportunity for pollution prevention improvements as here the separation process can be focused on for improvement in product recovery rather than the production process itself. Pollution prevention and closed-loop recycling opportunities exist in both configurations, however, as indicated by dashed lines in Figures 7.1 and 7.2, with either improved production process or separation efficiency. Any analysis of a waste treatment system should consider these potential opportunities or any uses of waste streams that can be considered feedstocks for other in-plant or external process needs.

1. OVERALL CONSIDERATIONS

Typically, when waste treatment options other than incineration are chosen, more than one unit process is required. For example, a typical process sequence that could serve as an alternative to incineration might be sedimentation with neutralization, biodegradation, followed by sludge separation, and finally landfilling the wastes. Before a specific treatment is chosen, the generator must determine the level of toxicity associated with the particular waste stream and whether treatment can be performed on-site or needs to be conducted off-site. The degree of toxicity associated with the waste will determine the level of need for detoxifying or destroying the waste

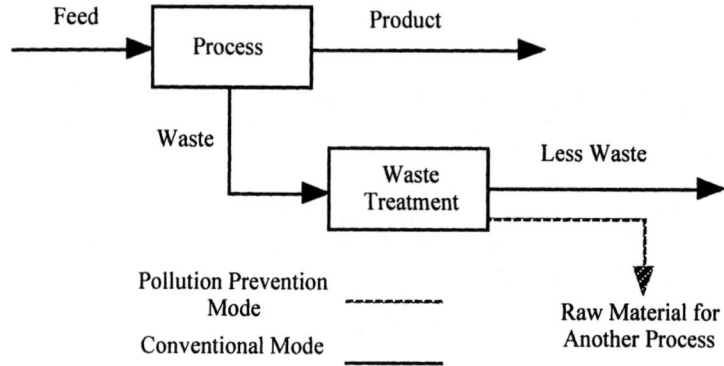

Figure 7.1. Common process/waste treatment system configuration with product/waste separation within the production process showing conventional and pollution prevention operating mode options.

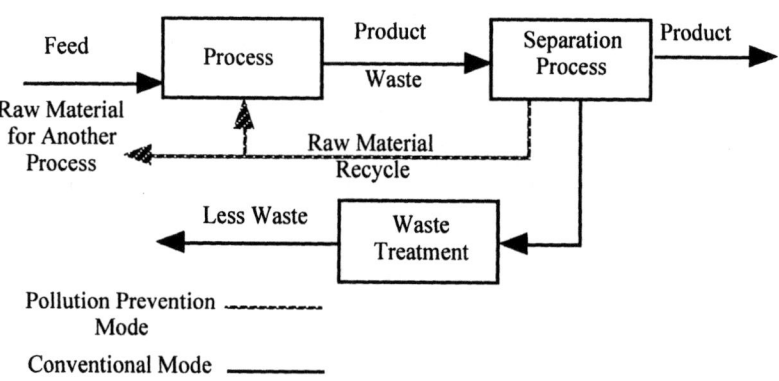

Figure 7.2. Common process/waste treatment system configuration with product/waste separation subsequent to the production process showing conventional and pollution prevention operating mode options.

and has a major impact on waste treatment costs. For on-site treatment, direct costs stem from: capital investment for treatment, raw material and labor costs, utility costs, sampling and monitoring costs, permitting costs, waste-end taxes, and insurance. The associated direct costs for off-site treatment include: treatment, transportation and storage costs, and waste-end taxes. The generator must make a decision to manage the waste directly or to employ the services of a commercial firm. Selection of an appropriate method for treatment of waste will be based not only on technical considerations, but also on economic and regulatory factors. If the waste is to be managed directly, technical considerations involved in the selection of a specific treatment process will include the characteristics of the waste, including its physical form, constituents present, concentration of each constituent, volume of waste; the applicability of available treatment processes; and the treatment objectives. Mixed waste streams will require more than one type of treatment process to meet applicable regulatory requirements.

Economic feasibility is obviously an important factor to be considered in treatment selection. Cost is a strong function of the amount of waste to be treated and the simplicity and degree of commercial usage of the treatment process. Environmental factors, in the form of state and local environmental regulations, also influence waste stream treatment selection. RCRA regulations, specifically, govern the type of waste treatment that may be used and the characteristics of the waste that is finally to be disposed of. The potential liabilities involved with improperly treated waste can include RCRA corrective action, remediation costs, and CERCLA cleanup costs.

As suggested above, the treatment of complex waste streams from industrial processes can be complex and expensive, and waste streams have inherent in them both short- and long-term liabilities. A discussion of some of the treatment processes appropriate for air, water and solid waste streams from industrial processes are described below so that the true cost of waste generation, and the associated benefits of waste and liability reduction provided by pollution prevention efforts, can be adequately captured in a pollution prevention opportunity assessment.

2. AIR POLLUTION CONTROL EQUIPMENT AND PROCESSES

A summary of air pollution control equipment and processes commonly used for the control of particulate and gaseous emissions from industrial facilities is provided in this section. The reader is referred to the voluminous literature on air pollution control equipment for more details. The references by Mycock, McKenna and Theodore[1], de Nevers,[2] and Wark, Warner, and Davis[3] are suggested as a starting point for this information.

2.1. Particulate Control Devices

The removal of particulates from a gas stream can be accomplished by a variety of physical and electrical methods, the choice of which is made based on the characteristics of the particles to be removed, and the removal efficiency that is required for a given situation. The devices described below are common in many industrial applications and include: cyclone separators, electrostatic precipitators, baghouses and fabric filters, and venturi scrubbers.

2.1.1. Cyclone Separators[2]

A common method of large particle removal from gas streams is by gravity in *gravity settlers*. *Cyclone separators* provide additional particle separation efficiency above that of gravity settlers by taking advantage of the density of particles to generate a centrifugal force used to carry them out of the gas stream and toward the wall of the separator. This centrifugal force is generated by inducing a vortex flow within the cyclone that forces gas entering an outer helix to spiral downward, then continue to spiral upward into a smaller inner helix as gas flows out of the cyclone. The particles that are separated move down the inner wall of the collector by gravity into a bottom conical hopper, where they are stored prior to removal from the collector.

High flow rates can be treated by combining multiple cyclones in parallel in a *multicyclone* configuration, and improved efficiencies can be realized through the placement of several cyclones in series. Cyclones are low headloss devices, but have a limited removal efficiency for particles with diameters smaller than 5 μm. They are used widely in industry for the removal of large particles as they are inexpensive both to build and operate, and they do not have a large space requirement. If high efficiency particle removal is required, particularly for gas streams containing small diameter particulates, a cyclone separator may be used for pretreatment of the gas stream, but additional gas removal devices such as those listed below will be required.

2.1.2. Electrostatic Precipitators

Electrostatic precipitators (ESPs) remove small particles from moving gas streams at high collection efficiency by imparting a charge on the particles and inducing them to move toward a collecting plate to neutralize this charge before the gas stream exits the unit. They have been used almost universally in power plants for removing fly ash from gases prior to discharge.

Electrostatic precipitators have the ability to efficiently remove fine particulates if a charge can be placed on the particles to be removed from the gas stream. The ability to transfer an electrical charge to the particles and have this charge remain stable long enough for the particle to migrate to the collecting plate depends to a large degree on the resistivity of the particles.

There are three classes of ESPs: *dry ESPs, wet ESPs (WEPs), and ionizing wet scrubbers (IWS)*. The dry units are by far the most popular. They have been used successfully to collect both solid and liquid particulate matter from many operations, including smelters, steel furnaces, petroleum refineries, and utility boilers. Capital costs for these units are not excessive, but high electrical demands make their operating costs high.

2.1.3. Baghouses and Fabric Filters

One of the oldest, simplest, and most efficient methods for removing solid particulates from gas streams is by filtration through fabric media. A *fabric filter* is capable of providing high collection efficiencies for particles as small as 0.5 µm and will remove a substantial quantity of particles as small as 0.01 µm. In its simplest form, the industrial fabric filter consists of a woven or felted fabric through which dust-laden gases are forced. A combination of factors results in the collection of particles on the fabric filters. When woven fabrics are used, a dust cake eventually forms; this, in turn, acts predominantly as a sieving mechanism. When felted fabrics are used, the dust cake is minimal or nonexistent.

As particles are collected, the pressure drop across the fabric filtering medium increases. Partly because of fan limitations, the filter must be cleaned at predetermined intervals. Dust is removed from the fabric by gravity and/or mechanical means. The fabric filters or bags are usually tubular or flat. The structure in which the bags hang is referred to as a baghouse; the number of bags in a baghouse may vary from several to several thousand. Quite often, when great numbers of bags are involved, the baghouse is compartmentalized to permit the cleaning of one compartment while others are still in service.

When the gas to be treated is at high temperature, high temperature resistant bags must be used to prevent fires and explosions within the baghouses. Glass fiber and Teflon™ filter bags can be used when process conditions demand temperature-resistant and reaction-resistant material. Capital costs of these units are high when specialty bags are required, and the need for on-going replacement of broken bags can make operation and maintenance costs high for these systems as well.

2.1.4. Venturi Scrubbers

Venturi scrubbers, classified as high-energy devices, are primarily used for high-efficiency particulate matter (including sub-micron level) collection and removal. Venturi scrubbers operate in a co-current flow fashion, with a liquid stream injected into a carrier gas stream within a venturi throat. The carrier gas reaches a maximum velocity at the venturi throat (typically 100 to 800 ft/s) where the liquid is introduced (typically 2 to 15 gpm liquid/1000 cfm gas). The liquid flashes within the throat, producing vast quantities of fine aerosols that are violently mixed with the carrier gas, and upon which fine particles impact and become entrained. Relatively poor gas absorption occurs within the venturi scrubber because this co-current flow regime limits absorption efficiency. Once past the venturi throat the carrier gas slows down, and much of the entrained aerosols will settle out of the gas stream by gravity separation, taking the particles they have entrained with them. Not all of the liquid separates from the gas stream, however, and an entrainment separator is almost always required to limit liquid emissions from the scrubber.

Since venturi scrubbers operate with very high carrier gas velocities and high pressure drops (5-in to 100-in water column, WC) they incur both high maintenance and operational costs. However, since the device itself is just a venturi, both the initial cost and space requirement are relatively small. Venturi scrubbers may also be used for gas absorption by simply increasing the amount of time the carrier gas and absorbent remain in contact. To accomplish this, the absorbent injection rate is usually increased to approximately 20 to 80 gpm liquid/1000 cfm gas. The increase in absorbent injection rate will obviously correspond to an increase in both energy consumption and absorbent processing costs, but may be warranted if both particulate and gaseous pollutant removal is required for a given gas stream.

2.2. Gaseous Control Devices

Gaseous contaminants of concern in industrial applications include organic compounds with low water solubility (hydrophobic organics), organic compounds with high water solubility (hydrophilic organics), and inorganic compounds, i.e., NOx, SOx, HCl, CO, etc., with variable water solubilities and reaction properties. The removal of gaseous contaminants from a gas stream can be accomplished by a variety of physical, chemical and thermal methods, the choice of which is made based on the characteristics of the gas(es) to be removed, and the removal efficiency that is required for a given situation. The devices described below are common in many industrial applications and include: absorbers for hydrophilic compounds, adsorbers for hydrophobic compounds, and vapor incinerators or flares for combustible gaseous components.

2.2.1. Absorbers

Absorption refers to transferring an objectionable gaseous component (absorbate) from a process gas stream (carrier gas) into a liquid (absorbent). Since absorption is a mass transfer operation, it can only occur if a concentration gradient exists between the carrier gas and the absorbent. To accomplish the removal of gaseous contaminants from gas streams, the absorption equipment is designed to maximize mass transfer rates. Mass transfer rates depend largely on the amount of surface area of each phase in contact with one another, as well as the amount of time the two phases remain in contact.

An absorber relies on relatively slow velocities between the carrier gas and absorbent to increase contact time and therefore enhance absorption of components out of a contaminated gas stream. The most widely used types of absorption equipment are *spray towers, venturi scrubbers* (discussed in Section 2.1.4), and *packed-bed units.*

Spray towers, which resemble cooling towers, are the simplest devices used for gas absorption and are classified as low-energy devices. They use a set of nozzles to spray a mist of liquid (absorbent) into an empty tower. Typically, the carrier gas enters the bottom of the tower and passes up through a liquid mist that is traveling downward to the bottom of the tower. Nozzles are usually selected and arranged to "mist" the entire tower cross section with fine liquid droplets, thereby maximizing the absorbent surface area and contact time. However, since the droplets tend to hit the sides of the tower after falling a short distance, spray towers are limited to applications where the gases to be removed are extremely soluble or where a high removal efficiency is not required. Spray towers are very inexpensive to purchase. O&M costs are low due to no complex internals; they can handle large volumetric flow rates (approximately 100,000 cfm), and they experience very low carrier gas system pressure drops (less than 2-in WC). They tend to have a relatively large footprint, however, making them a poor choice when high efficiency and small area requirements drive the selection of an absorber.

Packed-bed units, typically contained in a tower or column encasement, are the most common scrubbers used for gas absorption. Packed-bed units disperse the absorbent over packing material which provides a large surface area for continuous gas-liquid contact. Packed-bed units are classified according to the relative direction of gas and liquid flow. The most common arrangement, the *counter-current* design, has the carrier gas entering the bottom of the device and the absorbent entering the top. Therefore, the most dilute concentration of absorbate comes into contact with the purest absorbent providing a constant maximized concentration differential that theoretically achieves the highest efficiency possible. Although the carrier gas pressure drop is moderate (0.2 to 0.5-in WC/ft of packing) when compared to a venturi scrubber, a packed-bed unit is typically larger than a venturi scrubber and can handle a larger flow capacity. The counter-current packed-bed unit, however, can suffer from *"flooding"*, so care must be exercised when designing or specifying a packed-bed unit. The term flooding refers to the formation of a continuous absorbent layer above the packing due to excessive carrier gas velocity [pressure], which in turn does not allow absorbent to flow down through the tower properly to mix with the carrier gas. The carrier gas velocity typically used in a packed-bed unit is between 40 to 70% of the calculated flooding velocity.

Another packed-bed design uses *cross-flow*, where the absorbent flows perpendicular to the direction of carrier gas flow. This absorber system is smaller and experiences a lower pressure drop than other packed units for the same application. They are better suited to handle carrier gas exhaust streams with heavy particulate concentrations than other absorber designs. With high particulate loadings to a cross-flow absorber the absorbent spray flow rate is typically increased in the front half of the unit to increase particulate entrainment much like a venturi scrubber. The absorbent on the packing itself throughout the entire absorber is used for the gaseous contaminant removal. Finally, an unsprayed packing section could be left dry to act as an entrainment separator before the treated gas leave the unit.

2.2.2. Adsorbers

The process where one or more objectionable gaseous components (adsorbate) are removed from an effluent gas stream (carrier gas) by adhering to the surface of a solid (adsorbent) is termed *adsorption*. Typical adsorbents have a relatively large surface-to-volume ratio due to being highly porous. It is primarily on these internal surfaces that adsorption occurs. Adsorption is classified as either physical or chemical based on the manner by which the adsorbate is bound to the adsorbent. Physical adsorption describes adsorbate binding to the surface of the adsorbent via the weak forces of intermolecular cohesion (van der Waals' forces), a readily reversible process. Chemical adsorption refers to a much stronger bond being formed between the adsorbate and adsorbent via ionic or covalent chemical bonding which is not readily reversible.

Once an adsorbent has been classified, other distinguishing properties such as the nature of the chemical being treated, and the adsorbent surface area, pore distribution, particle size, and surface polarity are used in the selection of an optimal treatment media. Chemical characteristics such as aqueous solubility, polarity, octanol/water partition coefficient, vapor pressure, and molecular weight determine whether a given contaminant is capable of forming a chemical bond with a particular adsorbent surface. To evaluate process-specific treatment efficiency potential and to develop design data, adsorption isotherms may be conducted at a pilot scale to determine quantitatively the mass of adsorbate removed per mass of adsorbent.

The most common adsorbents used in industry are activated carbon, silica gel, activated alumina (alumina oxide), and molecular sieves (zeolites). Activated carbon, because it has a hydrophobic, non-polar surface, is used to control organic solvents, odors, toxic gases (e.g., chlorine and ammonia), and gasoline vapors, as well as a variety of other organic compounds. Silica gels are primarily used to remove moisture from carrier gases, but also have the ability to remove sulfur and other impurities as well. Silica gel is ineffective, however, above temperatures of 500°F. Activated alumina is used to dry gases under high pressure, prior to use in catalytic reactions. Molecular sieves can be used to capture or separate gases on the basis of molecular size and/or molecular configuration.

The main function of the adsorption equipment is to bring the gas and solid adsorbent into direct contact to facilitate adsorption. Complete package adsorption systems are available from a number of manufacturers for use in processes involving volatile solvents. Two or more adsorbers are required for continuous adsorption, one or more being in operation while others are being regenerated. Fluidized-bed and moving-bed equipment may be used for large-scale, continuous operation.

Processes that discharge gaseous pollutants that can be controlled by adsorption include: dry cleaning; degreasing; paint spraying; tank dipping; solvent extracting; metal-foil coating; plastic, chemical, pharmaceutical, rubber, linoleum, and transparent wrapping manufacturing; and fabric impregnation. In the manufacture of paints and varnishes, adsorption of the solvents followed by their recovery is not feasible because of fouling of the adsorbent with coating solids. Aqueous scrubbers to remove the paint solids and condensibles are often used prior to adsorption of the solvent by activated carbon.

2.2.3. Vapor Incinerators and Flares

One option for the high efficiency removal of a wide range of organic vapors is *vapor incineration* (also labeled *thermal oxidation*). This treatment process utilizes high temperature, oxygen-rich environments to oxidize chemical bonds in complex organic compounds to yield carbon dioxide, water, and other oxidized inorganic end products. Thermal oxidation processes can often be used for the destruction of suspensions of combustible solids and fumes of combustible liquids, and has also been successful for odor control.[4] Thermal oxidizers can be used both for concentrated, intermittent streams, such as for a relief valve from a process tank, or for continuous, dilute streams, such as from a paint drying oven. For intermittent, concentrated streams, elevated flares are typically used, while either direct thermal oxidation, or *catalytic oxidation* systems are used for removal of volatile organic compounds in dilute air streams.

Pollution prevention and energy reduction opportunities exist within processes utilizing thermal oxidation for organic vapor control. An opportunity exists to recover low grade heat from thermal oxidizers by installing heat exchangers in the incinerator exhaust (called a *recuperator*) to pre-heat the incoming waste gas prior to its combustion. With a recuperator, heat recoveries of 40 to 80% can be realized,[4] reducing the overall energy demand for the process.

Regenerative heat recovery systems are also now commonplace in industry for the recovery of heat from combustion gas. These systems involve the use of dual chambers filled with ceramic packing that are alternated to recover heat from the exhaust gas or provide heat to the inlet gas as flow direction through the heat recovery system is alternated by the use of simple valving. The regenerative systems are able to provide 80 to 95% heat recovery,[4] significantly improving energy utilization for waste gas treatment.

Alternatively, exhaust gas can be routed to a waste heat boiler for the production of process steam, or could be used directly or as a pre-heat gas stream to warm air that might be used for product drying. Many opportunities may exist within a process train and within a manufacturing facility to utilize the energy generated in the combustion of waste gases. If an organic waste stream cannot be eliminated from a production process, consideration should be given in the process design phase to the feasibility of using this waste stream as an energy source for the generation of low-grade energy elsewhere in the process.

3. WATER POLLUTION CONTROL EQUIPMENT AND PROCESSES

To control water pollution, a waste stream can be subjected to one or more physical, chemical, and biological treatment processes A great deal of additional detail on physical, chemical, and biological treatment is available in the literature, and the reader is referred to the following texts as a starting point for this review: Metcalf and Eddy;[5] McGhee;[6] Droste;[7] and Viesman and Hammer.[8] Some of the more common physical, chemical, and biological waste treatment processes found in industrial settings are described below.

3.1. Physical Treatment

There are more than 20 types of physical treatment processes known to be used in the handling of wastes; however, few of these are fully developed or commonly used in industry. Some treatment methods have been found to have little potential use, so that further research in these areas is unlikely. Zone refining, freeze drying, electrophoresis, and dialysis all fall into this category. The most common physical water treatment processes in industry are sedimentation, filtration, flocculation, and solar evaporation. Most other processes fall in between these two extremes; that is, they show some potential for future use but are not presently used to any great extent.

Physical treatment processes may be separated into two general categories based on whether they are designed to treat single-phase or multi-phase solutions. These process categories are: *phase separation* and *component separation*. In phase separation processes two distinct phases are separated based on their different physical properties, while in component separation processes a particular species is separated from a single-phase, multi-component system.

Sedimentation and *centrifugation* are examples of phase separation processes, liquid *ion exchange* and *freeze crystallization* are used for component separation, and *distillation* and *ultrafiltration* are used for both phase and component separation.

Phase separation processes are generally employed to reduce waste volume and to concentrate the waste into a single phase before further treatment and material recovery. Slurries, sludges, and emulsions contain more than one phase, and are the usual candidates for phase separation processes. Filtration, centrifugation, and flotation may be used on slurries that contain larger particles. If the slurry is colloidal, *flocculation* and ultrafiltration are generally used. If the slurry or sludge is known to contain volatile components, evaporation or distillation is used to remove them from the waste stream. Because emulsions are difficult to separate, the type of physical treatment required for these waste streams is usually selected on a case-by-case basis.

Component separation processes remove specific ionic or molecular species without the use of chemicals. Most of these are used in wastewater treatment and include such techniques as liquid ion exchange, *reverse osmosis*, ultrafiltration, *air stripping*, and *carbon adsorption*. The first three processes are used to remove ionic and inorganic components, while the last two techniques are used to remove volatile components, gases, and dissolved organic species from liquid waste streams.

This section includes discussions on the following physical wastewater treatment processes: flocculation and sedimentation; dissolved air flotation; filtration; centrifugation; air stripping; steam stripping; resin adsorption; and electrodialysis.

Many other physical treatment processes (carbon adsorption, distillation, evaporation, filtration, precipitation, ion exchange, reverse osmosis, etc.) are often employed in recycling operations and are discussed in more detail in Chapter 14, "Recycling."

3.1.1. Flocculation and Sedimentation

Flocculation and sedimentation are two processes used to separate waste streams that contain both a liquid and a solid phase. Both are well developed, highly cost-competitive processes, which are often used in the complete treatment of waste streams. They may also be used instead of, or in addition to, filtration. Some applications include the removal of suspended solids and soluble heavy metals from aqueous streams. Many industries use both processes in the removal of pollutants from their wastewaters. These processes work best when the waste stream contains a low concentration of the contaminating solids. Although they are applicable to a wide variety of aqueous waste streams, these processes are not generally used to treat nonaqueous or semisolid waste streams such as sludges and slurries.

Flocculation is a physical process in which fine suspended particles and colloids, which are difficult to settle out of the liquid, are encouraged to coalesce by the addition of coagulating chemicals to form larger, more easily settleable particles. The addition of the coagulating chemicals is referred to as *coagulation*, and is effective through the neutralization of charges on colloidal particles and the formation of *"sweep floc"* that enmeshes fine particles as it settles by gravity through the column of wastewater. Sedimentation is a physical process in which suspended solids are settled out through the use of gravitational forces acting on the solids. The solids removed by sedimentation form a sludge blanket that may be disposed of or treated to recover any valuable material.

3.1.2. Dissolved Air Flotation

Flotation, or *dissolved air flotation (DAF)*, is a physical process used to remove organic or inorganic solids suspended in wastewaters or slurries. The process is used to concentrate waste streams by forming a floatable sludge at the surface of a tank to remove valuable or toxic solids. DAF systems are commonly used to remove oils and greases from waste streams, and with coagulant addition, for the removal of heavy metal ions, cyanides, fluorides, and carbonyls from hazardous wastes. Flotation used to remove specific inorganics operates best when the desired product makes up less than 10 percent of the total solids in the waste stream.

Fine air bubbles are introduced to the bottom of a DAF tank to reduce the density of the suspended solids and carry them to the top of the tank. A froth forms at the surface of the tank, which is then removed by skimmers or scrapers. Individual materials or combinations of similar materials may be removed in this way if the proper coagulants are used. The sludge removed from the surface may be recovered or disposed of.

3.1.3. Filtration

Filtration is a popular liquid-solid separation process commonly used in treating wastewater and sludges. In wastewater treatment, it is used to purify the liquid by removing suspended solids. This is usually preceded by flocculation and sedimentation, with filtration used for further solids removal and polishing. In sludge treatment, filtration is used to remove liquid (sludge dewatering) and concentrate the solids, thereby lowering the sludge volume requiring further treatment or disposal. This method is highly competitive with other sludge-dewatering processes.

In the filtration process, a liquid containing suspended solids is passed through a porous medium. The solids are trapped within the medium resulting in the separation of solids from the waste stream. For large solids, a thick filter media such as sand may be used; for smaller particles, a fine filter such as a filter cloth is preferable. Fluid passage may be induced by gravity, positive pressure, or a vacuum. A few of the most popular filter types are the plate and frame filter press for sludge dewatering, and shell, leaf, and cartridge filters for liquid waste stream treatment. Refer to Chapter 14 for additional information concerning filtration options within a pollution prevention context.

3.1.4. Centrifugation

Centrifugation is another well-established liquid-solid separation process popular in commercial and municipal waste treatment facilities. It is usually used to reduce slurry and sludge volume and to increase the solids concentration in these waste streams. It is a technically and economically competitive process and is commonly used on waste sludges produced from water pollution control systems and on biological sludges produced in industrial and municipal treatment facilities.

Centrifugation is performed in a closed system and is therefore an excellent choice for treating volatile fluids. The liquid and solids are mechanically separated by centrifugal force. The removal of most of the liquid increases the solid concentration in, and reduces the volume of, the waste sludge. The collected solids may then be treated and disposed of or recovered. Three types of units are available for centrifugation: the solid bowl, the disk type, and the basket type centrifuge. The first two are used in large plants, while the third is primarily found in smaller plants. Refer to Chapter 14 for additional information concerning centrifugation applications for pollution prevention.

3.1.5. Air Stripping

Air stripping is generally used to remove volatile organics and other pollutant gases dissolved in liquid (usually aqueous) streams. This process works by providing contact between air and water to allow volatile compounds in the water to diffuse to the gaseous phase and leave the system with the injected air. The efficiency of air stripping depends upon a variety of conditions including: the volatility of the compounds to be removed from the water expressed by the chemical property *Henry's Law* (estimated by the vapor pressure/water solubility); the contact efficiency between the contaminated water and the injected air (affected by the configuration of the air stripper and the air/water ratio); and water temperature (affecting a compound's Henry's Law constant).

Air stripping is commonly used for the volatilization of ammonia from wastewater, for the stripping of hydrogen sulfide and carbon dioxide from drinking water, and in the removal of volatile dissolved organics (Henry's constants greater than 10 atm[9]) from contaminated groundwater supplies. The packed tower (see Chapter 5) is the most compact and efficient device for air stripping, producing greater than 99.9% removal efficiency for common volatile groundwater and wastewater contaminants.[9]

Air stripping systems are cost-effective liquid treatment processes, particularly if the stripped gas from the tower exhaust does not require further treatment. Capital and O&M costs double or triple if off-gas treatment is required. Tower fouling can also occur, however, if the contaminated water also contains high levels of iron or manganese due to chemical precipitation and/or biological growth on the tower packing material. This fouling can significantly reduce the removal efficiency of the air stripping units and will require additional O&M to ensure adequate stripper performance.

3.1.6. Steam Stripping

Steam stripping is designed to take advantage of the increase in Henry's Law constant and corresponding increased stripping efficiency with increased temperature. As with simple air stripping, steam stripping is used to remove dilute concentrations of volatile components from aqueous waste streams, i.e., ammonia, hydrogen sulfide, solvents, etc. The use of steam as the stripping gas phase improves efficiency to the point that compounds removed at marginal or unacceptable rates via air stripping can be effectively removed using the elevated water temperatures produced with the use of steam. For example, steam stripping has been used to recover sulfur from refinery waste and low volatility organics, such as phenol, from industrial wastes.

The stream-stripping process is carried out in a distillation column, which may be in either a packed or tray tower configuration. Steam is injected into the bottom of the column and is used to encourage the exchange of components from the waste stream, which is flowing downward in a countercurrent fashion. The product stream, rich in stripped components, is almost always further treated to remove these components. If the volatiles contain sulfur, the stream may need treatment for sulfur dioxide emissions. In a pollution prevention context, however, this concentrated stream could represent a feedstock for another process within the plant or in a related production facility.

Costs for steam stripping are much higher than for simple air stripping systems due to the cost of generating the steam required in the process. The quality of steam required in this process is low, however, and could easily be met using recovered heat from boiler or incinerator flue gas, providing additional pollution prevention opportunities when contaminants may not be effectively removed by air stripping alone.

3.1.7. Resin Adsorption

Resin adsorption is used in the removal of organic solutes from aqueous waste streams. Solute concentration in the wastewater may be as high as 8 percent. It is the preferred method when recovery of the adsorbate is desired since thermal regeneration required for activated carbon destroys the organic material recovered from the liquid stream. Resin adsorption is also useful when there is a high concentration of dissolved inorganic salts in the waste stream. Synthetic resins can be used to remove hydrophobic or hydrophilic solutes, which may then be recovered by chemical means. Their applications include phenol recovery, fat removal, and color removal from wastewaters. Possible additional uses include the removal of pesticides, carcinogens, and chlorinated hydrocarbons.

Resin adsorption is similar to carbon adsorption, except that two filter beds are used. One bed is used for adsorption while the other is being regenerated. The waste stream flows downward in the system at a rate of 1 to 10 gal/min-ft^2 of cross section, and adsorption terminates when the bed becomes saturated or the effluent concentration reaches the acceptable discharge level. Capital and O&M costs are generally high for these units, but they can still be cost-effective for some applications due to the selectivity they show for the adsorption of certain organics, and their ability to produce a recoverable concentrate stream following resin regeneration.

3.1.8. Electrodialysis

Electrodialysis is a fully developed process used in separating ionic components and is commonly used in the desalination of brackish waters. Separation of contaminants in an aqueous stream is achieved through the use of synthetic, selective membranes and an electrical field. The membranes allow only one type of ion to pass through them and may be chosen to be either anion or cation permeable. The electric field applied to the system causes the positive and negative anions to move in opposite directions and, hence, produces one stream rich in a particular ion and one depleted of that ion. Both streams may subsequently be internally or externally recycled or may be disposed of as necessary. It is important that pH and concentration be controlled for efficient operation of this system.

The potential uses of electrodialysis are in acid mine drainage treatment, the desalting of sewage treatment plant effluents, and in sulfite-liquor recovery. One disadvantage of this process is that it may produce low-level toxic or flammable gases at the electrodes during operation.

Capital and O&M costs can be high for this system, but, as described above, if ionic species of high value or high toxicity exist within the waste stream, the cost of this system can easily be justified. This system can also provide significant pollution prevention opportunities with the judicious use of selective membranes because of the quality of the concentrate stream that can be produced.

3.2. Chemical Treatment

Numerous chemical processes are used in the treatment of liquid waste streams. Most chemical treatment schemes are used in conjunction with other methods to achieve a desired end result. Selected treatment methods directly applicable to recycling efforts (for example precipitation) are discussed in Chapter 14 and are not repeated in this section. Here, the following chemical processes are discussed: calcination, chlorinolysis, electrolysis, hydrolysis, neutralization, oxidation, photolysis, and reduction.

Catalysis describes the modification of the mechanism and rate of a chemical reaction. In general, if a waste stream can be modified by a chemical reaction, the successful use of a catalyst will usually reduce operating costs and energy requirements of the process. Catalysts can be selective, such as in the detoxification of chlorinated pesticides by dechlorination, or versatile, as in the complete destruction of cyanides by air oxidation. Although catalytic oxidation is commonly used as an alternative to incineration in the decomposition of organic wastes, it is not widely used in the treatment of hazardous waste. The technology is not well developed, and because of catalyst specificity, progress in the development of one application does not necessarily mean progress in the development of another. In addition, the many components in a waste stream, in varying concentrations, can inhibit the effectiveness of a catalyst on the target reaction. Some laboratory research has been done on the following catalytic processes: oxidation of cyanides, sulfides, and phenols; and, decomposition of sodium hydrochloride solutions. A catalyst will normally reduce operating temperature and therefore conserve energy, a positive pollution prevention benefit generic to catalytic reactions. Catalytic processes may be higher in capital costs, but lower in operating costs, and thus result in lower overall expenditures for waste management.

3.2.1. Calcination

Calcination is a thermal decomposition process without any interaction with a gaseous phase. Operating temperatures of about 1800°F and atmospheric pressure can produce dry powder from slurries, sludges, and tars, as well as aqueous solutions, by driving off volatiles. Typical calciners are *open hearth*, *rotary kiln*, and *fluidized bed* configurations. Calcination is a well-established procedure and is recommended as a one-step process for the treatment of complex wastes containing organic and inorganic components. It is also one of the few processes that can satisfactorily handle sludges. A real advantage of calcination is that it can concentrate, destroy, and detoxify in a single step. In the process, organic components are usually destroyed while inorganics are reduced in volume and leachability, thus making them suitable for landfilling.

Treatment applications include recalcination of lime sludges from water treatment plants, coking of heavy residues and tars from petroleum refining operations, concentration and volume reduction of liquid radioactive wastes, and treatment of refinery sludges containing hydrocarbons, phosphates and compounds of calcium, magnesium, potassium, sodium, iron, and aluminum. The use of calcination is likely to expand in the future since it handles tars, sludges, and residues that are difficult to handle by other methods.

3.2.2. Electrolysis

Electrolysis is the reaction of either oxidation (loss of electrons) or reduction (gain of electrons) taking place at the surface of conductive electrodes immersed in an electrolyte under the influence of an applied electric potential. Electrolytic processes can be used for reclaiming heavy metals, including toxic metals from concentrated aqueous solutions. Electrolysis is not useful for organic waste streams or viscous, tarry liquids. Application to waste treatment has been limited because of cost factors. A frequent application is the recovery for recycling or reuse of metals, such as copper, from waste streams. Pilot applications include oxidation of cyanide waste and separation of oil-water mixtures. The energy costs of electrolysis can be 10 to 15 percent of operating costs. Gaseous emissions may occur and, if these are hazardous and cannot be vented to the atmosphere, further treatment (such as scrubbing) is required. Wastewater from the process may also require further treatment. Costs are dependent on the concentration of the feed and the desired output.

The most common waste treatment application of electrolysis is the partial removal of heavy metals from spent copper pickling solutions. When the typical concentration of the spent pickling solution is 2 to 7 percent copper, the system design is similar to that of a conventional electroplating bath. When recovering metal from more dilute streams, mixing and stirring are necessary to increase the rate of diffusion. It may also be necessary to use a large electrode surface area and a short distance between electrodes.

Another consideration in using this method is the removal of collected ions from the electrodes. Such removal may or may not be difficult, but must be addressed. The material collected must be ultimately disposed of if it is not suitable for reuse.

3.2.3. Hydrolysis

Hydrolysis is the reaction of a salt with water to produce both an acid and a base as shown in the following equation:

$$XY + H_2O \rightarrow HY + XOH \qquad (7.1)$$

These reactions generally require high temperature and pressure, acid or alkali addition, and sometimes catalysts. Feed streams can be aqueous, slurries, sludges, or tars. The process could be useful for a variety of wastes, but few applications have been employed. The petroleum industry uses hydrolysis to recover sulfuric acid from the sludge from the acid treatment of light oils. Hydrolysis has also been used to detoxify waste streams of carbamates, organophosphorus compounds, and other pesticides.

Energy requirements vary with application, but are generally high. Some products of hydrolysis are toxic and so require further treatment. An additional drawback is that, depending on the waste stream, the products may not be predictable and the mass of toxic substances discharged may be greater than that of the waste originally inputted for treatment. Operating costs are completely dependent upon application and can be extremely high if difficult to handle by-products are generated in the process.

3.2.4. Neutralization

Neutralization is a reaction that changes either low pH or high pH solutions to a pH of near 7 (neutrality). The treated stream undergoes essentially no change in physical form, except possibly precipitation or gas evolution. The process has extremely wide applications to aqueous and nonaqueous liquids, slurries, and sludges. It is very widely used in waste treatment. Some applications include treatment of pickle liquors, plating wastes, mine drainage, and for the disruption of oil emulsions. It is commonly used in the following industries: battery manufacturing; aluminum and coal production; inorganic and organic chemicals; photography; explosives; metal refining; pharmaceuticals; power plants; and textiles.

The reaction can take place in batch or continuous reactors (see Chapter 5 for details on reactors). The flow can be co-current or countercurrent depending on the desired results and the waste stream characteristics. The reaction can take place in a liquid or gaseous phase, or both. Limitations to the process include temperature dependence and residual formation.

Some common methods of neutralizing wastes include: mixing acidic and alkaline streams together, passing acid wastes through packed beds of limestone, mixing acid waste with lime slurries, adding solutions of concentrated bases such as caustic soda (NaOH) or soda ash (Na_2CO_3) to acid streams, passing waste flue gas from a boiler through alkaline waste liquids, adding compressed CO_2 to basic wastes, and adding acids such as H_2SO_4 or HCl to basic (alkaline) waste streams. The choice of the acid or alkali solution to be used is often based on process requirements, but cost is the primary consideration. Lime and sulfuric acid are relatively inexpensive. Lime will, in the process of neutralizing sulfate-bearing wastes, form calcium sulfate, which will precipitate and may be undesirable. Caustic soda and soda ash are more expensive but may be the better choice of bases for wastes containing sulfates. If sulfides or cyanides are present in the waste stream, toxic gas may be evolved during neutralization. These gases require special handling by gas treatment devices such as scrubbers.

3.2.5. Oxidation

Oxidation is a process in which one or more electrons are transferred from the chemical being oxidized to a chemical initiating the transfer. The main purpose of treating wastes by oxidation is detoxification. For example, oxidation is used to change cyanide to less toxic cyanate or to completely oxidize cyanide to carbon dioxide and nitrogen. Oxidation can also aid in the precipitation of certain ions in cases where the more oxidized ion has a lower solubility.

The following are some common oxidizing agents and their uses.

1. *Chlorine* as a gas or in *hypochlorite* salt is often used to oxidize cyanide. It can also be used on phenol-based chemicals, but this application is somewhat limited because of the formation of toxic chlorophenols when the process is not adequately controlled.
2. *Ozone* is used in the oxidation of cyanide to cyanate and to oxidize phenols to nontoxic compounds. Because ozone is an unstable molecule, it is highly reactive as an oxidizing agent. There are no inherent limitations to oxidation with ozone, but components of the wastewater should be examined to determine that enough ozone is being added to also oxidize other components that could compete with the toxic components for ozone.
3. Ozone with ultraviolet (UV) light can oxidize some compounds. Halogenated organic compounds are resistant to ozone, but in the presence of UV light can be completely oxidized.
4. Hydrogen peroxide (H_2O_2) is a powerful oxidizing agent and is successful with phenols, cyanides, sulfur compounds, and metal ions. The process is sensitive to pH, which must be controlled. The optimum pH is approximately 3 or 4, as process efficiency decreases at higher or lower pH values.
5. Potassium permanganate ($KMnO_4$) is an excellent oxidizing agent that reacts with aldehydes, mercaptans, phenols, and unsaturated acids. It has been used to destroy organics in wastewater and potable water. The reduced form of this compound is manganese dioxide, which is insoluble and can be easily removed by filtration. The process is pH-sensitive and is more rapid at high pH (up to 9.5) than at neutral to acid pH.

Oxidation is used in the treatment of wastes, the most common applications being for dilute waste streams. Energy consumption in the process is reactively low since only pumps and mixing equipment are needed. One disadvantage is that some oxidizing agents add other metal ions to the waste stream that may have to be subsequently removed. The costs are usually lower when continuous rather than batch processes are used. The oxidation process is not very successful with sludges. Incomplete reactions occur because of the difficulty in achieving good contact mixing with sludges.

3.2.6. Photolysis

Photolysis is the breakage of chemical bonds under the influence of UV or visible light. The extent of bond cleavage varies according to the compound and the light used in the reaction. Complete conversion of an organic compound to CO_2 and H_2O by this method is highly improbable and partially degraded components are likely to be hazardous. Large-scale waste treatment applications of photolysis have not been commercially developed. There are some UV-assisted ozonation and chlorination applications that have been successful. Specialized applications for the detoxification of pesticide-contaminated solvents are a possible future use.

3.2.7. Reduction

In chemical *reduction*, one or more electrons are transferred from the reducing agent to the chemical being reduced. This process can reduce the toxicity of a solution, making it safer for disposal, or encourage a particular chemical reaction. The first step in a reduction process is normally pH adjustment. The reducing agent can be a gas, solution, or finely divided powder. Mixing is critical in order to assure contact between the reducing agent and the waste for a complete reaction to occur. The reduction reaction is normally followed by a separation step, such as precipitation, to remove the reduced compound formed in the process.

Some common reducing agents are sulfur dioxide, sodium metabisulfite, sodium bisulfite, and ferrous salts for reducing chromium; sodium borohydride to reduce mercury; and alkali metal hydride to reduce lead. Sulfur dioxide is normally used as a gas. The reduction of chromium with sodium metabisulfite and sodium bisulfite is highly dependent on the pH and temperature of the waste solution. Sodium hydroxide must be added to control the pH during the reaction. Since the reduction of chromium with ferrous sulfate works best when the pH is below 3, acid must be added during this reaction.

The major application of chemical reduction is for the treatment of chromium waste. Chromium is very toxic in its hexavalent state (Cr^{6+}); trivalent chromium (Cr^{3+}) is much less toxic and will precipitate in an alkaline solution by forming an hydroxide. Chromium reduction is used for waste treatment in industries such as metal finishing, inorganic chemical manufacture, coil coating, battery manufacture, iron and steel manufacture, aluminum forming, electronic manufacture, porcelain enameling, and pharmaceutical manufacturing.

3.3. Biological Treatment

Biological processes involve chemical reactions, but are differentiated from the chemical category in that these reactions are biochemical in nature and take place in or around microorganisms. The most common use of biological processes is for the decomposition of biodegradable organic compounds and the removal of nutrients, primarily nitrogen, from liquid waste streams.

The biological waste treatment processes described here include: activated sludge, waste stabilization ponds and aerated lagoons, trickling filters, anaerobic digestion, composting, and enzyme treatment. All of these processes except enzyme treatment use microorganisms to decompose the waste. Enzyme treatment generally involves extracting the enzyme from the microorganism and using the purified enzyme to catalyze a particular reaction. With proper control, these processes are reliable and environmentally sound. Additional chemicals are usually not needed and operational expenses are relatively low for these biological treatment systems.

3.3.1. Activated Sludge

The *activated sludge* process uses microorganisms to decompose organic materials in aqueous waste streams. This is an aerobic, microbial process for degrading organic wastes that uses a high concentration of active microbial cells within a completely mixed or plug flow reactor, termed the *aeration tank*. The microorganisms in the aeration tank transform biodegradable organic substances enzymatically to produce energy, oxidized end products and new cellular material.

The liquid waste being treated is retained within the system for a period (the *hydraulic retention time*) of only 3 to 30 hours, before flowing into a *clarifier* used to separate the microbial biomass from the treated wastewater. This biomass, the activated sludge, is retained within the system for a period of 7 to 30 days (the *solids retention time*) through the recycling of settled solids (*return activated sludge*) from the clarifier underflow. To maintain proper growth conditions within the system, some solids are removed from the system (*waste activated sludge*) from the clarifier recycle line.

Besides removing soluble organics, the biomass with its large surface area also adsorbs organic wastes that are in a colloidal form. Volatile organics can be driven off during the aeration process, and this could be a significant source of VOC emissions from a facility. In addition metals and other recalcitrant, hydrophobic organics can concentrate in the sludge, potentially making treatment and disposal of the waste activated sludge problematic.

The standard activated sludge process has several modifications that are better suited for biologically treatable industrial wastes. *Step aeration* is a modification where the waste stream is added at several points along the aeration tank. This spreads out the waste loading, lessening the effect of a shock load on the system, and evening out the oxygen demand in the process. *Pure oxygen* can be used in place of air in an aeration tank. This increases the amount of oxygen available for the microorganisms, allowing higher strength wastes to be treated. *Sequencing Batch Reactors (SBRs)* are another modification of the activated sludge process. In SBRs aeration and settling take place in sequence within the same batch reactor. The advantage of an SBR is the ability to vary the retention time throughout various stages of the treatment process as needed to optimally stabilize the waste.[5]

3.3.2. Waste Stabilization Ponds and Lagoons

Waste stabilization ponds are shallow basins into which wastes are fed for biological decomposition. The chemical reactions involved are the same as those that occur in the other biological processes. Aeration is provided by the wind and diffusion across the pond surface, while anaerobic decomposition of settled solids takes place near the bottom of the pond, making them *facultative* (aerobic and anaerobic) systems. Because these systems have large surface areas and are not mechanically mixed, stabilization ponds support large populations of algae (an additional source of oxygen to the water) as well as bacteria. The ponds are commonly used for treatment of municipal sewage and dilute industrial wastes from industries such as steel, textiles, oil refineries, paper and pulp mills, and canneries.

Normally three or more waste stabilization ponds are used in series as each is only moderately efficient at organic removal efficiency. Some advantages of waste stabilization ponds include their low energy requirements and few effluent-handling problems when treating appropriate waste streams. Disadvantages include a high sensitivity to inorganic toxicants and high suspended solids loadings, and large land area requirements.

Aerated lagoons are earthen basins that are artificially aerated. Microbial reactions in aerated lagoons are the same as those that take place in the activated sludge process, i.e., primarily bacterial in nature, except that the biological solids are not recycled.

The aerated lagoon process has been used for petrochemical, textile, pulp and paper mill, cannery, leather tanning, gum and wood processing, and some other industrial waste streams. Aerated lagoons usually require less energy than activated sludge processes and are therefore less expensive to operate since recycle systems are not needed. More land area than activated sludge systems is required, however, as the retention time in an aerated lagoon is much longer than that for the activated sludge process. The removal efficiencies provided by aerated lagoons are also not as high as those for the activated sludge process since mixing and aeration are less effective in lagoons and they have no sludge recycle. One other disadvantage of the aerated lagoon is that they do not handle waste streams with variable organic and metal constituents as well as activated sludge systems do.

Occasionally, waste stabilization ponds and aerated lagoons must be lined to prevent groundwater contamination from pond/lagoon leachate.

3.3.3. Trickling Filters

Trickling filters are *fixed-film* biological systems used for the decomposition of liquid organic waste streams. In trickling filters, an inert packing material (originally rock but now plastic or ceramic packing) is contained within the bioreactor and waste is distributed over the media via rotating sprayer arms to encourage attachment and growth of microorganisms on this media. Over time the biofilm grows to a thickness that does not permit oxygen to penetrate through it and the lower layers of microbial film become anaerobic. With further microbial growth organic substrate penetration through the biofilm is limited and the lower layers of microbial mass die and lose their attachment to the media. With additional waste loading the biofilm is sheared (termed *sloughing*), the media surface is once again exposed, new microbes colonize the surface, and the process is repeated. In trickling filters treated effluent, not cell mass, is recirculated around the filter. This recirculation flow keeps the trickling filter wetted, dilutes incoming waste to moderate the waste load, provide additional contact time for the liquid to improve organic removal efficiency, and provides the energy to keep biomass sloughing and prevent clogging within the trickling filter.

Some industries that use trickling filters include refineries (with oil, phenol, and sulfide wastes), canneries, pharmaceutical manufacturers, and petrochemical manufacturers. The energy requirements for this process are low, although recycle pumps are required for recirculation of the filter effluent. Treatment efficiency is not as high as with activated sludge systems, typically 50 to 85 percent, and normally units are used in series and/or parallel to provide the required effluent quality.

3.3.4. Anaerobic Digestion

Anaerobic digestion utilizes anaerobic microorganisms (those that do not require oxygen) for the stabilization of high strength liquids and slurries. Anaerobic digestion is carried out by a consortia of microorganisms in a three step process. The first involves bacteria that break down complex organic compounds into simpler organic compounds (*acetogenic bacteria*). In the second step, other bacteria (*acid formers*) convert these simple organic compounds into organic acids. The third and final group (*methane formers*) converts these acids into methane and carbon dioxide, along with small amounts of water and ammonia. Balancing all populations to ensure that intermediate acids do not build up in the reactor, inhibiting the methane formers, is an important process control parameter.

Anaerobic digesters are operated with heating and mixing in the first of two closed reactors. Methane generated in the conversion of organic acids by the methane formers is used to heat the reactor to approximately 35°C. The second stage reactor is used for liquid/solids separation, and is normally fitted with a floating roof to also function as a gas storage vessel. If high strength organic waste is available in a plant, anaerobic digesters could be used to generate energy in the form of methane that could be recovered for use elsewhere in the plant. Anaerobic digestion is widely used for sewage sludge treatment, and has been highly successful for the treatment of brewery, alcohol distillery, meat packing, food processing, and cotton kiering waste streams.

3.3.5. Composting

Composting is a high-temperature, aerobic treatment process that is designed to stabilize high-solids content wastes and sludges. The primary microorganisms involved in the composting process are fungi and actinomycetes. In the composting process, wastes and sludges are piled in long rows, termed windrows. The solids content in these windrows is adjusted to 50 to 70 percent with the addition of an organic bulking agent such as sawdust, straw, manure, etc. The piled waste mixture must be maintained aerobic, so the pile is either turned mechanically, or is designed with internal air piping that can be used to supply oxygen to the pile without disturbing it in a *static pile* configuration. During decomposition, the interior of the pile can reach temperatures of 50 to 70°C, effectively destroying pathogenic organisms, weed seeds, etc., making the compost safe to subsequent uncontrolled use. The collection of leachate and runoff is normally required to prevent off-site migration of contaminants.

Complete waste stabilization in windrow or static pile systems requires approximately 3 or 4 months. Since there is a market for the soil and the decomposed organics (humus) in Europe, composting is frequently used there. Composting has been successful with municipal refuse and green yard waste, high concentration organic refinery sludges, and for the detoxification of some explosive and munitions-contaminated soils at DOD sites.

3.3.6. Enzyme Treatment

Enzyme treatment involves the application of specific proteins (simple or combined), which act as catalysts in degrading wastes. Enzymes work on specific types of compounds, specific molecules, or a specific bond in a particular compound. Enzymes are inhibited by the presence of insoluble inorganics, are sensitive to pH and temperature fluctuations, and do not adapt to variable contaminant concentrations.

Enzyme treatment may prove useful for specific industrial applications, but due to the specificity and fragile nature of current enzyme systems, enzyme treatment is not commonly used in waste treatment at present. The addition of enzymes in the past in an attempt to catalyze the activated sludge process has failed. Since it is not used on a large scale for waste treatment, little data are available on the economics of this process.

4. SOLID WASTE CONTROL EQUIPMENT AND PROCESSES

Solid wastes management is carried out using a variety of methods, most of which are employed by the facility responsible for the final disposal of this waste. At the generating facility solid waste is handled using two basic methods. The first is through physical processing with the ultimate objective being to reduce the volume of waste to be subsequently transported. Physical solid waste processing includes *air classification, shredding,* or *compaction.* The second method of solid waste handling is by *incineration* of the waste with the ultimate objectives of reducing the volume of waste to be transported for disposal, and perhaps recovering some of the energy within the solid waste stream. Both physical and thermal processes common in the handling of the solid waste stream are described below.

4.1. Physical Treatment[10]

Physical processes for the treatment of solid waste streams include methods for the reduction in particle size, for the separation of waste constituents based on their density, or for the increase in solid waste density. These processes are designed to reduce the volume of waste that must be transported to a location for final disposal, and do not result in the destruction of mass of waste constituents that are being handled. Several of these processes lead to the segregation of waste components, making them ideal for incorporation into a recycling/reuse program without large additional capital investments. The physical processes commonly used on-site in commercial facilities that are briefly discussed below include: *shredding, air classification*, and *compaction*.

4.1.1. Shredding

Generally the first step in the physical handling of solid waste that will be separated for recycling, composting, or incineration for energy production, or further processed for compaction, is waste *shredding*. Shredding is carried out using *grinders, chippers, shredders* or *hammer mills* to provide size reduction and to generate a final product that is relatively uniform in size to simplify subsequent waste handling steps. Waste that has been shredded has improved characteristics relative to energy recovery (more uniform particle size, more uniform composition, lower moisture content); subsequent sorting and classification (more uniform particle size, lower bulk density, lower moisture content), or subsequent compaction (more uniform particle size, lower moisture content, higher compactability), and may be routinely practiced, particularly for mixed solid waste streams. If a mixed waste stream is being handled, many abrasive materials in the waste make shredding expensive in terms of operation and maintenance, and may not warrant these high costs despite improvements in the physical quality of the waste stream. If recyclable components of a waste stream cannot be removed through pollution prevention techniques, some form of waste stream shredding may be justified to optimize recovery of these materials.

4.1.2. Air Classification

Air classification is used to separate low density, organic material from higher density, inorganic components within a solid waste stream. In a solid waste stream, the low density fraction primarily consists of paper and plastics, and is often used to separate components that will be further processed for recycling via composting, energy recovery via incineration, or for recycled feedstock preparation. Air classifiers are generally preceded by a mechanical shredder and are commonly configured to provide air flow upward, countercurrent to the downward flow of solid waste through a vertical throat or zig-zag channel. Heavy inorganics drop from the bottom of the classifier onto a materials conveyor, while the light, organic fraction of the waste is carried out of the classifier for subsequent separation from the classification air. Air classification is a relatively simple and inexpensive process that can provide efficient separation of light and heavy solid waste components. Air classification does require pre-shredding, however, making the overall process inclusive of shredding subject to the same constraints and high O&M costs as shredding systems alone.

4.1.3. Compaction

Compaction is a common solid waste process designed to reduce the volume of a solid waste stream (typically by a factor of 3:1 to 8:1) by increasing its bulk density. It is so common because transportation costs are a significant portion of the overall cost of managing solid waste, and compaction is effective in reducing transportation costs for both wastes and recycled materials. Compaction equipment can be either permanently located at a waste generation site (*stationary compactor*), or can be incorporated into mobile waste collection vehicles as is common for the majority of municipal, commercial, and industrial applications. Stationary compactors are common at large industrial and commercial facilities that must manage large volumes of paper waste, cardboard, and other compressible, non-peutresible waste materials. These stationary compactors can be effectively used in conjunction with shredding equipment to produce highly compacted and bailed bundles of recyclable paper and cardboard that minimize transportation costs. If waste disposal costs are high, and can be reduced through the bailing of waste to reduce the effective disposal volume of the material, the installation of a compaction system, even without the intent to provide component separation, may still be warranted.

4.2. Thermal Treatment

Thermal treatment or *incineration* is a viable method for reducing the volume of wastes, effectively destroying hazardous components of the waste, and in certain cases enables the recovery of useful energy leading to waste reduction and ultimately pollution prevention. Waste incineration has become a controversial topic despite its widespread use for the treatment of a variety of waste types, and despite statements by such entities as the Congressional Office of Technology Assessment that suggest that it "…is preferable to permanently reduce risks to human health and the environment by waste treatments that destroy or permanently reduce the hazardous character of the material, than to rely on long-term containment in land-based disposal structures."[11] The public perceives incineration to be an ineffective method for the destruction of waste constituents, releasing large amounts of untreated hazardous materials and hazardous incineration by-products to the atmosphere, despite data to the contrary.

Incineration generally reduces the volume of waste to be disposed of by 90 percent, and is the only treatment process that can recover energy from a mixed waste stream whose components are not otherwise recoverable.

Energy recovery systems can play a significant role in pollution prevention efforts, and a variety of incineration systems applicable to industrial processing of a solid waste stream are discussed below along with incinerator principles. A comprehensive treatment of incineration topics as particularly applied to hazardous waste incineration can be found in the literature.[12]

4.2.1. Incinerator Principles

The three critical factors for effective combustion of organic compounds include:

1. The *temperature* in the combustion chamber
2. The length of *time* the wastes are maintained at high temperatures
3. The *turbulence*, or mixing, of the waste with the air in the combustion zone

Virtually all burning in the combustion process occurs in what is defined as a *diffusion flame*. A diffusion flame consists of fuel molecules supplied from one side of the flame and oxygen molecules (from air) from the other side. In a narrow region where the two meet is a flame zone where almost all the oxidation combustion (in the form of a reaction) takes place. The temperature here is very high and most molecules are chemically transformed into atoms or small, highly reactive molecular species called *free radicals*.

Three zones can be distinguished in a diffusion flame: a *preheat zone*, a *reaction zone*, and a *recombination zone*. On the fuel side of the flame, considerable molecule degradation occurs in the preheat zone, and the fuel fragments leaving this zone are made up mainly of smaller molecular weight hydrocarbons, olefins, and hydrogen. In the reaction zone, the reactions are mainly of the free radical type. The major constituents in the reaction zone are similar irrespective of the composition of the original fuel/waste mixture. When certain smaller species are combined with oxygen or hydroxide radicals, compounds such as aldehydes (especially formaldehyde), ketones, alcohols, and acids can be produced as combustion intermediates. If the oxidation reactions are terminated as a result of flame quenching by cold air or water, or by a cold surface or lack of availability of additional oxygen, those intermediate species can escape from the flame zone. These are commonly referred to as *products of incomplete combustion (PIC)*. The carbon oxidation reactions do not reach true equilibrium because of the short retention time within the reaction zone, and the oxidation of the carbon proceeds mainly to carbon monoxide rather than carbon dioxide even though sufficient oxygen is present.

In the post-flame or recombination zone, slower recombination reactions occur, leading to further releases of heat. Here carbon monoxide is oxidized to carbon dioxide. Under certain conditions the organic radicals formed in the reaction zone will recombine with themselves and form olefins, aromatics, cycloolefins, and paraffins. Soot is also formed. Soot is the result of the reduction of organic carbon in the fuel to carbon particles in the reducing section of the flame. Since there are many possible routes of recombination, many different compounds in extremely low concentrations can be formed in any combustion process. In the case of the most simple fuel, methane, close to 100 reactions have been identified in the flame zone.

The major difference between conventional fossil fuels and chemical wastes is that, in addition to carbon and hydrogen, chemical wastes may also contain significant and varying amounts of other elements such as chlorine, sulfur, and nitrogen. Hence, the free radical composition in the reaction zone of a chemical waste flame can be different from that of a fossil fuel flame.

Adequate free oxygen must be present in the combustion zone to constantly mix with the gaseous waste forming a turbulent reaction environment within the incinerator. With proper supply and mixing of oxygen molecules and long enough residence time, the formation of PICs is minimized. Modern waste incinerators have a large combustion chamber (and often a secondary chamber) to provide adequate time for post-flame reactions to decompose virtually all of the small amount of organics that escape from the flame zone.

From the perspective of chemical thermodynamics and equilibrium, organic molecules cannot survive in the high-temperature oxidizing environment that exists in a typical incinerator. The theoretical or equilibrium combustion products are a function of the elemental composition of the fuel/waste mixture only and are not a function of the molecular structure of the mixture. In other words, a chemical waste that has the same elemental composition as that of a fossil fuel will produce precisely the same complete combustion products under equilibrium conditions. In fact, true equilibrium is seldom reached for two reasons: (1) reaction rates may be too slow in relation to the time available (kinetic limitation), and (2) mixing is less than perfect (mixing limitation). Mixing is the major limiting factor in a combustion device. However, in a well-designed and operated incinerator, equilibrium can be approached. This is especially true for high-temperature oxidation or organics that have a very fast reaction rate. Reactions associated with inorganics (for example, HCl, Cl_2, SO_2, SO_3, NO, and NO_2) are orders of magnitude slower, and the final combustion products may deviate significantly from equilibrium.

Under equilibrium conditions, almost all carbon in the organics will be converted into carbon dioxide, with a trace amount of carbon monoxide existing. Virtually all hydrogen will be converted into water or some inorganic acid, and all chlorine will be converted into hydrogen chloride, HCl, with some converted to chlorine gas, Cl_2. Most sulfur will be converted to sulfur dioxide, SO_2, with some to sulfur trioxide, SO_3. Metals within the waste may be converted into particulate metal oxides and exit the incinerator with either the fly ash or bottom ash. The combustion of nitrogen compounds is more complex. Depending on many factors, a portion of the nitrogen will be converted into different forms of oxides of nitrogen (NO and NO_2) and nitrogen gas (N_2).

Wastes are fed into an incinerator in batches or in continuous streams. A nozzle can be used to pump liquid waste feed into the incinerator in order to atomize the liquid to enhance its combustion efficiency. Solid waste feed is introduced into an incinerator either in bulk or through containers, using a conveyor or gravity feed system. The wastes are burned in a combustion chamber designed to withstand and maintain extremely high temperatures (between 1800°F and 2500°F) to ensure that all organic compounds are destroyed. The combustion ash is collected from the combustion chamber and disposed of as hazardous waste. Combustion gases are cooled, cleaned of gaseous and particulate contaminants, and released to the atmosphere through the incinerator stack.

The ability to recover energy from waste materials is a major advantage of incineration compared to other waste treatment options. Incinerators can generate 10 to 150 million Btu/h heat energy depending on the waste being combusted, and can be a significant source of energy within a manufacturing facility. Heat recovery is provided through the incorporation of a *waste heat boiler* into the incinerator secondary combustion chamber. These waste heat boilers are heat exchange systems that utilize the heat generated from the combustion of waste materials to convert boiler water to steam for process use or to drive turbines for power generation.

There are two types of waste heat boilers utilized in incinerator applications: *fire-tube boilers*, and *water-tube boilers*. Fire-tube boilers have hot gases passing on the inside of tubes, while the tubes are immersed in the boiler water. Water-tube boilers have water passing on the inside of tubes with hot gases flowing on the outside perpendicular to the tubes. Waste combustion can produce acid gases in the incinerator which may cause boiler corrosion problems especially in areas where temperatures drop below the dew point of HCl of approximately 300°F. In addition, slagging with alkali metal salts in the waste, and potential ash fouling of boiler tubes, can result from waste incineration. In general then, water-tube boilers are used in waste combustion applications to minimize these corrosion and fouling problems by moving only high-quality boiler feed water through the interior of the boiler tubes.

4.2.2. Rotary Kiln Incinerators

Rotary kiln incinerators were originally designed for lime processing. The rotary kiln is a cylindrical refractory-lined shell that is mounted at a slight incline from the horizontal plane to facilitate mixing the waste materials with circulating air. The kiln accepts solid and liquid waste materials of all types, with heating values between 1000 to 15,000 Btu/lb. Solid wastes and drummed wastes are usually fed by a pack-and-drum feed system, which may consist of a bucket elevator for loose solids and a conveyor system for drummed wastes. Pumpable sludges and slurries are injected into the kiln through a nozzle. Temperatures for burning vary from 1500°F to 3000°F. The kiln may be equipped with a lime or other caustic injection system to neutralize acid gas and combustion end products. Although liquid and even gaseous wastes may be incinerated in rotary kilns, these devices are designed primarily for combustion of solid wastes. They are exceedingly versatile in this regard, being capable of handling slurries, sludges, bulk solids of varying sizes, and containerized wastes. The only wastes that create problems in rotary kilns are aqueous organic sludges that become sticky on drying and form a ring around the kiln's inner periphery, and solids (e.g., drums) that tend to roll down the kiln and are not retained as long as the bulk solids. To reduce this problem, drums and other cylindrical containers are usually not introduced into the kiln when it is empty. Other solids in the kiln help to impede the rolling action.

Rotary kiln systems have a secondary combustion chamber to ensure complete combustion of the wastes. Air-tight seals close off the high end of the kiln; the lower end is connected to the secondary combustion chamber or mixing chamber. In some cases, liquid waste is injected into the secondary combustion chamber. The kiln acts as the primary chamber to volatilize and oxidize combustibles in the wastes. Inert ash is then removed from the lower end of the kiln. The volatilized combustibles exit the kiln, then enter the secondary chamber where additional oxygen is available and where ignitable liquid wastes or fuel can be introduced to ensure their complete combustion. It is in this secondary combustion chamber where energy recovery can take place using a waste heat boiler. Both the secondary combustion chamber and kiln are usually equipped with an auxiliary fuel firing system to bring units up to the desired operating temperatures. The auxiliary fuel system may consist of separate burners for auxiliary fuel, dual-liquid burners designed for combined waste-fuel firing, or single-fuel burners equipped with a premix system, whereby fuel flow is reduced and/or stopped and liquid waste flow is increased after the desired operating temperature is attained.

The published literature indicates that rotary kiln incinerators usually have a length-to-diameter (L/D) ratio between 2 and 10. Smaller L/D ratios often result in less particulate carryover, although carryover is a stronger function of the throughput velocity. Rotational speeds of the kiln are on the order of 0.2 to 1 inch/sec measured at the kiln periphery. Both the L/D ratio and the rotational speed are strongly dependent on the type of waste being combusted. In general, large L/D ratios along with slower rotational speeds are used when the waste requires longer residence time in the kiln for complete combustion. The residence time of the unpumpable wastes is controlled by the rotational speed of the kiln and the angle at which it is positioned. The residence times of liquids and volatilized combustibles are controlled by the gas velocity in the incineration system. Thus, the residence time of the waste material can be controlled to provide complete burning of the combustibles. This is a critical factor in limiting releases of some air pollutants from incineration devices.

Two types of rotary kilns are currently being manufactured in the United States today: *co-current* (burner at the front end with waste feed) and *countercurrent* (burner at the back end) systems. For a waste that easily sustains combustion, the positioning of the burner is arbitrary from an incineration standpoint, since both types will destroy a waste. However, for a waste having low combustibility (such as a high water-volume sludge), the countercurrent design offers the advantage of controlling temperature at both ends, which all but eliminates problems such as overheating the refractory lining. The countercurrent flow technique has been reputed to carry excessive ash over into the air pollution control system as a result of the associated higher velocities involved. However, this condition also increases the turbulence during combustion, which is generally a desirable factor.

Numerous hazardous wastes that previously were disposed of by potentially harmful methods (ocean dumping, landfilling, and deep-well injection) are currently being safely and economically destroyed by use of rotary kiln incinerators combined with proper flue gas treatment. Included in this list of primarily toxic wastes are polyvinyl chloride wastes, polychlorinated biphenyl (PCB) wastes from capacitors, obsolete munitions, and obsolete chemical warfare agents such as GB, VX, and mustard gas. Beyond these specific wastes, the rotary kiln incinerator is generally applicable in the destruction and ultimate disposal of any form of hazardous waste material that is at all combustible. Unlikely candidates are noncombustibles such as heavy metals, high moisture content wastes, inert materials, inorganic salts, and the general group of materials having a high inorganic content.

The primary advantages of the rotary kiln for waste incineration include the wide variety of wastes that can be incinerated simultaneously, the achievement of a high operating temperature, and the ability of obtaining a gentle and continuous mixing of incoming wastes. Another feature of the rotary kiln is that it can operate under sub-stoichiometric (oxygen-deficient) conditions to pyrolyze the wastes. The combustible off-gases with liquid wastes can then be incinerated in the secondary combustion chamber. This mode of operation also reduces the particulate matter carryover in the kiln gases. The rotating action allows better volatilization of the unpumpable wastes than stationary or fixed-hearth incinerators. Disadvantages of the rotary kiln include high capital and operating costs, the need for highly trained personnel to ensure proper operation, frequent replacement of the refractory lining if very abrasive or corrosive conditions exist in the kiln, and the generation of fine particulates (which become entrained in the exhaust gases) as a result of cascading action of the burning waste.

4.2.3. Fluidized Bed Incinerators

Fluidized bed incinerators burn finely divided solids, sludges, slurries, and liquids. These units have been used primarily in the petroleum and paper industries, as well as for processing nuclear wastes, spent cook liquor, wood chips, and in sewage sludge disposal. Wastes in any physical state can be applied to a fluidized-bed process incinerator. The fluidized-bed design takes its name from the behavior of a granular bed of nonreactive sand, through which a gaseous oxidizer (air, oxygen, or nitrous oxide) is passed at a high enough rate to cause the bed to expand in a highly turbulent state and act as a fluid above the combustion chamber floor. Preheating of the bed to startup temperature may be accomplished by a burner located above and impinging down on the bed. Waste is conveyed directly into the fluidized bed, where direct contact with the bed material improves heat transfer. After combustion, the exhaust and almost all the ash pass out of the top of the unit for cooling and further treatment. Ash caught in the bed material is eventually removed when the bed material is replaced. Auxiliary equipment for these fluidized incinerators includes a fuel burner system, an air supply system, and feed systems for liquid and solid wastes.

The two basic bed design modes, *bubbling bed* and *circulating bed*, are distinguished by the extent to which solids are entrained from the bed into the gas stream. The circulating bed incinerators have not been extensively used in waste treatment systems as the high gas velocities they use result in entrainment of a large portion of the solids in the exit gas that must be recovered and recycled back into the combustion unit. Higher gas velocities are used in this system to produce a high degree of turbulence throughout the incinerator to achieve associated higher heat and mass transfer rates and higher contaminant destruction efficiencies than would occur in the lower gas velocity bubbling bed incinerator.

Advantages of fluidized bed incinerators include a simple compact design, low cost, high combustion efficiency, low gas temperatures, little formation of nitrous oxide, and a large surface area for reaction. In addition, the bed can act as a heat-sink, requiring little or no preheat time between operating cycles. Disadvantages include problems with ash removal, carbon buildup in the bed resulting from increased residence times, high operating costs, and limited applicability to certain organic wastes for fear they will cause the bed to agglomerate.

4.2.4. Multiple Hearth Incinerators

The *multiple hearth incinerator* is a flexible unit that has been used to dispose of sewage sludges, tars, solids, gases, and liquid combustible wastes. Primarily used on sewage sludge, these units can be effective on residues from the manufacture of aromatic amines, bottoms from PVC manufacture, chemical sludge, oil refinery sludge, still bottoms, and pharmaceutical wastes. A typical multiple hearth furnace includes a refractory-lined steel shell, a central shaft that rotates, a series of solid flat hearths, a series of rabble arms with teeth for each hearth, an air blower, fuel burners mounted on the walls, an ash removal system, and a waste feeding system. Side ports for tar injection, liquid waste burners, and an afterburner may also be included. Sludge and solid combustible waste are fed into the unit through the furnace roof by a screw feeder or belt and flapgate. Liquids and gases are introduced to the unit through burner nozzles.

The rotating, air-cooled central shaft with air-cooled rabble arms and teeth distributes the waste material across the top hearth to drop holes. The waste falls to the next hearth and then the next until discharged as ash at the bottom. The waste is agitated as it moves across the hearths to make sure that new surface is exposed to hot gases as it passes from one hearth to the next. Normal incineration usually requires a minimum of six hearths. Temperatures in the top hearths are between 570°F and 1020°F to dry the wastes. Incineration occurs in the middle hearths, ranging between 1380°F and

1830°F. When combustion is complete, the remaining ash is screw-conveyed out of the unit.

Advantages to this system include a higher residence time for low volatility materials, ability to evaporate large quantities of water, ability to use many different fuels, and its high fuel efficiency. Disadvantages include slow temperature response throughout the unit owing to high residence times, difficulty in controlling the firing of fuel resulting from the slow temperature response, and high maintenance costs associated with the moving parts contained within this incinerator.

5. ILLUSTRATIVE EXAMPLES

Example 5.1. With respect to the treatment of toxic gases and particulates, briefly describe four commonly used adsorbents for control of gaseous pollutants.

SOLUTION. The following adsorbents are employed for gaseous pollution control:
1. Activated carbon is generally used to recover solvents, eliminate odors, and purify gases. Activated carbon also has the most versatility of all adsorbents in its ability to remove organics.
2. Activated alumina is generally used for the drying of gases.
3. Silica gel is also used for the drying of gases, but also has the ability to remove sulfur and other impurities.
4. Molecular sieves are used to remove gases whose molecules have the specific size and/or shape that will fit into the "pores" of the sieve.

Example 5.2. List the advantages and disadvantages of three commonly used devices for the control of particulates.

SOLUTION. The three commonly used air pollution devices for particulates are venturi scrubbers, electrostatic precipitators, and baghouses. Advantages and disadvantages are listed below.
1. *Venturi scrubbers, advantages*
    a. Low initial capital cost.
    b. Small space requirement.
    c. Efficiency not affected by resistivity of the particle.
    d. Able to remove gases and particulates at the same time.
    e. Can handle flammable, corrosive, and/or sticky materials.
   Venturi scrubbers, disadvantages
    a. Least efficient of the three particulate removal techniques.
    b. High operating costs due to large energy input to create a large pressure drop.
    c. Produces contaminated wastewater which requires further treatment and/or disposal.

2. *Electrostatic Precipitators, advantages*
   a. Low operating costs due to low fan energy input because of a low pressure drop.
   b. Capable of handling corrosive materials.
   c. May be designed to operate wet or dry.
   d. Dry precipitators can often operate at very high temperatures.

   Electrostatic Precipitators, disadvantages
   a. High initial capital cost.
   b. Addition of extra chemicals may be necessary to allow control of particles with high resistivity.
   c. Sensitive to changes in particulate concentrations in the waste stream.
   d. Combustible dust may create explosion hazards

3. *Baghouses, advantages*
   a. Often the most efficient of the three techniques.
   b. Not affected by the resistivity of the particles.
   c. Not affected by the changes in particulate concentration in the waste stream when either the "shaking" or "reverse-air-stream" methods of cleaning the bags is employed.

   Baghouses, disadvantages
   a. Sensitive to high (>550°F) temperatures.
   b. Very large.
   c. Efficiency is adversely affected by gas stream moisture.

Example 5.3. Consider a degreasing operation in a metal finishing process. Give an example of process modification that might be made to this part of the process to represent waste treatment.

SOLUTION. Several treatment methods are available, including:

1. Distillation of the solvent to remove impurities for reuse or use in another process area.
2. Inclusion of the solvent into a fuel stream in a steam generator or boiler.

Example 5.4. An air absorber for a vent line to a hydrochloric acid (HCl) production process must remove 99.9% of the HCl entering the absorber. The facility uses Tellerite packing that has a H_{OG} (height of a gas transfer unit) of 1.5 feet. Assuming that no equilibrium data are needed, calculate a rough estimate of the height of packing required for the tower.

SOLUTION. The height of packing in an absorber may be calculated from the equation:

$$Z = (N_{OG})(H_{OG})$$

120 Chapter 7

where Z is the column height in feet; N_{OG} is the number of transfer units; and H_{OG} is the height of a transfer unit in feet. In the absence of equilibrium data, assume that the slope of the equilibrium line is zero. For this condition,

$$N_{OG} = \ln(y_1/y_2)$$

where y_1 is the inlet concentration, and y_2 is the outlet concentration. For a basis of 100 moles/hr, 100 moles enter, and 0.1 moles are allowed to leave if there is 99.9% removal of HCl through the absorber. Solving for N_{OG} yields:

$$N_{OG} = \ln(100/0.1) = 6.908$$

The column packing height is then:

$$Z = (6.908)(1.5) = 10.362 \text{ ft}$$

The column height may be estimated by adding 10 feet to the packing height, for a total height of 20 ft.

Example 5.5. The Code of Federal Regulations, 40 CFR Part 60, Subpart XX, provides standards of performance for bulk gasoline terminals. This standard describes collection and processing of vapor displaced from tank trucks that are being filled. The emissions to the atmosphere are not to exceed 35 mg of total organic compounds per liter of liquid gasoline loaded. The tank transfer connections are to be vapor tight, which implies that the vapor processing system (unspecified) must meet the 35 mg limit.

Consider a tank whose vapor is air saturated with gasoline at 75°F so that 1 liter of vapor will be displaced for each liter of gasoline transferred. The following data applies: molecular weight of gasoline at 75°F ≈ 70 g/gmol; vapor pressure of liquid gasoline at 75°F = 6.5 psia; specific gravity of condensed vapor = 0.62.

a. Calculate the partial pressure of the gasoline vapor at this 35 mg/L limit.
b. What total system pressure would be required for the emission level limit to be obtained by condensation at 75°F?
c. What is the fraction of gasoline recovered by a vapor recovery system that reduces the gasoline concentration in the vapor from saturation at 75°F to 35 mg/L?

SOLUTION.
 a. Convert the vapor concentration, y_3, of 35 mg/L to a mole fraction basis:

$$y_3 = (35 \text{ mg/L})(22.4 \text{ L/gmol})(535°R/492°R)/[(1000 \text{ mg/g})(70 \text{ g/gmol})]$$
$$y_3 = 0.0122$$

This may be converted to a partial pressure based on 1 atmosphere since the system may be assumed to be open to the atmosphere.

$$p = y_3 P = (0.0122)(1 \text{ atm}) = 0.0122 \text{ atm}$$

b. Since the vapor pressure (p′) of gasoline at 75°F = 6.5 psia, then the total pressure (P) required for condensation to occur at 75°F is

$$P = p'/y_3 = 6.5/0.0122 = (533 \text{ psia})/(14.7 \text{ psia/atm}) = 36.3 \text{ atm}$$

c. To calculate the recovery, first calculate the mole fraction of the original vapor, y_1

$$y_1 = (6.5 \text{ psia})/(14.7 \text{ psia}) = 0.442$$

A total mass balance around the tank/ vapor recovery system yields:

$$w_1 = w_2 + w_3$$

where: w_1 = 1 gmol of vapor; w_2 = gmol of gasoline recovered; and w_3 = gmol of residual vapor stream.

A gasoline balance around the tank truck gives:

$$y_1 w_1 = y_2 w_2 + y_3 w_3$$

where $y_1 = 0.442$ = mole fraction of original vapor; $y_2 = 1.0$ = mole fraction of recovered stream; $y_3 = 0.0122$ = mole fraction of residual stream.

The equations above may be solved for w_2 as follows:

$$1.0 = w_2 + w_3$$
$$0.442 = w_2 + 0.0122 w_3 = w_2 + 0.0112 (1 - w_2)$$
$$0.442 = 0.0122 + 0.9878 w_2$$
$$w_2 = 0.435 \text{ gmol}$$

The term w_3 is therefore:

$$1.0 = 0.435 + w_3$$
$$w_3 = 0.565 \text{ gmol}$$

The recovery, R, may now be calculated as:

$$R = y_2 w_2 / (y_1 w_1) = (1.0)(0.435)/[(0.442)(1.0)]$$
$$R = 0.984 = 98.4\%$$

Example 5.6. Federal regulations require minimum performance standards for incinerators used for waste treatment. One of these standards consists of a specified destruction and removal efficiency (DRE) for principal hazardous organic constituents (POHCs) in the waste stream based on the mass of the POCH flowing into the incinerator and the mass of the POCH flowing out of the incinerator and its waste treatment systems. If the hazardous waste flow rate into an incinerator is 500 lb/hr, calculate the waste rate leaving the unit to achieve a DRE of 99.9999%.

122 Chapter 7

GIVEN: The equation defining the DRE of a hazardous waste is:

$$DRE = (M_{in} - M_{out})/M_{in} \, (100)$$

where M_{in} = mass flow rate in; and M_{out} = mass flow rate out.

SOLUTION. Solving the above equation for M_{out}, the following is obtained:

$$M_{out} = M_{in} \, [1 - (DRE/100)]$$

Calculate the mass flow rate out for a DRE of 99.9999%.

$$M_{out} = 500 \, [1 - (99.9999/100)] = 0.0005 \text{ lb/hr}$$

Example 5.7. Estimate the theoretical flame temperature of a waste mixture containing 25% cellulose, 35% motor oil, 15% water (vapor) and 25% inerts, by mass. Assume 5% radiant heat losses. The flue gas contains 11.8% CO_2, 13 ppm CO, and 10.4% O_2 (dry basis) by volume.

GIVEN: Additional data includes:

NHV of cellulose = 14,000 Btu/lb
NHV of motor oil = 25,000 Btu/lb
NHV of water = 0 Btu/lb
NHV of inerts (effective) = -1,000 Btu/lb

The flame temperature may be calculated using the Theodore-Reynolds equation:[12]

$$T = 60 + NHV/[\{0.325\}\{1 + (1 + EA)(7.5 \times 10^{-4})(NHV)\}]$$

where T = temperature, °F; NHV = net heating value of the inlet mixture, Btu/lb; EA = excess air on a fractional basis = 0.95 Y/(21-Y), where Y = dry mole % O_2 in the combustion (incinerated) gas. Assume the average heat capacity of the flue gas is 0.319 Btu/lb-°F.

SOLUTION. Determine the NHV for the mixture by multiplying the component mass fractions by their respective NHVs and taking the sum of the resulting product(s).

$$NHV = 0.25 \, (14,000 \text{ Btu/lb}) + 0.35 \, (25,000 \text{ Btu/lb}) + 0.15 \, (0.0 \text{ Btu/lb})$$
$$+ 0.25 \, (-1,000 \text{ Btu/lb}) = 12,000 \text{ Btu/lb}$$

Next, calculate the fractional excess air employed by the following:

$$EA = 0.95 \, Y/(21-Y) = 0.95 \, (10.40)/(21-10.4) = 0.932$$

Then calculate the flame temperature using the Theodore-Reynolds equation:

$$T = 60 + (12{,}000)/[\{0.325\} \{1 + (1 + 0.932)(7.5 \times 10^{-4})(12{,}000)\}] = 2068°F$$

Additional details on these equations can be found in Theodore and Reynolds.[12]

Example 5.8. A quench tower operates at a HCl removal efficiency of 65%. This is then followed by a packed tower absorber. What is the minimum collection efficiency that the absorber must have if an overall HCl removal efficiency of 99.0% is required?

SOLUTION. Select as a basis 100 lb/hr of HCl entering the unit. The mass of HCl leaving the spray tower is calculated from the following equation:

$$E = (W_{in} - W_{out})/W_{in}$$

Rearranging yields:

$$W_{out} = W_{in}(1-E) = 100(1.0 - 0.65)$$
$$= 35 \text{ lb/hr HCl leaving the spray tower}$$

Use the overall efficiency, E_o, to calculate the mass flow rate of HCL leaving the packed tower absorber.

$$W_{out} = W_{in}(1 - E_o) = 100(1.0 - 0.990)$$
$$= 1.0 \text{ lb/hr leaving the packed tower}$$

The efficiency of the packed tower, E_p, can now be calculated:

$$E_p = (W_{in} - W_{out})/W_{in} = (35.0 - 1.0)/5.0 = 0.971 \text{ or } 97.1\%$$

6. SUMMARY

1. Recognizing that not all wastes can be eliminated through source reduction methods or recycling efforts, viable treatment processes are available for managing the remaining wastes. These waste management processes should be understood so that the consequences of waste generation, and the true benefits of waste elimination, can be quantified.
2. Waste treatment options can be divided into four general categories:
    - *Physical treatment*
    - *Chemical treatment*
    - *Biological treatment*
    - *Thermal treatment*

3. These treatment options are applied using a variety of methods for the removal of contaminants from air, liquid, and solid process streams. Many of these methods involve separation processes that can be taken advantage of for in-process recycling and pollution prevention (Figures 7.1 and 7.2).

4. Air pollution control equipment is often employed at plants for the removal of particulate and gaseous contaminants from air streams. Examples of air pollution control equipment for particulate control include: *cyclone separators, electrostatic precipitators, baghouses and fabric filters*, and *venturi scrubbers*. Gaseous contaminant control devices include: *absorbers* for hydrophilic compounds, *adsorbers* for hydrophobic compounds, and *vapor incinerators or flares* for combustible gaseous components. Pollution prevention and energy reduction opportunities exist within processes utilizing thermal oxidation for organic vapor control with the addition of heat recovery devices (*recuperators* or *regenerative heat recovery systems*) in the incinerator off-gas

5. To control water pollution, a waste stream can be subjected to a variety of *physical, chemical*, and/or *biological* processes depending on the nature of the waste materials being treated.

6. Physical treatment processes include: *phase separation* and *component separation*. In phase separation two distinct phases are separated based on their different physical properties, while in component separation a particular species is separated from a single-phase, multi-component system. Examples of physical treatment processes for liquid waste treatment include: *flocculation and sedimentation, dissolved air flotation, filtration, centrifugation, air stripping, steam stripping, resin adsorption*, and *electrodialysis*.

7. A variety of chemical processes are used in the treatment of liquid waste streams, most of which must be used in conjunction with other methods to achieve a desired end result. Chemical processes discussed in this chapter include: *calcination, chlorinolysis, electrolysis, hydrolysis, neutralization, oxidation, photolysis*, and *reduction*.

8. Biological treatment processes utilize biochemical processes taking place primarily within bacterial cells, to degrade organic compounds to stabilized endproducts. The predominant form of biological treatment of liquid industrial wastes is aerobic treatment in the form of *activated sludge, waste stabilization ponds and aerated lagoons*, and *trickling filters*. *Anaerobic digestion* takes place in the absence of dissolved oxygen, and is commonly used for municipal sludge treatment and in the stabilization of high strength industrial waste from breweries, distilleries, meat packing plants, and food processing facilities. *Composting* is a high-temperature, aerobic treatment process designed to stabilize high-solids content wastes and sludges.

9. Solid wastes at a plant site are usually managed using two basic methods. One is some type of physical process to reduce the volume of waste requiring disposal. These physical processes include: *shredding, air classification*, or *compaction*. Physical processes for solid waste management often result in the separation of waste components, making them suitable for

in-process recycling methods. The second solid waste stream handling method is *incineration*, and is used to reduce the mass and volume of solid waste requiring final disposal. The three critical factors for effective combustion of organic compounds in incineration systems include: *temperature, time*, and *turbulence*.

10. *Rotary kiln, fluidized bed*, and *multiple hearth incinerators* can be used for solid waste combustion, and are designed and operated to optimize the time, temperature, and turbulence of combustion reactions in a variety of ways. All of these incineration systems provide opportunities for energy recovery and subsequent pollution prevention benefits that result through the incorporation of *waste heat boilers* for steam and/or power generation.

7. PROBLEMS

1. The four main types of gas phase air pollution control include: condensation (contact or surface condensers), and the three control approaches discussed in this chapter, i.e., adsorption, absorption, and incineration. During the drilling of a gas well in a formation containing a mixture of natural gas with hydrogen sulfide, which of these control methods should be used to reduce H_2S emissions from the drilling rig?

2. List five factors that should be considered when selecting a solvent for a gas absorption column to be used as an emission control device.

3. Activated carbon adsorption equipment is commonly used for the control of gaseous pollutants. List and briefly discuss several design or process factors that affect its effectiveness in controlling gaseous pollutants.

4. Figures 7.1 and 7.2 show two possible process diagrams for the treatment of wastes in a process. Figure 7.2 shows a process where the waste is removed from the product stream before treatment. Suggest some processes that might achieve this product/waste stream separation. Also, for each process, describe how the process achieves the separation, and what conditions are required for it to occur.

5. Of the many chemical treatment methods available, oxidation is frequently used when inexpensive oxidants are available. Some of these include air, hydrogen peroxide and common bleach (sodium hypochlorite). Ozone, though expensive, is occasionally used in wastewater treatment applications.

Ammonium sulfide is a noxious substance that can be used for the generation of sulfide ion in inorganic qualitative analysis. The substance is also used in the photographic and textile industries. An aqueous solution of ammonium sulfide has a high pH (approximately 11 to 12), has a foul odor, and reacts with acids to form hydrogen sulfide, a toxic and foul smelling gas. It cannot be disposed of directly but must be treated before final disposal. Fortunately, ammonium sulfide can be oxidized quite easily with a hypochlorite solution such as household bleach, according to the following equation:

$$(NH_4)_2S + NaOCl \rightarrow NH_4Cl + Na_2SO_4$$

DATA: One container of aqueous $(NH_4)_2S$ weighs 2.3 kg; the solution is 20.1% $(NH_4)_2S$. Household bleach NaOCl contains 5.25% active ingredient, weighs about 8.7 lb/gal, and costs $0.89 (retail). One "lab pack" (55 gal drum) will hold about 12 containers of aqueous ammonium sulfide and costs $350 to dispose of through a qualified disposal company.

    a. Balance the above equation.

    b. Calculate the cost of disposal of 1 gallon of ammonium sulfate using household bleach (sodium hypochlorite) as an oxidizing agent.

    c. Compare the economics of this disposal by oxidation with the cost of simply calling in the waste disposal company to place it in a "lab pack" and take it away.

6. Regan Chemical Company transports solid hazardous waste to a disposal site. On average, Regan's hauling trucks carry 4 tons of waste per trip for a total of 32,000 tons per year. In the event a truck overturns, it can be assumed that 2 tons of the waste are spilled. DOT statistics indicate that 1 out of 4,000 waste hauling trucks overturn during an average trip. Industry studies and Chevron Corporations's SMART (Save Money and Reduce Toxics) data indicate that cleanups resulting from transportation spills cost as much as $10,000 per ton. Calculate the total potential liability of producing and "disposing " of the waste in this matter. Express the answer on both an annual and per ton basis.

7. Explain what role physical treatment processes play in biological waste treatment systems.

8. The general formula for the alkyldichlorobenzenes is $C_nH_{2n-8}Cl_2$, where $n > 6$. Write a balanced, general chemical equation for the combustion of alkyldichlorobenzenes.

9. Medical sludge containing mercury is burned in an incinerator at a mercury feed rate 4.6 lb/hr. The resulting 500°F product (40,000 lb/hr of gas; MW=32) is quenched with water to a temperature of 150°F. The resulting stream is filtered to remove all particulates. What happens to the mercury? Assume the process pressure is 14.7 psi and that the vapor pressure of Hg at 150°F is 0.005 psi.

10. Polychlorinated biphenyls (PCBs) are mixed with waste oil for burning in a rotary kiln incinerator. What will happen as the percentage of PCBs is increased in the waste mixture assuming that all other variables (e.g., excess air, heat loss, feed rate) are kept constant?

11. A hazardous waste incinerator is burning an aqueous slurry of soot (i.e., carbon), producing a small amount of fly ash. The waste is 70% water by mass and is burned with 0% excess air (EA). The flue gas generated contains 0.30 grains of particulates in each 8.0 cubic feet (actual) at 580°F.

Calculate the particulate concentration in the flue gas in grains/acf, in grains/scf, in grains/dscf, and in grains/dscf corrected to 50% EA. If the regulations require the particulate emissions to be less than 0.08 grains/dscf corrected to 50% EA, is this incinerator in compliance or must additional particulate control measures be taken?

Assume that when the flue gas passes through a waste heat boiler no water condensation occurs.

12. A state incinerator emission limit requires 99% HCl control and allows 0.07 grains/dscf at 68°F, corrected to 50% EA. An incinerator is to burn 5 tons/hr hazardous waste containing 2% Cl, 80% C, 5% inerts, and the balance H_2O by weight.
   a. Calculate the maximum mass emission rate of equivalent HCl in lb/hr that may be emitted.
   b. Calculate the maximum mass emission rate of particulates in lb/hr that may be emitted. What is the actual particulate emission rate if all the inerts are emitted from the stack as fly ash?
   c. Determine the combustion efficiency of this incinerator if a stack test indicates that the flue gas contains 12% CO_2 and 20 ppm CO.

13. A hazardous waste incinerator has been burning a certain mass of mixed tetrachlorobyphenyls ($C_{12}H_6C_{14}$) per hour and the HCl produced was neutralized with soda ash (Na_2CO_3). If the incinerator switches to burning an equal mass of dichlorobenzene ($C_6H_4C_{12}$), by what factor will the consumption of soda ash be increased?

14. Estimate the theoretical flame temperature of a waste mixture containing 25% cellulose, 35% motor oil, 15% water (vapor), and 25% inerts, by mass. Assume 5% radiant heat losses. The flue gas contains 11.8% CO_2, 13% CO, and 10.4% O_2 (dry basis) by volume.
   GIVEN:
   NHV of Cellulose = 14,000 Btu/lb
   NHV of Motor Oil = 25,000 Btu/lb
   NHV of Water = 0 Btu/lb
   NHV of Inerts (effective) = -1,000 Btu/lb
   Assume the average heat capacity of the flue gas = 0.325 Btu/lb-°F.

8. REFERENCES

1. Mycock, J., McKenna, J., and Theodore, L., *Handbook of Air Pollution Control Engineering and Technology*, CRC Press, Lewis Publishers, Boca Raton, Florida, 1995.
2. de Nevers, N., *Air Pollution Control Engineering*, McGraw-Hill Book Company, San Francisco, California, 1995.
3. Wark, K., Warner, C. F., and Davis, W. T., *Air Pollution, Its Origin and Control*, Addison-Wesley, Menlo Park, California, 1998.
4. Cooper, C. D. and Alley, F. C., *Air Pollution Control, A Design Approach*, 2nd ed., Waveland Press, Inc., Prospect Heights, Illinois, 1994.
5. Metcalf & Eddy, Inc., *Wastewater Engineering: Treatment, Disposal, Reuse*, 3rd ed., McGraw-Hill Book Company, San Francisco, California, 1991.
6. McGhee, T. J., *Water Supply and Sewerage*, 6th ed., McGraw-Hill Book Company, San Francisco, California, 1991.
7. Droste, R. L., *Theory and Practices of Water and Wastewater Treatment*, John Wiley & Sons, Inc., New York, New York, 1997.
8. Viessman, W. and Hammer, M. J., *Water Supply and Pollution Control*, 6th ed., Addison-Wesley, Menlo Park, California, 1998.

9. U.S. EPA, *Cleanup of Releases from Petroleum USTs: Selected Technologies*, EPA/530/UST-88/001, Office of Underground Storage Tanks, Washington, D.C., 1988.
10. Tchobanoglous, G., Theisen, H., and Vigil, T., *Integrated Solid Waste Management: Engineering Principles and Management Issues*, McGraw-Hill, San Francisco, California, 1993.
11. Office of Technology Assessment, Report to U.S. Congress, *Technologies and Management Strategies for Hazardous Waste Control*, Washington, D.C., March 1985.
12. Theodore, L. and Reynolds, J., *Introduction to Hazardous Waste Incinera-tion*, Wiley-Interscience, New York, New York, 1988.

CHAPTER 8

Process Diagrams
Contributing Author: Mary Wrieden

The complete design specification for a medium to large-sized chemical process includes process diagrams; energy and material balance tables; chemical, mechanical, electrical, metallurgical, and civil engineering design considerations and plans and specification; and process flow sheets. The process flow sheet is the key instrument for defining, refining, and documenting a chemical process. The process flow diagram is the authorized process blueprint, serves as the framework for specifications used in equipment designation and design, and is the single, authoritative document employed to define, construct, and operate a chemical process. This chapter explains the different types of flow sheets encountered in process operation and design, and the role of these flow sheets in process development and improvement.

1. FLOW SHEETS

Beyond equipment symbols and process stream flow lines, there are several essential components contributing to a *process flow sheet*. These include equipment identification numbers and names, temperature and pressure designations, utility designations, volumetric or molar flow rates for each process stream, and a material balance table pertaining to process flow lines. The process flow sheet may show additional information such as energy requirements, major instrumentation, and physical properties of the process streams. When properly assembled and employed, this type of process schematic provides a coherent picture of the overall process. The flow sheet symbolically and pictorially represents the interrelations among the various flow streams and equipment, and permits easy calculations of material and energy balances. A number of symbols are universally employed to represent equipment, equipment parts, valves, piping, and so on. These symbols obviously reduce, and in some instances replace, detailed written descriptions of the process.

A flow sheet usually changes over time with respect to degree of sophistication and details it contains. A crude flow sheet may initially consist of a simple, freehand block diagram offering information about the equipment only; a later version (described in the next section) may include line drawings with pertinent process data, such as overall and componential flow rates, utility and energy requirements, and instrumentation. During the later stages of a design project, the flow sheet consists of a highly detailed piping and instrumentation diagram (P&ID), which is covered in a later section of this chapter. For information on aspects of the design procedure, which are beyond the scope of this text, the reader is referred to the literature.[1,2]

130 Chapter 8

As discussed in Chapter 4, one can conceptually view a (chemical) plant as consisting of a series of interrelated building blocks that are defined as units or unit operations. The process flow sheet ties together the various pieces of equipment that make up the process. Flow schematics follow the successive steps of a process by indicating where the pieces of equipment are located and where the material streams enter and leave each unit.

There are five basic types of schematic diagrams in general use in process engineering. These are

1. *Block diagrams.*
2. *Graphic flow diagrams.*
3. *Process flow diagrams.*
4. *Process piping and instrumentation diagrams.*
5. *Tree diagrams.*

1.1. Block Diagrams

The *block diagram* is the simplest but least descriptive of the schematic diagrams in common use. As the name implies, it consists of neat rectangular blocks which usually represent a single unit operation in a plant or an entire section of the plant. These blocks are connected by arrows indicating the flow sequence. The block diagram is extremely useful in the early stages of a process design and is particularly valuable in presenting the results of economic or operating studies since the significant data can be placed within the blocks. Four different types of operations can be described by block diagrams. As indicated in Figure 8.1, these include: *pure inventory operations, separation processes, assembly processes,* and *chemical change.*

Increases in inventory represent accumulation; decreases represent depletion. An operation such as this is defined as a *pure inventory operation.* The flow diagram for this operation is shown in its most generalized form in Figure 8.1a. If no chemical change takes place, a starting material can be physically separated into more than one product. This is by definition a *separation operation.* The generalized flow diagram for this operation is shown in Figure 8.1b. It would apply equally well to the recovery of the two components in an apparatus or the partial separation of an entering feed water into exiting steam and exiting hot water. Two or more material streams that are combined or assembled without chemical reaction is, by definition, an *assembly operation.* A generalized flow diagram for this operation is presented in Figure 8.1c. With more material streams involved, it would apply equally well to an automobile assembly line or the creation of an emulsion by the intimate mixing of oil, water, and an emulsifying agent. These three types of operation have one important common characteristic, the absence of chemical change. This sets them apart from the fourth type, the *chemical operation.* In Figure 8.1d two reactants, A and B, are shown entering a chemical reaction producing products C and D. The flow diagram will vary with the number of reactants and products involved.

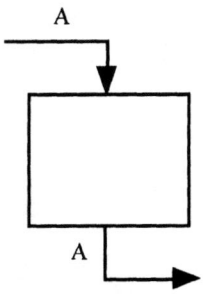

a. *Pure inventory operation.*
   No chemical change
   No phase change
   No separation or combination
   of materials.

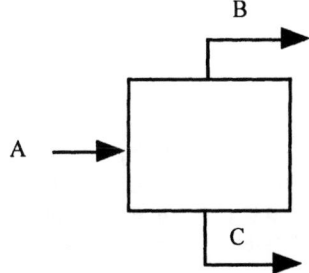

b. *Separation.*
   No chemical change
   Physical separation of
   materials occurs with or
   without phase change.

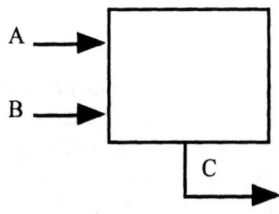

c. *Assembly.*
   No chemical change
   Physical combination of
   materials occurs with or
   without phase change.

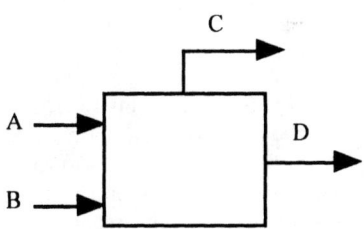

d. *Chemical change.*
   Reactants (A and B) undergo
   molecular rearrangement to
   form products (C and D),
   with or without phase
   change.

Figure 8.1. Fundamental operations described by block diagrams.

1.2. Graphic Flow Diagrams

Graphic flow diagrams are used most frequently in advertising, company financial reports, and technical reports in which certain features of the flow diagram require extra emphasis. It should present the desired information clearly and in an eye-catching fashion that is both novel and informative.

1.3. Process Flow Diagrams

The process flow diagram, or PFD, is a pictorial description of the process. It provides the basic processing scheme, the basic control concept, and process information from which equipment can be specified and designed. As described earlier, it provides the basis for the development of the P&I

diagram (P&ID), equipment design and specifications, and serves as a guide for the design, construction, and operation of a chemical process. The process flow diagram usually includes:

1. Material balance data (may be on separate sheets).
2. Flow scheme, equipment, and interconnecting streams.
3. Basic control instrumentation.
4. Temperature and pressure at various points.
5. Any other important parameters unique to each process.

Data on spare and parallel equipment are often omitted as is valving. A valve is shown only where its specification can aid in understanding intermittent or alternate flows. Instrumentation is indicated to show the location of variables being controlled and the location of the actuating device, usually a control valve. To help the reader better understand the process flow sheet, a list of commonly used symbols for valves is presented in Figure 8.2.[3]

1.4. Process Piping and Instrumentation Flow Diagrams

The P&ID, which provides the basis for detailed design, offers a precise description of piping, instrumentation, and equipment. This key drawing defines the plant system, describes equipment, and shows all instrumentation, piping, and valving. It is used to train personnel and aids in troubleshooting during start-up and operation. The P&ID assigns item numbers to all equipment (e.g., towers, reactors, and tanks); gives dimensions of equipment and vessel elevations; and shows all piping, including line numbers, sizes, specifications, and all valves. All instrumentation is covered, giving numbers, function, types, and indicating whether the instrumentation is actuated electronically or pneumatically.

A general knowledge of the symbols for flow, level, pressure, and temperature controllers, as given in Figure 8.3 to 8.6, is needed to comprehend flow diagrams like the simple example presented in Figure 8.7.[3] In this vessel, with an inlet feed on top of the tank equipped with a flow controller, the level in the tank is maintained by a level-controlling device. When the level rises above the high level point, the level controller sends a signal to a valve actuator and the valve is opened, dropping the liquid level within an acceptable range within the vessel. When the level approaches a specified lower limit value, the valve is closed. More complicated systems can be analyzed in a similar manner used in this simple example.

Certain essential information is often concisely provided via notations adjacent to the representation of each piece of process equipment. Experience has dictated the information required for common items such as pumps and vessels. For special equipment, overall dimensions and significant process and operating characteristics are often given. The notations provided in Table 8.1 are suggested for several common items. Any consistent system is satisfactory.

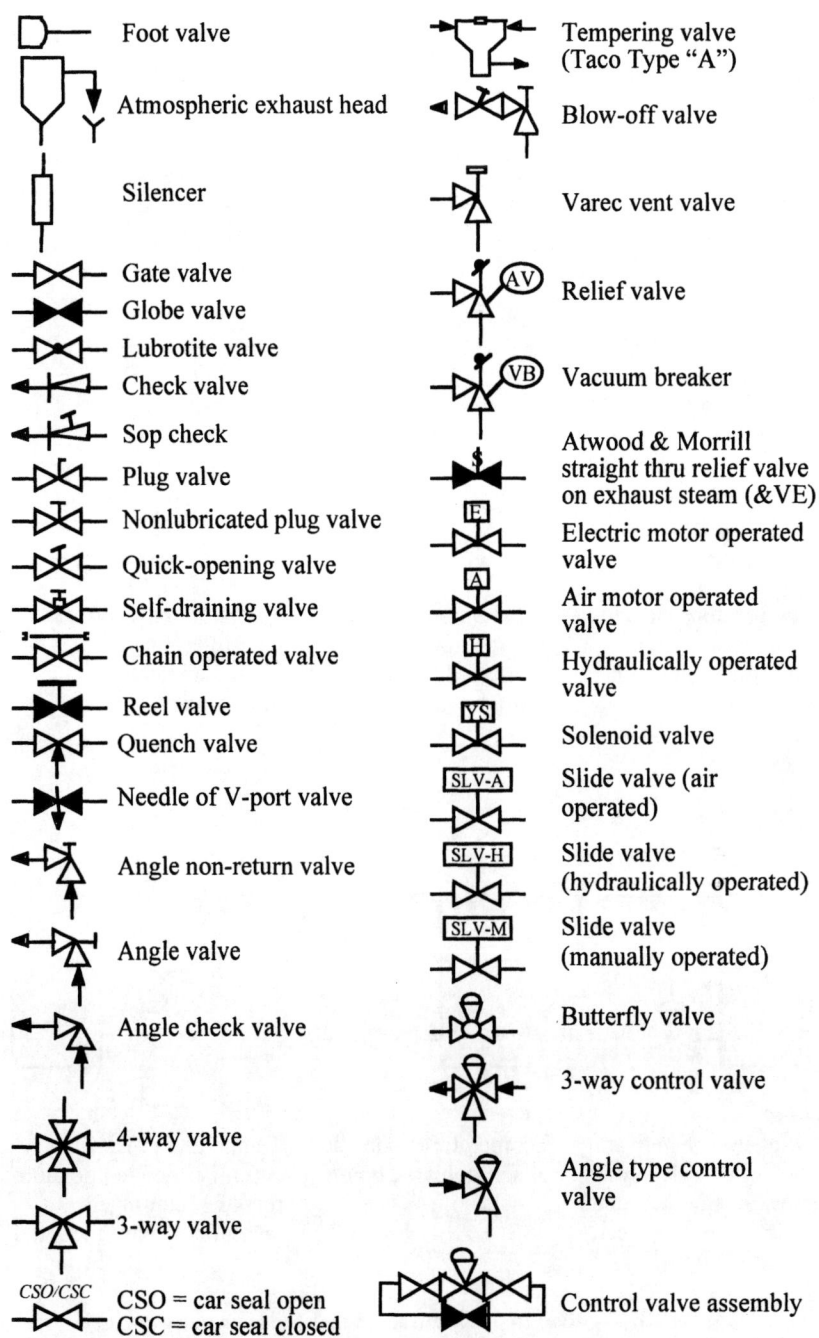

Figure 8.2. Commonly used valve symbols.

134 Chapter 8

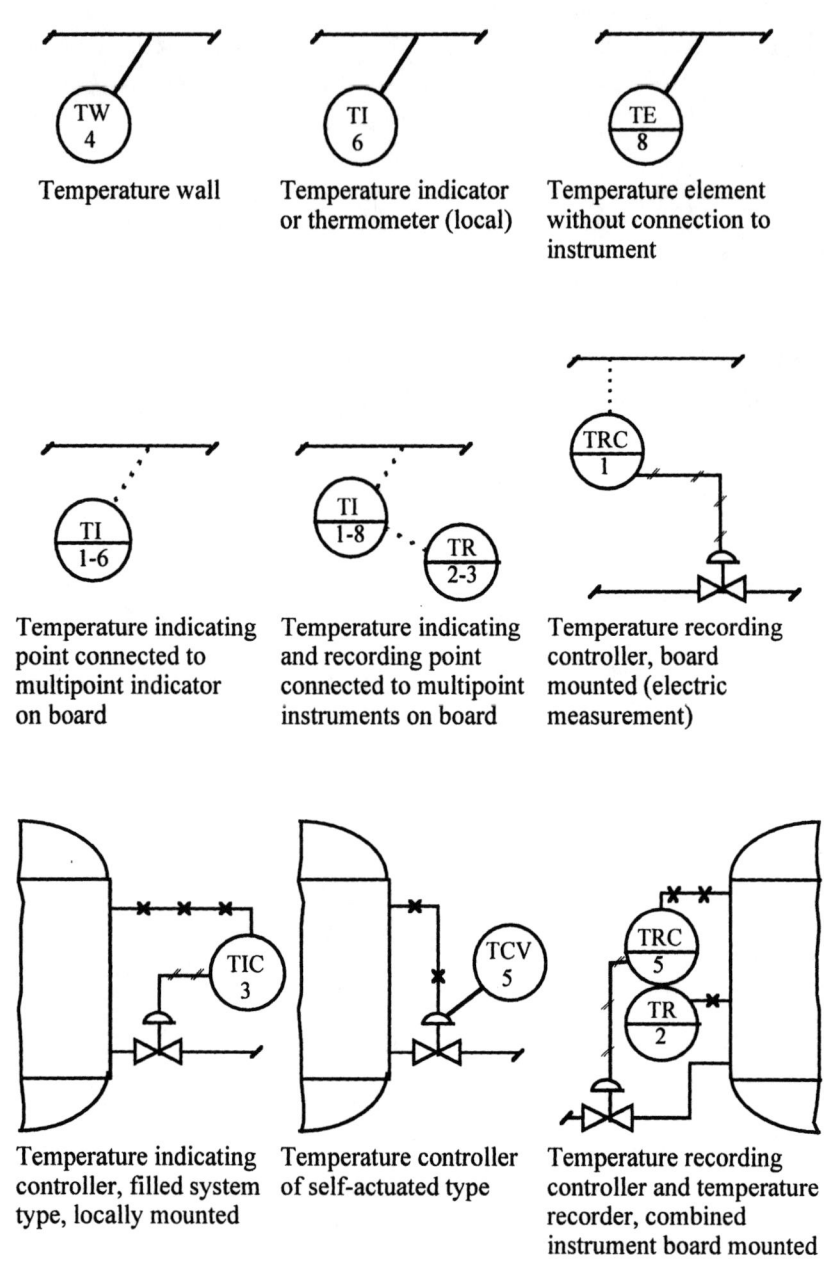

Figure 8.3. Typical instrumentation symbols for temperature.

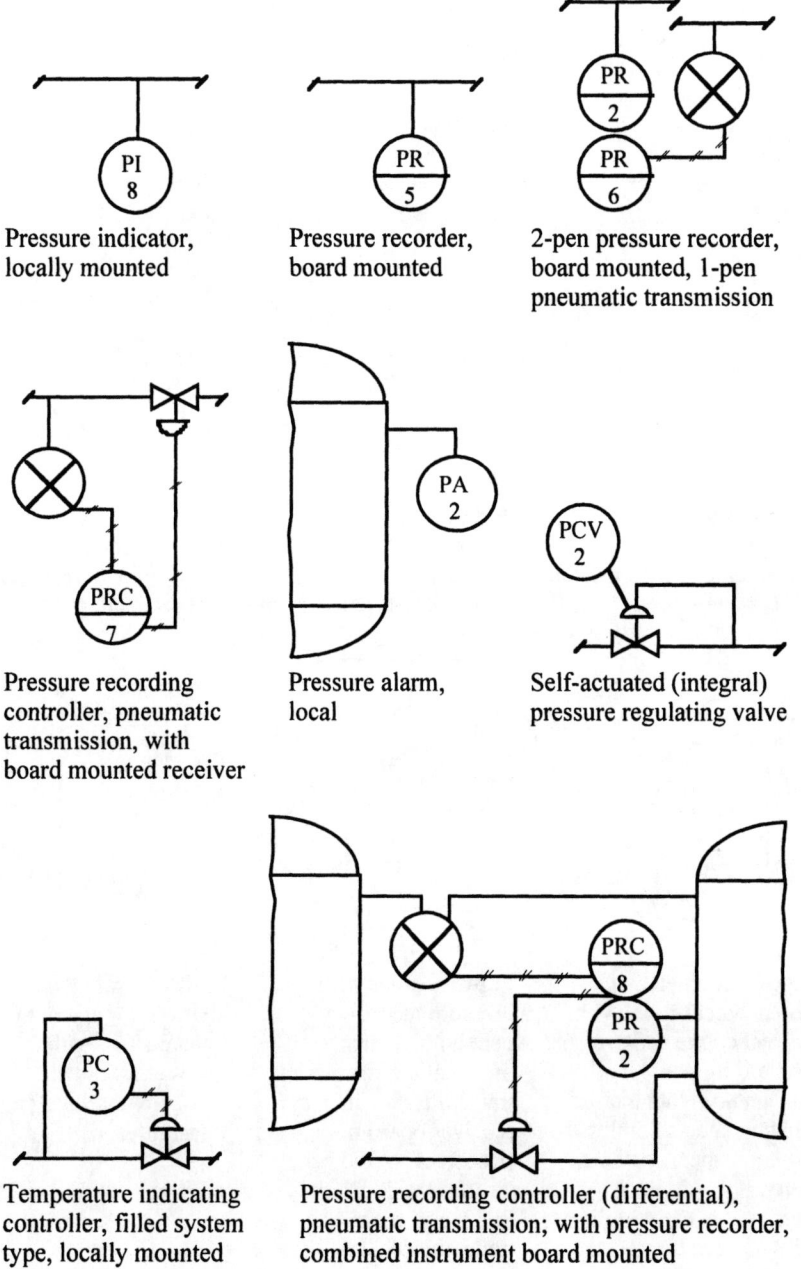

Figure 8.4. Typical instrumentation symbols for pressure.

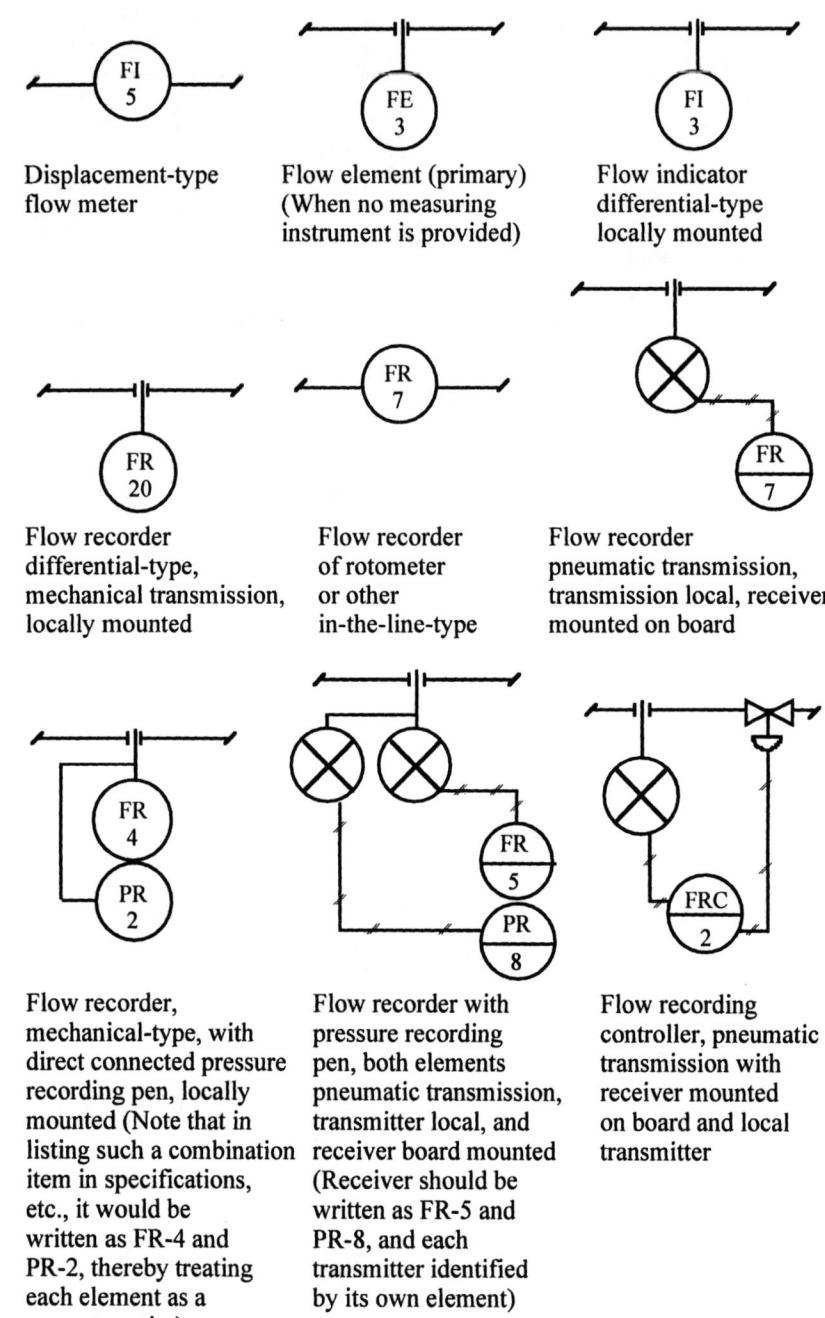

Figure 8.5. Typical instrumentation symbols for flow.

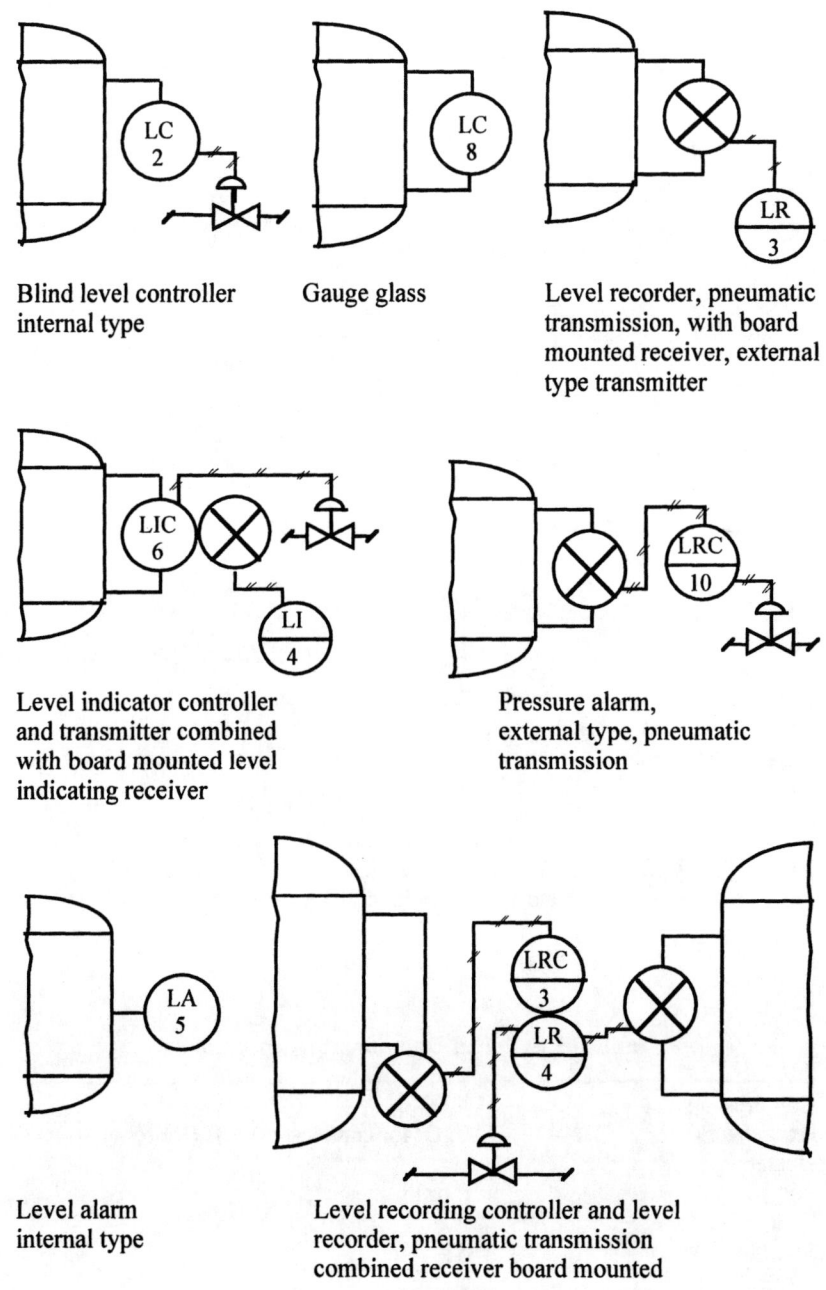

Figure 8.6. Typical instrumentation symbols for liquid level.

138 Chapter 8

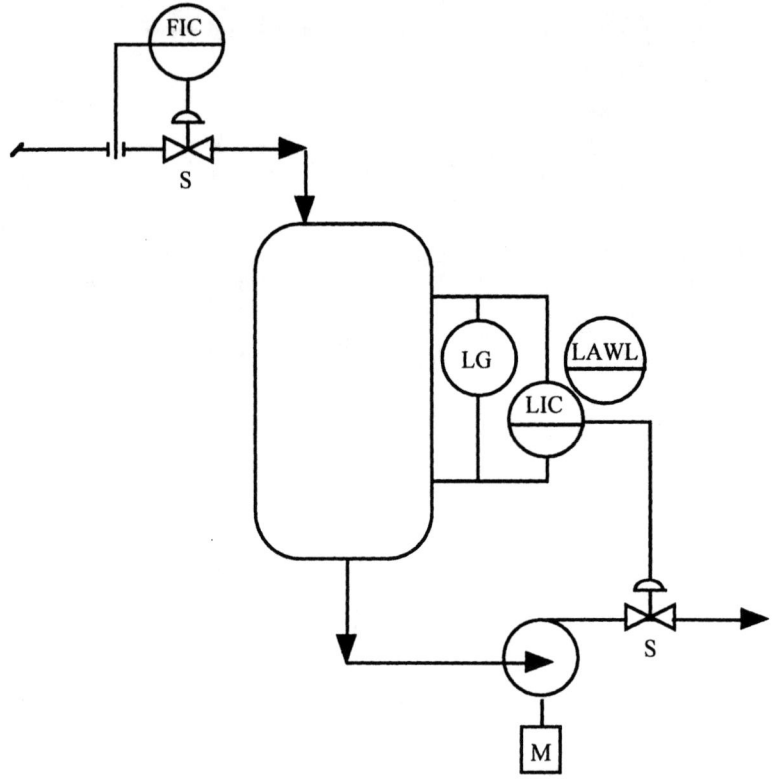

Figure 8.7. Vessel with level control on outlet. FIC, flow indicator controller; LG, level gauge; LIC, level indicator controller; LAWL, level alarm for water level; M, motor.

Table 8.1. Typical Equipment Designations.

Compressors	K, C
Exchangers	E, C (for condensers), RB (for reboilers)
Heaters	H
Pumps	P or PU
Reactors	R
Storage tanks	ST
Towers	T
Vessels	V

Process information that is often necessary includes the following:

Compressors
1. Service
2. Stages
3. Suction conditions
4. First stage suction
5. Second stage suction
6. Second stage discharge

Heat Exchangers
1. Service
2. Differential pressure across shell and tubes
3. Heat transfer area
4. Duty (heat transfer rate)
5. Design conditions
6. Temperature and pressure at inlet and outlet

Pumps
1. Service
2. Size and type
3. Fluid being handled
4. Pump operating temperature
5. Fluid density at pump operating temperature
6. Design flow rate at pump operating temperature
7. Design operating service pressure and net positive suction head (NPSH)

Vessels
1. Service
2. Diameter, height, wall thickness
3. Special features (lining, etc.)
4. Design conditions
5. Operating conditions

1.5. Tree Diagrams

Another type of flow diagram applicable to the chemical process industry is the *tree diagram*. Tree diagrams are used primarily in the study of hazards and/or chemical accidents and have become an integral part of any study or analysis involving accident and emergency management.[3,4] In addition, and as described earlier in this book, accident and emergency management is an integral part of pollution prevention. The following discussion details the two main types of analysis employed in the chemical process industry, *fault tree analysis (FTA)* and *event tree analysis (ETA)*.

1.5.1. Fault Tree Analysis

A *fault tree* analysis is carried out using a flow diagram that highlights conditions that cause system failure. Fault tree analysis attempts to describe how and why an accident or other undesirable event has occurred or could occur. Fault tree analysis shows the relationship between the occurrence of the undesired event, the "top event," and one or more antecedent events, called "basic events." The top event may be, and usually is, related to the basic events via certain intermediate events. A fault tree diagram depicts the casual chain linking the basic events to the intermediate events and the latter to the top event. In this chain, the logical connection between events is indicated by so-called "logic gates." The principal logic gates are the AND gate, symbolized on the fault tree by ⌒, and the OR gate, symbolized by ⋀.

1.5.2. Event Tree Analysis

An *event tree* analysis uses a flow diagram that represents the possible steps leading to a failure or accident. The sequence of events begins with an initiating event and terminates with one or more undesirable consequences. In contrast to a fault tree, which works backward from an undesirable consequence to possible causes, an event tree works forward from the initiating event to possible undesirable consequences. The initiating event may be equipment failure, human error, power failure, or some other event that has the potential to adversely affect the environment or an ongoing process and/or equipment. The event tree may limit hazard analysis because it cannot quantify the potential of the event occurrence. In addition, all the initial occurrences must be identified for a complete analysis. The event tree should be used to examine, rather than to evaluate, the possibilities and consequences of a failure. A fault tree analysis can be used to establish the probabilities of the event tree branches.

2. PROCESS SIMULATIONS

A *process simulation* develops in the same manner as the process flow sheet. A simulation requires a block diagram. The streams and equipment shown on the simulation block diagram represent the streams and equipment shown on both the process flow diagram and the piping and instrumentation diagram of the process. Like the process flow diagram and the piping and instrumentation diagram, the simulation and simulation block diagram necessitate a uniform system of numbers and symbols for designating streams and equipment.

At the earliest stage of simulation development a technical individual prepares a pressure profile of the process depicted on the process flow diagram. The pressure profile establishes operating pressures for the various units involved in the process. Conservative estimates of pressure drops through lines, valves, and process equipment allow for the inevitable changes that come with refinement of the process.

For the purpose of estimating flow rates and process conditions, the engineer and/or scientist represents equipment such as valves, drums, towers, dryers, coalescers, and reactors with flashes, splitters, and black boxes. As the process matures, specific unit operations replace most of these modules.

Simulations provide process data for streams and equipment. Stream process data include items such as pressure, temperature, flow rate, composition, phase, density, surface tension, and viscosity. Equipment process data includes items such as pressure, temperature, duty, efficiency, and tray loading. Valve sizing and line sizing calculations require stream process data. Select stream process data for inlet and outlet streams and equipment process data appear on equipment data sheets. Equipment sizing calculations also require process data.

A simulation verifies the integrity of the process depicted on the process flow diagram. Performance data obtained from an operating plant or pilot studies should resemble that predicted by the simulation. The reader is referred to the literature[5,6] for more information about process simulation techniques.

3. POLLUTION PREVENTION AND FLOW SHEETS

As described earlier, a flow sheet depicts all of the equipment and streams in a process. Boundaries drawn around portions of the flow sheet form systems. These systems can consist of the entire plant, areas of the plant, multiple pieces of equipment, or individual pieces of equipment. Examination of the mass and energy balances associated with these systems yields process deficiencies and pollution prevention opportunities. However, the viability of a pollution prevention system depends upon the location of these boundaries.[7]

No system is an island. Systems work together. All systems must be considered when taking steps to reduce waste, recycle, treat wastes, and dispose of wastes. Additional "systems" affected by the process that do not appear on the flowsheet, such as raw material sources and ultimate disposal sites, should also be considered.

As stated earlier, the process flow sheet matures from crude to sophisticated. Material substitution, modifications to process chemistry, changes in raw materials supplies, etc., are often involved in pollution prevention options. Because of the significant effect of these measures on process chemistry and process configuration, they often limit the number of possible process scenarios that move forward during design. Hence, pollution prevention suggestions and modifications should be incorporated into the flow sheets as these options are identified, and they should be considered at all stages of process development. The reader is referred to Appendix E for information on available pollution prevention software packages to aid in the incorporation of pollution prevention ideas into the design process.

4. ILLUSTRATIVE EXAMPLES

Example 4.1. As part of a plant's pollution prevention program, flue gas from a process is mixed with recycled gas from an absorber (A), and the mixture passes through a waste heat boiler (H) which uses water as the heat transfer medium. It then passes through a water spray quencher (Q) in which the temperature of the mixture is further decreased and, finally, through an absorber (A) in which water is the absorbing agent (solvent) for one of the species in the flue gas stream. Prepare a simplified flow diagram for the process.

SOLUTION.
1. Prepare a line diagram of the process.

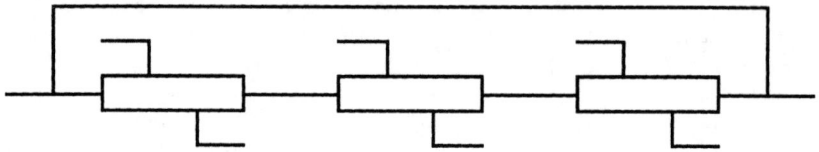

2. Label the equipment.

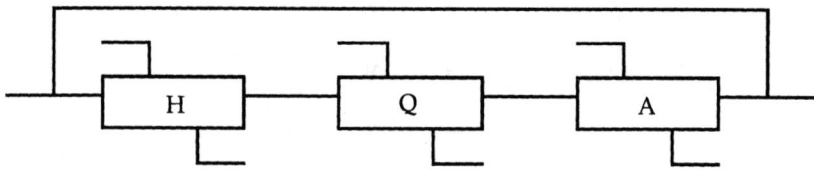

3. Label the flow streams.

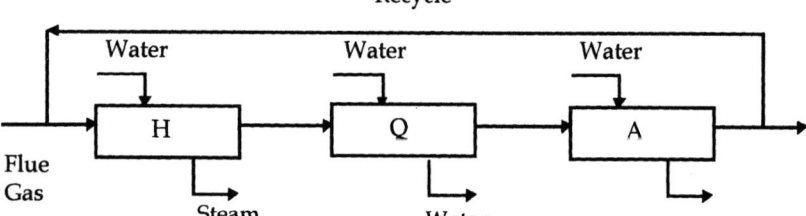

Example 4.2. The heat-generating unit in a coal-fired power plant may be simply described as a continuous-flow reactor into which fuel (mass flow rate F) and air (mass flow rate A) are fed, and from which effluents ("flue gas," mass flow rate E) are discharged.
  a. Draw a flow diagram representing this process. Show all flows into and out of the unit.
  b. Write a mass balance equation for this process.
  c. Suppose the fuel contains a mass fraction y of incombustible component C (for example, ash). Assume that all of the ash is carried out of the reactor with the flue gas (note that in reality, a fraction of the ash generated will remain within the heat-generating unit as bottom ash and must be removed periodically). Write a mass balance equation for component C.
  d. What is the mass fraction (z) of C in the exit stream E and what is the effect of increasing the combustion air flow upon z?

SOLUTION.
  a. Draw a flow diagram of the process. See figure on the next page.
  b. Write an overall mass balance equation for the system.

$$F + A = E$$

Component C is not combustible and is carried out with the exhaust, therefore $C_{in} = C_{out}$.

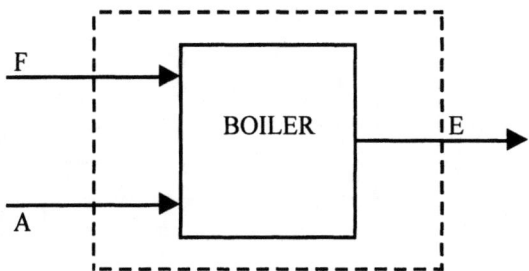

c. Write the component mass balance equation of C.

$$yF + (0)A = zE$$

d. Combining both equations to solve for z leads to:

$$E = F + A$$
$$yF = z(F + A)$$
$$z = y\ (F/(F + A))$$

Therefore, increasing the combustion air flow (A) decreases the mass fraction of C in the exhaust (a dilution effect). For this reason, effluent concentrations are given at a specified excess air value (fraction or percent).

Example 4.3. If a building fire occurs, a smoke alarm sounds with probability = 0.9 (i.e., it is expected to have a 10% failure rate). The sprinkler system functions with a probability = 0.7 whether or not the smoke alarm sounds. The consequences are minor fire damage (alarm sounds, sprinkler works), moderate fire damage with few injuries (alarm sounds, sprinkler fails), moderate fire damage with many injuries (alarm fails, sprinkler works), and major fire damage with many injuries (alarm fails, sprinkler fails). Construct an event tree and indicate the probabilities for each of the four consequences.

SOLUTION. Determine the first consequence(s) of the building fire and list the probabilities of the first consequence.

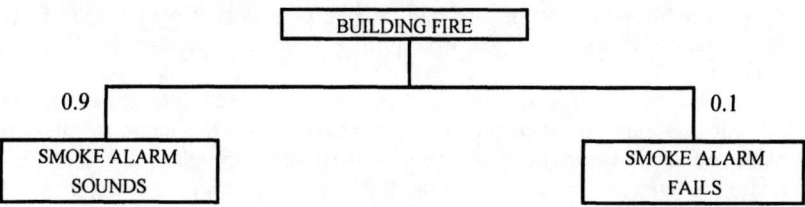

Determine the second consequence(s) of the building fire and list the probabilities of the consequence (s).

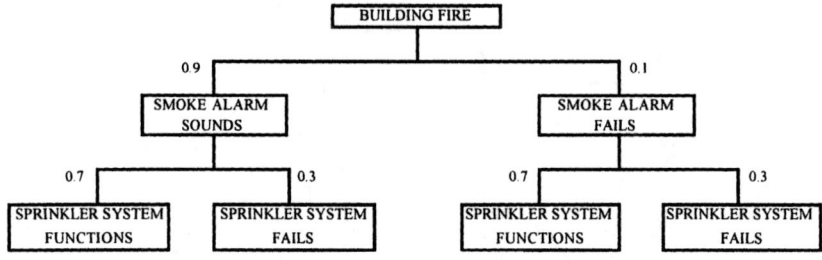

Determine the final consequences and calculate the probabilities of minor fire damage, moderate fire damage with few injuries, moderate fire damage with many injuries, and major fire damage with many injuries.

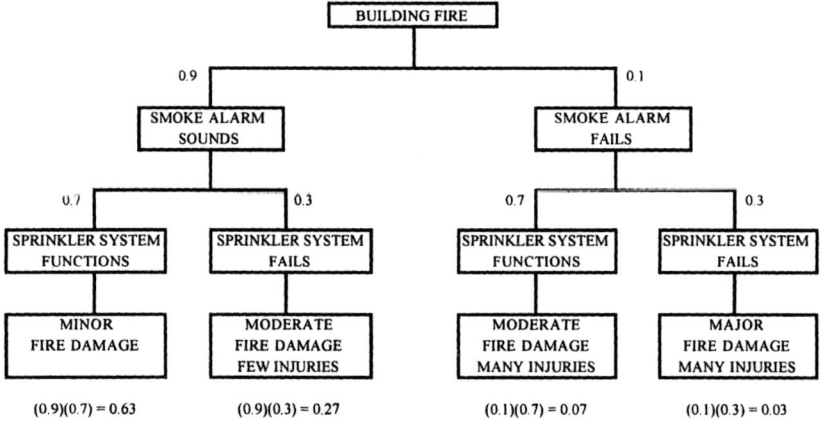

5. SUMMARY

1. *Flow sheets* essentially define, refine, and document a process.
2. A flow sheet usually changes over time with respect to its degree of sophistication and details. A crude flow sheet may initially consist of a simple, freehand block diagram offering information about the equipment only; a later version may include line drawings with pertinent process data such as overall and componental flow rates, utility and energy requirements, and instrumentation. During the later stages of a design project, the flow sheet consists of a highly detailed piping and instrumentation diagram (P&ID).
3. The *block diagram* is the simplest but least descriptive of the schematic diagrams used in the process industry. As the name implies, it consists of neat rectangular blocks which usually represent a single unit operation in a plant or an entire section of a plant. These blocks are connected

by arrows indicating the flow sequence. The block diagram is extremely useful in the early stages of a process change and is particularly valuable in presenting the results of economic or operating studies since significant process data can be placed within the blocks.

4. The *graphic flow diagram* is used most frequently in advertising, company financial reports, and technical reports in which certain features of the flow diagram require extra emphasis. It should present the desired information clearly and in an eye-catching fashion that is both novel and informative.

5. The *process flow diagram*, or *PFD*, is a pictorial description of the process. It is the authorized process blueprint, and the single, authoritative document employed to define, construct, and operate the chemical process. It gives the basic processing scheme, the basic control concept, and the process information from which equipment can be specified and designed. It provides the basis for the development of the *P&I diagram (P&ID)* and also serves as a guide for the plant operator. The process flow diagram usually includes:

- Material balance data (may be on separate sheets)
- Flow scheme equipment and interconnecting streams
- Basic control instrumentation
- Temperature and pressure at various points
- Any other important parameters unique to each process

6. The P&ID, which provides the basis for detailed design, offers a precise description of piping, instrumentation, and equipment. This key drawing defines the plant system, describes equipment, and shows all instrumentation, piping, and valving. It is used to train personnel and aids in troubleshooting during start-up and operation. The P&ID assigns item numbers to all equipment (e.g., towers, reactors, and tanks), gives dimensions of equipment and vessel elevations, and shows all piping, including line numbers, sizes, and specifications, and all valves. All instrumentation is covered, giving numbers, function, types, and indicating whether it is electronically or pneumatically actuated.

7. The two main types of accident analyses employed in the chemical process industry are *fault tree analysis (FTA)* and *event tree analysis (ETA)*. A fault tree analysis is a flow diagram that spotlights conditions that cause system failure. Fault tree analysis attempts to describe how and why an accident or other undesirable event has occurred or could occur. An event tree analysis is a flow diagram that represents the possible steps leading to a failure or accident. In contrast to a fault tree, which works backward from an undesirable consequence to possible causes, an event tree works forward from the initiating event to possible undesirable consequences.

8. The *process simulation* develops in the same manner as the process flow sheet. Like the process flow diagram and the piping and instrumentation diagram, the process simulation and its block diagram necessitate a uniform system of numbers and symbols for designating streams and equipment. Simulations provide process data for verifying the integrity of the process depicted on the process flow diagram.

146 Chapter 8

9. Boundaries drawn around portions of the flow sheet form systems. These systems consist of the entire plant, areas of the plant, multiple pieces of equipment, and individual pieces of equipment. Examination of the mass and energy balances associated with these systems can yield process deficiencies and opportunities for pollution prevention. However, the viability of a pollution prevention system depends upon the location of these boundaries. Pollution prevention suggestions and modifications should be incorporated into the flow sheet at all stages of process development.

6. PROBLEMS

1. A mixture of water and ethyl alcohol, 40% alcohol by mass, is fed to a flash unit where it is separated into two streams. The vapor and liquid output streams contain 80 and 28 mass percent alcohol, respectively. Draw a flow diagram of the process.

2. Potassium nitrate is obtained from an aqueous solution of 20% KNO_3. During the process, the aqueous solution of potassium nitrate is evaporated, leaving an outlet stream with a concentration of 50% KNO_3, which then enters a crystallization unit where the outlet product is 96% KNO_3 (anhydrous crystals) and 4% water. A residual aqueous solution which contains 0.55 g of KNO_3 per g of water also leaves the crystallization unit and is mixed with the fresh solution of KNO_3 at the evaporator inlet.
   a. Draw a flow diagram of the process.
   b. Calculate the feed rate to the evaporator and the recycle rate to the evaporator in units of kg/hr when the feedstock flow rate is 5000 kg/hr.

3. Construct a decision tree given the following events:
GIVEN:
   Date: a husband's anniversary
   Decision: buy flowers or do not buy flowers
   Consequences (buy flowers): domestic bliss or suspicious wife
   Consequences (do not buy flowers): status quo or wife in tears/
       husband in doghouse

4. The sketch below shows a simplified process flow diagram for the manufacture of ammonia from natural gas, which can be considered to be pure methane. In the first stage of the process, natural gas is partially oxidized in the presence of steam to produce a synthesis gas containing hydrogen, nitrogen and carbon dioxide. The carbon dioxide is removed. The remaining mixture of hydrogen and nitrogen is reacted over a catalyst at high pressure to form ammonia.

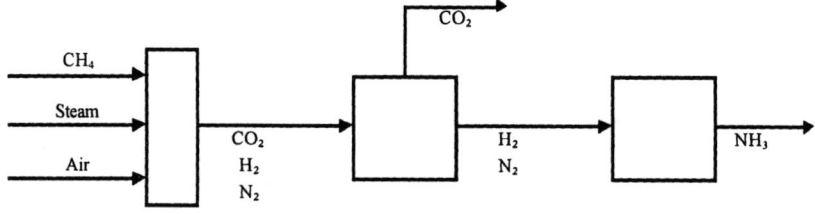

The two reactions for the process, neglecting side reactions and impurities, are

$$7\ CH_4 + 10\ H_2O + 8\ N_2 + 2\ O_2 \rightarrow 7\ CO_2 + 8\ N_2 + 24\ H_2$$
$$N_2 + 3\ H_2 \rightarrow 2\ NH_3$$

Calculate the amount of methane gas, both in pounds per day and in standard cubic feet (60°F, 1 atm) per day, required for a global-scale ammonia plant producing 1200 tons of ammonia per day.

5. A plant operator calibrates instruments to correct for inaccuracy and imprecision. Accuracy deals with the correctness of a value obtained from an instrument or the regularity of an offset amount, e.g., how much different is the instrument reading from the actual process condition. Precision deals with the reproducibility of the readings, e.g., how much variation is there in the instrument reading for identical process conditions? For the flow meter data given below, construct a calibration curve to correct for accuracy and account for imprecision. The "plant" readings are from a new instrument; the "correct" readings are from a carefully calibrated test instrument temporarily installed in series with the new instrument.

	Plant Instrument, ft³/min	Correct Value, ft³/min
Set 1	9.8	9.6
	11.0	10.4
	18.9	18.5
	20.2	19.5
	30.5	30.1
Set 2	30.0	29.4
	40.2	39.6
	39.7	39.5
	49.8	49.4
	50.4	50.0
Set 3	58.5	58.0
	61.0	60.5
	72.5	72.1
	70.2	69.7
	81.0	80.2
Set 4	79.8	79.2
	90.7	90.3
	89.5	89.1
	101.4	101.0
	99.0	98.2

6. A runaway chemical reaction can occur if coolers fail (A) or there is a bad chemical batch (B). Coolers fail only if both Cooler #1 fails (C) and cooler #2 fails (D). A bad chemical batch occurs if there is a wrong mix (E) or there is a process upset (F). A wrong mix occurs only if there is an operator error (G) and instrument failure (H).
   a. Construct a fault tree.

b. If the following annual probabilities are provided by the plant engineer, calculate the probability of a runaway chemical reaction occurring in a year's time if

$$P(C) = 0.05$$
$$P(D) = 0.08$$
$$P(F) = 0.06$$
$$P(G) = 0.03$$
$$P(H) = 0.01$$

7. REFERENCES

1. Treybal, R.E., *Mass-Transfer Operations*, 1st ed., McGraw-Hill, New York, New York, 1955.
2. Coulon, J., Richardson, J., and Skinnott, R., *An Introduction to Chemical Engineering Design*, Pergamon Press, Elmsford, New York, 1983.
3. Theodore, L., Reynolds, J., and Taylor, F., *Accident and Emergency Management*, Wiley-Interscience, New York, New York, 1989.
4. Theodore, L., Reynolds, J., and Morris, K., *A Theodore Tutorial, Accident and Emergency Management: Industrial Applications*, Theodore Tutorials, East Williston, New York, 1992.
5. Sowell, R., Why a Simulation System Doesn't Match the Plant, *Hydrocarbon Processing* (3):102-107, 1998
6. Schad, R., Make the Most of Process Simulation, *Chemical Engineering Progress* (1):21-26, 1998.
7. Allen, D. and Rosselot, K., *Pollution Prevention for Chemical Processes*, Wiley-Interscience, New York, New York, 1997.

CHAPTER 9

Economic Considerations
Contributing Author: Keith Colacioppo

Pollution prevention has been recognized as potentially one of the lowest-cost options available for waste/pollutant management. Hence, an understanding of the economics involved in pollution prevention options is quite important in making decisions at both the engineering and management levels. Every engineer should be able to execute an economic evaluation of a proposed project. If the project is not profitable, it should obviously not be pursued, and the earlier such a project can be identified, the fewer are the resources that will be wasted in its development.

Before the cost of a pollution prevention program can be evaluated, the factors contributing to the cost must be recognized. There are two major contributing factors to the overall cost of a project: *capital costs* and *operating costs*. These major cost categories are discussed in this chapter. Once the total cost of the various process and pollution prevention options has been estimated, the engineer must then determine whether or not a specific project will be profitable. This involves converting all cost contributions to an annualized basis, a method that is also discussed later in this chapter. If more than one project proposal is under study, this method provides a basis for comparing alternate proposals and for choosing the best option on a total cost basis. Since project optimization is another subject of concern, a brief description of a perturbation analysis is presented. Other factors that can be important in an overall analysis of a process change for minimizing waste generated are plant siting and the layout of the plant, which are also considered later in this chapter.

Detailed cost estimates are beyond the scope of this chapter and this text. Such procedures are capable of producing accuracies in the neighborhood of ± 5 percent; however, such estimates generally require many months of engineering work and project evaluation. This chapter is designed to give the reader a basis for a preliminary cost analysis only, with an expected accuracy of approximately ± 20 percent.

In addition, economic subject areas such as simple interest, compound interest, present worth, depreciation, perpetual life, rate of return, payback period, bonds, etc., are either not presented or only superficially reviewed. Details on these topics are available in the literature.[1]

An economic evaluation is carried out using standard measures of profitability. Each company and organization has its own economic criteria for selecting projects for implementation. In performing an economic evaluation, various costs and savings must be considered. As in any project, the cost elements of a pollution prevention project can be broken down into capital costs and operating costs. The economic analysis described in this chapter represents a preliminary, rather than detailed cost analysis. For smaller facilities with only a few (and perhaps simple) processes, the entire pollution prevention assessment procedure will tend to be much less formal. In this situation, several obvious pollution prevention options, such as installation of flow controls and good operating practices, may be implemented with little or no economic evaluation. In these instances, no complicated analyses are necessary to demonstrate the advantages of adopting the selected pollution prevention options. A proper perspective must be maintained between the magnitude of savings that a potential option may offer and the amount of manpower required to carry out the technical and economic feasibility analyses.

1. ECONOMIC AND COSTING PROCEDURES[2]

The purpose of this section is to outline the basic elements of a pollution prevention cost-accounting system that incorporates both traditional and less tangible economic variables. The intent is not to present a detailed discussion of economic analysis, but to help identify the more important elements that must be considered to properly quantify costs and benefits of pollution prevention options.

The greatest driving force behind any pollution prevention plan is the promise of economic opportunities and cost savings provided by the waste reduction effort over the long term. Hence, an understanding of the costs and benefits involved in pollution prevention programs/options is important in making decisions at both the engineering and management levels. If a project cannot be justified economically after all factors have been taken into account, it should obviously not be pursued.

Besides the traditional cost factors involved in pollution prevention projects, there are also other important costs and benefits associated with pollution prevention that need to be quantified if a meaningful economic analysis of a pollution prevention project is going to be performed. Table 9.1 summarizes various cost accounting methods that have evolved over time to begin to attempt to capture the intangible costs of waste production and the benefits that accrue with the minimization or elimination of waste streams.

The *Total Systems Approach (TSA)* referenced in Table 9.1 attempts to quantify not only the economic aspects of pollution prevention but also the social costs associated with the production of a product or service from cradle to grave – often referred to as *life cycle analysis*. (Life cycle costing and analysis is treated in more detail in Chapter 11 of Part 2 of this book.) TSA also attempts to quantify less tangible benefits such as the reduced risk derived from not using a hazardous substance. More emphasis will certainly be placed on the TSA approach in future pollution-prevention assessments. For example, at present, a utility considering the option of converting from a gas-fired boiler to coal-firing is usually not concerned with the environmental effects and implications associated with such activities as mining, transporting, and storing the coal prior to its usage as an energy feedstock. More complete life cycle considerations will be incorporated into future pollution prevention assessments and business decisions.

Table 9.1. Developments in Economic Analysis Approaches.

Prior to 1945	Capital costs only
1945 - 1960	Capital and some operating costs
1960 - 1970	Capital and operating costs
1970 - 1975	Capital, operating, and some environmental control costs
1975 - 1980	Capital, operating, and environmental control costs
1980 - 1985	Capital, operating, and more sophisticated environmental control costing
1985 - 1990	Capital, operating, and environmental controls, and some life-cycle analysis (Total Systems Approach)
1990 - 1995	Capital, operating, and environmental control costs and life-cycle analysis (Total Systems Approach)
1995 - 2000	Widespread acceptance of Total Systems Approach
>2000	Total Systems Approach with Activity Based Costing

2. ENVIRONMENTAL ACCOUNTING[3]

In December 1993, a national workshop of experts drawn from business, professional groups, government, non-profits, and academia produced an *Action Agenda*[4] "to encourage and motivate businesses to understand the full spectrum of environmental costs and integrate these costs in decision making." The *Agenda* identifies four overarching issue areas that require attention to advance environmental accounting: (1) better understanding of terms and concepts, (2) creation of internal and external management incentives, (3) education, guidance, and outreach, and (4) development and dissemination of analytical tools, methods, and systems.

Through EPA's *Design for the Environment Program*, accounting tools, resource directories, and detailed case studies of environmental accounting practices have been developed and disseminated as part of the Agency's *Environmental Accounting Project*. Many of these resources can be found on the Internet at http://www.epa.gov/opptintr/acctg/pdf/. The following material is excerpted from EPA's introductory document on environmental accounting[3] that can be found on the Internet at the URL above under the document name: busmgt.pdf.

2.1. What Is Environmental Accounting?

Different uses of the umbrella term *environmental accounting* arise from three distinct contexts: national income accounting, financial accounting, and management accounting.

National income accounting is a macro-economic measure. Gross Domestic Product (GDP) is an example. The GDP is a measure of the flow of goods and services through the economy. It is often cited as a key measure of our society's economic well-being. The term environmental accounting may refer to this national economic context. For example, environmental accounting can use physical or monetary units to refer to the consumption of the nation's natural resources, both renewable and nonrenewable. In this context, environmental accounting has been termed *natural resources accounting*.

Financial accounting enables companies to prepare financial reports for use by investors, lenders, and others. Publicly held corporations report information on their financial condition and performance through quarterly and annual reports, governed by rules set by the U.S. Securities and Exchange Commission (SEC) with input from industry's self-regulatory body, the Financial Accounting Standards Board (FASB). Generally Accepted Accounting Principles (GAAP) are the basis for this reporting. Environmental accounting in this context refers to the estimation and public reporting of environmental liabilities and financially material environmental costs.

Management accounting is the process of identifying, collecting, and analyzing information principally for internal purposes. A key purpose of management accounting is to support a business's forward-looking management decisions, and in the context of this textbook, it is the most relevant of the environmental accounting contexts. Management accounting can involve data on costs, production levels, inventory and backlog, and other vital aspects of a business. The information collected under a business's management accounting system is used to plan, evaluate, and control in a variety of ways:

1. Planning and directing management attention,
2. Informing decisions such as purchasing (e.g., make vs. buy), capital investments, product costing and pricing, risk management, process/product design, and compliance strategies, and
3. Controlling and motivating behavior to improve business results.

Unlike financial accounting, which is governed by GAAP, management accounting practices and systems differ according to the needs of the businesses they serve. Some businesses have simple systems; others have elaborate ones. Just as management accounting refers to the use of a broad set of cost and performance data by a company's managers in making a myriad of business decisions, environmental accounting refers to the use of data about environmental costs and performance in business decisions and operations. Table 9.2 lists many types of internal management decisions that can benefit from the consideration of environmental costs and benefits.

2.2. What Is an Environmental Cost?

Uncovering and recognizing *environmental costs* associated with a product, process, system, or facility is important for good management decisions. Attaining such goals as reducing environmental expenses, increasing revenues, and improving environmental performance requires paying attention to current, future, and potential *environmental costs*. How a company defines an environmental cost depends on how it intends to use the information (e.g., cost allocation, capital budgeting, process/product design,

Table 9.2. Types of Management Decisions Benefiting from Environmental Cost Information.

Product Design	Capital Investments
Process Design	Cost Control
Facility Siting	Waste Management
Purchasing	Cost Allocation
Operational	Product Retention and Mix
Risk Management	Product Pricing
Environmental Compliance Strategies	Performance Evaluations

other management decisions) and the scale and scope of the exercise. Moreover, it may not always be clear whether a cost is "environmental" or not. Whether or not a cost is "environmental" is not critical; the goal is to ensure that relevant costs receive appropriate attention.

2.3. Identifying Environmental Costs

Environmental accounting terminology uses such words as *full, total, true,* and *life cycle* to emphasize that traditional approaches are incomplete in scope because they overlook important environmental costs (and potential cost savings and revenues). In looking for and uncovering relevant environmental costs, managers may want to use one or more organizing frameworks as tools. There are many different ways to categorize costs. Accounting systems typically classify costs as: *direct materials and labor, manufacturing or factory overhead* (i.e., operating costs other than direct materials and labor), *sales, general and administrative (G&A) overhead*, and *research & development (R&D)*.

Environmental expenses may be classified in any or all of these categories in different companies. To better focus attention on environmental costs for management decisions, the *EPA Pollution Prevention Benefits Manual* and the Global Environmental Management Initiative (GEMI) environmental cost primer use similar organizing frameworks to distinguish costs that generally receive management attention, termed the "usual" costs or "direct" costs, from costs that may be obscured through treatment as overhead or R&D, distorted through improper allocation to cost centers, or simply overlooked. These latter costs are termed "hidden," "contingent," "liability" or "less tangible" costs. Figure 9.1 lists examples of these costs under the labels "conventional," "potentially hidden," "contingent," and "image/relationship" costs.

Conventional Costs. The costs of using raw materials, utilities, capital goods, and supplies are usually addressed in cost accounting and capital budgeting, but are not usually considered environmental costs. However, decreased use and less waste of raw materials, utilities, capital goods, and supplies are environmentally preferable, reducing both environmental degradation and consumption of nonrenewable resources. It is important to factor these costs into business decisions, whether or not they are viewed as

	Potentially Hidden Costs	
Regulatory	**Upfront**	**Voluntary** (Beyond Compliance)
Notification	Site studies	Community relations/
Reporting	Site preparation	outreach
Monitoring/testing	Permitting	Monitoring/testing
Studies/modeling	R & D	Training
Remediation	Engineering and	Audits
Recordkeeping	procurement	Qualifying suppliers
Plans	installation	Reports (e.g., annual
Training		environmental reports)
Inspections	**Conventional Costs**	Insurance
Manifesting		Planning
Labeling	Capital equipment	Feasibility studies
Preparedness	Materials	Remediation
Protective Equipment	Labor	Recycling
Medical surveillance	Supplies	Environmental studies
Environmental	Utilities	R & D
insurance	Structures	Habitat and wetland
Financial assurance	Salvage value	protection
Pollution control		Landscaping
Spill response	**Back-End**	Other environmental
Stormwater	Closure/decommissioning	projects
management	Disposal of inventory	Financial support to
Waste management	Post-closure care	environmental groups
Taxes/fees	Site survey	and/or other researchers
	Contingent Costs	
Future Compliance Costs	Remediation	Legal expenses
Penalties/fines	Property damage	Natural resource damages
Response to future releases	Personal injury damage	Economic loss damages
	Image and Relationship Costs	
Corporate image	Relationship with professional	Relationship with lenders
Relationship with customers	staff	Relationship with host
Relationship with investors	Relationship with workers	communities
Relationship with insurers	Relationship with suppliers	Relationship with regulators

Figure 9.1. Examples of environmental costs incurred by firms.[3]

"environmental" costs. The dashed line around these *conventional costs* in Figure 9.1 indicates that even these costs (and potential cost savings) may sometimes be overlooked in business decision-making.

Potentially Hidden Costs. Figure 9.1 shows several types of environmental costs that may be potentially hidden from managers. First are *upfront environmental costs*, which are incurred prior to the operation of a process, system, or facility. These can include costs related to siting, design of environmentally preferable products or processes, qualifications of suppliers, evaluation of alternative pollution control equipment, and so on. Whether classified as overhead or R&D, these costs can easily be forgotten when managers and analysts focus on operating costs of processes, systems, and facilities. Second are *regulatory* and *voluntary environmental costs* incurred in operating a process, system, or facility; because many companies traditionally have treated these costs as overhead, they may not receive appropriate attention from managers and analysts responsible for day-to-day operations and business decisions. The magnitude of these costs also may be more difficult to determine as a result of their being pooled in overhead accounts. Third, while upfront and current operating costs may be obscured by management accounting practices, *back-end environmental costs* may not be entered into management accounting systems at all. These environmental costs of current operations are *prospective,* meaning they will occur at more or less well defined points in the future. Examples include the *future* cost of decommissioning a laboratory that uses licensed nuclear materials, closing a landfill cell, replacing a storage tank used to hold petroleum or hazardous substances, and complying with regulations that are not yet in effect but have been promulgated. Such back-end environmental costs may be overlooked if they are not well documented or accrued in accounting systems.

Figure 9.1 contains a lengthy list of "*potentially hidden*" *environmental costs*, including examples of the costs of upfront, operational, and back-end activities undertaken to: comply with environmental laws (i.e., regulatory costs), or go beyond compliance (i.e., voluntary costs). In bringing these costs to light, it also may be useful to distinguish among costs incurred to respond to *past pollution* not related *to ongoing operations*; to control, clean up, or prevent pollution from *ongoing operations*; or to prevent or reduce pollution from *future operations*.

Contingent Costs. Costs that may or may not be incurred at some point in the future – here termed "*contingent costs*" – can best be described in probabilistic terms: their expected value, their range, or the probability of their exceeding some dollar amount. Examples include the costs of remedying and compensating for future accidental releases of contaminants into the environment (e.g., oil spills), fines and penalties for future regulatory infractions, and future costs due to unexpected consequences of permitted or intentional releases. These costs may also be termed "contingent liabilities" or "contingent liability costs." Because these costs may not currently need to be recognized for other purposes, they may not receive adequate attention in internal management accounting systems and forward-looking decisions.

Image and Relationship Costs. Some environmental costs are called "less tangible" or "intangible" because they are incurred to affect subjective (though measurable) perceptions of management, customers, employees, communities, and regulators. These costs have also been termed "*corporate image*" and "*relationship*" costs. This category can include the costs of annual environmental reports and community relations activities, costs incurred voluntarily for environmental activities (e.g., tree planting), and costs incurred for pollution prevention award/recognition programs. The costs themselves are not "intangible," but the direct benefits that result from relationship/corporate image expenses often are.

Is It An "Environmental" Cost? Costs incurred to comply with environmental laws are clearly environmental costs. Costs of environmental remediation, pollution control equipment, and noncompliance penalties are all unquestionably environmental costs. Other costs incurred for environmental protection are likewise clearly environmental costs, even if they are not explicitly required by regulations or go beyond regulatory compliance levels.

There are other costs, however, that may fall into a gray zone in terms of being considered environmental costs. For example, should the costs of production equipment be considered "environmental" if it is a "clean technology?" Is an energy-efficient turbine an "environmental" cost? Should efforts to monitor the shelf life of raw materials and supplies in inventory be considered "environmental" costs (if discarded, they become waste and result in environmental costs)? It may also be difficult to distinguish some environmental costs from health and safety costs or from risk management costs.

The success of environmental accounting does not depend on "correctly" classifying all the costs a firm incurs. Rather, its goal is to ensure that relevant information is made available to those who need or can use it. To handle costs in the gray zone, some firms use the following approaches: allowing a cost item to be treated as "environmental" for one purpose but not for another, treating part of the cost of an item or activity as "environmental," or treating costs as "environmental" for accounting purposes when a firm decides that a cost is more than 50% environmental.

There are many options. Companies can define what should constitute an "environmental cost" and how to classify it, based on their goals and intended uses for environmental accounting. For example, if a firm wants to encourage pollution prevention in capital budgeting, it might consider distinguishing environmental costs that can be avoided by pollution prevention investments, from environmental costs related to remedying contamination that has already occurred. But for product costing purposes, such a distinction might not be necessary because both are costs of producing the good or service.

2.4. Applying Environmental Accounting to Cost Allocation

An important function of environmental accounting is to bring environmental costs to the attention of corporate stakeholders who may be able and motivated to identify ways of reducing or avoiding those costs while at the same time improving environmental quality. This can require, for example, pulling some environmental costs out of overhead and allocating those environmental costs to the appropriate accounts. By *allocating* environmental costs to the products or processes that generate them, a company can motivate affected managers and employees to find creative pollution prevention alternatives that lower those costs and enhance profitability. For example, Caterpillar's East Peoria, Illinois, plant no longer dumps waste disposal costs into an overhead account; rather, the costs of waste disposal are allocated to responsible commodity groups, triggering efforts to improve the bottom line through pollution prevention.[5]

Overhead is any cost that, in a given cost accounting system, is not wholly attributed to a single process, system, product, or facility. Examples can include supervisors' salaries, janitorial services, utilities, and waste disposal. Many environmental costs are often treated as overhead in corporate cost accounting systems. Traditionally, an overhead cost item has been handled in either one of two ways: it may be allocated on some basis to specific products, or it may be left in the pool of costs that are not attributed to any specific product.

If overhead is allocated incorrectly, one product may bear an overhead allocation greater than warranted, while another may bear an allocation smaller than its actual contribution. The result is poor product costing, which can affect pricing and profitability. Alternatively, some overhead costs may not be reflected at all in product cost and price. In both instances, managers cannot perceive the true cost of producing products and thus internal accounting reports provide inadequate incentives to find creative ways of reducing those costs.

Separating environmental costs from overhead accounts where they are often hidden and allocating them to the appropriate product, process, system, or facility directly responsible reveals these costs to managers, cost analysts, engineers, designers, and others. This is critical not only for a business to have accurate estimates of production costs for different product lines and processes, but also to help managers target cost reduction activities that can also improve environmental quality. The axiom "one cannot manage what one cannot see" pertains here. There are two general approaches to allocating environmental costs: build proper cost allocation directly into cost accounting systems, or handle cost allocation outside of automated accounting systems. Companies may find that the latter approach can serve as an interim measure while the former option is being implemented.

Figure 9.2 highlights the misallocation of environmental costs using conventional cost allocation procedures. Suppose Widget B is solely responsible for toxic waste generation and the subsequent management costs in a given facility, and Widget A creates no toxic waste during its production. The misallocation occurs because the toxic waste management cost is lumped together in an overhead cost pool that is misallocated to both Widgets A and B, even though none of the toxic waste management cost results from the production of Widget A. The effect is to distort the actual costs of producing Widgets A and B.

Figure 9.3 illustrates a cost accounting system that correctly attributes the environmental costs of Widget B only to Widget B. By breaking environmental costs out of overhead and directly attributing them to products, managers will have a much clearer view of the true costs of producing Widget A and B. Alternatively, environmental costs can be allocated to responsible processes,

158 Chapter 9

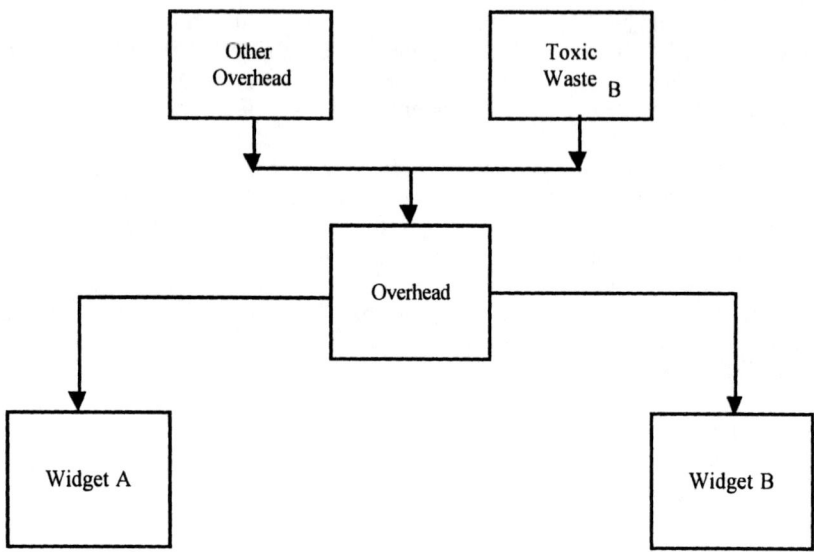

Figure 9.2. Example of misallocation of environmental costs under traditional cost system.[3]

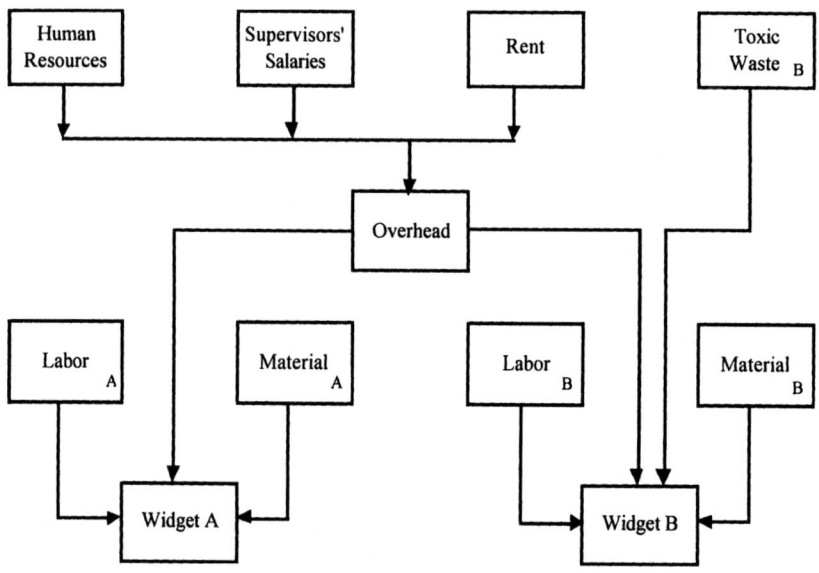

Figure 9.3. Example of revised cost accounting system.[3]

systems, or departments. Environmental costs resulting from several processes or products may need to be allocated based on a more complex analysis, and future costs (e.g., toxic waste disposal) may need to be amortized and allocated to proper cost centers.

The preceding discussion applies equally to the appropriate crediting of revenues derived from sales or use of by-products or recyclables (e.g., raw materials and supplies). Although the focus of cost allocation is on environmental costs, environmental revenues should be treated in a similar fashion.

2.5. Applying Environmental Accounting to Capital Budgeting

Capital budgeting includes the process of developing a firm's planned capital investments. It typically entails comparing predicted cost and revenue streams of current operations and alternative investment projects against financial benchmarks in light of the costs of capital to a firm.[6,7] It has been quite common for financial analysis of investment alternatives to exclude many environmental costs, cost savings, and revenues. As a result, corporations may not have recognized financially attractive investments in pollution prevention and "clean technology." This is beginning to change. When evaluating a potential capital investment it is important to fully consider environmental costs, cost savings, and revenues to place pollution prevention investments on a level playing field with other investment choices. To do this, identify and include the *types* of costs (and revenues) (i.e., the "cost inventory") that will help to demonstrate the financial viability of a cleaner technology investment. Analyze qualitatively those data and issues that cannot be easily quantified, such as the potential less tangible benefits of pollution prevention investments. Figure 9.1 may help in identifying potentially relevant costs (and savings).

After collecting or developing environmental data (either from the accounting system or by manual means), allocate and project costs, cost savings, and potential revenues to the products, processes, systems, or facilities that are the focus of the capital budgeting decision. Begin with the easiest to estimate costs and revenues and work toward the more difficult to estimate environmental costs and benefits such as contingencies and corporate image. The benefit of improved corporate image and relationships due to pollution prevention investments can impact costs and revenues in ways that may be challenging to project in dollars and cents. Some of these less tangible benefits of pollution prevention investments include: increased sales due to enhanced company or product image; better borrowing access and terms; equity more attractive to investors; health and safety cost savings; increased productivity and morale of employees; greater retention, reduced recruiting costs; faster, easier approvals of facility expansion plans or changes due to increased trust from host communities and regulators; enhanced image with stakeholders such as customers, employees, suppliers, lenders, stockholders, insurers, and host communities; and improved relationships with regulators. Information about past expenditures on corporate image also may be helpful in estimating future benefits (e.g., potential savings or reductions in those outlays resulting from the investment) for companies that want to go beyond the qualitative consideration of these benefits. For more information on integrating environmental costs into capital budgeting, see EPA.[7]

2.6. Applying Environmental Accounting to Process/Product Design

The design of a process or product significantly affects its environmental costs and performance. The design process involves balancing cost, performance, cultural, legal, and environmental criteria.[8]

Many companies are adopting "design for the environment" or "life cycle design" programs to take environmental considerations into account at an early stage of product development. To do so, designers need information on the environmental costs and performance of alternative product/process designs, much like the information needed in making capital budgeting decisions. Thus, making environmental cost and performance information available to designers can facilitate the design of environmentally preferable processes and products.

For example, the Rohm and Haas Company has developed a model to estimate in R&D the environmental cost of new processes. The model includes conventional, hidden, contingent, and relationship costs. In early phases of process development, the cost model prompts process researchers to select and justify process chemistries, operating conditions, and equipment that embody the principles of pollution prevention. As the project progresses, the model identifies environmental cost reduction opportunities. The model can provide financial analysts with an economic picture of the potential environmental risk of a new process prior to its commercialization.[9]

3. CAPITAL EQUIPMENT COSTS

Equipment cost is a function of many variables, one of the most significant of which is equipment capacity. Other important variables include operating temperature and/or pressure conditions, and degree of equipment sophistication. Preliminary capital equipment costs estimates are often made using simple cost-capacity relationships that are valid when the other variables are confined to narrow ranges of values. These cost-capacity relationships can be represented by approximate, linear (on log-log coordinates) cost equations of the form[10]

$$C = aQ^b \qquad (9.1)$$

where C = cost, Q = some measure of equipment capacity, and a, b = empirical "constants" that depend mainly on equipment type.

It should be emphasized that this procedure is suitable for rough estimation only; actual estimates from vendors are more preferable. If more accurate values are needed and if old price data are available, the use of an indexing method is better, although a bit more time-consuming. The method consists of adjusting earlier cost data to present values using factors that correct for inflation. A number of such indices are available. One of the most commonly used is the Chemical Engineering Fabricated Equipment Cost

Index (FECI).[11] Past and current values of this index are listed in Table 9.3. Other indices for construction, labor, building, engineering, and so on are also available in the literature.[11]

Table 9.3. Fabricated Equipment Cost Index.

Year	Index	Year	Index
1999*	388.5	1986	318.4
1998	389.5	1985	325.3
1997	386.5	1984	334.1
1996	381.7	1983	327.4
1995	381.1	1982	326.0
1994	368.1	1981	321.8
1993	359.2	1980	291.6
1992	358.2	1979	261.7
1991	361.3	1978	238.6
1990	357.6	1977	216.6
1989	355.4	1976	200.8
1988	342.5	1975	192.2
1987	323.8		

*estimated as of April, 1999
Source: *Chemical Engineering*, McGraw-Hill, New York.

Generally, it is not wise to use cost data older than 5 to 10 years, even with the use of the cost indices. Within that time span, technologies used in the processes may have changed drastically. Use of the indices could cause estimates to be much greater than actual costs. Such an error might lead to the choice of an alternative proposal other than the least costly.

The usual technique for determining the capital costs (i.e., total capital costs, which include equipment design, purchase, and installation) for the facility is based on the factored method of establishing direct and indirect installation costs as a function of the known equipment costs. This is basically a modified Lang method, whereby cost factors are applied to known equipment costs.[12,13]

The first step is to obtain from vendors (or, if less accuracy is acceptable, from one of the estimation techniques previously discussed) the purchase prices of the primary and auxiliary equipment. The total base price, designated by X, which should include instrumentation, control, taxes, freight costs, and so on, serves as the basis for estimating the direct and indirect installation costs. The installation costs are obtained by multiplying X by the cost factors, which are available in the literature.[12-17] For more refined estimates, the cost factors can be adjusted to model more closely the proposed system by using adjustment factors that take into account the complexity and sensitivity of the system.[12,13]

The second step is to estimate the direct installation costs by summing all the cost factors involved in the direct installation costs, which include piping, insulation, foundation and supports, and so on. The sum of these factors is designated as the DCF (direct installation cost factor). The direct installation costs are then the product of the DCF and X.

The third step consists of estimating the indirect installation costs. The procedure here is the same as for the direct installation costs; that is, all the cost factors for the indirect installation costs (engineering and supervision, start-up, construction fees, etc.) are added; the sum is designated by ICF (indirect installation cost factor). The indirect installation costs are then the product of ICF and X.

Once the direct and indirect installation costs have been calculated, the total capital cost (TCC) may be evaluated as:

$$TCC = X + (DCF)(X) + (ICF)(X) \qquad (9.2)$$

This is then converted to annualized capital costs with the use of the Capital Recovery Factor (CRF), which is described later in Equation 9.3. The annualized capital cost (ACC) is the product of the CRF and TCC and represents the total installed equipment cost distributed over the lifetime of the facility.

Guidelines for improving equipment purchasing are listed here:

1. Do not buy or sign any documents unless certified independent test data are provided.
2. Contact and visit facilities of previous clients of the vendor.
3. Obtain prior approval from local regulatory officials.
4. Require a guarantee from the vendors involved. Start-up assistance is usually needed and assurance of prompt technical assistance should be obtained in writing. A complete and coordinated operating manual should be provided.
5. Vendors should provide key replacement parts if necessary.
6. Finally, 10 to 15 percent of the cost should be withheld until the installation is completed.

4. OPERATING COSTS

Operating costs can vary from site to site since these costs, in part, reflect local conditions (e.g., staffing practices, labor, and utility costs). Operating costs, like capital costs, may be separated into two categories: direct and indirect costs. Direct costs are those that cover material and labor and are directly involved in operating the facility. These include labor, materials, maintenance and maintenance supplies, replacement parts, waste (i.e., residues after incineration) disposal fees, utilities, and laboratory costs. Indirect costs are those operating costs associated with, but not directly involved in, operating the facility. Costs such as overhead (e.g., building and land leasing and office supplies), administrative fees, local property taxes, and insurance fees fall into this category.

The major direct operating costs are usually those associated with labor and materials. Materials costs for systems involve the cost of chemicals needed for the operation, replacement parts, and so on. Labor costs differ greatly, depending primarily on the degree of controls and/or instrumentation. Typically, there are three working shifts per day with one supervisor per shift. On the other hand, at some manufacturing plants, the system may be manned by a single operator for only one-third or one-half of each shift; at some sites only an operator, supervisor, and site manager are necessary to run the facility. Salary costs vary from state to state and depend significantly on the location of the facility. The cost of utilities generally consists of that for electricity, water, fuel, and steam. Annual utility costs are estimated with the use of material and energy balances. Costs for waste disposal can be estimated on a per ton basis. Costs of landfilling can range from $10 per ton if the waste is nonhazardous to $100 per ton or more if the material is hazardous. The costs of handling any scrubber effluent can vary depending on the method of disposal. For example, if conventional sewer disposal is used, the effluent probably has to be cooled and neutralized before disposal, with the cost for this depending on the solids concentration in the scrubber discharge. Annual maintenance costs can be estimated as a percentage of the capital equipment costs. The annual costs of replacement parts can be computed by dividing the costs of the individual part by its expected lifetime. Such life expectancies can be found in the literature.[12] Laboratory costs depend on the number of samples tested and the extent of the tests. These costs can be estimated as 10 to 20 percent of the operating labor costs.

Indirect operating costs consist of overhead, local property tax, insurance, and administration, less any credits. Overhead comprises payroll, fringe benefits, social security, unemployment insurance, and other compensation that is indirectly paid to plant personnel. This cost can be estimated as 50 to 80 percent of the operating labor, supervision, and maintenance costs.[17,18] Local property taxes and insurance can be estimated as 1 to 2 percent of the total capital cost (TCC), and administration costs can be estimated as 2 percent of the TCC. Refer above to *Section 2.3. Applying Environmental Accounting to Cost Allocation* regarding limitations in conventional accounting practices in the allocation of environmental costs to the overhead cost category.

The total operating cost is the sum of the direct operating costs and the indirect operating costs less any credits that may be recovered (e.g., the value of recovered steam). Unlike capital costs, operating costs are always calculated on an annual basis.

5. PROJECT EVALUATION

In comparing alternate pollution prevention processes or different options of a particular process from an economic point of view, it is recommended that the total capital cost be converted to an annual basis by distributing it over the projected lifetime of the facility. The sum of both the annualized capital costs (ACC) and the annual operating costs (AOC) is known as the total annualized cost (TAC) for the facility. The economic merit of the proposed facility, process, or scheme can be examined once the total annual cost is available. Alternate facilities or options (e.g., two different processes for accomplishing the same degree of pollution prevention, or a baghouse versus an electrostatic precipitator for air pollution control) may also be compared. Note: a small flaw in this procedure is the assumption that the operating costs remain constant throughout the lifetime of the facility. However, since the analysis is geared to comparing different alternatives, the changes with time should be somewhat uniform among the various alternatives, resulting in little loss of accuracy.

The conversion of the total capital cost to an annualized basis involves an economic parameter known as the capital recovery factor (CRF). These CRFs can be found in any standard economics text[19,20] or can be calculated directly from Equation 9.3:

$$CRF = [(i)(1+i)^n]/[(1+i)^n - 1] \qquad (9.3)$$

where n = projected lifetime of the system, yr; and, i = annual interest rate expressed as a fraction.

The CRF is a positive, fractional number. The ACC is computed by multiplying the TCC by the CRF. The annualized capital cost reflects the cost associated with recovering the initial capital outlay over the depreciable life of the system. Investment and operating costs can be accounted for in other ways, such as a present-worth analysis. However, this capital recovery method is preferred because of its simplicity and versatility. This is especially true when comparing somewhat similar systems having different depreciable lives. In such decisions there are usually other considerations besides economic ones, but if all other factors are equal, the proposal with the lowest total annualized cost should be the most viable one.

Consider the following waste management example. If an on-site (internal) incineration system is under consideration for construction, the total annualized costs should be sufficient to determine whether or not the proposal is economically attractive as compared with other proposals. If, however, a commercial incineration process is being considered, the profitability of the proposed operation becomes an additional factor. The method presented below assumes a facility lifetime of 10 years and that the land is already available.

One difficulty in this analysis is estimating the revenue generated from the facility, because both technology and costs can change from year to year. Also affecting the revenue generated is the amount of waste to be processed by the facility. Usually, the more waste processed, the greater are the revenues. If a reasonable estimate can be made as to the revenue that will be generated from the facility, a rate of return can be calculated.

This method of analysis is known as the discounted cash flow method using an end-of-year convention; that is, the cash flows are assumed to be generated at the end of the year, rather than throughout the year (the latter obviously being the real case). An expanded explanation of this method can be found in any engineering economics text.[19] The data required for the analysis are the TCC, the annual after-tax cash flow (A), and the working capital (WC). For example, for hazardous waste incineration facilities WC includes the on-site fuel inventory, caustic soda solution, maintenance materials (spare parts, etc.), and wages for approximately 30 days. The WC is expended at the start-up of the plant (time = 0 yr) and is assumed to be recoverable after the life of the facility (time = 10 yr). For simplicity, it is assumed that the WC is 10 percent of the TCC and that the TCC is spread evenly over the number of years used to construct the facility. (For this example, the construction period is assumed to be 2 years.)

Usually, an after-tax rate of return on the initial investment of at least 30 percent is desirable. The method used to arrive at a rate of return will be discussed briefly. An annual after-tax cash flow can be computed as the annual revenues (R) less the annual operating costs (AOC) and less income taxes (IT). Income taxes can be estimated at 50 percent (this number may be lower with the passage of new tax laws) of taxable income (TI).

$$IT = 0.5 \, (TI) \tag{9.4}$$

The taxable income is obtained by subtracting the AOC and the depreciation of the plant (D) from the revenues generated (R) or:

$$TI = R - AOC - D \tag{9.5}$$

For simplicity, straight-line depreciation is assumed; that is, the plant will depreciate uniformly over the life of the plant. For a 10-year lifetime, the facility will depreciate 10 percent each year:

$$D = 0.1 \, (TCC) \tag{9.6}$$

The annual after-tax cash flow (A) is then:

$$A = R - AOC - IT \tag{9.7}$$

This procedure involves a trial-and-error solution. There are both positive and negative cash flows. The positive cash flows consist of A and the recoverable working capital in Year 10. Both should be discounted backward to time = 0, the year the facility begins operation. The negative cash flows consist of the TCC and the initial WC. In actuality, the TCC is assumed to be spent evenly over the 2-year construction period. Therefore, one-half of this flow is adjusted forward from after the first construction year (time = - 1 yr) to the year the facility begins operating (time = 0). The other half, plus the WC, is assumed to be expended at time = 0. Forward adjustment of the 50 percent TCC is accomplished by multiplying by an economic parameter known as the single-payment compound amount factor (F/P), given by:

$$F/P = (1 + i)^n \tag{9.8}$$

where i = rate of return, fraction; and n = the number of years, in this case 1 yr.

For positive cash flows, the annual after-tax cash flow (A) is discounted backward by using a parameter known as the uniform series present worth factor (P/A). This factor is dependent on both interest rate (rate of return) and the lifetime of the facility and is defined by:

$$P/A = [(1 + i)^n - 1]/[i(1 + i)^n] \tag{9.9}$$

where n = lifetime of facility, for this example, 10 yr. It should be noted that the P/A is the inverse of the CRF (capital recovery factor). The recoverable working capital at Year 10 is discounted backward by multiplying WC by the single payment present worth factor (P/F), which is given by:

$$P/F = 1/(1 + i)^n \tag{9.10}$$

where n = lifetime of the facility, in this case, 10 yr.

The positive and negative cash flows are now equated and the value of i, the rate of return, may be determined by trial and error from Equation 9.11:

$$\text{Term 1} + \text{Term 2} = \text{Term 3} + \text{Term 4} \tag{9.11}$$

where Term 1 = $((1 + i)^{10} - 1)/(i(i + i)^{10})$ A = worth at Year = 0 of annual after-tax cash flows; Term 2 = $WC/(1 + i)^{10}$ = worth at Year = 0 of recoverable WC after 10 yr; Term 3 = (WC + 0.5 TCC) = assumed expenditures at Year = 0; and Term 4 = $0.5 (TCC)(1 + i)^1$ = worth at Year = 0 of assumed expenditures at Year = -1. As indicated above, this is defined as the discounted cash flow method using an end-of-year conversion.

This analysis has dealt with only obvious economic considerations, that is, cash up front (capital costs) and operating and maintenance costs. Strictly speaking, one needs to consider the more subtle (but potentially substantial) financial impacts of pollution prevention and/or waste minimization as discussed above in *Section 2 Environmental Accounting*.

6. PERTURBATION STUDIES IN OPTIMIZATION[21]

Once a pollution prevention scheme for a particular process has been selected, it is common practice to optimize the process from a capital cost and operation and maintenance (O&M) standpoint. There are many optimization procedures available, most of them too detailed for meaningful application to a waste generating facility. These sophisticated optimization techniques, some of which are routinely used in the design of conventional chemical and petrochemical plants, invariably involve computer calculations. Use of these techniques in pollution prevention analysis, however, is usually unwarranted.

One simple optimization procedure that is recommended is the perturbation study. This involves a systematic change (or perturbation) of variables, one by one, in an attempt to locate the optimum design from a cost and operation viewpoint. To be practical, this often means that the engineer must limit the number of variables by assigning constant values to those process variables that are known beforehand to play an insignificant role in the performance of the process under evaluation. Reasonable guesses and simple or short-cut mathematical methods can further simplify the procedure. Much information can be gathered from this type of study since it usually identifies those variables that significantly impact on the overall performance of the process and also helps to identify the major contributors to the total annualized cost.

7. PLANT SITING AND LAYOUT

Plant siting and layout can also play an important role in pollution prevention and waste reduction. Because of economic factors, details on this design consideration and its potential impact on pollution prevention are considered below.

The proper location of a plant is as important to its success as the selection of a process. Not only must many tangible factors such as labor supply and raw material sources be carefully considered, so also must a number of intangible factors which are more difficult to evaluate. The selection of a plant site must be based on a very detailed study in which all factors are weighed as carefully as possible. Such a study often requires a substantial outlay of capital, but false economies at this point may lead to great losses in the future.

For many processes, one or more predominant factors effectively minimize the number of possibilities for a plant location. Raw material and transportation costs may be such that a plant must be located near a material supply source. Thus only the sites near to sources of raw material need be studied, and these may be few in number. Similarly, labor requirements may be heavy enough to eliminate cities below a certain size. These and other factors serve as effective screening agents that save both time and money in the plant siting process.

Important factors to be considered in the study of areas and sites for plant location include raw materials, transportation, process water, waste disposal, fuel and power, labor, and weather.[22] These are discussed individually below.

7.1. Raw Materials

Although the source of raw materials need not be at the plant site, it is an extremely important factor in the ultimate location of the plant. Process development work and economic studies will indicate the minimum standards for raw materials selection. When these standards have been determined, all possible sources of acceptable raw materials can be located and a detailed analysis can proceed.

The size of each raw materials source must be determined in the light of existing and estimated future requirements for the raw materials. An attempt must be made to estimate the life of the raw material source based on future requirements. Alternate sources or substitutes in the area should also be located and evaluated. The cost of raw materials delivered to the plant site can then be determined for all sources meeting the process quality and quantity specifications.

7.2. Transportation

It is not possible to present a complete discussion of freight rates in this text. The engineer in charge of obtaining information related to plant location needs only realize that experts must be consulted in establishing freight charges and optimum location with respect to transportation. Such advice is available from freight agents and traffic experts of railroads and other transportation facilities. In addition, the traffic manager of the company can be of great assistance in obtaining the necessary information and aiding in its interpreta-tion.

The effect of transportation facilities and rates on plant location can be a controlling factor in plant siting. The plastic industry, for example, which must deliver many small shipments to various users in the minimum amount of time, finds a location near the majority of users is mandatory. Less-than-carload-lot rates are very high. Therefore, the distance that the material must be shipped should be kept to a minimum. Each of the four major transporta-tion methods in use today (railroad freight, trucking, water transport, and air shipping) has its benefits and drawbacks which should all be considered carefully during the plant siting evaluation.

7.3. Process Water

Process industries rank above all others as users of water. No process plant could operate without water as a cooling medium and as a direct raw material in certain phases of a process. The local water supply, therefore, must be studied before an area can even be considered as a possible site for a processing facility. A detailed estimate of present and future water require-ments must precede the plant siting study. Then the availability of water in the region being considered should be carefully investigated. If well water is to be used, a complete study of the past history of the groundwater table is necessary. If groundwater supplies are adequate, they are preferred, because of their lower temperature and generally higher water quality as compared to surface water sources.

Surface waters from streams or lakes also require careful consideration since both their quantity and quality can vary greatly on a seasonal basis. Streams discharging into the ocean during times of low runoff can become saline as a result of seawater intrusion. Under such conditions, plant design may have to include large storage facilities for fresh water to be collected during periods of high runoff and used during dry seasons when the stream becomes saline.

Companies moving into relatively non-industrialized areas often fail to consider the possibility of other plants following suit. The size of the water supply should be adequate not only for the future needs of the proposed plant but also for the anticipated needs of other industries that might move into the area. It is also desirable to consider alternate sources of supply that may be required as the preferred water source becomes depleted. Water quality must be studied as well as water quantity. Chemical and bacteriological examin-ation will indicate the extent of treatment required and can aid in the development of water cost figures for comparison with other locations. The possible contamination of the water source by other industries in the area should be anticipated. Note that "contamination" may consist of raising the temperature of the water to a level that renders its use as a cooling medium impossible.

7.4. Waste Disposal

The forward-looking engineer must also consider waste disposal in any plant siting and layout decisions. It may be just as important in sparsely populated areas as in large metropolitan regions having special regulations addressing these concerns. Waste disposal/control may be accomplished either on-site or off-site. The decision regarding these two options is primarily dictated by economic considerations. However, legal, environmental, public relations, health and safety, and other considerations can also play important roles in the decision-making process. Plant location takes on added significance if a location for off-site disposal/control of wastes must be incorporated into the plant siting process.

7.5. Fuel and Power

All process plants require both steam and electric power in their operations. Power is either purchased from local utility companies or generated at the plant site. Even if power is generated by the process plant, arrangements must be made for standby power from the local utility for emergencies. Steam is rarely purchased but is generated at the plant for use in the process and as a driving medium for pumps and compressors.

A detailed knowledge of the quantity of power and steam required for the operation of the proposed plant must be obtained before the siting study can proceed. The costs of all fuels available in the area should be carefully analyzed.

7.6. Labor

A large portion of the cost of any manufactured item is represented by labor. Although labor rates are increasingly becoming similar in most parts of the country, factors such as skill, labor relations, and the general welfare of the work force affect labor productivity and efficiency. Each region being considered for plant location must be surveyed to determine the availability and the skills of the labor market. The skills need not exactly match those required by the process plant. A stable labor force is valuable in successful plant operation.

7.7. Weather

Weather data for a number of years should be assembled for each community being studied. Particular attention should be given to such natural disasters as hurricanes, earthquakes, and floods, which can often be predicted from meteorological data. In certain locations these catastrophic events must be assumed to be probable, and this increases construction costs. Extremely cold weather often hampers process plant operation and requires special construction features to protect equipment from freezing. Predominantly warm weather permits cheaper construction, but may also reduce the efficiency of the labor force.

8. FACTORS IN PLANNING LAYOUTS

Rational design must include the arrangement of processing areas, storage areas, and handling areas with regard to such factors as:

1. New site development or addition to a previously developed site.
2. Future expansion.
3. Economic distribution of services (water, process steam, power, and gas).
4. Weather conditions.
5. Building code requirements.
6. Waste disposal problems.
7. Sensible use of floor and elevation space.
8. Safety considerations (possible hazards).
9. Existing environmental regulations.
10. Anticipated changes in environmental regulations.

8.1. Methods of Layout Planning

In starting a detailed planning study, space requirements must be known for various products, by-products, and raw materials, as well as for process equipment. A starting or reference point, together with a directional schematic flow pattern, will enable design engineers to make a trial plot plan, as explained below. A number of such studies will be required before a suitable plot and elevation plan can be chosen.

The basic blocks used in building an arrangement for plot plans are often used in the unit area concept. This method of planning is particularly well adapted to large plant layouts. Unit areas are often delineated by means of distinct process phases and operational procedures, by the presence or absence of contamination, and by safety requirements. Thus, the determination of the shape and extent of a unit area and the interrelationships of each area in a master plot plan is one of the first tasks of layout planning. Figure 9.4 demonstrates an example of this type of planning.

8.2. Modeling System Layouts

Years ago plastic scale models were fabricated for each plant under construction, providing excellent three-dimensional representations of the actual facilities. Today, in the age of the microcomputer, it is quicker, easier, and much cheaper to generate models by means of computer graphics.

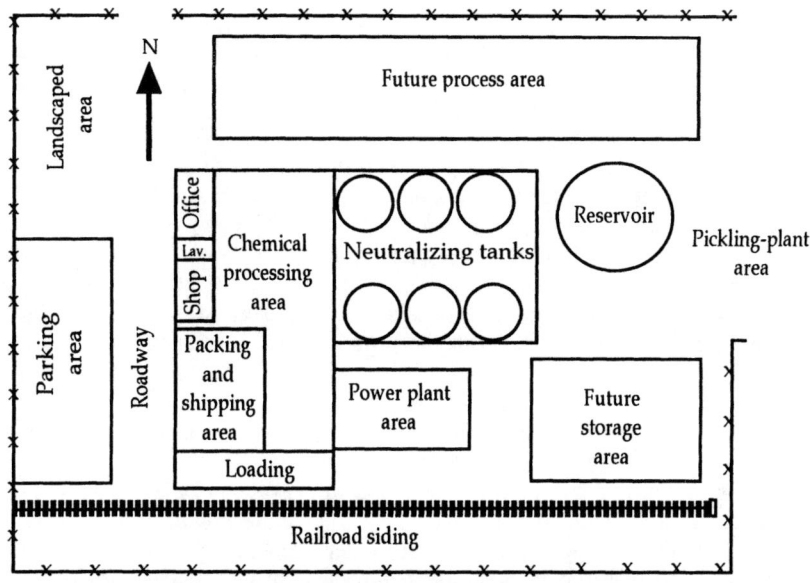

Figure 9.4. Typical master plot plan.

8.3. Principles of Plant Layout

Storage facilities for raw materials and intermediate and finished products may be located in isolated or adjoining areas. Hazardous materials should be isolated because they can be a risk to life and property when stored in large quantities. Storage in adjoining areas initially, to reduce materials handling, may introduce an obstacle to future expansion of a plant.

Arranging the storage of materials to facilitate or simplify handling is also a point to be considered in design. Where it is possible to pump a single material to an elevation so that subsequent handling can be accomplished by gravity into intermediate reaction and storage units, materials handling costs may be reduced. In making a layout, ample space should be assigned to each piece of equipment as accessibility is an important factor in maintenance and long-term pollution prevention.

It is extremely poor economic practice to fit the equipment layout too closely into a building. A slightly larger building than appears necessary will cost little more than one that is overcrowded, yet the extra cost will be small in comparison with the penalties that will result if, to eliminate bottlenecks in maintenance, processing, etc., the building must be expanded.

Operations that constitute a process are essentially a series of unit operations that may be carried on simultaneously. Since these operations are repeated several times in the flow of materials, it should be possible to arrange the equipment into groups of like pieces. Such a layout will make it possible the equipment into groups of like pieces. Such a layout will make it possible for one or two operators to tend all equipment in a given group.

Access for initial construction and maintenance is a necessary part of planning. Space should be provided for repair and service equipment, such as cranes and fork lifts, and specialized items such as snow removal equipment, as well as access ways around doors and underground hatches.

Possible expansion must always be kept in mind. The question of multiplying the number of units or increasing the size of the prevailing unit or units merits more study than can be given here. It suffices to say that one must exercise engineering judgment regarding these types of process layout decisions.

Whether floor space is a major factor in the design of a particular plant depends on the value of land near the proposed site. The engineer should, however, allot floor space economically, consistent with good housekeeping in the plant, and should be sure to consider linear flow of materials, access to equipment, space to permit working on parts of equipment that need frequent servicing, and the safety and comfort of the operators.

The distribution of gas, air, water, steam power, and electricity is not always a major consideration in plant layout, since the flexibility of distribution of these services permits designing to meet almost any condition. Some regard for the proper replacement of each of these services, however, helps to ensure ease of operation and orderliness and aids in reducing maintenance costs.

After a complete study of quantitative factors, the selection of the building or buildings must be considered. Standard factory buildings are to be desired but, if none can be found satisfactory to handle the space and process requirements of the engineer, an architect specializing in this area should be consulted to design a building around the process as opposed to a beautiful structure into which a process must fit.

Consideration of equipment for materials handling is only a minor factor in most cases of arrangement, owing to the multiplicity of devices available for this purpose. Where materials handling is paramount in a process, however, serious thought must be given to it. Again, engineering judgment must be exercised. Whenever possible, one should take advantage of the topography of the site when moving materials around a plant site.

All existing or possible future railroads and highways adjacent to the plant must be known in order to plan rail sidings and access roads within the plant. Railroad spurs and roadways of the correct capacity and at the right locations should be provided for in a traffic study and overall master track and road plan of the plant area.

It must be recognized that there is no single solution to the problem of equipment layout. There are many rational designs. A plan must be decided on after exercise of engineering judgment, striking a balance between the advantages and disadvantages of each possible choice. A great deal of planning is governed by local and national safety and fire code requirements. Fire protection, consisting of reservoirs, mains, hydrants, hose houses, fire pumps, sprinklers in buildings, explosion barriers and directional routing of explosion forces to clear areas, and dikes for combustible product storage tanks, must be incorporated to comply with such codes, to protect costly plant investment, and to reduce insurance rates. Many safety considerations can be better understood by performing a Hazard and Operability (HAZOP) study.[22]

9. PLANT AND PROCESS DESIGN

It is obviously difficult to treat the general subject of plant design from an economic point of view in any great detail. However, some of the key factors that need to be considered early in a design project are briefly discussed below.

9.1. Research Considerations

At this early stage, researchers (probably at the bench-scale level) should be concerned with:

- Raw material quality
- Product quality
- All potential environmental problems that may arise throughout the development of the product or process

9.2. Process Development Considerations

At this stage, engineers (usually at the pilot-plant level) should be concerned with:

- Flexibility in the selection of feed or raw materials to minimize (any) waste generated
- Methods of improving process reliability to minimize losses, spills, and off-specification production
- Ability to monitor and control all pollutant/waste streams
- Potential impacts of the process on both the public and plant employees

9.3. Process Design

The engineer should be concerned with pollution prevention for economic reasons since eliminating or reducing any pollutants generated is preferred to modifying the operation once the plant is designed, built, and running. For example, it is sometimes more cost effective to employ a more pure feedstock than to manage a pollutant/waste that is generated with a less pure raw material. Construction materials should be selected that will prevent or reduce the possibility of equipment failure, which can result in waste discharges to the environment. Instrumentation and controls should be included to help maximize production efficiency and reduce the amount of waste generated. Piping, gaskets, and valves should be designed and constructed to minimize any fugitive emissions; problems with mechanical seals are commonplace.

9.4. Design for the Environment (DfE) Initiative

The EPA Office of Pollution Prevention and Toxics created the Design for the Environment (DfE) program in 1991 to help businesses incorporate environmental considerations into the design and redesign of products, processes, and technical and management systems. The program has three goals: (1) to encourage voluntary reduction of the use of specific hazardous chemicals by businesses, governments, and other organizations through actual design or redesign of products, processes, and technical and management systems; (2) to change the way businesses, governments, and other organizations view and manage for environmental protection by demonstrating the benefits of incorporating environmental considerations into the up-front design and redesign process; and (3) to develop effective voluntary partnerships with businesses, labor organizations, government agencies, and environmental and community groups to implement DfE projects and other pollution prevention activities. DfE projects include broad institutional efforts aimed at changing general business practices, as well as voluntary, cooperative projects with trade associations and businesses in specific industries.

A typical industry project includes developing a Cleaner Technologies Substitutes Assessment (CTSA), a methodology for evaluating the comparative risk, performance, cost, and resource conservation of alternatives to chemicals and processes currently used by specific industry sectors. The CTSA methodology has grown out of DfE industry projects, which are cooperative, joint efforts with trade associations, businesses, public-interest groups, and academia to assist businesses in specific industries to select more environmentally sound products, processes and technologies. A CTSA document is the repository for the technical information developed by a particular DfE project, including detailed environmental, economic, and performance information on traditional and alternative chemicals, manufacturing methods, and technologies. The goal of a CTSA is to provide businesses with information to make environmentally informed choices and design for the environment.

More information on the DfE program and tools for use in various product and process design efforts aimed at pollution prevention and waste minimization can be obtained on the Internet through the U.S. EPA web site at the following URL: http://www.epa.gov/opptintr/dfe.

10. ILLUSTRATIVE EXAMPLES

Example 9.1. When considering the cost of air toxics control technologies it is often important to know how long it will take to recover the cost of an investment in control equipment. The time required to recover this cost is called the payout time. The commonly accepted formula for payout time is:

$$\text{Payout Time} = \frac{\text{Fixed Capital Investment}}{(\text{Annual Profit} + \text{Annual Depreciation})}$$

spends $10,000 for new scrubber packing for a tower that strips air toxics out of a gas stream. The manager decides to depreciate the equipment at $1430/yr (7-year straight-line method depreciation) and estimates that the equipment will generate $1500/yr in annual profit.

SOLUTION. Since the payout time is calculated as the fixed capital investment divided by the sum of the annual profit plus the annual depreciation, the following result is obtained:

$$\text{Payout Time} = \frac{\$10,000}{(\$1500 + \$1430)} = 3.44 \text{ yr}$$

Example 9.2. The equipment (adsorber) costs for a proposed pollution prevention project were estimated to be $852,644 in 1982. Your company has now proposed a similar type of pollution prevention project for your facility in 1998. Assume an annual rate of return of 10% and a operating period of 10 years.
    a.   What is the cost of the "new" project?
    b.   If the total installation cost is 60% of the total equipment cost, what is the annualized capital cost of this project?

SOLUTION.
    a.   The Fabricated Equipment Cost index, FECI, for 1975 through 1999 is provided in Table 9.2. Estimate the capital cost, CC, of the adsorber in 1998 using the following approach:

$$\text{Cost}_{1998} = \text{Cost}_{1982} \left(\frac{\text{FECI}_{1998}}{\text{FECI}_{1982}} \right) = \$852,644 \left(\frac{389.5}{326} \right) = \$1,018,726$$

    b.   Calculate the installation cost, IC, as:

$$IC = 0.60 (\$1,018,726) = \$611,236$$

Calculate the total capital cost, TCC, in 1998 dollars as:

$$TCC = CC + IC = \$1,018,726 + \$611,236 = \$1,629,962$$

Calculate the capital recovery factor, CRF, using the following equation:

$$CRF = \frac{i(1+i)^n}{(1+i)^n - 1}$$

For i = 0.1, and n=10, the CRF is found to be:

$$CRF = \frac{0.1(1.1)^{10}}{(1.1)^{10} - 1} = 0.1627$$

The annualized capital cost, ACC, of the equipment in $/yr, is then calculated as follows:

$$ACC = TCC\,(CRF) = \$1{,}629{,}962\,(0.1627) = \$272{,}530/yr$$

Example 9.3. An engineer has compiled the data below for two different project options – one that includes a comprehensive pollution prevention program option, and one that does not. From an economic point of view, which project should the engineer select? The lifetime of the equipment is 10 years and the interest rate is 10%.

	Project with Pollution Prevention	Project w/o Pollution Prevention
Equipment cost	$1,294,000	$1,081,000
Installation cost	$786,000	$659,000
Operating Labor	$39,900/yr	$8500/yr
Maintenance	$43,000/yr	$17,000/yr
Utilities	$958,000/yr	$821,000/yr
Overhead	$51,300/yr	$13,900/yr
Taxes, insurance and administration	$86,200/yr	$72,600/yr
Credits	$380,000/yr	$0

SOLUTION. For the project with pollution prevention, calculate the total capital cost, TCC(w), as follows:

$$TCC(w) = \$1{,}294{,}000 + \$786{,}000 = \$2{,}080{,}000$$

Calculate the Capital Recovery Factor, CRF(w), using the following expression:

$$CRF(w) = \frac{i(1+i)^n}{(1+i)^n - 1} = \frac{0.1(1.1)^{10}}{(1.1)^{10} - 1} = 0.1627$$

Calculate the annualized capital cost, ACC(w), of the process with pollution prevention options as follows:

$$ACC(w) = (\$2{,}080{,}00)\,(0.1627) = \$338{,}500$$

For the project without pollution prevention, calculate the total capital cost, TCC(w/o), as follows:

$$TCC(w/o) = \$1,081,000 + \$659,000 = \$1,740,000$$

Calculate the annualized capital cost, ACC(w/o), for the option without pollution prevention measures as follows noting that the CRF for this option is identical to that calculated for the option with pollution prevention options:

$$ACC(w/o) = (\$1,740,000)(0.1627) = \$283,200$$

The better choice for the project can then be chosen based on the following summary of costs:

	Project with Pollution Prevention	Project w/o Pollution Prevention
Annual capital cost	$338,500/yr	$283,200/yr
Operating Labor	$39,900/yr	$8500/yr
Maintenance	$43,000/yr	$17,000/yr
Utilities	$958,000/yr	$821,000/yr
Overhead	$51,300/yr	$13,900/yr
Taxes, insurance and administration	$86,200/yr	$72,600/yr
Credits	-$380,000/yr	$0
Total annual cost	$1,136,900/yr	$1,216,200/yr

The project with pollution prevention is the better choice on a total annual cost basis by $79,300/yr.

11. SUMMARY

1. Pollution prevention has been recognized as one of the lowest-cost options for waste management. Hence, an understanding of the economics involved in pollution prevention options is quite important in making decisions at both the engineering and management levels. Every engineer and/or scientist should be able to execute an economic evaluation of a proposed project. If the project is not profitable it should obviously not be pursued, and the earlier such a project can be identified, the fewer are the resources that will be wasted during the evaluation process.

2. Besides the traditional cost factors involved in pollution prevention projects, there are also other important costs and benefits associated with pollution prevention that need to be quantified if a meaningful economic analysis of a pollution prevention project is going to be performed. A *Total Systems Approach (TSA)* or *life cycle analysis* attempts to quantify not only the economic aspects of pollution prevention but also the social costs associated with the production of a product or service from cradle to grave, and the less tangible benefits that result from the reduced risk derived from not using a hazardous substance.

3. *Environmental accounting* is a relatively new term that applies to the modification of standard accounting practices to attempt to capture the true

environmental costs of waste production, and the benefits that result from pollution prevention efforts. In a *management accounting* context, environmental accounting refers to the use of data about environmental costs and performance in business decisions and operations.

4. *Environmental costs* can be classified in many ways, and include *conventional costs* (raw materials, utilities, capital goods, etc.), *potentially hidden costs* (siting, qualification of suppliers, regulatory, decommissioning, costs for future compliance), *contingent costs* (remediation and compensation for future releases, fines for future violations, consequences of permitted releases, etc.), and *image/relationship costs* (annual environmental reports, community relations costs, voluntary environmental activities costs, etc.).

5. The emphasis of environmental accounting is to quantify these various non-conventional costs and allocate them to each process/product option as appropriate so that the true cost and impact of management and engineering decisions can reflect the environmental burden that each option represents. This proper allocation of cost should be carried out in a company's cost allocation process as well as in its capital budgeting, and process/product design activities.

6. Before the cost of a pollution prevention program or option can be evaluated, the factors contributing to the cost must be recognized. There are two major contributing factors: *capital costs* and *operating costs*. Once the total costs of the "change" have been estimated, the engineer must determine whether or not the project will be profitable. This involves converting all cost contributions to an annualized rate. If more than one project proposal is under study, this method provides a basis for comparing alternate proposals and for choosing the best proposal.

7. An economic evaluation is carried out using standard measures of profitability. Each company and organization has its own economic criteria for selecting projects for implementation. In performing an economic evaluation, various costs and savings must be considered. The economic analysis described in this chapter represents a preliminary, rather than detailed, analysis. For smaller facilities with only a few (and perhaps simple) processes, the entire pollution prevention assessment procedure will tend to be informal.

8. *Equipment (capital) cost* is a function of many variables, one of the most significant of which is equipment capacity. Other important variables include operating temperature and/or pressure conditions and degree of equipment sophistication. Preliminary estimates are often made from simple cost-capacity relationships that are valid when the other variables are confined to narrow ranges of values. The Chemical Engineering Fabricated Equipment Cost Index (FECI) is a commonly used index that allows earlier cost data to be adjusted to present values using factors that correct for inflation.

9. *Operating costs* can vary from site to site since these costs, in part, reflect local conditions (e.g., staffing practices, labor, and utility costs). Operating costs, like capital costs, may be separated into two categories: direct and indirect. Direct costs are those that cover material and labor and are directly involved in operating the facility. These include labor, materials,

maintenance and maintenance supplies, replacement parts, waste (i.e., residues after incineration) disposal fees, utilities, and laboratory costs. Indirect costs are those operating costs associated with but not directly involved in operating the facility; costs such as overhead (e.g., building-land leasing and office supplies), administrative fees, local property taxes, and insurance fees fall into this category.

10. Once a particular process pollution prevention scheme has been selected, it is common practice to optimize the process from a capital cost and operation and maintenance (O&M) standpoint. There are many optimization procedures available, most of them too detailed for meaningful application to a waste-generating facility. These sophisticated optimization techniques, some of which are routinely used in the design of conventional chemical and petrochemical plants, invariably involve computer calculations. Use of these techniques in pollution prevention analyses is usually not warranted, however.

11. The proper location of a plant is as important to its success as the selection of a process. Not only must many tangible factors such as labor supply and raw material sources be carefully considered, but also a number of intangible factors, which are more difficult to evaluate. The selection of a plant site must be based on a very detailed study in which all factors are weighed as carefully as possible. Such a study often requires a substantial outlay of capital, but false economies at this point may lead to great losses in the future.

12. EPA's Design for the Environment (DfE) program began in 1991 to help businesses incorporate environmental considerations into the design and redesign of products, processes, and technical and management systems. More information on the DfE program and tools for use in various product and process design efforts aimed at pollution prevention and waste minimization can be obtained on the Internet through the U.S. EPA web site at http://www.epa.gov/opptintr/dfe.

12. PROBLEMS

1. Discuss the two major factors contributing to the cost of a pollution prevention program.

2. Define the various methods of analyses that are employed in calculating depreciation allowances.

3. An engineer or scientist must often decide if it is truly worthwhile to invest in new equipment. One calculation that is often used to make this decision is called the percent rate of return on investment which is given by:

$$\text{Percent rate of return} = \frac{\text{Annual Profit}}{\text{Initial Investment Cost}} \times 100$$

Calculate the percent rate of return on investment for the plant manager's $10,000 investment described in Illustrative Example 9.1.

4. From an economic point of view, the break-even point of a process operation is defined as that condition when the costs (C) exactly balance the

income (I). The profit (P) is therefore, $P = I - C$. At break-even, the profit is zero. The cost and income (in $) for a particular operation are given by the following equations:

$$I = \$60,000 + 0.021 \, N$$
$$C = \$78,000 + 0.008 \, N$$

where N is the yearly production rate of the item being manufactured. Calculate the break-even point for this operation.

5. An engineer's total equipment cost on a given project is $1,049,600 in 1999. If the total installation cost of the equipment for an engineer's project is 60% of the total equipment cost, what is the annualized capital cost for this project? Assume an annual rate of return of 10% and an operating period of 10 years.

6. Three different control devices are available for the removal of a hazardous contaminant from a process stream. The service life is 10 years for each device. Capital and annual operating costs are as follows:

Device	Initial Cost	Annual Operating Cost	Salvage value in Year 10
A	$300,000	$50,000	$0
B	$400,000	$35,000	$0
C	$450,000	$25,000	$0

Which device is the most economical assuming straight-line depreciation?

13. REFERENCES

1. Theodore, L. and Kneus, K., *A Theodore Tutorial, Engineering Economics and Finance*, Theodore Tutorials, East Williston, New York, 1996.
2. Perry, R. H and Green, D., *Perry's Chemical Engineers' Handbook*, 7th edition, McGraw-Hill Publishing Co., New York, New York, 1998.
3. U.S. EPA, *An Introduction to Environmental Accounting As A Business Management Tool: Key Concepts And Terms*, Office of Pollution Prevention And Toxics (MC 7409), Washington, D.C., EPA 742-R-95-001, 1995.
4. U.S. EPA, *Stakeholder's Action Agenda: A Report of the Workshop on Accounting and Capital Budgeting for Environmental Costs,* December 5-7, 1993, Office of Research and Development, Washington, D.C., EPA 742-R-94-003, 1994.
5. Owen, Jean V., Environmental Compliance: Managing the Mandates, *Manufacturing Engineering* (3), 1995.
6. White, A. and Becker, M., Total Cost Assessment: Catalyzing Corporate Self Interest in Pollution Prevention," *New Solutions,* (Winter):34, 1992.
7. U.S. EPA, *Total Cost Assessment: Accelerating Industrial Pollution Prevention through Innovative Project Financial Analysis, With Applications to the Pulp and Paper Industry*, Office of Pollution

Prevention And Toxics (MC 7409), Washington, D.C., EPA-741-R-92-002, 1992.
8. U.S. EPA, *Life Cycle Design Guidance Manual: Environmental Requirements and the Product System*, Office of Pollution Prevention And Toxics (MC 7409), Washington, D.C., EPA-600-R-92-226, 1993.
9. Thomas, S. T., Weber, V., Berger, S. A., and Klawiter, I. L., *Estimate the Environmental Cost of New Processes in R&D*, Presented at the American Institute of Chemical Engineers (AIChE) Spring National Meeting, April, 1994.
10. U.S. EPA, *Report to Congress*, Office of Solid Waste, Washington, D.C., EPA/530-SW-86-033, 1986.
11. McGraw-Hill Companies, *Chemical Engineering*, New York, New York, 1999.
12. Neveril, R. B., *Capital and Operating Costs of Selected Air Pollution Control Systems*, U.S. EPA, Office of Air Quality, Programs, and Standards, Research Triangle Park, North Carolina, EPA 450/5-80-002, 1978.
13. Vatavuk, W. M. and Neveril, R. B., Factors for Estimating Capital and Operating Costs, *Chemical Engineering*, (November 3):157-162, 1980.
14. Vogel, G. A. and Martin, E. J., Hazardous Waste Incineration. Part 1. Equipment Sizes and Integrated-Facility Costs, *Chemical Engineering*, (September 5):143-146, 1983.
15. Vogel, G. A. and Martin, E. J., Hazardous Waste Incineration. Part 2. Estimating Costs of Equipment and Accessories, *Chemical Engineering*, (October 17):75-78, 1983.
16. Vogel, G. A. and Martin, E. J., Hazardous Waste Incineration. Part 3. Estimating Capital Costs of Facility Components, *Chemical Engineering*, (November 28):87-90, 1983.
17. Ulrich, G. D., *A Guide to Chemical Engineering Process Design and Economics*, Wiley-Interscience, New York, New York, 1984.
18. Vogel, G. A. and Martin, E. J., Hazardous Waste Incineration. Part 4. Estimating Operating Costs, *Chemical Engineering*, (January 9):97-100, 1984.
19. DeGarmo, E. P., Canada, J. R., and Sullivan, W. G., *Engineering Economics*, 6th edition, Macmillan, New York, New York, 1979.
20. Hodginan, C., Selby, S., and Weast, R.C., ed., *CRC Standard Mathematical Tables*, 12th Edition. The Chemical Rubber Company, Cleveland, OH, 1961.
21. ICF Technology Incorporated, *New York State Waste Reduction Guidance Manual*, ICF Technology Incorporated, Alexandria, Virginia, 1989.
22. Theodore, L., Reynolds, J., and Taylor, F., *Accident and Emergency Management*, Wiley-Interscience, New York, New York, 1989.

PART 2
POLLUTION PREVENTION PRINCIPLES

As discussed earlier in Part 1 of this text, pollution prevention has been defined by the U.S. Environmental Protection Agency (EPA) to include both *source reduction* and *closed-loop recycling*. These two approaches represent the major components of pollution prevention and the highest priority options in EPA's hierarchy of integrated waste management options. Treatment has been included in EPA's integrated waste management hierarchy as the next option for handling wastes after all source reduction and recycling efforts have been employed. Recognizing that not all waste will be eliminated or treated, ultimate disposal remains the last option in the waste management hierarchy. Additional aspects of pollution prevention not generally considered in a traditional waste management hierarchy that are relevant to current and future practices and that are covered in this part of the text include related areas of energy conservation; and health, safety, and accident management.

Part 2 of this text is devoted to specific pollution prevention principles and details of available pollution prevention options. An introduction to the concept of pollution prevention, EPA's pollution prevention strategy, pollution prevention regulations, and Federal, state and local pollution prevention initiatives are reviewed in Chapter 10. Energy conservation, accident and emergency management, health risk assessments, multimedia approaches, life-cycle analysis, sustainable development, ISO 14000, environmental justice, and ethical considerations are presented in Chapter 11. Before a discussion of actual methods for minimizing the generation of wastes is given, the reader is instructed on how to identify specific areas within a plant that may have potential for pollution prevention and how to assess the technical and economic feasibility of implementation. This process of identification, defined as the *pollution prevention assessment*, is discussed in Chapter 12. The chapters that follow in Part 2 contain material on strategies for applying specific pollution prevention options. *Source reduction* techniques, including product changes and source control measures, are discussed in Chapter 13, while recycling options and technologies are highlighted in Chapter 14. Specific industrial applications of pollution prevention are presented along with industrial-specific case studies in Part 3.

Pollution prevention clearly has become a top priority for both industries and regulators. As the commitment to pollution prevention increases, the problems created by the generation of wastes will begin to lessen. For industry, this means that pollution prevention must be an integral part of a company's overall operational strategy. Industry must also view pollution prevention as a continuous process of searching for new areas to assess, investigating new methods of waste reduction and recycling, and analyzing and implementing these methods, before treatment and disposal are needed.

CHAPTER 10

From Pollution Control to Pollution Prevention

1. OVERVIEW

Traditionally, environmental protection efforts have emphasized control of pollution after it has been generated (Table 10.1). Although this approach may, in many circumstances, be effective in protecting human health and the environment, this method of waste management has certain disadvantages. Specifically, this type of pollution control does not always solve the problem of pollution; rather, it often transfers pollution from one medium to another, resulting in no net environmental benefit. In addition, management of waste after it is generated requires investment in pollution control equipment and expenditures of materials and energy that would not be required if the waste is not generated. The underlying principle of pollution prevention is based on limiting the amount of waste produced "up-front," rather than developing extensive treatment processes "downstream" to ensure that the waste poses no threat to human health or to the environment. Because of the growing appreciation of the benefits of waste avoidance, the overall approach to waste management in the United States has begun to shift from pollution control driven activities to pollution prevention activities as suggested in Table 10.1.

Table 10.1. Evolution of Waste Management Approaches in the United States.

Prior to 1945	No control
1945 - 1960	Little control
1960 - 1970	Some control
1970 - 1975	Greater control, U.S. EPA established
1975 - 1980	More sophisticated control
1980 - 1985	Beginnings of waste-reduction management
1985 - 1990	Waste-reduction management
1990 - 1995	Formal pollution prevention programs (Pollution Prevention Act of 1990)
1995 - 2000	Widespread acceptance of pollution prevention
>2000	Green manufacturing, sustainable development, design for the environment, ...

The concept of pollution prevention was first defined as "waste minimization" and was introduced when the U.S. Congress specifically stated the following in the *Hazardous and Solid Waste Amendments of 1984* to the *Resource Conservation and Recovery Act (RCRA)*:

> The Congress hereby declares it to be the national policy of the United States that, wherever feasible, the generation of hazardous waste is to be reduced or eliminated as expeditiously as possible. Waste that is nevertheless generated should be treated, stored, or disposed of so as to minimize the present and future threat to human health and the environment. According to the United States Environmental Protection Agency (EPA) 250 million tons of solid waste are generated annually in the U.S. Thus far the majority of the environmental protection efforts are centered around treatment and control of pollution.

In its 1986 report to Congress (EPA/530/SW86-033), EPA detailed the concept of waste minimization as:

> The reduction to the extent feasible, of hazardous waste that is generated or subsequently treated, stored or disposed of. It includes any source reduction or recycling activity undertaken by a generator that results in either (1) the reduction of total volume or quantity of hazardous waste, or (2) the reduction of toxicity of hazardous waste, or both, so long as such reduction is consistent with the goal of minimizing present and future threats to human health and the environment.

The most important piece of legislation enacted to date, however, is the *Pollution Prevention Act of 1990*. The Pollution Prevention Act, signed into law in November 1990, established pollution prevention as a "national objective." The Act notes that:

> There are significant opportunities for industry to reduce or prevent pollution at the source through cost-effective changes in production, operation, and raw materials use.... The opportunities for source reduction are often not realized because existing regulations, and the industrial resources they require for compliance, focus upon treatment and disposal, rather than source reduction.... Source reduction is fundamentally different and more desirable than waste management and pollution control.
> The Act establishes the pollution prevention hierarchy as national policy, declaring that pollution should be prevented or reduced at the source wherever feasible, while pollution that cannot be prevented should be recycled in an environmentally safe manner. In the absence of feasible prevention or recycling opportunities, pollution should be treated; disposal or other release into the environment should be used as a last resort.

Source reduction is defined in the law to mean any practice which reduces the amount of any hazardous substances, pollutant or contaminant entering any waste stream or otherwise released into the environment (including fugitive emissions) prior to recycling, treatment or disposal; and which reduces the hazards to public health and the environment associated with the release of such substances, pollutants, or contaminants.

In the current working definition used by the EPA, source reduction and closed-loop recycling are considered to be the most viable pollution prevention techniques, preceding treatment and disposal. In its original "Pollution Prevention Policy Statement" published in the January 26, 1989, *Federal Register*, EPA encouraged organizations, facilities, and individuals to fully utilize source reduction and recycling practices and procedures to reduce risk to public health, safety, and the environment.

Figure 10.1 depicts the EPA hierarchy of preferred approaches to integrated waste management, and ultimately pollution prevention. As one proceeds down the hierarchy, more and more waste is potentially generated, and more energy is expended in the management of this waste. As illustrated by Figure 10.1, integrated waste management may be broken down into the following four major components:

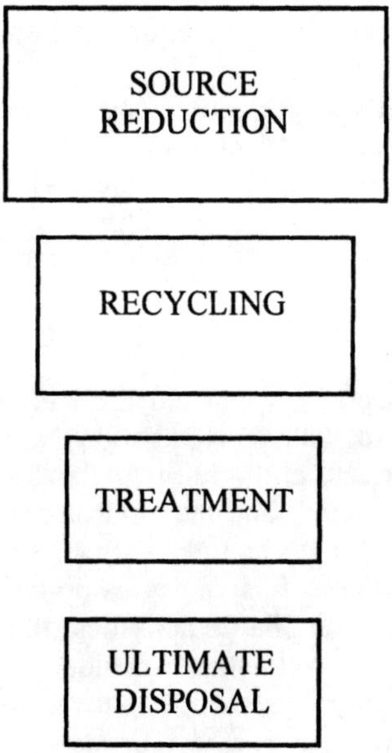

```
┌─────────────────────┐
│     SOURCE          │
│     REDUCTION       │
└─────────────────────┘

    ┌───────────────┐
    │   RECYCLING   │
    └───────────────┘

      ┌───────────┐
      │ TREATMENT │
      └───────────┘

        ┌─────────┐
        │ULTIMATE │
        │DISPOSAL │
        └─────────┘
```

Figure 10.1. U.S. EPA's integrated waste management and pollution prevention hierarchy.

1. *Source reduction*, consisting of technologies to reduce the amount of wastes initially generated, is the primary approach emphasized by EPA for waste management and pollution prevention. The techniques involved are applied to the production process prior to the point of waste generation. Methods that eliminate or reduce the amount of waste generated by a particular process, either through process modification, equipment alteration, or material substitution are part of source reduction.
2. *Recycling*, the secondary approach, attempts to recover a usable material from a waste stream. The methods involved can take place within (closed-loop) or at the end of the process and can be implemented either on- or off-site.
3. For the waste remaining after all possible source reduction and recycling techniques have been employed, the next approach in the waste management hierarchy is the use of physical, biological, and chemical *treatment* methods, including incineration. This results in a reduction of the toxicity and volume of waste requiring ultimate disposal.
4. The last approach for managing wastes is *ultimate disposal* as defined by the U.S. EPA, consisting of landfilling, landfarming, deep-well injection, and ocean dumping.

A more detailed flow diagram, providing a more in-depth evaluation of waste management and pollution prevention options, is provided in Figure 10.2.

The extension of pollution prevention activities to include not only waste reduction but also energy conservation and accident and emergency management is provided in Chapter 11, along with a host of other related pollution prevention topics. Source reduction is discussed in Chapter 13, while recycling approaches are highlighted in Chapter 14.

2. FEDERAL REGULATIONS

In the mid-1970s, Congress enacted the *Resource Conservation and Recovery Act (RCRA)* to regulate the generation and disposal of waste. Since then, numerous amendments and additional laws have been passed in an attempt to protect public health and the environment. Pollution prevention practices may enable a firm to comply with these regulations without an investment in control devices. In some cases pollution prevention measures may be required as part of compliance activities stipulated in the regulations. The most important of these regulations include: *the Clean Air Act (CAA); the Clean Water Act (CWA); the Comprehensive Environmental Response, Compensation, and Liability Act (CERCLA); the Superfund Amendments and Reauthorization Act (SARA); the Hazardous and Solid Waste Amendments (HSWA); the Safe Drinking Water Act (SDWA); the Toxic Substance Control Act (TSCA); the Pollution Prevention Act; and the Emergency Planning and Community Right-to-Know Act (EPCRA).*

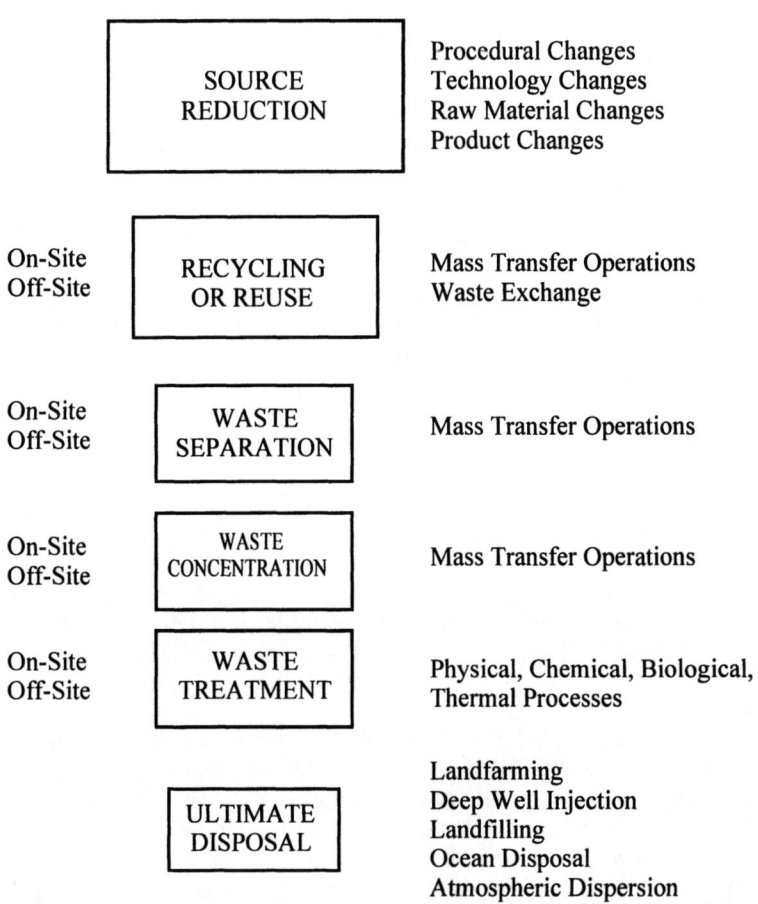

Figure 10.2. Pollution prevention and waste management options.

Prior to the passage of the Pollution Prevention Act of 1990, the main regulatory driver for pollution prevention was RCRA, as amended by HSWA. As indicated in its name, the Resource Conservation and Recovery Act is intended to restrict the improper disposal of more than 400 discarded commercial chemical products and specific chemical constituents of industrial waste steams destined for treatment or disposal on land so that waste reduction and recycling become attractive options for waste management. RCRA was revised by the HSWA in 1984. One of the most important provisions of HSWA related to pollution prevention was the restriction of the land disposal of RCRA hazardous wastes in order to minimize the potential for future risk to human health and the environment. This provision, called the *land disposal restriction (LDRs)*, or "land ban," directed EPA to establish treatment standards for each of seven groups of RCRA hazardous wastes by specific dates. The intent was to restrict the land disposal of all RCRA hazardous wastes by these statutory deadlines. The standards for the sixth group of wastes, which comprise the final third of the listed RCRA wastes, were promulgated on May 8, 1990. The seventh and last waste category, used for newly identified wastes, has an associated statutory deadline within 6 months of identification of these new wastes as hazardous. These LDRs provide a significant incentive for industry to minimize its waste production, specifically through source reduction.

In addition to the LDRs, EPA has implemented the RCRA waste minimization mandate by establishing waste *manifests* and requiring reporting each time a hazardous waste shipment is sent to an off-site storage, treatment, or disposal facility. When a hazardous waste is created, the generator must sign a certification or manifest. Additionally, a generator must complete a biennial report that describes the efforts made at reducing the volume and toxicity of wastes they generated, and that compares the current year's volume and toxicity of wastes to those of previous years.

The *Clean Water Act* includes *National Pollutant Discharge Elimination System* (NPDES) *permits, effluent guidelines*, and zero discharge goals designed to control releases to surface waters of 126 individual chemicals released into the water. This pollutant list includes volatile organic substances such as benzene, chloroform, and vinyl chloride; acid compounds such as phenol and its derivatives; pesticides such as chlordane, DDT, and toxaphene; heavy metals such as lead and mercury; PCBs; and other organic and inorganic compounds.

The *Clean Air Act* includes *National Emission Standards for Hazardous Air Pollutants (NESHAPs)* designed to control emissions of six specific chemicals (asbestos, beryllium, mercury, vinyl chloride, benzene, and arsenic) and one generic category (radionuclides) released into the air from specific industrial sources. Amendments to the Clean Air Act in 1990 extended the control of air emissions to a more extensive list of hazardous air pollutants *(HAPs)*, and established requirements for development of *Risk Management Plans* that provide documentation of the past chemical accident and release history from a facility, assessment of the impact of the uncontrolled releases of selected toxic and explosive materials from manufacturing facilities on surrounding communities, and mitigation and control of impacts of these releases if they exceed acceptable risk levels.

The *Emergency Planning and Right to Know Act (EPCRA)*, another name for the *Superfund Amendments and Reauthorization Act (SARA) Title III*, includes the *Toxics Release Inventory (TRI)* reporting requirement by facilities. Section 313 of SARA applies to more than 320 chemicals and chemical categories released into all environmental media. TRI reports help the public understand the volume and nature of toxic pollutants legally released into the environment on a on-going basis. Section 302 of SARA regulates *Extremely Hazardous Substances,* and applies to more than 360 chemicals for which facilities are required to prepare emergency management plans if these chemicals are present at the facility above certain threshold quantities. Releases of these chemicals to the air, land, or water trigger required reporting by the facility to the State Emergency Response Committee

(SERC) and the Local Emergency Planning Committee (LEPC) under SARA Section 304. As indicated above, EPCRA provides the platform and mandate for public involvement in the management of hazardous materials within their communities, and gives the public a role in influencing future regulations and industrial performance.

While these various regulations have been designed to place increasingly strict controls on the production and management of hazardous wastes, and the handling and use of hazardous materials in the manufacturing process, it was the Pollution Prevention Act that clearly defined and set a unified strategy for pollution prevention practices and activities to be carried out in the U.S. This Act, EPA's original pollution prevention strategy, and laws specifically addressing hazardous and toxic chemicals are discussed below.

2.1. The Pollution Prevention Act of 1990

The most important federal regulation regarding pollution prevention in the U.S. is the Pollution Prevention Act of 1990. This Act was signed into law in November, 1990, and established pollution prevention as a "national objective." The 1990 Act established a waste management and pollution prevention hierarchy as national policy, declaring that pollution should be prevented or reduced at the source wherever feasible, while pollution that cannot be prevented should be recycled in an environmental safe manner. In the absence of feasible prevention or recycling opportunities, pollution should be treated; disposal or other releases(s) into the environment should be used as a last resort. The Pollution Prevention Act also formalized the establishment of EPA's Office of Pollution Prevention, independent of the single-medium programs, to carry out the functions required by the Act and to develop and implement a strategy to promote source reduction. Among other provisions, the law directs EPA to:

1. Facilitate the adoption of source reduction techniques by businesses and by other federal agencies.
2. Establish standard methods of measurement for source reduction.
3. Review regulations to determine their effect on source reduction.
4. Investigate opportunities to use federal procurement to encourage source reduction.
5. Develop improved methods for providing public access to data collected under federal environmental statutes.
6. Develop a training program on source reduction opportunities, model source reduction auditing procedures, a source reduction clearinghouse, and an annual award program.

Under the Act, facilities required to submit TRI reports to EPA must now also provide information on pollution prevention and recycling for each facility and for each toxic chemical that they manage. The information that must be provided includes the quantities of each toxic chemical entering the waste stream and the percentage change from the previous year, the quantities recycled and percentage change from the previous year, source reduction practices, and changes in production from the previous year. Finally, the Act requires EPA to report to Congress within 18 months (and biennially afterward) on the actions needed to implement the strategy to promote source reduction.

Local governments have also played a significant role in promoting pollution prevention in the industrial, consumer, transportation, agricultural, and public sectors of their communities. Many local governments have already taken the lead in putting successful recycling programs into place. A variety of tools are available to promote prevention. Local governments (including cities, counties, sewer and water agencies, planning departments, and other special districts) can and have provided:

1. Educational programs to raise awareness in businesses and the community of the need to reduce waste and conserve resources.
2. Technical assistance programs that provide on-site help to companies and organizations in reducing pollution at the source.
3. Regulatory programs that promote prevention through mechanisms such as codes, licenses, and permits.
4. Government procurement policies regarding the purchase of recycled products, reusable products, and products designed to be recycled.

In addition, many local governments have passed resolutions and ordinances relating to waste reduction, energy conservation, automobile use, procurement policies, and so on. Such resolutions and ordinances can be useful steps in signaling a public commitment to operate using environmentally sound principles. These actions also help to define environmental performance goals and targets and to delineate the specific responsibilities of different local agencies in waste management and waste reduction efforts.

2.2. EPA's Pollution Prevention Strategy

EPA's *Pollution Prevention Strategy*, released in February 1991, was developed by the Agency in consultation with all program and regional offices. The strategy provides guidance on incorporating pollution prevention principles into EPA's ongoing environmental protection efforts and includes a plan for achieving substantial voluntary reductions of targeted high risk industrial chemicals. The strategy is aimed at maximizing private sector initiative while challenging industry to achieve ambitious prevention goals. A major component of the strategy has been the *Industrial Toxics Project*, known more generally as the *33/50 Program*. The overall goals of the 33/50 program were to reduce national pollution releases and off-site transfers of priority toxic chemicals and to encourage pollution prevention. The specific program goal was to voluntarily reduce releases and transfers of 17 toxic chemicals by 33 percent by the end of 1992 and 50 percent by the end of 1995. Both goals were reached 1 year ahead of schedule.

The 17 pollutants identified as targets of the industrial toxics project presented both significant risks to human health and the environment and significant opportunities to reduce such risks through prevention. The list (Table 10.2) was drawn from recommendations submitted by program offices, taking into account such criteria as health and ecological risk, potential for multiple exposures or cross-media contamination, technical or economic opportunities for pollution prevention, and limitations for their effective treatment using conventional technologies.

All of the targeted chemicals are included on EPA's Toxic Release Inventory (TRI); thus, reductions in their releases can be measured in each year's TRI reports. Despite company participation being strictly voluntary, almost 1,300 parent companies, operating more than 6,000 facilities in the U.S., have participated in the 33/50 program. Companies examined their own industrial processes and identified and implemented cost-effective pollution prevention practices for toxic chemicals. Information on a variety of "success stories" is available on EPA's web site at: http://www.epa.gov/opptintr/3350/. These success stories detail not only technical innovations, but new managerial and cost-accounting approaches that have led to significant reductions in hazardous materials releases by participating companies.

Table 10.2. The 33/50 Program Target Chemicals.

Benzene	Methyl Ethyl Ketone (MEK)
Cadmium	Methyl Isobutyl Ketone (MIBK)
Carbon Tetrachloride	Nickel
Chloroform	Tetrachloroethylene (PCE)
Chromium	Toluene
Cyanide	1,1,1-Trichloroethane (TCA)
Dichloromethane	Trichloroethylene (TCE)
Lead	Xylene
Mercury	

EPA's pollution prevention strategy also provides guidance on incorporating pollution prevention into the Agency's existing program, emphasizing the need for continued strong regulatory and enforcement programs. At the same time, the strategy favors flexible, cost-effective approaches that involve market-based incentives where practical. For example, the strategy calls for the use of "regulatory clusters," through which EPA will categorize the rules it intends to propose over the next several years for certain chemicals and their sources. The clusters are intended to foster improved cross-media evaluation of the cumulative impact of standards, provide more certainty for industry, and encourage early investment in pollution prevention activities.

The strategy outlines several short-term measures that will address various institutional barriers within the Agency's own organization that limit its ability to develop effective prevention strategies. Such measures include designating special assistants for pollution prevention in each assistant administrator's office, developing incentives and awards to encourage Agency staff to engage in pollution prevention efforts, incorporating prevention into the comprehensive 4-year strategic plans by each program office, and providing pollution prevention training to Agency staff.

The 33/50 Program for the manufacturing sector represents the first focus of a comprehensive Agency strategy. After the completion of this project in 1995, EPA staff of the program focused on dissemination of results from the program in the form of: company profiles that summarize successful programs at industrial facilities; a generic waste reduction manual, the Facility Pollution Prevention Guide; numerous industry-specific Guides to Pollution Prevention; supplementary information in reports, handbooks, reference manuals, bibliographic reports, and videos; the development of seven Manufacturing Technology Centers across the nation for research and training purposes; state programs for grants and awards; and several hot lines described later in the text. One current initiative is the *Design for Environment (DfE) Program* which is a cooperative, voluntary effort by EPA, industry, professional organizations, state and local governments, other federal agencies (including the Small Business Administration), and the public aimed at developing specific pollution prevention information in a number of small business-dominated industries including printing (screen printing and lithography), dry cleaning, and electronics/printed wiring boards. Specifically, the DfE Program uses its industry projects to bring together comparative information on the environmental and human health risks, exposures, performance and costs of alternative products and technologies so that small businesses can make more informed environmental management decisions, realizing that small businesses often do not have the resources or technical expertise to develop this kind of technical information on their own.

EPA continues to seek to work with the Departments of Agriculture, Energy, and Transportation to develop strategies for preventing pollution from agricultural practices and energy and transportation uses. EPA has already begun several joint initiatives, including a cooperative grants program for sustainable agriculture research with the Department of Agriculture and a joint program with the Department of Energy to demonstrate energy efficiency and waste reduction in key sectors that has led to the *Industries 2000 Project*.

Another important goal of EPA's Pollution Prevention Program is to ensure that pollution prevention training and education are available to government, industry, academic institutions, and the general public. Training and education are needed to help institutionalize prevention as the strategy of choice in all environmental decision-making activities. The U.S. EPA created the *National Pollution Prevention Center for Higher Education (NPPC)* in 1991 to collect, develop, and disseminate educational materials on pollution prevention; the University of Michigan was selected as the site. The NPPC represents a collaborative effort between business and industry, government, non-profit organizations, and academia and has as its mission to: promote sustainable development by educating students, faculty, and professionals about pollution prevention; create educational materials; provide tools and strategies for addressing relevant environmental problems; and establish a national network of pollution prevention educators. In addition to developing educational materials and conducting research, the NPPC also offers an internship program, professional education and training, and conferences. The NPPC web site address is: http://www.umich.edu/~nppcpub/.

The *Pollution Prevention Information Clearinghouse (PPIC)* is a multimedia clearinghouse of technical, policy, programmatic, legislative, and financial information dedicated to promoting pollution prevention through efficient information transfer. The Clearinghouse is operated by EPA's Office of Pollution Prevention and Toxics, and can be reached on-line at: http://www.epa.gov/opptintr/library/libppic.htm. The Clearinghouse serves as a repository for hard copy reference materials and provides on-line information retrieval and ordering system. In addition, the Clearinghouse provides on-line interactive access to a wide range of pollution prevention topics that include: general pollution prevention information, enforcement and compliance issues, environmentally preferable purchasing, pollution prevention incentives for states, design for the environment, environmental accounting, hospitals for a healthy environment, voluntary standards network, education and training, environmental labeling, and information on persistent and bioaccumulative chemicals. Finally, the Clearinghouse provides a national calendar of conferences and workshops relating to pollution prevention, as well as a PPIC technical assistance hot line (202/260-1023) to answer pollution prevention questions, access information in the PPIC, and assist in document searches and ordering.

3. PROGRESS TOWARD POLLUTION PREVENTION[1]

The progress that has been made in pollution prevention has been summarized in a U.S. EPA publication, *Pollution Prevention 1997: A National Progress Report*.[1] This report highlights activity in the pollution prevention area, updating the first such report issued by EPA in 1991. Each of the chapters in this report summarizes pollution prevention activities of a particular sector of society. The document is highlighted by guest commentaries representing organizations as diverse as the National Association of Counties, the Dow Chemical Company, the Department of Defense, and the North Carolina Department of Environment, Health and Natural Resources to provide a diverse perspective on the progress and future challenges of waste management in this country.

The progress in waste reduction and pollution prevention in six sectors is summarized in the report. These six sectors include: EPA, federal agencies, industry, state and tribal entities, educational institutes, and non-profit organizations. A brief overview of these findings is provided below.

3.1. Promoting Pollution Prevention at EPA

The progress in seven theme areas that have provided the focus for the Agency's activities in pollution prevention as identified by the 1993 U.S. EPA Pollution Prevention Policy statement[2] given below:

1. *Incorporating pollution prevention into the mainstream work of EPA.* The Agency has undertaken a concerted effort since 1990 to find the best ways to incorporate prevention into regulations and permitting, through such efforts as the *Source Reduction Review Project* and EPA's *Common Sense Initiative.*
2. *Building a national network of prevention programs.* EPA realizes that it must work with state and local governments to develop a national network of programs that will assist regulators at all levels of government in promoting pollution prevention. EPA is providing funding support, technical assistance and information dissemination, and forming government partnerships at the federal, state, and local levels to focus efforts on pollution prevention as the national goal for environmental management.
3. *Pioneering cross-media prevention programs as new models for government/industry interaction.* Such voluntary programs have been developed by EPA to explore ways to formalize interactions between EPA and industry in pollution prevention efforts. These partnerships include the *33/50 Program*, *Climate Wise*, the *Green Lights Program*, *WasteWise*, *Design for the Environment*, *Project XL*, *Environmental Accounting*, *WAVE (Water Alliances for Voluntary Efficiency)*, and the *Pesticide Environmental Stewardship Program*.
4. *Establishing new federal partnerships.* EPA is working with other federal agencies to promote pollution prevention across the federal government.
5. *Generating environmental information on pollution prevention.* One key to pollution prevention is access to information on pollutant sources, types of pollution generated, and technologies that can help prevent pollution at the source. Programs such as the *Toxics Release Inventory (TRI)* yield information for industry, government, and communities on major types of releases, the industrial and government facilities that are releasing them, and the environmental media into which the pollutants are being released. EPA's *Consumer Labeling Initiative* is examining ways to provide consumers with better environmental information, including improved product labels.
6. *Developing partnerships for technological innovation in pollution prevention.* EPA's partnerships with industry and universities are developing new technologies for future pollution prevention efforts. These partnerships include the *Green Chemistry Challenge* and an *Environmental Leadership Program*, both of which support facilities that have volunteered to demonstrate innovative approaches, and incorporate pollution prevention concepts into their corporate culture.
7. *Changing existing federal laws to encourage pollution prevention as the preferred method for reducing risks to health and the environment.*

3.2. Promoting Pollution Prevention at Other Federal Agencies

The federal government is the largest buyer of goods and services and is the largest property owner in the U.S. In its varied roles as purchaser of products, facility manager, regulator, and policy maker, the federal government is uniquely situated to encourage pollution prevention through the example of its own actions. Federal agencies have become substantially more active in pollution prevention since 1990, under the guidance of legislation and a number of Executive Orders. Some of these activities include:

1. *The U.S. Department of Agriculture's Sustainable Agriculture Research and Education Program.* This program provides competitive grants for research, education, and extension projects in four regions of the country. The projects are focused on reducing farmers' pesticide use, improving their waste management capabilities, and helping them reduce energy consumption.
2. *The U. S. Agency for International Development Environmental Pollution Prevention Project.* This project focuses on pollution prevention programs for urban and industrial waste in developing countries by providing technical assistance for diagnosing problems, training, information dissemination, and assistance in program development.
3. *The Department of Commerce's National Institute of Standards and Technology (NIST) Programs.* NIST is assisting industry in technology development through four programs: the *Advanced Technology Program*, which provides cost-shared grants for high-risk technologies with commercial potential; the *Manufacturing Extension Partnership,* providing manufacturing and energy efficiency and waste management assistance to small and mid-sized companies; collaborative laboratory research with industry; and the *Malcolm Baldrige National Quality Award*, an outreach program honoring quality in manufacturing.
4. *The Department of Defense (DOD).* DOD engages in numerous pollution prevention activities affecting both military installations and weapon systems. As a major user and generator of hazardous substances, DOD has focused on reducing the use of these chemicals in its own facilities and by its suppliers. *Life-cycle assessment* is an integral part of these projects.
5. *The Department of Energy (DOE).* DOE also uses, generates, and releases a large amount of hazardous substances; its recent successes in addressing this problem have earned DOE an "Environmental Champion" award. Each facility is responsible for developing pollution prevention goals and determining the best method for achieving them.
6. *The General Services Administration (GSA).* GSA is one of the largest purchasing units of the government and, with EPA, is piloting several projects to evaluate and distribute information on environmentally preferable products. EPA has a web site

devoted to environmentally preferable purchasing located at the following URL: http://www.epa.gov/opptintr/epp/.
7. *The Department of the Interior (DOI)*. DOI is approaching pollution prevention and waste minimization at the Bureau level. Organizations such as the *National Park Service* and the *U.S. Geological Survey* are proceeding with plans to reduce the amount of toxic materials used, stored, and disposed.
8. *The National Aeronautical and Space Administration's (NASA)*. NASA's pollution prevention strategy has resulted in a significant reduction in releases of TRI reportable substances over the last few years. NASA is using facility-specific plans to promote and implement pollution prevention goals.
9. *The U.S. Postal Service's Waste Minimization/Pollution Prevention Program*. This program has resulted in a 76 percent decrease of solid hazardous waste generation by the Postal Service since 1992. Changes have occurred in the painting of service vehicles, the use of dry cell batteries, recycling of mail trays and pallets, and numerous other areas.
10. *The Department of Transportation (DOT)*. DOT is moving on several fronts to integrate pollution prevention into its activities. Reductions in energy use by encouraging walking and bicycling, use of recycled materials in asphalt, wetlands mitigation, and decreased use of polluting substances are some of DOT's initiatives.
11. *The "Greening of the White House" Project*. Through this project, President Clinton has instituted numerous changes in the operations of the White House to transform it into a model for energy efficiency, waste reduction, and environmental protection.

3.3. Promoting Pollution Prevention in Industry

Although pollution prevention has spread to a wider audience, industry remains at the center of pollution prevention activities as industry is the largest user of energy and hazardous materials, and is subsequently the largest generator of hazardous waste in the U.S. economy. Studies have shown that the economic benefits of waste avoidance can be compelling arguments in favor of pollution prevention, but only when managers are able to clearly identify the cost savings that pollution prevention efforts produce. Environmental accounting is a key factor in demonstrating to businesses the value of waste minimization and pollution prevention.

One starting place for considering progress in pollution prevention in the industrial sector is the TRI maintained by EPA. TRI data, which are collected and published annually, show a steady decline in the volume of toxic chemicals released to the environment by the manufacturing sector. Since 1988, the year TRI reporting was first required, releases of hazardous substances have decreased by 44 percent, although the volume of waste generated has increased. This increase in overall waste generation can be attributable, at least in part, to an improving economy and, therefore, increased production. One of EPA's best-known voluntary programs, the 33/50 Program, had a goal of reducing releases of 17 selected chemicals by 33 percent as of 1992, and 50 percent by 1995. This program achieved both the 1992 and 1995 goals a year ahead of schedule.

Companies that serve as models for pollution prevention responses have common elements, beginning with strong management support and commitment to the pollution prevention programs they initiate. Five large corporations that have been leaders in industrial pollution prevention efforts, and that are highlighted in EPA's 1997 pollution prevention progress report[1] include: the leading chemical manufacturers *Monsanto* and *Union Carbide*; *Public Service Electric and Gas*, a utility that used materials management to yield pollution prevention returns; *AT&T*, which has made innovative use of environmental accounting methods to further its pollution prevention goals; and *Home Depot*, one of the most active retailers promoting a pollution prevention agenda among its clients and staff.

While larger companies frequently have both the financial and technical resources and expertise in-house to implement pollution prevention practices, smaller businesses may have even a greater need for assistance than their larger counterparts. These businesses may find it difficult to identify opportunities for pollution prevention in their processes and products and may also have fewer resources available to implement the changes, whether in equipment, accounting practices, or other areas, than larger firms. Five examples of successful small businesses are described in EPA's pollution prevention progress report,[1] along with information on federal and state programs that are available to assist small businesses, including EPA's Small Business Compliance Assistance Centers, state Small Business Development Centers, and NIST-supported state Manufacturing Extension Partnership centers.

Industry pollution prevention initiatives go beyond changes in manufacturing processes to include product stewardship programs to reach suppliers and customers with a pollution prevention message; working with communities and stakeholders to create more sustainable products and expand market share; and selling "green" or environmentally preferable products. Innovative ideas and technologies in pollution prevention, ranging from new soldering process for circuit boards to using ultraviolet light to coat beer cans (thereby eliminating emissions of volatile organic compounds), are some of the approaches being used by industry to improve their environmental and manufacturing performance, and reduce their compliance burdens and environmental liabilities.

3.3.1. Waste Reduction Always Pays (WRAP)

Several *Chemical Process Industry (CPI)* companies have compiled an outstanding record in minimizing waste generated at their facilities. One such company is Dow Chemical. Although there is no attempt here to endorse (or condemn) any one company's waste management program, this example is included only to provide a brief outline of what the authors of this book believe to be a successful program.[3]

Dow's *Waste Reduction Always Pays (WRAP)* program takes a balanced approach to reducing the environmental impact of operations with a focused effort on source reduction and recycling activities. The environmental policy places a priority on waste and emission reduction, and environmental guidelines support the hierarchy of waste management, that is, source reduction, recycling, treatment, and land disposal as a last option. When efforts to reduce waste at the source or to recycle a waste stream have not been successful, Dow still has the environmental responsibility to manage its waste stream. The primary treatment methods used for Dupont's wastes are state-of-the-art incineration and biological treatment where feasible (see Chapter 7 for details on these treatment methods).

Dow has five operating divisions in the United States. Each division has an individual contact who is responsible for the development and implementation of the WRAP program in a manner that fits the particular division's needs. There is also an individual who is responsible for the overall WRAP effort in the United States. The division contacts design their activities around the following broad waste management goals:
1. Reduce waste releases to the environment.
2. Give recognition for excellence in pollution prevention efforts.
3. Develop a pollution prevention mentality.
4. Provide support for pollution prevention projects.
5. Measure and track the progress of pollution prevention efforts.
6. Strive for continuous improvement.
7. Reduce long-term costs.

The following iterative process is utilized by operating facilities to implement the WRAP program at Dow:
1. Inventory all process losses to air, water, and land.
2. Identify the sources of those losses.
3. Prioritize emission reduction efforts.
4. Allocate resources and implement emission reduction projects.
5. Document and report progress toward emission reduction.
6. Communicate progress internally and externally.
7. Plan for future reductions.

The process of identifying waste streams and determining how those streams are produced in a facility has been best accomplished by one or more brainstorming sessions which utilize the expertise of plant engineers and operations personnel. Prioritization of pollution prevention efforts involves considering several criteria which include the volume and toxicity of individual waste streams, and cost-effective projects are always sought out. In order to reduce waste, projects have to be implemented, and the implementation of projects has required the proper allocation of resources. The development by Dupont of a mechanism for obtaining necessary manpower and capital to engage in pollution prevention efforts has played a key role in completing projects. By documenting and communicating accomplishments, creative pollution prevention ideas are shared and waste reduction progress is verified within the WRAP program. This documentation also builds a base of information that Dow can use to communicate its pollution prevention activities to surrounding communities and regulatory agencies. Finally, periodic reviews of waste inventories and a reevaluation of the prioritization process allows for adjustments and flexibility within a facility's pollution prevention action plan.

Waste reduction projects are recognized at an annual WRAP Outstanding Achievement Awards luncheon. The U.S. area president presents an award to an outstanding waste reduction project from each division and to one outstanding project from a product department. The award is presented to a representative (normally an engineer or an operations technician who played a major part in the project) of the team that completed each project.

The following are some of the ideas that have been implemented to reduce waste at Dow's facilities:
- *Raw Materials* - Improved the quality of feed streams, utilized inhibitors, switched to reusable containers or bulk shipments.
- *Reactors* - Improved physical mixing in reactors, ensured better distribution of feed streams, improved reactant injection, utilized alternate catalysts, provided for separate reactor recycle streams, improved heating and/or cooling techniques, implemented different reactor designs.
- *Heat Exchangers* - Used lower pressure steam, desuperheated plant steam, installed a thermocompressor, used staged heating, utilized on-line cleaning techniques, improved the monitoring of exchanger fouling.
- *Pumps* - Recovered seal flushes and purges, installed sealless pumps.
- *Process Control* - Improved on-line control; automated start-ups, shutdowns, and product changeovers; optimized daily operations.

In summary, Dow's WRAP program has utilized ongoing efforts involving quality initiatives, environmental initiatives, and even capital contests as vehicles to promote and expand its waste reduction activities. In the final analysis, however, waste reduction is accomplished only through the completion of pollution prevention projects. Providing the mechanisms, support, and capital required to implement pollution prevention projects has been an important management responsibility and a critical component of the success of Dow's WRAP program.

3.3.2. Chemical Manufacturers Association (CMA) Activities[4]

The *Chemical Manufacturers Association (CMA)* is a nonprofit trade association whose member companies represent more than 90 percent of the productive capacity for basic industrial chemicals produced in the U.S. Since 1986 the CMA has implemented several formal programs to encourage and support the chemical industry's ongoing efforts to reduce air, water, and solid waste releases to the environment. In the past, CMA's efforts have focused on three separate, but connected, areas: waste minimization (loosely termed pollution prevention), air quality, and SARA Title III emergency management planning. *CMA's Waste Minimization Program*, established in 1986, has four programmatic elements – industry support (workshops, resource manuals, etc.), communications, measurement and tracking, and advocacy and outreach. *CMA's Air Quality Control Policy* of 1986 covers both process emissions and accidental releases. For process emissions, the policy encouraged companies to implement four steps: develop an inventory of existing emissions of toxic air pollutants; assess the impact of these emissions on employees and the surrounding community and determine the adequacy of the current control technology; reduce these emissions as needed to safeguard employees, public health, and the environment; and communicate the results of these actions to appropriate communities and governmental agencies. CMA's Title III program is designed to help assure industry-wide compliance, strengthen the chemical industry's position in communities, and make Title III work at the local level. CMA instituted a massive education and awareness campaign through newsletters, meetings and seminars, booklets, and video-taped materials (some of which are currently used by one of the authors in a graduate course at Manhattan College).

In 1989, these three programmatic elements were consolidated into CMA's *Waste and Release Reduction Program*. This consolidation was affirmed with the adoption of a new policy, the *Chemical Release Reduction Policy*. This policy, which is designed for local implementation, combines the previous policies and programs into a consolidated policy. The goal of the Chemical Release Reduction Policy includes extending reduction efforts to all environmental media and striving for long-term, continuous improvement in emission reductions. The Waste and Release Reduction Program expanded each of the existing program activities and initiated several new activities. The Chemical Release Reduction Policy encourages members to participate in activities to obtain ambient air quality data, to work toward building consensus on the use of scientifically sound methods to assess and evaluate the effects of chemicals on public health, and to provide technical assistance to customers and smaller companies.

The Chemical Release Reduction Policy is designed for voluntary implementation at the local level by each member company. Four broad elements are included in the Policy:

1. Provide inventory releases to all environmental media, starting with the pollutants listed in SARA Section 313. Members are encouraged to include other chemicals in their inventories as appropriate.
2. Establish reduction priorities for the inventoried releases based on factors such as their potential health and environmental impact and the level of community concern.
3. Develop a release reduction plan based on the reduction priorities developed in the previous step. Although reductions may be achieved by many methods, the following order in making the reductions is preferred: source reduction, recycling/reuse, and treatment.
4. Communicate release reduction progress and future plans to employees and the public.

Technical experts of member companies volunteer their time to produce comments, develop resource and educational materials, and develop policies and positions related to environmental controls and pollution prevention efforts in the chemical industry. Under the Waste and Release Reduction Program, including its component activities, CMA has developed a wide variety of guidance materials, some of which are described below:

1. *Waste minimization resource manual* - To assist companies in implementing waste reduction programs, CMA publishes a manual entitled *CMA's Waste Minimization Resource Manual* in order to share industrial experiences. This manual helps companies to expand existing or develop new waste minimization programs.
2. *Regional air monitoring* - To better understand the relationship between releases to the air and the impact on ambient air quality, CMA published a guidance document entitled *Chemicals in the Community: Implementing Regional Air Monitoring Programs*. This document provides a basic, yet comprehensive, guide for planning and implementing regional monitoring programs for toxic air pollutants.
3. *Evaluating ambient atmospheric chemical concentrations* - To identify the types of methods used by industry health professionals, CMA published *Chemicals in the Community: Methods to Evaluate Airborne Chemical Levels*. This document presents some options that companies may consider when developing a formal decision process to select methods for assessing the impact of their chemical emissions on the adjacent community.
4. *Atmospheric fate* - To assist nontechnical and technical readers who have limited knowledge of atmospheric chemistry, CMA published *Chemicals in the Community: Understanding Atmospheric Fate*. This document gives the reader a general foundation and understanding of the terms used in chemical fate assessment and modeling.
5. *Risk communication* - To educate members about providing and explaining information about chemical risks, CMA commissioned several notable experts to write the document *Risk Communication, Risk Statistics, and Risk Comparisons: A Manual for Plant Managers*.
6. *Fugitive emissions* - To assist companies to better estimate releases from equipment, CMA developed the *Plant Organization Software System for Emissions from Equipment (POSSEE)*, published a series of training videotapes, and produced the guidance document *Improving Air Quality: Guidance for Estimating Fugitive Emissions from Equipment*.
7. *Secondary emissions* - To assist companies to better estimate releases from wastewater ponds and lagoons, CMA developed *Programs to Assist Volatile Emissions (PAVE)*, a secondary emissions modeling system and the guidance manual *Improving Air Quality: A Guide to Estimate Secondary Emissions*.

This is only a partial list of some of the CMA documents available to assist companies in implementing the CMA Chemical Release Reduction Policy at their facilities. Other CMA efforts are ongoing.

Under another of CMA's environmental initiatives, *Responsible Care*^{SM}, the chemical industry is committing to continuously improve its performance in health, safety, and protection of the environment. Under the Responsible Care^{SM} initiative, CMA member companies signed 10 guiding principles that form the framework for improving industry's performance. These guiding principles are listed in Table 10.3.

Two aspects of Responsible Care^{SM} make it unique. First, CMA member companies must participate in Responsible Care as an obligation of membership in CMA. Second, through its Public Advisory Panel, the public is directly involved in shaping the Responsible Care initiative.

To implement the guiding principles, CMA has developed a series of implementation codes. One, the *Community Awareness and Emergency Response (CAER) Code*, assures that members:
- Initiate and maintain a community outreach program to communicate openly and respond to public concerns, and
- Have an emergency response program to respond rapidly and effectively to emergencies

The CAER code builds on the success of the voluntary CAER program initiated in 1985 to respond to public concern about accidental chemical releases. Within 2 years, more than 1100 facilities had initiated CAER activities to improve joint community/industry emergency response efforts. Other codes that have been developed include: pollution prevention, distribution, process safety, employee health and safety, and product stewardship.

3.4. Promoting Pollution Prevention at the State and Tribal Level

Across the country, from corporate boardrooms to individual households, great strides have been made in pollution prevention over the past 9 years since the nation's Pollution Prevention Act was signed. Pollution Prevention concepts now permeate all aspects of society, and new ideas for source reduction are being explored and applied to deal with the expanding complexity of the environmental problems. Many state pollution prevention programs have been in the forefront of this innovation, and represent a great deal of this progress. State activities have shifted since 1990, from legislation and program development, to implementation issues dealing with integration of pollution prevention into existing regulatory programs, and to efforts to measure progress in pollution prevention.

For almost a decade, EPA has been providing grants to states to support their pollution prevention efforts. The *Pollution Prevention Incentives for States (PPIS)* grant program (http://www.epa.gov/opptintr/p2home/ppis.htm) provides matching funds to states to support pollution prevention activities and develop state programs. EPA designed the grant program to give the states flexibility to address local needs. Because states have closer, more direct contact with industry and hence are more aware of local needs, EPA believes that state-based environmental programs can make a unique contribu-

Table 10.3. Guiding Principles for Responsible Care^{SM}
A Public Commitment.

Our industry creates products and services that make life better for people around the world - both today and tomorrow. The benefits of our industry are accompanied by enduring commitments to Responsible Care^{SM} in the management of chemicals worldwide. We will make continuous progress toward the vision of no accidents, injuries or harm to the environment and will publicly report our global health, safety and environmental performance. We will lead our companies in ethical ways that increasingly benefit society, the economy and the environment while adhering to the following principles:

- To seek and incorporate public input regarding our products and operations.
- To provide chemicals that can be manufactured, transported, used and disposed of safely.
- To make health, safety, the environment and resource conservation critical considerations for all new and existing products and processes.
- To provide information on health or environmental risks and pursue protective measures for employees, the public and other key stakeholders.
- To work with customers, carriers, suppliers, distributors and contractors to foster the safe use, transport and disposal of chemicals.
- To operate our facilities in a manner that protects the environment and the health and safety of our employees and the public.
- To support education and research on the health, safety and environmental effects of our products and processes.
- To work with others to resolve problems associated with past handling and disposal practices.
- To lead in the development of responsible laws, regulations and standards that safeguard the community, workplace and environment.
- To practice Responsible Care^{SM} by encouraging and assisting others to adhere to these principles and practices.

---

tion to pollution prevention. Since the grant program began in 1989, almost every state has established a pollution prevention program, and many of these provide long-term funding for pollution prevention activities. Over half of the states have enacted pollution prevention laws, and many of the state pollution prevention programs have increasingly become institutionalized within their state environmental protection departments, partnering with their air, waste and water programs to integrate pollution prevention into core environmental regulatory activities. Now that the states have developed basic capabilities, EPA has shifted responsibility for implementing the grant program from Headquarters to the Regions, giving the Regions flexibility to focus resources on regional priorities.

In 1997 EPA's Office of Pollution Prevention and Toxics initiated a new grant program to establish a national network of pollution prevention information centers, four existing centers and five new ones, designated as the *Pollution Prevention Resource Exchange (P2Rx)* . The P2Rx (Internet accessible at http://www.epa.gov/p2/p2rx.htm) is a network of regional pollution prevention information centers that supports state and local technical assistance programs by: providing access to high quality, synthesized, peer reviewed pollution prevention information and expertise; minimizing duplication of effort in developing information, organizing outreach efforts, and producing publications; leveraging resources by improving collaboration among assistance providers; building distribution channels for hard copy materials and access to electronic pollution prevention information; supplying referrals to pollution prevention experts; and promoting consistent and predictable formats for sharing pollution prevention information. The P2Rx group is collaborating on two new sectors, hospitality and metal fabrication, and four mature sectors: dry cleaning, automotive repair, metal finishing, and printing.

Even though state agencies incorporate pollution prevention ideals and aspects into several of their programs, a large majority of state agencies have a separate pollution prevention bureau. These bureaus emphasize pollution prevention techniques for a wide range of domestic to industrial applications. Pollution prevention programs focus on areas such as agriculture, the automotive industry, the printing industry, mercury prevention, waste reduction, and the health industry. They also address issues that concern small business. In an attempt to maximize their efforts, pollution prevention bureaus provide newsletters, fact sheets, pollution prevention resource libraries, on-site technical assistance, and waste material clearinghouses, among other ventures.

Some states administer their pollution prevention programs through regulatory agencies with media-specific offices such as air, water, or solid waste. Other states also involve nonregulatory agencies, such as university-based technical assistance programs, small business assistance programs, manufacturing extension agents, and technology transfer foundations. Implementing these programs involves a variety of approaches including technical assistance and outreach, mandatory facility planning, and regulatory integration. At least 40 states offer confidential, on-site pollution and waste assessments for small to medium-sized businesses. Over 30 states operate information clearinghouses on pollution prevention and 30 states have some form of a pollution prevention facility planning program. States also offer hot lines to provide specific information and answer questions; computer searches to provide up-to-date information; support research on specific pollution prevention techniques; provide workshop and training seminars; produce pollution prevention technology transfer publications; and provide pollution prevention grants and loans, particularly to small businesses. States are also incorporating pollution prevention into regulatory activities such as enforcement settlements, permitting, and compliance inspections.

As state pollution prevention programs look ahead, they face two primary challenges. The first is to evaluate and measure the effectiveness of their technical assistance and outreach efforts in terms of actual pollution prevention results at the company level. The second is the ongoing need to integrate pollution prevention into both local and state regulatory programs. The continued success of state pollution prevention programs will be dependent upon their ability to develop long-lasting partnerships with their inspection and compliance programs and to help these programs truly integrate pollution prevention and source reduction techniques into solutions for environmental health and safety and compliance problems.

Tribal governments have only recently been able to consider, both economically and technically, pollution prevention as an alternative to end-of-pipe controls. Although most tribal governments are still in the nascent phase of environmental management, since 1992, tribes have received 18 PPIS grants and 14 *Environmental Justice through Pollution Prevention (EJP2)* grants. Several tribes have taken steps to integrate pollution prevention into their regulatory and voluntary programs. Critical issues for integration of pollution prevention into tribal activities are the lack of communication among tribes, and the need for education and outreach on general pollution prevention principles. Tribes, EPA, and state agencies are hoping to overcome these barriers by increased tribal participation in national conferences and membership in organizations such as the National Pollution Prevention Roundtable.

3.5. Promoting Pollution Prevention at Educational Institutions

Most formal pollution prevention education is available through university and graduate school-level coursework. However, there has been a recent explosion of interest in pollution prevention in educational institutions at all levels.

In kindergarten through high school, pollution prevention is being added to educational curricula in order to encourage children to practice pollution prevention at school and at home. Educational partnerships and organizations have created materials that engage children's imagination and enable them to see the practical results of pollution prevention. Examples of these educational programs include the Texas *Learning to Be Water Wise and Energy Efficient*, and the curriculum *Environmental ACTION*.

A number of universities and non-profit organizations are developing curricular materials incorporating pollution prevention into courses in business, accounting, engineering, chemistry, finance, and environmental sciences. For example, the *Management Institute for Environment and Business* is working with business schools to encourage an understanding of how source reduction and waste minimization can improve the profitability of a company through environmental accounting, design for the environment, life-cycle analysis, and quality management. As indicated above, the U.S.

EPA-funded *National Pollution Prevention Center for Higher Education (NPPC)* represents a valuable resource for educational materials for use in college and university-level institutions, as well as for professional education and training.

Universities are leading in the research and development of new pollution prevention concepts, such as life-cycle analysis and industrial ecology. Often spurred by students, some universities have become activists in implementing pollution prevention on their campuses. Universities are also a vital source of information for industry and communities. There at least 35 university-centers for pollution prevention in existence throughout the U.S., most of which are detailed in Appendix E. These centers work with industry on technology development and information dissemination, data collection, audits, and training and conferences. Many states have established their compliance assistance/pollution prevention coordinators at a regulatory agency with the technical assistance program located at a university. Universities are forming partnerships with federal and state agencies, industry, and local community organizations to solve real world environmental problems on a local, regional, and national scale.

3.6. Promoting Pollution Prevention through Community and Non-Profit Organizations

Community involvement has been crucial in achieving many of the pollution prevention successes that have been achieved since 1990. *Community-Based Environmental Protection (CBEP)* projects focus on local conditions and problems, recognizing that each community is unique and that solutions for one locale are not necessarily applicable to another. CBEP also encourages partnerships between public and private entities to address local environmental issues and make effective use of local resources. While community concerns over industrial pollution are a primary focus for pollution prevention programs, pollution prevention at the local and regional level is often interwoven with issues of transportation, land use, and building design/indoor air quality.

Pollution prevention and waste minimization concerns have been a primary impetus for several new professional associations, including the *American Institute for Pollution Prevention* and the *National Association of Physicians for the Environment*. Established local government organizations, including the *National Association of Counties* and the *Center for Neighborhood Technologies*, have found a new role in helping to advance prevention among their members. Pollution prevention has been the driver for established environmental groups, such as the *Environmental Defense Fund* and the *Natural Resources Defense Council*, to take on new and expanded roles in collaborative projects. Other groups promote prevention by supplying the public with data developed under the TRI.

4. THE FUTURE OF POLLUTION PREVENTION

What does the future hold for pollution prevention? What are likely to be the greatest challenges in the years ahead and is society prepared to meet them? A variety of views from long-time pollution prevention practitioners are summarized below.[1]

1. Joseph Ling, retired from 3M where he served as Vice President for Environmental Engineering and Pollution Control, suggests that while we may not have all the answers, it's time to forge ahead. As he puts it "we need to take that step today [toward sustainability], and not worry about stumbling tomorrow."
2. David Thomas, director of the Waste Management and Research Center at the Illinois Department of Natural Resources, suggests that pollution prevention is one aspect of a larger environmental revolution that is shaping a new, more sustainable future. Challenges that lie ahead include properly accounting for the true cost of waste and incorporating pollution prevention into the global marketplace.
3. Harry Freeman, executive director of the Louisiana Environmental Leadership Pollution Prevention Program at the University of New Orleans, argues that "pollution prevention is a process rather than an end" and suggests that the focus of pollution prevention may shift to clean products rather than industrial processes and waste streams.
4. Joanna Underwood, president of INFORM, Inc., argues that the concept of pollution prevention has taken center stage in environmental thinking, but industry progress in source reduction "has only been marginal." Underwood urges business to find innovative answers to source reduction, that better data be made available to the public through improved materials accounting techniques, and that the burden of proof should be placed on manufacturers to show that new proposed chemicals are safe for intended uses. "Exposure prevention" should be one of the new guiding principles for environmental management. This argument is shared by the authors of this text.
5. Gerald Kotas, co-director of the National Climate Wise Program and senior environmental scientist with the Office of Energy Efficiency and Renewable Energy of the DOE, calls for partnerships to be formed to develop creative solutions that will lead to fundamental changes in our lifestyle that are necessary for a sustainable future.

Although it is difficult to predict the future, the authors of this text unanimously agree with the practitioners above that pollution prevention will become the standard environmental management option of the new millennium, i.e., the 21st Century.

210 Chapter 10

5. ILLUSTRATIVE EXAMPLES

Example 5.1. Traditionally pollution abatement emphasized pollution control. Today, the shift is towards pollution prevention. What are the differences between the two terms? When was the concept of pollution prevention formally introduced by the EPA?

SOLUTION. Pollution control is an end-of-pipe process. It is performed after the waste is generated. It often transfers the pollutant from one environmental medium to another. Pollution prevention focuses on the reduction of pollution at the source. This implies "up-front" reduction or elimination of waste from a process.

Pollution prevention was introduced when the Congress included "waste minimization" in the 1984 Hazardous and Solid Waste Amendments of the Resource Conservation and Recovery Act. The most formal statement by the Congress of pollution prevention as a "national objective" was made in the passage of the Pollution Prevention Act of 1990.

Example 5.2. Explain the following waste reduction techniques that are used in industry:
    a. Good housekeeping
    b. Material substitution
    c. Equipment design modification
    d. Recycling
    e. Waste exchange
    f. Detoxification

SOLUTION.
    a. *Good housekeeping*: Improper labeling, storage and dumping of hazardous chemicals often increase the risk of spillage, groundwater contamination and waste treatment plant upsets. Educating the employees on the proper handling and disposal of hazardous chemicals reduces such events tremendously.
    b. *Material substitution.* This method involves replacing the polluting ingredients of a product with a less toxic one. Material substitution is most economical when a product is being developed for the first time. With an existing product, new chemicals often require an additional investment in new equipment. However, the additional investment may be less expensive than the cost of additional control systems necessary to meet new emission standards. For example, replacement of an organic based solvent as a carrier for tablet coating with a water based coating costs a pharmaceutical company $60,000. Alternatively, the air pollution control system to meet emission standards for the organic solvent had a capital cost of $180,000 and $30,000 annual operating costs. Clearly, this pharmaceutical company will replace the organic based coating with a water based one to realize the economic and pollution prevention benefit this material substitution provides.
    c. *Equipment design modification.* Old equipment may contribute to the production of harmful by-products and excessive emission of pollutants.

Replacing such process equipment might eliminate or minimize the generation of harmful by-products. Low interest federal loans may be available to assist in these changes.

   d. *Recycling.* On-site waste recycle and reuse programs are very effective and reduce both raw material consumption and waste production. The use of off-site recyclers may be a practical option for smaller industries. Some of the popular on-site recovery processes include: distillation, adsorption, filtration and electrolysis. Distillation is mainly used to recover up to 85 to 90 percent of the original organic solvent. The spent organic compound is distilled and the solvent vapor is condensed to recover the pure solvent. Adsorption processes generally use activated carbon beds to adsorb the solvent and can achieve up to 90 percent solvent recovery. The compound is recovered in a concentrated form during regeneration of the bed either by steam or inert gas regeneration. Filtration separates solids from the liquid stream via simple filtration or use of a fabric membrane. The separated solid and liquid streams could be reused with or without further treatment depending on their purity and their ultimate use. Electrolysis is commonly used for metal recovery from waste streams.

   e. *Waste exchange.* Waste from one industry may actually be a valuable resource as a raw material in another industry. Avoided disposal costs for the waste generator and inexpensive raw material for the receiving industry are the inherent advantages of this waste reduction technique. Most of the chromium used in U.S. industries is imported. Recovery and reuse of chromium will not only minimize the amount of waste but also make the U.S. less dependent on imports.

   f. *Detoxification.* After considering all other options for waste reduction, detoxification of waste by neutralization or other techniques may be considered. Thermal, biological and chemical treatments are some of the detoxification processes commonly used. However, energy costs, along with the potential of generating new compounds and end product streams, must be evaluated thoroughly to justify the perceived environmental benefits of these proposed detoxification processes.

Example 5.3. Consider a degreasing operation in a metal finishing process. This process involves the use of solvents to remove compounds, such as oils, from metal parts to prepare them for further processing. Give an example of modifications that might be made to this part of the process that represent:
   a. Source reduction
   b. Recycling
   c. Waste treatment
   d. Waste disposal

SOLUTION. Examples of methods that might be used include:
   a. *Source reduction.* Substitution of a different, nontoxic or non-polluting solvent (e.g., citric acid-based) for a toxic solvent such as TCE. Note that when using a water soluble citric acid-based solvent, the grease and oil from the metal parts will float on top of the solvent and can be skimmed

off so that the solvent can be reused over and over again. This is an available option on many citric acid-based degreaser systems.

  b. *Recycling.* Filtering and reuse of solvent for noncritical cleaning.

  c. *Waste treatment.* Distillation of spent solvent to recover high-purity solvent that can be reused in the process.

  d. *Waste disposal.* Disposal of still bottoms from a solvent distillation process by shipping them to an off-site landfill.

6. SUMMARY

1. Most environmental protection efforts have traditionally emphasized control of pollution from waste substances after it has been generated. However, this tradition of pollution control does not solve the problem of pollution; rather it alters the problem by transferring pollution from one medium to another, resulting in no net environmental benefit. This has caused interest and focus to shift from ways to deal with the amounts of waste generated toward efforts in reducing the amounts of waste produced.

2. The concept of pollution prevention was first defined as "waste minimization" and was introduced when the U.S. Congress passed the *Hazardous and Solid Waste Amendments of 1984*. However, the formal identification of pollution prevention as a "national objective" did not take place until passage of the *Pollution Prevention Act of 1990*.

3. Pollution prevention options may be broken down into four major categories based on EPA's waste management/pollution prevention hierarchy. These options include:

> *Source reduction*, use of technologies to reduce the volume of wastes initially generated, is the primary approach. The techniques involved are applied to the production process prior to the point of waste generation. Methods that eliminate or reduce the amount of waste generated by a particular process, either through procedure modifications or material substitution, are employed.
>
> *Recycling*, the second approach, attempts to recover a usable material from a waste stream. The methods involved can take place within the process or at the end of the process and can be implemented either on- or off-site.
>
> *Treatment* is used for the management of wastes remaining after all possible source reduction and recycling techniques have been employed. Treatment can include the use of physical, biological, chemical, and thermal treatment methods for the reduction of the toxicity and volume of waste requiring ultimate disposal.
>
> *Ultimate disposal* is the last option for waste management, and can include landfilling, landfarming, deep-well injection, and ocean disposal.

4. Pollution prevention activities can help a firm to comply with many federal regulations, including:

- *Clean Air Act (CAA)*
- *Clean Water Act (CWA)*

- *Comprehensive Environmental Response, Compensation, and Liability Act (CERCLA)*
- *Superfund Amendments and Reauthorization Act (SARA)*
- *Hazardous and Solid Waste Amendments (HSWA)*
- *Safe Drinking Water Act (SDWA)*
- *Toxic Substance Control Act (TSCA)*
- *The Pollution Prevention Act of 1990*

5. The problems associated with the handling of wastes have become a major concern for industry as a whole, and for the chemical industry specifically. There are real incentives for avoiding the production of wastes up-front, and they will increase as more regulations are passed and treatment capabilities for waste substances remain limited.

6. *EPA's Pollution Prevention Strategy* provides guidance on incorporating pollution prevention principles into EPA's ongoing environmental protection efforts and is aimed at maximizing private sector initiative while challenging industry to achieve ambitious prevention goals. One successful program reflecting this Agency strategy is the *33/50 Program* which has resulted in the voluntary reduction in releases and transfers of 17 toxic chemicals by 33 percent by the end of 1991 and 50 percent by the end of 1994, both goals being reached 1 year ahead of schedule.

7. Additional voluntary and Agency/industry partnering programs include: *Climate Wise*, the *Green Lights Program, WasteWise, Design for the Environment, Project XL, Environmental Accounting, WAVE (Water Alliances for Voluntary Efficiency)*, and the *Pesticide Environmental Stewardship Program*.

8. Progress toward pollution prevention goals has been achieved by promotion of pollution prevention concepts in the following groups:
- *EPA* through such Agency programs as the *Common Sense Initiative*, Agency/industry partnerships listed above, partnerships with other federal agencies, the *Consumer Labeling Initiative*, the *Green Chemistry Challenge* and an *Environmental Leadership Program*.
- *Other Federal Agencies* via a number of Executive Orders through such programs as the *U.S. Department of Agriculture's Sustainable Agriculture Research and Education Program*, the *U. S. Agency for International Development Environmental Pollution Prevention Project*, the NIST *Manufacturing Extension Partnership Program*, the *U.S. Postal Service's Waste Minimization/Pollution Prevention Program*, and the *"Greening of the White House" Project*.
- *Industry* through involvement in voluntary EPA/industry and industry-promoted waste reduction and pollution prevention initiatives.
- *State and Tribal Level* through the EPA-funded *Pollution Prevention Incentives for States (PPIS)* grant program, the EPA-funded *Pollution Prevention Resource Exchange* network, and tribal involvement in PPIS and *Environmental Justice through Pollution Prevention* grants.

214 Chapter 10

- *Education Institutions* via education program development such as the Texas *Learning to Be Water Wise and Energy Efficient*, and the active participation curriculum *Environmental ACTION*, and higher education curriculum resource materials produced by the *National Pollution Prevention Center for Higher Education* and the *Management Institute for Environment and Business*.
- *Community and Non-Profit Organizations* through *Community-Based Environmental Protection (CBEP)* projects, professional associations such as the *American Institute for Pollution Prevention* and the *National Association of Physicians for the Environment*, and through pollution prevention focused activities by established environmental groups such as the *Environmental Defense Fund* and the *Natural Resources Defense Council*.

9. With improved dissemination of pollution prevention information, and with the growing appreciation for the significant economic and liability-reduction benefits generated from waste minimization and pollution prevention efforts, pollution prevention is becoming the preferred environmental management option as we move into the 21st Century.

7. PROBLEMS

1. EPA's pollution prevention hierarchy includes the following components: source reduction, recycling, treatment, and ultimate disposal. Explain the nature of activities involved in each of these approaches and explain each component briefly.

2. Consider the two pollution problems presented below, briefly discuss which of the options discussed in this chapter would be most applicable, and discuss the limitations of the others.
   a. Radioactive wastes from a nuclear power plant
   b. Sulfuric acid waste from the pharmaceutical industry

3. Define and discuss the differences between the following terms:
   a. Pollution prevention
   b. Pollution control
   c. Waste minimization

4. One of the major driving forces behind pollution prevention programs is liability concerns. In addition, the term liability is very often used when discussing environmental regulations. Related terms are retroactive liability and cradle-to-grave liability. Briefly explain these terms. What are their implications related to pollution prevention?

5. Discuss activities a local government can undertake to promote pollution prevention in their jurisdiction.

6. The major components of a company's waste reduction program include the following: management commitment, communication of the program to the rest of the company, waste audits, cost/benefit analysis, implementation of the program, and follow-up. Describe each of these general principles that are necessary for a successful pollution prevention program.

8. REFERENCES

1. U.S. EPA, *Pollution Prevention 1997: A National Progress Report*, Office of Pollution Prevention and Toxics, Washington, D.C., EPA/742/R-97/00, 1997.
2. U.S. EPA, *Pollution Prevention Policy Statement: New Directions for Environmental Protection--Memorandum*, Office of the Administrator, Washington, D.C., EPA/742/F-93/009, 1993. (Internet access at URL: www.epa.gov/opptintr/p2home/p2policy.htm).
3. Lindsly, J. A., *Waste Reduction – Methods Utilized to Make it Happen*, Paper #90-41.6, Presented at the Air and Waste Management Association Annual Meeting, Pittsburgh, Pennsylvania, 1990.
4. Mason, A. M., *Pollution Prevention – A Chemical Industry Perspective*, Paper #90-46.1, Presented at the Air and Waste Management Association Annual Meeting, Pittsburgh, Pennsylvania, 1990.

CHAPTER 11

Additional Pollution Prevention Components

1. INTRODUCTION

There are numerous areas of environmental concern that are incorporated under the pollution prevention umbrella. The primary pollution prevention components are *source reduction*, highlighted in Chapter 13, and *closed-loop recycling*, highlighted in Chapter 14 of this book. Another major component of pollution prevention is *energy conservation*. Energy utilization and energy conservation issues have significantly affected American life since the Arab oil embargo in 1973. Energy conservation is directly related to pollution prevention since a reduction in energy demand generally corresponds to less energy production and, consequently, less pollutant output. The third major area of concern related to pollution prevention is that of *accident and emergency management*. Events like Chernobyl and Bhopal have increased public awareness of the risk associated with hazardous materials use, and have stimulated the development of regulatory policies concerned with chemical release emergency response planning and general requirements imposed on industry for the proper handling of both hazardous wastes and hazardous raw materials used in production processes

Energy conservation, and accident and emergency management topics are discussed in the sections that follow along with a variety of other pollution prevention topics ranging from *health risk assessment* and *life cycle analysis*, to *ISO 14000* and *environmental justice* considerations.

2. ENERGY CONSERVATION

The environmental impacts of *energy conservation* and consumption are far-reaching, affecting air and water (as well as land) quality and public health. Combustion of coal, oil, and natural gas is responsible for air pollution in urban areas, acid rain that is damaging lakes and forests, and some of the nitrogen pollution that is harming estuaries. Although data show that for the period from 1977 to 1989 annual average ambient levels of all *criteria air pollutants* were down nationwide, 96 major metropolitan areas still exceed the national health-based standard for ozone, and 41 metropolitan areas exceed the standard for carbon monoxide, two criteria pollutants associated with energy production.

Energy consumption also appears to be the primary man-made contribution to global warming (the greenhouse effect). EPA has concluded that energy use through the formation of carbon dioxide during combustion processes has contributed approximately 50 percent to the global warming that has occurred in the last 20 years. As a response to now overwhelming evidence that greenhouse gas emissions will lead to a rise in global mean temperature, and because industrial energy use accounts for nearly 30% of total greenhouse gas emissions, *Climate Wise* was formed April 1994 to help reduce greenhouse gas emissions in the industrial sector. Climate Wise (Climate Wise Hot Line: (800) 459-WISE (800-459-9473), or web site location: http://www.epa.gov/climatewise) is a joint U.S.EPA/DOE program that was formed in response to President Clinton's "Climate Change Action Plan," to help the U.S. honor its international commitment to reduce greenhouse gas emissions to 1990 levels by the Year 2000. Climate Wise is a voluntary partnership program designed to encourage U.S. industry to take advantage of the environmental and economic benefits associated with energy efficiency improvements and greenhouse gas emissions reductions. Pollution prevention in this area can best be achieved through energy conservation and increased energy efficiency (in combustion, transmission, distribution, etc.).

Climate Wise encourages comprehensive, cost-effective industrial energy efficiency and pollution prevention actions and allows companies to tailor their programs to meet the needs and opportunities of their operations. Climate Wise targets the industrial sector, providing partners support for a variety of cost-effective emissions reduction opportunities, from improving boiler and air compressor system efficiency to fuel switching and cogeneration. The program provides partners with access to technical resources and technical assistance related to energy conservation, pollution prevention, and economic development; provides opportunity for regional business-to-business exchange workshops related to company profitability, productivity, and environmental performance; provides positive public recognition; and access to financial resources including Small Business Administration guaranteed loans, low-interest buy-downs, State tax credits, utility programs, and private-sector financing opportunities. As part of this partnership, companies must submit an Action Plan within 6 months of joining the program that details ways to reduce greenhouse gas emissions by implementing energy efficiency and environmental management practices. The program also provides companies with guidance on how they can quantify energy savings and emission reductions.

As of May 1998, Climate Wise Partners represented about 12% of the total U.S. industrial energy use base, with Climate Wise Partners collectively projected to save more than $240 million annually by the Year 2000 in energy and waste management related costs through participation in the Climate Wise program. The distribution of types of Climate Wise projects reported in 1998[1] is shown in Figure 11.1. Most are energy conservation related, with typical projects Climate Wise participants undertake including:

- Optimizing boiler efficiency, improving air compressors and steam traps, and insulating piping and heating equipment.
- Cogenerating, using renewable energy, reusing waste heat, and reducing vehicle fleet and employee trips.
- Engineering energy-efficient production processes, designing energy-efficient products, using recycled materials in products, and developing products that are easy to recycle.

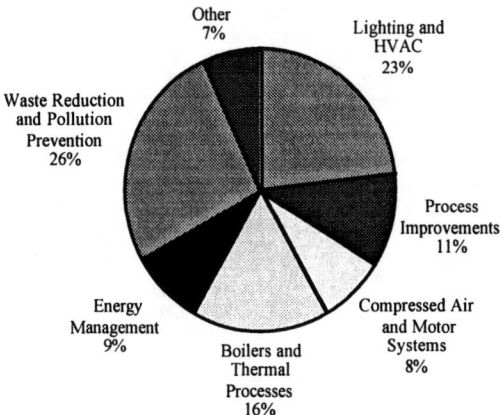

Figure 11.1. Distribution of U.S. EPA Climate Wise project types reported in 1998.[1]

The Climate Wise program has published a "First Tier" checklist for the evaluation of industry energy use to identify areas of opportunity where potential energy and waste production savings might exist. This checklist is provided in Figure 11.2, and indicates that energy conservation measures that can be taken in the manufacturing and process industries include basic actions such as: implementation of sound operation, maintenance, and inspection (OM&I) programs; implementation of pollution prevention programs; recovering energy from hot water/liquids; recovering energy from hot (flue) gases; conducting formal energy conservation training programs for all employees, etc. In addition, specific energy efficiency options should be considered relative to industrial heat exchangers, ovens and furnaces, insulation, boilers, chillers, compressed air systems, lighting, motors and drives, and power factor issues. These specific energy issues are discussed based on recommendations provided by the Michigan Manufacturing Technology Center.[2]

2.1. Heat Exchangers

Heat exchangers can be a source of waste, especially with products that are temperature-sensitive. Reducing tube-wall temperature is the key to pollution prevention for heat exchangers. Table 11.1 highlights the pollution prevention options available for heat exchanger units and the impact each has on heat exchanger performance.

Boilers
- Optimize boiler size and boiler loading
- Analyze flue gas and optimize air/fuel ratio
- Install over-fire draft control
- Convert to atomizing burner
- Install characterizable fuel valve
- Clean boiler tubes
- Establish burner maintenance schedule
- Install stack dampers
- Recover waste heat from flue gas or blowdown to pre-heat combustion air or feedwater
- Minimize boiler blowdown with better feedwater treatment
- Automate blowdown control
- Turn off hot water circulation pump when boilers are not in use
- Fuel-switch to less carbon-intensive fuel

Steam System
- Implement steam trap maintenance program
- Shut off steam traps on super-heated steam lines when not in use
- Install correctly sized steam traps
- Repair steam leaks in lines, valves and reducing stations
- Improve insulation of steam lines, condensate lines, and condensate tanks
- Recover and recompress vented steam for low-pressure applications
- Flash condensate to lower steam pressure
- Increase condensate return to boiler
- Install de-aerator in place of condensate tank
- Replace barometric condensers with surface condensers
- Clean steam coils in process tanks
- Close off unused steam lines
- Use minimum steam operating pressure

Furnaces, Ovens, and Kilns
- Minimize warm-up time and temperature
- Use optimum temperature and minimum safe ventilation rates
- Automate controls
- Recover waste heat for use in other processes
- Optimize combustion and heat transfer
- Improve insulation, seals, and refractories
- Implement direct firing or direct electric heating in place of indirect heating

Waste Heat Recovery and Heat Containment
- Recover waste heat in other applications
- Clean fouled heat-exchanger surfaces (filter) contaminated streams if fouling is heavy
- Install or improve equipment insulation
- Isolate hot equipment from air-conditioned areas

Cogeneration and Renewables
- Install cogeneration equipment
- Generate electricity with waste heat
- Generate electricity with renewable resources, (e.g., biomass, photovoltaics, wind turbines)

Process Cooling
- Use cooling tower water instead of refrigeration or chiller
- Use outside air cooling
- Reduce refrigeration system operating pressure
- Raise cooling water temperature
- Use waste heat for absorption refrigeration
- Clean condensers and coils
- Improve insulation
- Use continuous freezing in place of batch freezing

Compressed Air Systems
- Use cooler air for compressor intakes
- Install, upgrade or adjust compressor controls
- Right-size compressors/optimize loading
- Reduce pressure
- Eliminate compressed air use
- Repair air leaks
- Recover waste heat
- Change dryer filters
- Clean intercoolers
- Adjust operating schedules to minimize equipment idle time
- Remove or close off unused compressed air lines

Process Controls
- Optimize temperature, pressure, flow, and material movement
- Install automated systems

Other Technologies
- Next generation technologies:
  _____
  _____
  _____
  _____
  _____
  _____

Figure 11.2. Action plan checklist from EPA's Climate Wise program.[1]

2.2. Ovens and Furnaces

Ovens and furnaces utilize more energy than all other plant operations. Avoiding hot tube-wall temperatures is a major pollution prevention and energy efficiency issue for furnaces as was indicated above for heat exchangers. A direct heat furnace can be replaced with a high temperature intermediate heat exchanger so that a heat transfer fluid can be used to eliminate direct heating. In addition, energy can be conserved by making use of existing steam superheat if available to heat a process stream, avoiding exposure of the fluid to the hot tube-wall temperature of a furnace. Installation of insulation (below) or implementation of heat recovery from a furnace flue gas can also be an effective means of reducing energy demands associated with furnace operations.

Table 11.1. Pollution Prevention Options and Their Effects on Heat Exchanger Performance and Energy Efficiency.

Pollution Prevention Technique	Impact on System
Use lower pressure steam	Reduces tube-wall temperature
Desuperheat steam	Reduces tube-wall temperatures and increases the effective surface area of the exchanger because the heat transfer coefficient of condensing steam is ten times greater than that of superheated steam
Install a thermocompressor	Reduce tube-wall temperature by combining high and low pressure steam
Use staged heating	Minimize degradation, staged heating can be accomplished first using waste heat, then low pressure steam and finally, desuperheated high pressure steam
Use on-line cleaning techniques for exchangers	Recirculating sponge balls and reversing brushes can be used to reduce exchanger maintenance, and also to keep the tube surface clean so that lower temperature heat sources can be used
Use scraped-wall exchanger	To recover saleable products from viscous streams, e.g., monomers from polymer tar
Monitor exchanger fouling	Sometimes an exchanger fouls rapidly when plant operating conditions are changed too fast or when a process upset occurs; monitoring can help to reduce such fouling
Use noncorroding tubes	Corroded tube surfaces foul more quickly than noncorroded ones

Burner and combustion controls may be upgraded to improve furnace and oven energy efficiency by replacing open age-type burners with sealed-in burners, replacing atmosphere burners with power burners, and/or installing a fuel/air ratio control system. Installation of furnace pressure controls can also increase the energy efficiency of ovens or furnaces. Optimization of fuel burning equipment is vital to energy savings and efficiency, and inefficient burning of fuel is more likely if the facility does not have good documentation of operational changes, or if the operations and maintenance manual does not reflect the true operations of the facility.

2.3. Insulation[3,4,5]

Upgrading or installing insulation on pipes, buildings, and hot and cold process equipment may be a significant source of energy and cost savings.

Insulation is important to any operation where the transfer of fluids occurs at other than ambient temperatures. Insulation can be classified as fibrous, cellular, polyisocyanate, and granular based on the material it iscomposed of.

Fibrous insulation is made up of a variety of materials including small diameter fibers that create insulating air space. Examples are glass, silica, rock wool, slag wool, or alumina silica. Fiber-glass is an inexpensive option for applications of 50°F to 850°F but has low resistance to abuse and readily absorbs moisture. Mineral Wool: has the lowest cost for higher temperature applications, 1200°F to 1800°F and is most appropriate for non-austenitic stainless steel and low abuse situations. *Cellular insulation* is made up of small individual cells of air separated from each other. Examples are glass or foamed plastic.

Polyisocyanate insulation is appropriate for applications of 140°F to 300°F and -100°F to -300°F. It has similar costs to calcium silicate, but has better moisture resistance and is good for dual temperature applications. Foam glass is good for temperatures of 300°F to 900°F. *Granular insulation* is composed of small nodules of material with voids or hollow spaces. This material may be loose, pourable, or combined with binders to be rigid. Examples of granular insulation include calcium silicate, expanded vermiculite, perlite, cellulose, and diatomaceous earth. Calcium silicate is relatively inexpensive and best for applications between 300°F and 1200°F.

2.4. Boilers

Boilers are used in many industrial applications, both for process heating and in-plant heating during the winter months. The major energy use issue with boilers is their efficiency of fuel combustion. As described above for ovens and furnaces, burner and combustion controls should be considered for upgrading to improve boiler energy efficiency. The combustion performance of boilers should be monitored at least twice per year; the air/fuel ratio should be tuned as necessary to ensure maximum boiler energy conservation. This energy conservation measure can potentially provide a 10% to 15% reduction in boiler energy demands.

Heat exchangers installed in boiler vent stacks will provide low-grade heat recovery that can be used to pre-heat combustion air, or for other process or space-heating needs within the facility. This energy conservation measure can potentially provide 10% to 25% energy savings.

As indicated in Chapter 6, steam production and use can represent a significant energy in manufacturing facilities, and more careful management of steam pressures, along with repair of leaks in steam lines, can provide significant energy and cost savings to a facility. Both higher than necessary steam operating pressures and leaks in the steam lines that allow steam to be wasted result in higher steam production requirements from the boiler than is absolutely necessary to meet basic system requirements. In addition, returning steam condensate to the boiler to recover the energy remaining in the condensate rather than feeding ambient temperature feed water can add an additional 5% to 8% energy savings without compromising boiler operation.

2.5. Chillers and Cooling Towers

Chillers are used at many industrial facilities to provide process and in-plant cooling, function to lower ambient temperatures much as boilers are used to raise process temperatures. These chillers consist of a compressor, an evaporator, an expansion valve, and a condenser, and are classified as *reciprocating chillers*, *screw chillers*, or *centrifugal chillers*, depending on the type of compressor used. Reciprocating chillers are usually used in smaller systems, but can be used in systems as large as 800 tons (2800 kW). Screw chillers are normally used in the 200 tons to 800 tons range (700 kW to 2800 kW), while centrifugal chillers are available in the 200 tons to 800 tons range, and are used for very large systems (greater than 800 tons [2800 kW]).

Since outdoor wet-bulb temperature affects cooling tower performance, significant power savings can be realized in chiller operations if a two-speed or ideally a variable-speed fan is used for the cooling tower fan to reduce fan motor power consumption. During periods of lower outdoor wet-bulb temperature, the design amount of cooling can be obtained with lower airflow rates. As the air flow rate decreases, the fan speed and the motor power requirements also decrease. The power draw by a fan is proportional to the cube of the fan speed, and even accounting for lower motor efficiency at lower speed, lower by approximately 15%, significant savings occur because at low speed, the power required to deliver a unit of cooling air is significantly less than at high speed.

Chiller energy consumption is also affected by the condensing water temperature. As the condensing water temperature increases, the pressure rise across the compressor increases increasing the work that must be done by the compressor for a given heat duty. Normal condensing water temperature setpoints range between 65°F and 85°F. If the setpoint temperature is in the middle of the range, significant energy savings can be realized by lowering this temperature to perhaps 60°F, as the rule of thumb suggests a 0.5% improvement in chiller efficiency for each degree Fahrenheit decrease in the condenser water setpoint temperature. Dropping the condensing water setpoint to 60°F from 75°F would be expected to yield a nearly 7.5% reduction in chiller energy demand.

Finally, chiller efficiency increases as the delivered chilled water temperature increases, approximately a 1% efficiency increase for each degree Fahrenheit increase in the chilled water setpoint temperature. As with steam production discussed above, consideration should be given to actual process needs so that chiller process water temperature can be adjusted as feasible to decrease overall chiller energy requirements.

2.6. Compressed Air Systems

Compressed air is perhaps the most expensive form of energy used in an industrial setting because approximately 90% of the energy used is converted into heat while only 10% is actually converted to compressed air energy.[6] As indicated above for chillers, the work required by a compressor is proportional to the absolute temperature between the inlet and outlet flow streams. If inlet compressor air is taken from inside a manufacturing facility it is typically 20°F to 30°F warmer than actual outside air. In lowering the intake air temperature by collecting intake air from outside the facility, significant energy savings can be obtained.

Because of the high inefficiency of producing compressed air, higher than required air pressures and seemingly small air leaks can represent significant energy expenditures. If, as in a typical manufacturing firm, about 10% of the compressed air generated is lost through leaks, this loss corresponds to large energy requirements due to the original inefficiency in the production of the compressed air. Since most process equipment requires only 80 to 100 psi operating pressures, any increase in line pressures above this minimum level represents high energy costs in compression requirements, and in increased potential for line leaks. Means other than line pressure increases should be used to increase compressed air volumes. Significant energy savings can be realized then by reducing the line pressure to the minimum required, and by routinely identifying and fixing air leaks with an effective OM&I program.

2.7. Lighting

Significant advances in lighting technology have been made in recent years, and many lighting systems that represented good practice several years ago are inefficient in view of these new lighting innovations. Since a facility's lighting generally represents approximately 20% of its total electric bill, evaluation of lighting requirements and potential lighting savings is often quite cost effective. The energy-efficient lighting technologies available today can dramatically reduce energy consumption and prevent pollution while delivering comparable or better lighting. If energy-efficient lighting were used everywhere it were profitable, the electricity required for lighting would be cut by 50 percent, and aggregate national electricity demand would be reduced by more than 10 percent.[7]

As with the Climate Wise program, the U.S. EPA sponsors a voluntary, non-regulatory program, the *Green Lights program*, that is specifically focused on encouraging major U.S. corporations to install energy-efficient lighting technologies wherever they are profitable and maintain or improve lighting quality. The underlying principle of the Green Lights program is "environmental protection at a profit." Over 1,900 organizations participate in this program, with projected annual cost savings from current participants exceeding $72 million annually.[7]

Through Green Lights, EPA is aiding U.S. companies to realize their potential to prevent pollution at a profit by installing energy-efficient lighting designs and technologies. Improved lighting technologies are identified and implemented that lead to improved corporate profit by lowering their electricity bills, improving lighting quality, and increasing worker productivity. Reduced energy demands for lighting also reduce air pollution caused by electricity generation, which includes carbon dioxide, sulfur dioxide, and nitrogen oxides.

Lighting energy can be wasted in several ways including transmission losses (lighting is blocked by dirt or other obstructions), over-lighting and use of inefficient light sources. A comprehensive Lighting Survey should be used to identify current lighting conditions and lighting system performance. When combined with a lighting assessment tool, such as ProjectKalc[7] or MEDS,[2] the most energy efficient and best quality lighting for specific areas within an industrial facility can be identified. ProjectKalc offers full analysis of potential lighting upgrades. It provides comprehensive energy and economic analysis of upgrades involving controls, relamping, delamping, tandem wiring, etc., and includes user-modifiable databases of costs, labor time, and performance for over 8,000 common hardware applications applicable for a variety of industrial applications including shop, task and office lighting, exit signs, outdoor lighting, and low-use areas like conference rooms and warehouses. Similarly, simple calculations and recommendations for lighting retrofits may be made using the economic feasibility and payback period estimation features of the MEDS web site related to Lighting.[2]

A number of energy saving opportunities incorporated into these Lighting Survey/Retrofit Evaluation programs that exist for industrial lighting applications are summarized below.

High-intensity discharge (HID) lamps have long service lives and high efficiency, providing on order of magnitude greater lumens per watt than standard incandescent lamps. HID lamps are often the most energy efficient option for many lighting applications. HID lamps contain a sealed arc tube of mercury, sodium, and halide compounds inside a glass bulb. A continuous electric arc between two electrodes in an atmosphere of inert gas produces light. Tubes may also have additional starting electrodes.

One of the most obvious and beneficial steps to conserve energy is to turn off lights when they are not needed. *Occupancy sensors* detect when an employee has left, and after a set period of time, turns off the lamp. The sensor turns lights back on when movement is once again detected in the room. *Photocell sensors* automatically turn off electric lights when natural light is sufficient for a particular lighting purpose. Savings of 10% to 30% of lighting costs may be achieved by the installation of these lighting control measures. Significant lighting-related energy savings can also be realized by purchasing the newest technology lighting. High-efficiency lamps can reduce power consumption by 5% to 15%, while electronic ballasts can save 10% to 25% of lighting energy costs. Significant savings of energy (up to 95%) and labor may be achieved with the replacement of incandescent exit signs with *LED exit signs*. Energy savings are about $35 per year, while due to LED lamp lives of up to 20 years, labor costs to change lamps and maintain equipment are practically eliminated. Inexpensive retrofit exit sign light kits are available from several vendors.

2.8. Motors and Drives

Motor systems, including motors, drives, pumps, fans, compressors, and their control-system and mechanical-load components, account for nearly 75 percent of the electricity used by industry. The U.S. EPA and DOE have created as part of the Climate Wise program a voluntary, non-regulatory program, the *Motor Challenge*, that provides technical assistance (Motor Challenge Hotline, 1-800-862-2086) and independent performance validation to industrial end users, motor and drive manufacturers and distributors, utilities, research institutes, and State energy offices when converting to energy-efficient, electric motor systems. More than 900 companies have become involved in the Motor Challenge program as of 1998.

Several generic energy saving opportunities are available related to industrial uses of motors and drives. *Synthetic lubricants* have demonstrated an ability to provide energy savings due to reduced friction and heat on bearings, gearboxes, and other related parts. They also increase the service life of these components.

Facilities should should consider replacing worn out motors with high efficiency motors, as they are never as efficient after being rewound as they were originally. Payback of the extra cost of high-efficiency motors is usually less than 2 years for motors operated for at least 4,000 hours and at at least 75% load. The greatest potential for energy savings (up to 10% of energy costs) occurs for motors in the 1 to 20 hp range where motor energy efficiency can vary greatly from standard to high efficiency motors. Above 20 hp, efficiency gains become smaller, as motors over 200 hp are already relatively efficient.

Facilities should also consider replacing standard V-belts with cogged V-belts. Substitution of these notched cog belts for the conventional V-belt offers energy savings (approximately 2.5% of energy costs) because the standard V-belt is subjected to large compression stresses when conforming to the sheave diameter. The notched V-belt has less material in the compression section of the belt, thereby minimizing rubber deformation and compression stresses. The result is higher operating efficiency for the notched V-belt.

2.9. Power Factor Issues

The ratio of real, usable power (kW) to apparent power (kVA) is known as the *Power Factor (PF)*. For example, assume that a manufacturing plant has an average annual PF of 0.78. A PF of 0.78 means that for every 78 kW of usable power that the plant requires, the utility must supply 78 kW/PF or 100 kVA. If the plant's PF is changed from 0.78 to 0.95, then for every 78 kW demanded by the plant, the utility need only supply 78 kW/0.95 or 82 kVA. The utility supplying electricity to a facility will assess a power factor charge when the PF falls below a specified level because more apparent power must be supplied as the user's PF decreases. To avoid penalties from being incurred from the local utility company, the PF should be maintained as close to unity (1.0) as is practical for the entire manufacturing plant. Capacitors can be installed at any point in the electrical system and will improve the power factor between the point of application and the power source. Capacitors can be added at each piece of equipment, ahead of groups of small motors, or at main services. Electrical contractors should be consulted for the optimal approach for reducing power loss within a given facility using capacitor control.

3. ACCIDENT AND EMERGENCY MANAGEMENT

Accidents can occur in many ways. There may be a chemical spill, an explosion, or a nuclear plant out of control. There are often accidents in transport: trucks overturning, trains derailing, or ships capsizing. There are "acts of God" such as earthquakes and storms. It is painfully clear that accidents are a fact of life. The one common thread through all of these situations is that accidents are rarely expected and, unfortunately, they are frequently mismanaged.

Accident and emergency management issues are relevant in the context of pollution prevention as both are focused on eliminating, or at worst, reducing, the release of hazardous materials into the environment to protect public health and the environment. Both pollution prevention and accident and emergency management approaches rely on the evaluation of existing conditions to identify opportunities to reduce releases and/or the impact of these releases on the receptor population.

Development of plans for handling accidents and emergencies must precede the actual occurrence of these events. In recent years hazardous chemical release incidents from the chemical, petrochemical, and refinery industries have caused particular concern. Since the products of these industries are essential in a modern society, every attempt must be made to identify and reduce the risk of accidents or emergencies in these areas.

Risk evaluation of accidents serves a dual purpose. It estimates the *probability* that an accident will occur and also assesses the severity of the *consequences* of an accident. Consequences may include damage to the surrounding environment, financial loss, or injury to life. This section outlines the methods used to identify hazards, and the causes and consequences of accidents. Risk assessment of accidents provides an effective way to help ensure either that a mishap does not occur or that the likelihood of an accident is reduced. The result of the risk assessment allows concerned parties to take precautions to prevent an accident and mitigate its consequences before it happens.

An *accident* is an unexpected event that has undesirable consequences. The causes of accidents have to be identified in order to help prevent accidents from occurring. Any situation or characteristic of a system, plant, or process that has the potential to cause damage to life, property, or the environment is considered a *hazard*. A hazard can also be defined as any characteristic that has the potential to cause an accident. The severity of a hazard plays a large part in the potential amount of damage a hazard can cause if it occurs. *Risk* is the probability that human injury, damage to property, damage to the environment, or financial loss will occur upon the occurrence of an accident. An *acceptable risk* is a risk whose probability is unlikely to occur during the lifetime of the plant or process. An acceptable risk can also be defined as an accident that has a high probability of occurring, with negligible consequences. Risks can be ranked qualitatively in categories of high, medium, and low. Risk can also be ranked quantitatively as the annual number of fatalities per million affected individuals. This is normally denoted as a number times one millionth. For example, a fatality rate of 3×10^{-6} indicates that, on the average, for every million individuals, three individuals will be at risk of death every year.

Another quantitative approach that has become popular in industry is the Fatal Accident Rate (FAR) concept. The FAR describes the number of fatalities over the lifetime of 1,000 workers. The lifetime of a worker is defined as 10^5 hours, which is based on a 40-hour work week for 50 years. A reasonable FAR for a chemical plant is 3.0 with 4.0 usually taken as a maximum. The FAR for an individual at home is approximately 3.0. A FAR of 3.0 means that there are 3 deaths for every 1,000 workers over a 50-year period.

There are several steps in evaluating the risk of an accident as indicated in Figure 11.3. These are detailed below for the example where the system in question is a chemical plant.[8]

1. A brief description of the equipment and chemicals used in the plant is needed.
2. Any hazard in the system has to be identified. Hazards that may occur in a chemical plant include: fire, explosions, toxic vapor releases, rupture of a pressurized vessel, slippage, runaway reactions, and corrosion.
3. The event or series of events that will initiate an accident has to be identified. An event could be a failure to follow correct safety procedures, improperly repaired equipment, or failure of a safety mechanism.
4. The probability that the accident will occur has to be determined. For example, if a chemical plant has a 10-year life, what is the probability that the temperature in a reactor will exceed the specified temperature range? The probability can be ranked from low to high. A low probability means that it is unlikely for the event to occur in the life of the plant. A medium probability suggests that there is a possibility that the event will occur. A high probability means that the event will probably occur during the life of the plant.
5. The severity of the consequences of the accident must be determined.
6. If the probability of the accident and the severity of its consequence are low, then the risk is usually deemed acceptable and the plant should be allowed to operate. If the probability of occurrence is too high or the damage to the surroundings is too great, then the risk is usually unacceptable and the system needs to be modified to minimize these effects.

The heart of the hazard risk assessment algorithm provided is enclosed in the dashed box in Figure 11.3. The algorithm allows for reevaluation of the process if the risk is deemed unacceptable (the process is repeated either with Step One or Step Two).

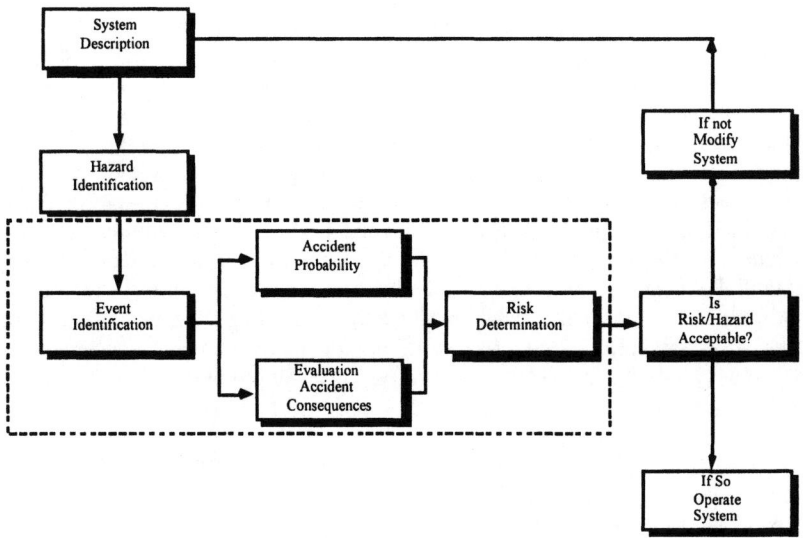

Figure 11.3. Hazard risk assessment flowchart.[8]

4. HEALTH RISK ASSESSMENT

Since 1970 the field of *health risk assessment* has received widespread attention within both the scientific and regulatory communities. It has also attracted the attention of the public. Studies on cancer caused a turning point in the world of risk because it opened the eyes of risk scientists and health professionals to the world of risk assessments. There are many definitions for the word risk. As described in the previous section, it is a combination of uncertainty and damage; a ratio of hazards to safeguards; a triplet combination of event, probability, and consequences; or even measure of economic loss or human injury in terms of both the incident likelihood and the magnitude of the loss or injury.

Properly conducted, risk assessments have received fairly broad acceptance, in part because they put into perspective the terms *toxic*, *hazard*, and *risk*. Toxicity is an inherent property of all substances. The study of toxicity of substances (toxicology) reveals that all chemical and physical agents can produce adverse health effects at some dose or under specific exposure conditions. In contrast, exposure to a chemical that has the capacity to produce a particular type of adverse effect represents a hazard. Risk, however, is the probability or likelihood that an adverse outcome will occur in a person or a group that is exposed to a particular concentration or dose of the hazardous agent. Therefore, risk is generally a function of exposure or dose. Consequently, health risk assessment is defined as the process or procedure used to estimate the likelihood that humans or ecological systems will be adversely affected by a chemical or physical agent under a specific set of exposure conditions.[9]

Health risk assessments provide an orderly, explicit, and consistent way to deal with scientific issues in evaluating whether a hazard exists and what the magnitude of the hazard may be. This evaluation typically involves large uncertainties because the available scientific data are limited, and the mechanisms causing adverse health impacts or environmental damage are only imperfectly understood. When one examines risk, how does one decide how safe is safe, or how clean is clean? To begin with, one has to look at both sides of the risk equation–that is, both the toxicity of a pollutant and the extent of public exposure. Information is required under both the current and potential exposure conditions, considering all possible exposure pathways. In addition to human health risks, one needs to look at potential ecological or other environmental effects. In conducting a comprehensive risk assessment, one should remember that there are always uncertainties and that the impact of these uncertainties and the effect of assumptions on the risk calculation must be included in the analyses.[10]

In recent years, several guidelines and handbooks have been produced to help explain approaches for doing health risk assessments. As discussed by a special National Academy of Sciences committee convened in 1983, most human or environmental health hazards can be evaluated by dissecting the analysis into four parts: *hazard identification, dose-response assessment or hazard assessment, exposure assessment*, and *risk characterization* (see Figure 11-4). For some perceived hazards, the risk assessment process might stop with the first step, hazard identification, if no adverse effect is identified or if an agency elects to take regulatory action without further analysis.[9] Hazard identification involves an evaluation of various forms of information in order to identify the different hazards. Dose-response or toxicity assessment is required in an overall assessment; responses/effects can vary widely since all chemicals and contaminants vary in their capacity to cause adverse effects. This step frequently requires the extrapolation of experimental data for animals to humans. Exposure assessment is the determination of the magnitude, frequency, duration, and routes of exposure of human populations and ecosystems. Finally, in risk characterization, toxicology and exposure information are combined to obtain a qualitative or quantitative expression of risk.

Risk assessment involves the integration of the information and analysis associated with the above four steps to provide a complete characterization of the nature and magnitude of risk, and the degree of confidence associated with this characterization. A critical component of the assessment is a full elucidation of the uncertainties associated with each of the major steps in the process. Under this broad concept of risk assessment are encompassed all of the essential problems of toxicology. Risk assessment takes into account all of the available dose-response data. It should treat uncertainty not by the application of arbitrary safety factors, but by stating them in quantitatively and qualitatively explicit terms, so that they are not hidden from decision makers. Risk assessment, defined in this broad way, forces an assessor to confront all the scientific uncertainties, and to set forth in explicit terms the means used in specific cases to deal with these uncertainties.[11]

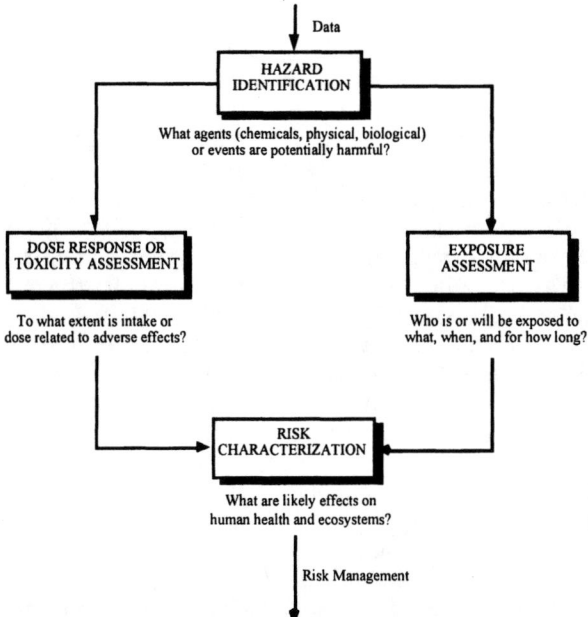

Figure 11.4. Hazard risk assessment flowchart.[8]

5. MULTIMEDIA ANALYSIS

It is now increasingly clear that some treatment technologies, while solving one pollution problem, have created others. Most contaminants, particularly toxic compounds, present problems in more than one medium. Since nature does not recognize neat, jurisdictional compartments, these same contaminants are often transferred across media. Air pollution control devices or industrial wastewater treatment plants prevent wastes from going into the air or water, but the toxic ash and sludge that these systems produce can become hazardous waste problems themselves. For example, removing trace metals from a flue gas usually transfers the products to a liquid or solid phase. Does this exchange an air quality problem for a liquid or solid waste management problem? Wastes disposed of on the land or in deep wells may contaminate groundwater, and evaporation from ponds and lagoons can convert solid or liquid wastes into air pollution problems.

The Great Lakes offer an excellent example of how our understanding of environmental problems has expanded, revealing the need for a multimedia focus and a total systems approach to adequately protect human health and our interdependent ecosystem. Two decades ago, a study by the International Joint Commission (IJC) identified nutrients and toxics problems in the five Great Lakes and found that Lake Ontario and Lake Erie, in particular, were afflicted with eutrophication problems. Since then, the U.S. and Canada have undertaken cooperative efforts that have successfully reduced nutrient loadings, particularly phosphorous, and helped to reverse eutrophication in the most severely affected areas. Since 1972, the U.S. government has spent over $7.6 billion on pollution problems in the Great Lakes, mostly for more than 1,000 municipal sewage treatment plants. With point source contributions of phosphorus increasingly under control, the importance of controlling toxic contamination is becoming more evident. Although some progress has been made, concentrations of persistent toxic substances such as mercury, PCBs, and lead remain unacceptably high in some parts of the Great Lakes, both in water and in sediments. Interestingly, the IJC has found atmospheric deposition to be a major pathway to contamination and has observed airborne sources for 10 or 11 "critical" toxic pollutants. For example, an estimated 50% of the heavy metals found in the Great Lakes results from atmospheric deposition, not water (or other liquid) discharges. Studies have registered deformities in fish and wildlife exposed to contaminated sediments and other sources of toxic chemicals in the Great Lakes. Although the decline in conventional pollutants has encouraged an increase in fish populations in some areas, all Great Lakes states still advise residents to limit or, in some cases, to eliminate their consumption of popular sport fishing species, such as perch, walleye, brown trout, and chinook salmon, because of their contamination by a variety of toxic contaminants.

EPA's own single-media offices, often created sequentially as individual environmental problems were identified and responded to in legislation, have played a role in impeding the development of cost-effective multimedia prevention strategies. In the past, innovative cross-media agreements involving or promoting pollution prevention, and voluntary arrangements for overall reductions in releases, had not been encouraged. However, new initiatives are characterized by their use of a wide range of tools, including market incentives, public education and information, small business grants, technical assistance, research and technology applications, as well as the more traditional regulations and enforcement, as has been described for some of the EPA/DOE energy conservation programs earlier in this chapter.

A *multimedia approach* can be applied to any type of system, product, or process. In the development to follow, a relatively simple example involving a chemical plant is first discussed. This is followed by a more detailed analysis of the manufacture, use, and ultimate disposal of a product. The discussion then concludes with a review of (potential) emissions to the environment from a hazardous waste incineration (HWI) facility.

Regarding a chemical plant, the reader is referred to Figure 11.5. This relatively simple method of analysis, as applied to a chemical plant alone, enables one to understand and quantify how various chemical inputs to the plant, i.e., chemicals, fuel, additives, and so forth, are chemically transformed within the facility and partitioned through plant components into various chemical outputs, i.e., the gaseous, aqueous, and solid discharge streams. The principles of the conservation law for mass (see Chapter 3, Part 1) clearly show that if a substance is removed from one discharge stream without chemical change (as opposed to one with a chemical, biological, or thermal reaction), the source and fate of all the chemicals in the process streams of a chemical plant may be determined and quantified. Chemical changes can be accounted for through chemical kinetic considerations.

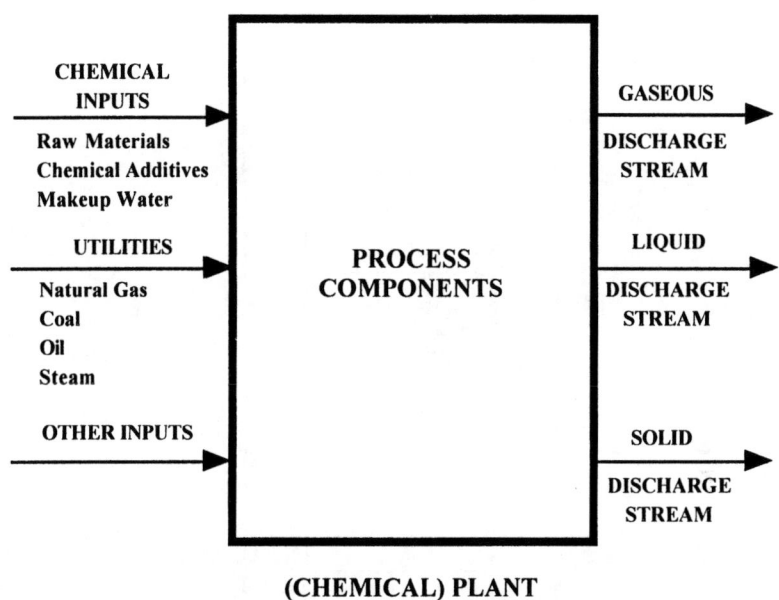

Figure 11.5. Multimedia approach for a hypothetical chemical plant.

Perhaps a more meaningful understanding of the multimedia approach can be obtained by examining the production and ultimate disposal of a product or service. A flow diagram representing this situation is depicted in Figure 11.6. Note that each of the ten steps in the overall "process" has potential inputs of mass and energy and may produce an environmental pollutant and/or a substance or form of energy that may be used in a subsequent or later step. Traditional partitioned approaches to environmental control can provide some environmental relief, but a total systems approach is required if optimum improvements, in terms of pollution/waste reduction, are to be achieved. This type of analysis has come to be described as the *life cycle* approach. Various components of the life cycle analysis methodology are reviewed in the next section of this chapter.

It should be obvious to the reader that a multimedia approach that includes energy conservation considerations requires a total systems approach. Much of the environmental engineering work in the future will focus on this area, since it appears to be the most cost-effective way of solving many environmental problems, and for evaluating new products and processes to minimize environmental releases and impacts to human health and the environment during all stages of a product's life.

234 Chapter 11

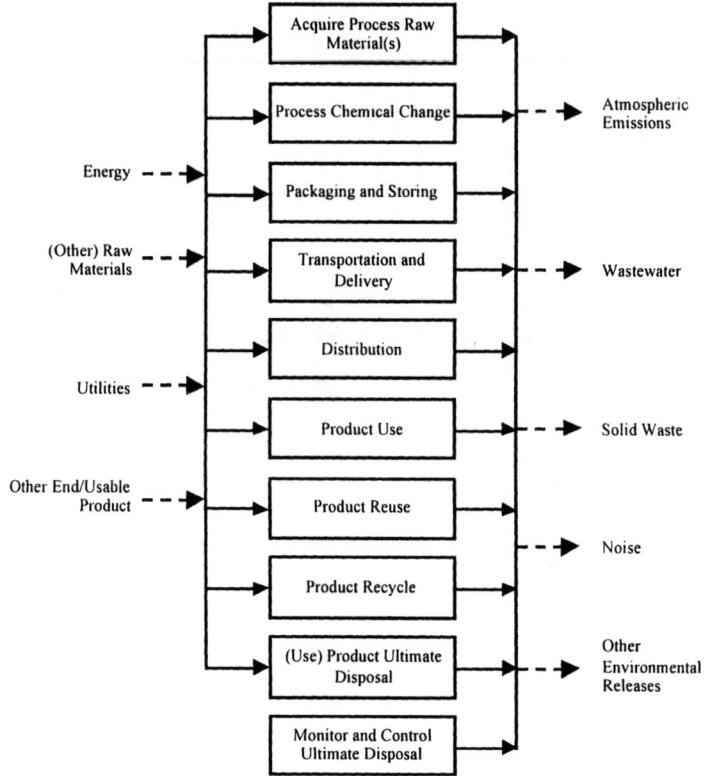

Figure 11.6. Overall multimedia process flow diagram for a hypothetical product.

6. LIFE CYCLE ANALYSIS[12]

Historically, the environmental review of products or processes has considered only those impacts that could be easily translated into financial terms (e.g., permitting costs and pollution control equipment costs). Consequently, these financially-based analyses often do not fully capture the benefits of pollution prevention opportunities, particularly those that reduce environmental concerns for the present and future.

Without the tools to completely document environmental benefits, pollution prevention opportunities have often been difficult to support when competing against more easily-quantified environmental projects such as end-of-pipe controls, and non-environmental investments such as remodeling or plant expansion.

Life cycle analysis (LCA) has developed over the past 20 years to provide decision makers with analytical tools that attempt to accurately and comprehensively account for the environmental consequences and benefits of competing projects. LCA is a procedure to identify and evaluate "cradle-to-grave" natural resource requirements and environmental releases associated with processes, products, packaging, and services. LCA concepts can be particularly useful in ensuring that identified pollution prevention opportunities are not causing unwanted secondary impacts by shifting burdens to other places within the life cycle of a product or process. LCA is an evolving tool undergoing continued development. Nevertheless, LCA concepts can be useful in gaining a broader understanding of the true environmental effects of current practices and of proposed pollution prevention opportunities.

LCA is a tool to evaluate all environmental effects of a product or process throughout its entire life cycle. This includes identifying and quantifying energy and materials used and wastes released to the environment, assessing their environmental impact, and evaluating opportunities for improvement.

LCA can be used in various ways to evaluate alternatives including in process analysis, material selection, product evaluation, product comparison, and policy-making. The method is also appropriate not only for process engineering evaluations, but also for use by a wide range of facility personnel ranging from materials acquisition staff, to new product design staff, to staff involved in investment evaluation.

Figure 11.7 illustrates the possible life stages that can be considered in a LCA and the typical inputs/outputs measured. Note the similarity of this figure with that presented above for the multimedia approach for process flow analysis in Figure 11.6. The unique feature of the LCA assessment is its multimedia focus on the entire "life cycle," rather than a single manufacturing step or environmental emission.

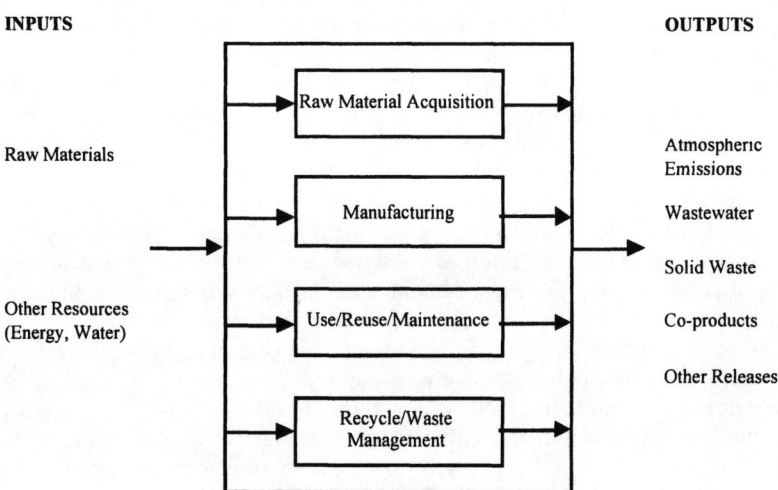

Figure 11.7. Life cycle stages for a hypothetical product.[12]

The theory behind this approach is that operations occurring within a facility can also cause impacts outside the facility's gates that need to be considered when evaluating project alternatives. Examining these "upstream and downstream" impacts can point out benefits or drawbacks to a particular opportunity that otherwise may have been overlooked. For example, examining whether to invest in washable/reusable cloth towels or disposable paper towels in a vehicle maintenance facility should include a comparison of all major impacts, both inside the facility (e.g., disposing of the paper towels) and "outside the gate" (e.g., wastewater discharges from the off-site washing of the reusable towels).

In addition, the LCA approach makes it possible to verify that a project that effectively solves one particular pollution problem does not result in cross-media shifting of pollution to another media (e.g., from waterborne to atmospheric releases). By examining all resource inputs (e.g., energy, materials, water) and environmental releases (e.g., air, water, and solid waste) across the entire life cycle of the product, process, or activity, the LCA can identify unacceptable cross-media transfers and transfers of pollutants to other life cycle stages.

6.1. The LCA Process

LCA generally consists of the following components: *Goal Definition and Scoping, Inventory, Impact Analyses*, and *Improvement Analyses*. *Goal Definition and Scoping* is a screening process which involves defining and describing the product, process or activity; establishing the context in which the assessment is to be made; and identifying the life cycle stages to be reviewed for the assessment. *Inventory Analysis* involves identifying and quantifying energy, water and materials usage, and environmental releases (e.g., air emissions, solid waste, wastewater discharge) during each life cycle stage. *Impact Assessment* is used to assess the human and ecological effects of material consumption and environmental releases identified during the inventory analysis. *Improvement Assessment* involves evaluating and implementing opportunities to reduce environmental burdens as well as energy and material consumption associated with a product or process.

Gaining a complete understanding of a proposed project's environmental effects requires identifying and analyzing inputs and releases from every life cycle stage. However, securing and analyzing this data can be a daunting task. In many cases facility decision makers may not have the time or resources to examine each life cycle stage or to collect all pertinent data.

Before beginning to apply LCA concepts to projects under review, facility managers must first determine the purpose and the scope of the study. In determining the purpose, facility managers should consider the type of information needed from the environmental review (e.g., Does the study require quantitative data or will qualitative information satisfy the requirements?). Once the purpose has been defined, the boundaries or the scope of the study should then be determined. What stages of the life cycle are to be examined? Are data available to study the inputs and outputs for each stage of the life cycle to be reviewed? Are the available data of an acceptable type and quality to meet the objectives of the study? Are adequate staff and resources available to conduct a detailed study?

These questions should be answered during the goal definition and scoping component of the project. Goal definition and scoping links the purpose and scope of the assessment with available resources and time and allows reviewers to outline what will and will not be included in the study. In some cases, the assessment may be conducted for all stages of the life cycle (i.e., raw materials acquisition, manufacturing, use/reuse/maintenance, and recycling/waste management). In many cases, the analysis may begin at the point where equipment and/or materials enter the facility. In other cases, primary emphasis may be placed on a single life cycle stage, such as identifying and quantifying waste and emissions data. In all cases, managers should ensure that the boundaries of the LCA address the purpose for which the assessment is conducted and the realities of resource constraints. At a minimum, all life cycle stages in which significant environmental impacts are likely to occur should be included in the LCA.

Determining the purpose and scope of the study will help to identify the type of environmental analysis that should be conducted when applying LCA concepts, i.e., either a *Life Cycle Checklist*, or a *Life Cycle Assessment Worksheet*.

6.2. The Life Cycle Checklist

If resources are limited and an in-depth, quantitative life cycle analysis is not practical, a facility manager may consider using a simple checklist to identify and highlight certain environmental implications associated with competing projects. A checklist using qualitative data instead of quantitative inputs can be very useful when available information is limited or as a first step in conducting a more thorough LCA. In addition, a *Life Cycle Checklist* should include questions regarding the environmental effects of current operations and/or potential projects in terms of materials and resources consumed and wastes/emissions generated. Figure 11.8 provides a sample checklist. The checklist used by an individual facility can be tailored to emphasize areas of specific concern. For example, a facility in an area of the country where landfill space is limited may want to emphasize the collection and evaluation of solid waste generation data. Similarly, a facility located in arid or semi-arid areas may want to collect and evaluate information relating to water consumption.

A Life Cycle Checklist is relatively easy to use and requires limited resources. On the other hand, it does not provide a detailed or complete assessment of the environmental consequences associated with the activity under review. If a more detailed analysis is required, a more comprehensive *Life Cycle Assessment Worksheet* may be required.

6.3. The Life Cycle Assessment Worksheet

A *Life Cycle Assessment Worksheet* provides a codified method of collecting and analyzing more detailed information and conducting a more in-depth analysis than when using a Life Cycle Checklist to identify and evaluate the resource and material inputs and the environmental releases associated with each life cycle stage of a process evaluation or pollution prevention project. Life cycle-based worksheets (Figure 11.9) are generally organized into three sections. The first section asks for a flowchart of the process steps or activities to be included in the analysis. The second section asks for inputs (i.e., raw materials, energy, and water), and the third section asks for outputs (i.e., products, air, water, and land releases). Electronic versions of these worksheets are available on disk or on-line in Lotus 1-2-3 or Excel format from EPA's Pollution Prevention Information Clearinghouse (PPIC technical assistance hot line 202/260-1023).

Issue	Question	Yes	No
Material Usage	Does the project minimize the use of raw materials?		
Resource Conservation	Does the project minimize energy usage?		
	Does the project minimize water usage?		
Local Environmental Impacts	Does the project eliminate or minimize impacts to the local environment, i.e., air, water, land?		
Global Environmental Impacts	Does the project eliminate or minimize impacts known to cause global environmental concerns, e.g., global warming, ozone depletion, acid rain?		
Toxicity Reduction	Does the project improve the management of toxic materials and/or processes which result in human/ecological exposure?		

Figure 11.8. An example life cycle checklist.[12]

6.4. The Life Cycle Inventory Analysis

The first step in identifying and evaluating the inputs and outputs associated with life cycle stages under review is to describe and understand each step in the process. One common method to do this is to construct a system flow diagram for the product, process, or activity being studied. Each step within the relevant life cycle stages is represented by a box. Each box is connected to other boxes that represent the preceding and succeeding step. A simple example of a process flow diagram is illustrated below. In this example, the life cycle stages covered within the diagram begin at the point a solvent is purchased for use and enters an industrial facility property. Each of these boxes can be further divided into detailed process flow steps.

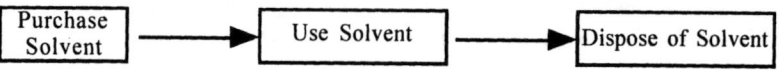

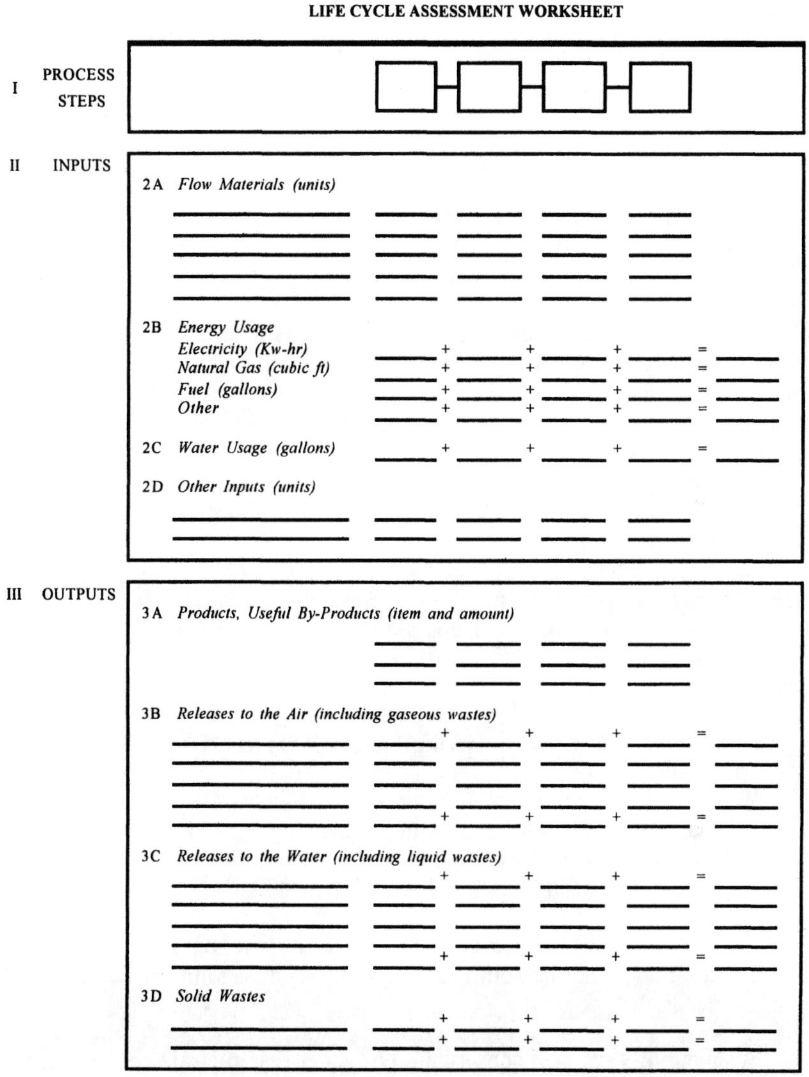

LIFE CYCLE ASSESSMENT WORKSHEET

Figure 11.9. An example life cycle assessment worksheet.[12]

When all relevant steps for each stage of the product, process, or activity under review have been identified, the flow diagram should be expanded to identify the specific energy and material inputs, and the specific environmental releases associated with each box on the diagram. This step is of crucial importance, because data on these identified inputs and releases will be collected later, and will form the basis for all findings and conclusions. The diagram below illustrates the inputs and releases for each step in the sample flow diagram for solvent purchasing at an industrial facility.

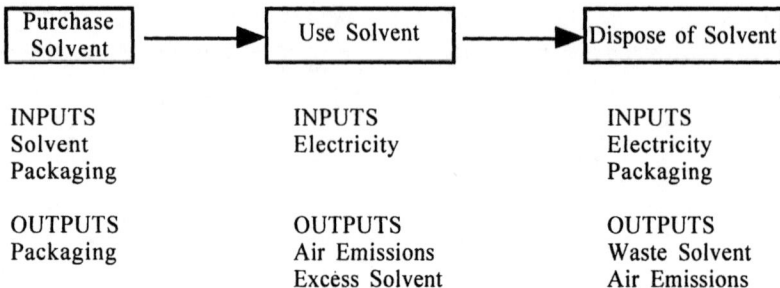

INPUTS	INPUTS	INPUTS
Solvent	Electricity	Electricity
Packaging		Packaging

OUTPUTS	OUTPUTS	OUTPUTS
Packaging	Air Emissions	Waste Solvent
	Excess Solvent	Air Emissions

Once a flow diagram has been developed, personnel conducting the LCA should identify sources of information that will describe and quantify the material and energy inputs, and the environmental releases associated with each box in the process flow diagram to complete the Inventory portion of the LCA.

Following the *Inventory*, *Impact Analysis* to assess the human and ecological effects of material consumption and environmental releases is carried out as described for the *Risk Characterization* in the *Health Risk Assessment* section above. If unacceptable impacts are indicated from the *Impact Analysis*, some alternative options must be selected or some mitigation must be carried out to cost effectively reach acceptable risk levels. This implementation of mitigation activities is incorporated into the LCA *Improvement Assessment* step.

7. SUSTAINABLE DEVELOPMENT

Sustainability involves simultaneous progress in four major areas: human, economic, technological, and environmental. The United Nations (World Commission on Environment and Development 1987 Report, *Our Common Future*) defined "sustainable development" as:

> Development that meets the need of the present without compromising the ability of future generations to meet their own needs.

Sustainability requires conservation of resources, minimizing depletion of nonrenewable resources, and using sustainable practices for managing renewable resources. There can be no product development or economic activity of any kind without available resources. Except for solar energy, the supply of resources is finite. Efficient designs conserve resources while also reducing impacts caused by material extraction and related activities. Depletion of nonrenewable resources and overuse of otherwise renewable resources limits their availability to future generations.

Another principal element of sustainability is the maintenance of ecosystem structure and function. Because the health of human populations is connected to the health of the natural world, the issue of ecosystem health is a fundamental concern to sustainable development. Sustainability requires that the health of all diverse species as well as their interrelated ecological functions be maintained. As only one species in a complex web of ecological interactions, humans cannot separate their survivability from that of the total system.

Sustainable development demands change. Consumption of energy, natural resources, and products must eliminate waste. The manufacturing industry can develop green products that can meet the sustainability requirements. Life cycle analysis, design for environment and toxic use reduction are elements that help sustainability. Sustainable manufacturing, for example, extends the responsibility of industry into material selection, facility and process design, marketing, cost accounting, and waste disposal. Extending the life of a manufactured product is likely to minimize waste generation. Design engineers must consider many aspects of the product including its *durability, reliability, remanufacturability* and *adaptability*.

Designing a product that can withstand wear, stress, and degradation extends its useful life. This, in many cases, reduces the cost and impact on the environment. Reliability is the ability of a product or system to perform its function for the length of an expected period under the intended environment. Reducing the number of components in a system and simplifying the design can enhance the reliability. Screening out potentially unreliable parts and replacing with more reliable parts helps to increase the system reliability.

Adaptable designs rely on interchangeable parts. For example, consumers can upgrade components as needed to maintain state-of-the-art performance. In remanufacturing, used worn products are restored to "like-new" condition. Thus remanufacturing minimizes the generation of waste. Products that are expensive but not subject to rapid change are the best candidates for remanufacturing. Design continuity between models in the same product line increases interchangeable parts. The parts must be designed for easy disassembly to encourage remanufacturing.

Design of products that emphasizes efficient use of energy and materials reuse and recycling reduces waste and supports sustainability. By effective recycling, material life can be extended. Materials can be recycled through open loop or closed loop pathways. For example, post-consumer material is recycled in an open loop one or more times before disposal. (This topic is discussed in more detail in Chapter 14.) However, in a closed-loop pathway, such as with solvents, materials within a process are recovered and used as substitutes for virgin material. Minimizing the use of virgin materials supports sustainability. Thus, resource conservation can reduce waste and directly lower environmental impact. Manufacturing a less material-intensive product not only saves materials and energy but will also be lighter, thus reducing energy and costs related to product transportation. Process modifications and alterations specifically focused on replacing toxic materials with more benign ones minimize the health risk and the environmental impact of material used and product manufacturing. It also improves the health and safety of employees. Process redesign may also yield "zero discharge" by completely eliminating waste discharges. Thus, sustainability can be accomplished through several different approaches. Evaluating these options up-front will aid in developing truly sustainable processes and products, and

242 Chapter 11

is much more desirable than implementing control measures after unacceptable waste releases occur.

Finally, responsible businesses can begin moving toward sustainability by taking six steps:

1. Foster a company culture of sustainability.
2. Initiate voluntary performance improvements.
3. Apply eco-efficiency (material and energy conservation, toxic use reduction, recycling, etc.) concepts.
4. See opportunities for sustainable business growth.
5. Invest in creativity, innovation, and technology for the future.
6. Reward employee commitment and action.

8. ISO 14000[13,14,15]

ISO 14000 is a series of environmental standards, developed outside of the regulatory community, that has the potential to revolutionize the way industry, government, and the public respond to environmental issues. ISO 14000 is designed and managed by the International Organization for Standardization (ISO), a private sector body. The group was founded in 1947, and is responsible for standardizing everything from paper size to film speeds. The American National Standards Institute (ANSI) is the U.S. representative to ISO.

The 14000 standards are a voluntary series of guidelines designed to address environmental management. They focus on management, rather than specific performance based on the belief that correct management will lead to better performance. ISO 14000 asks industry leaders to accept their responsibility for the environment, and consider this in corporate decision making. It also encourages companies to be proactive, and incorporate pollution prevention ideals into their environmental management activities.

ISO 14000 is actually made up of many individual standards, which fall under two broad categories: organizational evaluation and product development evaluation. Individual standards are assigned numbers within the 14000 series. Some of these standards have been finalized, but many are still in the process of being developed.

8.1. Company Concerns

Corporations are recognizing the benefits of ISO 14000 certification. Certification allows companies to reduce the cost of doing business, create consistency, improve public image and to be recognized on a global level as environmental leaders. Many experts say that it is quickly becoming a requirement of many international customers, and expect it to become a Federal requirement in the near future.

Proper environmental management should reduce the cost of doing business as greater efficiency is achieved, allowing revenue to be used for growth and capital projects instead of waste management and pollution control. As indicated in Chapter 9, historically, only waste disposal has been included in determining the cost of environmental management. However, many other costs must be looked at such as OSHA compliance, DOT shipping costs, permitting fees, protective equipment, hazardous materials storage, training, staffing, insurance, spill cleanup, fines, long term liability costs and more. By considering this 'total' cost, companies have begun to realize the importance of looking at alternatives which are more environmentally sound and likely to cost less on a life cycle basis.

Having a proper life cycle management plan in place will help companies focus on prevention, reduce the costs of maintaining compliance, and reduce liability. As described above, life cycle management considers at all stages of a material's life. This includes raw material use, manufacturing, transportation, and potential waste generation and disposal. Pollution prevention activities are compatible with life cycle analysis efforts, as for example, facilities may consider non-hazardous alternatives to hazardous solvents, or use smaller amounts of water or materials by processes such as countercurrent rinsing when evaluating the outcome of a life cycle assessment for their products or processes.

Commitment to environmental protection and continuous improvement will help businesses consider environmentally sound options when they become available. By involving employees and the community in crafting an environmental policy, companies can improve their corporate image, and open the door for further improvements.

The environmental policy of an organization in compliance with ISO standards must contain several elements including:

- Environmental objectives and targets for the organization
- Advocacy of the use of best technology and management practices
- The commitment to meet or exceed any legal and regulatory requirements related to environmental compliance for the company
- The commitment to continual improvement and prevention of pollution
- A framework for setting and reviewing environmental objectives and targets of the company
- A commitment to consider the views of interested parties

The policy must be documented, implemented and maintained, and must be communicated to all employees. The policy should also be available to the public.

8.2. The Certification Process

Certification is a process that begins with upper management commitment. Once the decision is made to improve, an *Environmental Management System (EMS)* must be developed. This is a system to ensure that environmental decisions are made holistically. It includes a written program, education and training, and knowledge of relevant laws. Development often involves teams from all corporate departments, and helps companies realize how far-reaching environmental considerations are in their business. Through this process, areas of environmental concern are identified, allowing the company to begin developing and prioritizing environmental improvements. Once prioritized target areas are identified, a corporate environmental policy must be designed. Again, it is important to remember that pollution prevention is critical in redesigning the corporate approach to the environment where a company commits to trying to produce the best product, with the smallest environmental impact.

Part of the certification process requires that businesses quantify the impact they are having on the environment. This evaluation should include factors such as air emissions, solid waste, hazardous waste, potential contamination, and use of raw materials. A full understanding of a company's impact is critical to identifying opportunities for change and improvement. After baseline studies are completed, a full-scale environmental audit must be completed. This process can be performed in-house, or by an outside party, and includes a mass balance of all raw materials and wastes generated through the processes used at a facility. Companies must consider raw materials, operational wastes, and disposal in assessing their products and processes. These steps help provide guidelines for change, and should be performed at least annually. As discussed above, life cycle assessments are an integral part of the process.

The process described above represents a company's approach to self-certification of compliance with ISO standards, without a third party audit. An organization can also be certified through third party auditing. This procedure involves identifying an accredited registrar firm for certification. The certified auditors from the registrar's office then will plan to visit the facility and audit their EMS. The audit team then completes its audit report and submits its findings to the registrar. The registrar, based on the findings and recommendations of the audit report, makes a registration decision and acts accordingly, and communicates the decision to the organization that they audited. Effective communication between the registrar and the facility to be audited well before the time of audit will alleviate many problems and make the audit go smoothly and successfully. A pre-audit by the facility is essential to identify problem areas and correct them before the third party audit begins.

8.3. Benefits and Pitfalls of ISO 14000

Some of the potential benefits of the ISO 14000 standards are listed below.

1. The ISO 14000 standards provide industry with a structure for managing their environmental problems which presumably will lead to better environmental performance.
2. It facilitates trade and minimizes trade barriers, harmonizing different national standards. As a consequence, multiple inspections, certifications, and other conflicting requirements could be reduced.
3. It expands possible market opportunities.

4. In developing countries, ISO 14000 can be used as a way to enhance regulatory systems that are either nonexistent or weak in their environmental performance requirements.
5. A number of potential cost savings can be expected, including:
    - Increased overall operating efficiency
    - Minimized liability claims and risk
    - Improved compliance record (avoided fines and penalties)
    - Lower insurance rates
6. ISO 14000 registration can demonstrate an organization's commitment and credibility regarding environmental issues (e.g., corporate image and community goodwill).

Some of the potential pitfalls of the ISO 14000 standards include:

1. Implementation of ISO 14000 standards can be a tedious and expensive process.
2. ISO 14000 standards can indirectly create a technical trade barrier to both small businesses and developing countries due to their limited knowledge and resources (e.g., complexity of the process and high cost of implementation, lack of a registration and accreditation infrastructure, etc.).
3. ISO 14000 standards are a voluntary standard. However, some countries may make ISO 14000 standards a regulatory requirement which can potentially lead to a trade barrier for foreign countries who cannot comply with the standards.
4. Certification/registration issues, including:
    - the role of self-declaration versus third party auditing
    - accreditation of the registrars
    - competence of ISO 14000 auditors
    - harmonization and worldwide recognition of IS0 14000 registration certificates worldwide

9. MISCELLANEOUS POLLUTION PREVENTION CONSIDERATIONS

9.1. Small Quantity Generators

Small Quantity Generators (SQGs) are defined as facilities generating between 100 and 1,000 kg of hazardous waste per calendar month, while a *Conditionally Exempt Small Quantity Generator* generates by definition less than 100 kg of hazardous waste per calendar month. These sources represent a wide variety of industrial groups and SIC codes (Table 11.2), and present unique challenges in terms of regulation and technical assistance. These generators generally have fewer than 5 to 10 employees and are managed and staffed by individuals with limited training in identification and management of hazardous wastes or pollution prevention.

While these approximately 600,000 to 700,000 SQGs in the U.S. do not individually contribute significant quantities of materials to the waste stream, the nature of their waste materials (Table 11.3) and their aggregate are a significant part of the total hazardous waste stream, generating as much as 1 million metric tons of hazardous wastes annually.

Table 11.2. Primary SIC Groups Likely to Contain SQGs.[16]

Pesticide End-Users	Pesticide Application Services
Chemical Manufacturing	Wood Preserving
Formulators	Laundries
Photography	Textile Manufacturing
Equipment Repair	Metal Manufacturing
Construction	Motor Freight Terminal
Printing/Ceramics	Paper Industry
Analytical and Clinical Laboratories	Educational and Vocational Shops
Wholesale and Retail Sales	
Cleaning Agents and Cosmetic Manufacturers	
Furniture/Wood Manufacturing and Refinishing	
Vehicle Maintenance (with and without lead-acid batteries)	
Other Services (Funeral, Cleaning, Maintenance, etc.)	
Other Manufacturers (Leather, Flat Glass, Cotton Ginning, Asbestos, Abrasives, etc.)	

Table 11.3. Major Waste Types Generated by SQGs.[17]

Spent solvents
Strong acids and alkalis
Photographic wastes
Dry cleaning filtration residues

Management of the large number and diffuse nature of waste from SQGs is a challenge, as the common SQG description includes small size, concern for short-term costs, apprehension about regulations and audits, and limited technical training, or any training for that matter, waste management problems and solutions. In addition, many of the firms are not aware of new manufacturing and production technologies and approaches that could provide them with improved materials use and increased productivity. These firms are fertile ground for *Technical Assistance Programs* of various types which take a nonregulatory posture in their assistance to small businesses. Many of these Technical Assistance Programs provide both technical manufacturing and waste management information, provide complementary or reduced-cost energy and/or manufacturing efficiency audits, provide referrals to consultants, provide free hotline emergency information, training and seminars, and even may have available economic incentives for energy, manufacturing, and waste management improvement project implementation for small business clients. The U.S. EPA has a Small Business Assistance Program accessible through their home page on the Internet (http://www.epa.gov), and EPA provides funding associated with the Clean Air Act for individual state small business assistance and voluntary compliance audit programs as well. The U.S. Department of Commerce funds the Manufacturing Extension Partnership (MEP) program (http://www.mep.nist.gov/) throughout the 50 states which provide technical assistance, through a network of technical field agents, to small manufacturing entities with 500 employees or less. The MEP program focuses on providing manufacturing technical assistance in the areas of process layouts, energy efficiency, lean manufacturing, environmental compliance, environmental management systems, etc., all directed at improving the competitiveness of small U.S. manufacturers. As suggested by these technical assistance areas, the MEP program support is intimately connected to improving the small manufacturers' environmental performance as well. The reader should note that the bulk of the material presented in Part 3, on applications and case studies, is primarily directed toward SQGs.

9.2. Domestic Activities

Many of the ideas and techniques discussed above for industrial settings have relevance to individuals and activities that take place at home. The use of pollution prevention principles on the home front clearly do not involve the use of high-tech equipment or major lifestyle changes. Success of pollution prevention at the domestic level depends only on the active and willing participation of the public in energy and waste saving activities. Some examples of individual actions that can be taken to reduce waste, conserve energy, or improve health and safety at home or in the office are identified below, while an extensive discussion of domestic pollution prevention opportunities can be found in Theodore and Theodore:[8]

At Home
- *Waste Reduction* - Purchase products with the least amount of packaging; Borrow items used infrequently.
- *Energy Conservation* - Use energy efficient lighting, e.g., fluorescent bulbs; Install water-flow restriction devices on sink faucets and shower heads.
- *Accident, Health, and Safety* - Handle materials to avoid spills, trips, and slips; Keep hazardous materials out of reach of children.

At the Office
- *Waste Reduction* - Pass on verbal messages when written correspondence isn't required, email when a letter isn't needed; Reuse paper before recycling it.
- *Energy Conservation* - Don't waste utilities just because you are not directly paying for them; Take public transportation to the office.
- *Accident, Health, and Safety* - Know building evacuation procedures; Follow company medical policies, e.g., annual physicals.

The public plays an important role in supporting pollution prevention, waste minimization, and life cycle assessment approaches of businesses by purchasing environmentally friendly goods and services from business and industry. What is evident, however, is that a desire to use "green" products and services by the public will be of no avail unless these goods and services are available in the marketplace. This is another role that the public can play in stimulating pollution prevention and sustainable development, communicating to business organizations and government institutions the support of green products and legislation, and the disapproval of products and/or activities that have negative impacts on the environment. Through this communication and feedback, individuals can become intimately involved in improving the environmental impact of our current society to provide a safer and cleaner environment for future generations.

9.3. Ethical Considerations

Some would argue that the ethical behavior of engineers is more important today than at any time in the history of the profession. The engineers' ability to direct and control the technologies they master has never been stronger. In the wrong hands, the scientific advances and technologies of today's engineer could become the worst form of corruption, manipulation, and exploitation. Engineers, however, are bound by a code of ethics that carry certain obligations associated with the profession. Some of these obligations include:

1. Support one's professional society.
2. Guard privileged information.
3. Accept responsibility for one's actions.
4. Employ proper use of authority.
5. Maintain one's expertise in a state of the art world.
6. Build and maintain public confidence.
7. Avoid improper gift exchange.
8. Practice conservation of resources and pollution prevention.
9. Avoid conflicts of interest.
10. Apply equal opportunity employment.
11. Practice health, safety, and accident prevention.
12. Maintain honesty in dealing with employers and clients.

There are many codes of ethics that have appeared in the literature. The preamble for one of these codes is provided below as a typical code of engineering ethics:

> Engineers in general, in the pursuit of their profession, affect the quality of life for all People in our society. Therefore, an Engineer, in humility and with the need for Divine guidance, shall participate in none but honest enterprises. When needed, skill and knowledge shall be given without reservation for the public good. In the performance of duty and in fidelity to the profession, Engineers shall give utmost.[18]

As individuals and as organization members, we may agree that our primary ethical responsibility is to protect public health, with accompanying responsibility to protect the environment. In practical daily work, however, we are usually consumed with serving our employers and customers or clients. Within a company, conflicts may arise for individuals depending on how the responsibility for environmental compliance is allocated. To encourage individual responsibility for costs associated with production, a company may assign pollution control costs to each production unit. With this approach, each line manager may be inclined to minimize pollution control costs to show higher profits. The corporate entity should encourage ethical behavior in this case to strive for pollution prevention and waste minimization efforts rather than illegal activities to avoid required pollution control costs. In general this method of localized responsibility is preferred over the situation where the responsibility for pollution control and related costs is centralized so that production units have less incentive to prevent pollution and maintain pollution control equipment at their individual units.

Although the environmental movement has grown and matured in recent years, its development is far from stagnant. To the contrary, change in individual behavior, corporate policy, and governmental regulations are occurring at a dizzying pace. Because of the Federal Sentencing Guidelines, the Defense Industry Initiative, as well as a move from compliance to a values-based approach in the marketplace, corporations have inaugurated company-wide ethics programs, hotlines, and senior line positions responsible for ethic training and development. The Sentencing Guidelines allow for mitigation of penalties if a company has taken the initiative in developing ethics training programs and codes of conduct. In the near future, these same Guidelines will inevitably apply to infractions of environmental law, requiring regulated businesses to consider ethics integration into engineering activities as a means of mitigating adverse effects from unintentional compliance violations.

9.4. Environmental Justice

The issue of *environmental justice* has come to mean different things to different people. EPA indicates that environmental justice is the fair treatment and meaningful involvement of all people, regardless of race, color, national origin, or income, with respect to the development, implementation, and enforcement of environmental laws, regulations, programs, and policies. Fair treatment means that no racial, ethnic, or socioeconomic group should bear a disproportionate share of the negative environmental consequences resulting from industrial, municipal, and commercial operations, or from the execution of federal, state, local, or tribal programs and policies.

The concept of environmental justice is composed of four, key, interrelated elements, namely: environmental racism, environmental health, environmental equity, and environmental politics. Overconsuming resources and polluting the ecosystem in such a way that it enjoins future generations from access to reasonable comforts irresponsibly transfers problems to the future in exchange for short-term gain. Beyond this intergenerational conflict, enormous inequi-ties exist in the distribution of resources between developed and less developed countries. Inequities also occur within national boundaries, and it is these inequities that have led to a growing environmental justice movement in the U.S.

Environmental justice can be achieved, in part, with a concerted effort on the part of grassroots and mainstream activists. Minorities have a responsibility to exercise their rightful political and legal power. At the same time, federal protection policy needs to devote attention to specific minority environmental concerns, to monitor the implementation and enforcement of environmental regulations, and to incorporate considerations of equity into policy making procedures.

According to a U.S. General Accounting Office study examining population ethnicity and location of off-site hazardous waste landfills in the southeastern region of the United States, African Americans comprise the majority of the population of three out of every four communities with such hazardous waste landfills.[19] While siting decisions supposedly result from technical concerns, there are no geological reasons to site waste disposal facilities in low-income, minority areas. Political and economic reasons provide a partial explanation for this concentration. Residents are often unaware of the negative effects of environmental hazards and of the available recourses to opposition. Further, they are rarely politically organized or influential, and they often lack the economic and legal resources needed to oppose such unfavorable land uses.

Even though low-income, minority communities have historically lacked political, legal, and economic power, community activism and mobilization have been effective in combating certain environmental problems. Further, existing environmental legislation and the power of the state and federal bureaucracy can actually be used in achieving the goals of community activists. *People United for a Better Oakland (PUEBLO)*, a multiracial organization centered in Oakland, California, successfully employed community activism though political empowerment and legal and economic pressure in the pursuit of environmental justice. Alarmed by a study claiming 20% of Oakland children had lead blood levels high enough to cause permanent brain damage, PUEBLO activists began a campaign to "Get the Lead Out" in 1990. Activists demanded lead screening for low-income children, locally mandated changes of city lead ordinances, and lead cleanup in existing contaminated sites. Even though lead screening was legally mandated by the state of California, many clinics and doctors were not offering the service. Further, PUEBLO activists were concerned that many families did not realize that lead in their children was a problem. In a 1991 class-action lawsuit Matthews v. Coye, the NAACP, ACLU, and NRDC all rallied for Medicaid to provide for the testing of lead in Californian children. A settlement required the state of California to screen about 500,000 low-income children under the age of six for lead poisoning at a cost of about $15 to $20 million. In addition, Oakland imposed a 10 dollar tax levy on existing homes. This tax was use to finance projects of home abatement, park testing, and parental education to prevent and protect against lead poisoning. This was a substantial victory for a poor minority community, and demonstrates that environmental activism can work to reduce risk and to challenge the inequitable pattern of waste distribution, and implementation and enforcement of environmental laws.

To date, environmental legislation has not fully addressed the inequities inherent in capitalism; economic factors predispose certain segments of the population to increased risk. The process of policy making, implementation, and enforcement of environmental regulations has redistributed and concentrated risks in low-income, minority communities. Even though these communities seem to lack political, legal, and economic power, community activism and mobilization have been effective in some instances in combating specific environmental problems. The example of the PUEBLO activism illustrates the possibility of success when minorities actively exercise their rightful political and legal power in combination with the power of federal and state governments and existing environmental legislation. This type of activism promises to increase in the future.

Although success incorporates mainstream participation, the environmental justice movement should be led by the voice and concerns of low-income, minority victims globally. Future eradication of the disproportionate exposure of environmental pollution will hopefully challenge wider systemic inequalities that plague minority communities worldwide.

10. ILLUSTRATIVE EXAMPLES

Example 10.1. To make informed, intelligent decisions regarding pollution prevention, one must have a firm grasp of the intricate relationships among energy, technology and pollution. Pollution can often be traced back to the misuse of energy and thus one of the most potent weapons in minimizing pollution is to reduce the use of energy. Technology should be directed toward energy efficiency, i.e., toward getting more "bang for the Btu." The greatest strides in pollution reduction since 1970 have been achieved through energy conservation with little or no reduction in the standard of living. In fact, in most cases, large savings in cash expenditures have been achieved. It has been calculated that if the entire U.S. were to switch to compact fluorescent bulbs overnight, it would be transformed from an energy importing nation to an energy exporting nation.

As a consumer society, we have been encouraged to use more material goods as a matter of course. All of these materials require energy at every step of production. Energy, regardless of its form, produces pollution in its generation and use. Below are three simple activities that we engage in everyday. Choose one of these activities and develop a "pollution tree" which traces the basic activity back to the original source of raw material, listing the type of waste/pollution at each step. You may expect to generate one main pathway and several side paths. Estimate the impact of each (large, moderate, small). Generate alternatives and or minimization techniques. Think globally!
    a.   Mowing the lawn.
    b.   Washing clothes.
    c.   Having a hamburger, soda and fries at the local fast food outlet.

SOLUTION. Mowing the lawn. Mowing one lawn is a trivial activity but when the impact is multiplied by the millions of lawns in the U.S., the impact may be considerable. This would be more likely to be a problem in cities such as Los Angeles and New York City which are already under a heavy pollution load. The "pollution tree" for mowing a lawn is shown below.

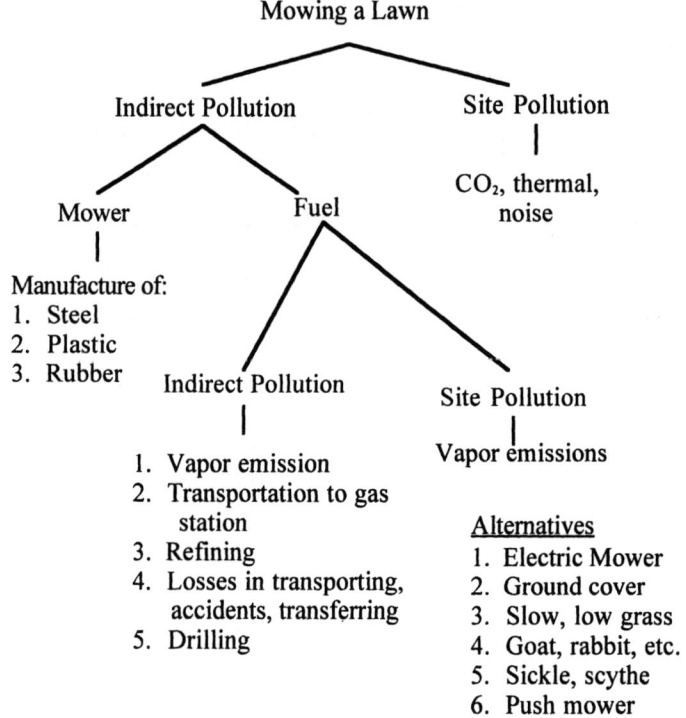

As can be seen, each of the items under indirect pollution (manufacture, refining, etc.) will require a pollution tree of their own. Also, the use of an electric mower, of course, brings in the pollution generated at the power generating station (CO_2, other emissions) as well as the pollution caused by the fuel used at the power plant. It is up to the individual to determine how extensive a tree is required as it quickly becomes very complex. Overall, the pollution due to lawn mowing is very small, relative to say, the automobile. This problem does not consider the ultimate disposal of the lawn mower nor of the clippings, if they are collected. Nor does it consider the impact of having a lawn in the first place. An extensive lawn (e.g., a golf course) which is maintained with fertilizers, herbicides and pesticides presents a very significant pollution source, and would require a modified "pollution tree."

Example 10.2. Dry cleaning operations represent a common Small Quantity Generator that can pose significant health, safety, and environmental impact risks if not managed properly. At a small dry cleaning operation, an employee uses trichloroethylene (TCE) to remove stains from clothing. TCE has been identified as a carcinogen. Also, since it is a volatile organic compound (VOC), emissions of TCE to the atmosphere contribute to photochemical smog formation, and EPA has placed TCE on its toxics list.

Estimate allowable worker exposure and emissions rates from dry cleaning operations by answering the following questions. The information below is needed for this estimation.

GIVEN: Vapor pressure of TCE (from CRC Handbook)
log P (torr) = (-1816.8/T) + 7.95642, where T is in Kelvin;
molecular weight of TCE = 131.39 g/gmol;
TLV data: TLV TWA = 50 ppm, TLV STEL = 200 ppm
*NOTE: The "TLV" is the threshold limit value or the exposure concentration below which there are not adverse health effects to the exposed individual; "TWA" is the maximum allowable, 8-hr time-weighted average exposure concentration; "STEL" is the short-term exposure limit defined as the 15-min TWA exposure which should not be exceeded at any time during an 8-hr work day.

The time variation of the concentration of a species emitted into a volume V may be calculated from:

$$\frac{dC}{dt} = \frac{G(t)}{V} - kQ\left(\frac{C}{V}\right)$$

where C = concentration of species at time t, kg/m^3; G(t) = emission rate of species at time t, kg/s; V = volume of room, m^3; Q = room ventilation rate, m^3/s; and k = mixing constant for imperfect mixing, with k = 0.3 at height for stagnant conditions. This equation assumes that the concentration of the species = 0 in the ventilation air.

   a. Suppose no ventilation of the workroom occurs. Assuming a constant usage rate, how much TCE can evaporate during a typical 8-hr workday before the recommended maximum exposure is exceeded? Assume the room is 4 m wide, 5 m long, and 4 m in height.
   b. Assume that actual evaporation rates of TCE average 50 g/d. What minimum ventilation rates are needed to maintain the steady-state room concentration of TCE below acceptable values?
   c. Assume that, on the average, there are two dry cleaning stores per urban mi^2, and that the stores operate 6 d/wk. For the ventilation rates computed in Part b, what annual emission rates can be attributed to dry cleaning in a major metropolitan area, which has an area of approximately 100 mi^2?
   d. Since the evaporation rate is proportional to vapor pressure, the emissions might be controlled by lowering the temperature at which the TCE is held. What reduction in temperature (from the ambient temperature of 25°C) would be required to halve the emissions rate? What temperature is required to reduce the vapor pressure to 10% of that at 25°C?

e. As part of a pollution prevention strategy, it is suggested that the TCE vapors might be recovered from the ventilation stream by chilling at constant (atmospheric) pressure. Evaluate the feasibility of this suggestion by computing the temperature to which the stream must be cooled. Assume the room concentration of TCE is equal to 25% of the TWA value.

SOLUTION.
a. If G_{tot} is the total vapor emitted during the day, the equation given may be solved to obtain

$$C = G_{tot}/V$$

NOTE: It has been assumed that the concentration in the room is spatially homogeneous.

Since the vapor concentration is cumulative, it is appropriate to compute the mass of vapor which represents the short-term exposure limit value (the concentration to which a person should be exposed no longer than 15 minutes).

Using C = STEL = 200 ppm maximum concentration = 200×10^{-6} atm, and V = 80 m³, the total vapor mass allowed to be emitted daily is:

$$\frac{n}{V} = \frac{p}{RT} = \frac{(200 \times 10^{-6} \text{ atm})}{(82.057 \text{ cm}^3\text{-atm/gmol-K})(298 \text{ K})} = 8.179 \times 10^{-9} \text{ gmol/cm}^3$$

$$= 8.179 \times 10^{-3} \text{ gmol/m}^3$$

$$G_{tot} = (8.179 \times 10^{-3} \text{ gmol/m}^3)(131.39 \text{ g/gmol})(80 \text{ m}^3)$$
$$= 85.97 \text{ g/8-h working day.}$$

b. At steady state, C = G/kQ. Under these conditions, the TWA value should be used, since it represents the average concentration to which a worker can be continuously exposed over an 8-hour workday.

Using the ideal gas law, 50 ppm = 0.269 g/m³. Substituting into the steady-state solution,

$$0.0269 \text{ g/m}^3 = 50 \text{ g/d} \left(\frac{1}{0.3 \, Q}\right) (1 \text{ d/8hr})(1 \text{ hr/3600 s})$$

Rearranging to solve for Q yields:

$$Q = 0.022 \text{ m}^3/\text{s} = 1291 \text{ L/min}$$

as the minimum ventilation rate required to ensure that the TWA is not exceeded.

c. The total number of emissions sources is:

$$100 \text{ mi}^2 \, (2 \text{ dry cleaners/mi}^2) = 200 \text{ sources}$$

Each source emits 50 g/d, so the total is

10 kg/d = 3210 kg/yr = 6864 lb/yr operating 312 d/yr.

d. First compute the vapor pressure at 25°C:

$$\log P = -\frac{1816.8}{298} + 7.95642 = 1.860; P = 72.5 \text{ torr}$$

Then use 50% of this value in the vapor pressure equation and solve for temperature:

$$\log (36.2) = -1816.8/(T_2) + 7.95642; T_2 = 11°C$$

This temperature can reasonably be achieved. Similarly for 1% of this value:

$$\log (7.24) = -1816.8/(T_3) + 7.95642; T_3 = -17.0°C$$

e. Condensation occurs if the partial pressure in the cooled mixture equals the component vapor pressure at that temperature (assuming an ideal gas mixture):

$$50 \times 10^{-6} \text{ atm } (760 \text{ torr/atm}) (0.25) = 9.5 \times 10^{-3} \text{ torr}$$
$$\log (0.0095) = -1816.8/T + 7.95642$$
$$T = 182 \text{ K} = -90.9°C$$

This is not a realistic temperature for an economical condensation unit. The vapors must first be concentrated, or compressed, or a different method of vapor recovery might be used. For example, it may be possible to cost-effectively separate the vapors using membrane separation techniques, and alternative cleaning solvents should be considered.

Example 10.3. You are Director of Environmental Affairs for Sludge Chemical Co. Your job is to implement the environmental policies established by executive management. As a middle manager, you have no say in setting these policies, but you agree with the policy decisions that have been made. Thanks to these policies and your effective implementation, Sludge Chemical Co. has an exemplary environmental record and has never been cited for non-compliance with a regulation.

The CEO of Sludge Chemical Co. says that he wants to continue to assure compliance with all state and federal rules and has announced the formation of a new department, Internal Environmental Audits (IEA). This group will serve as a company-wide watchdog and will report directly to Sludge Chemical's outside legal counsel who will then report to the CEO. Restating what a wonderful job you have done, the CEO now grants you complete dominion over all things environmental. Henceforth, any problems will be reported by IEA via Sludge's attorney, so there is no need for you to call with details. What reasons can you give for the new department and its unconventional structure? Are you uncomfortable with this situation? If so, why?

SOLUTION. State and federal governments are showing an increasing propensity towards criminal prosecution of violators of environmental regulations. The courts have generally upheld the concept of a "corporate veil" that protects everyone except those with direct knowledge of environmental violations. In most cases where corporate executives have been successfully prosecuted, they have been directly involved in the day-to-day management of hazardous materials. Recently, a new tactic has been employed whereby the top executives are notified of violations by corporate counsel, preferably outside counsel. These communications are shielded by the attorney-client privilege and, as such, are not subject to review by the courts. As Director of Environmental Affairs, you should be concerned about this type of arrangement since it may build an effective legal barrier leaving you on the outside. A corporate structure that prevents direct reporting of violations to management should be questioned.

Example 10.4. A New Jersey utility has determined that for every 10,000 kW it generates, it must burn approximately 1 ton of coal. The coal used at this particular facility is Illinois #6, a form of bituminous coal with the approximate chemical formula $C_{100}H_{85}S_{2.1}N_{1.5}O_{9.5}$. The company is embarking on a new public relations project to demonstrate to consumers the impact of wasted energy on the environment. As the utility's engineer, you have been asked to calculate the amount of CO_2 and SO_2 discharged into the atmosphere for every kW of energy produced.

SOLUTION. First the molecular weight of the coal is calculated using the following atomic weight data: C = 12.00 lb/lbmol, H = 1.00 lb/lbmol, S = 32.06 lb/lbmol, N = 14.00 lb/lbmol, O = 16.00 lb/lbmol

The molecular weight is then determined from the molecular formula as follows:

$$MW = (100)(12.00) + (85)(1) + (2.1)(32.06) + (1.5)(14) + (9.5)(16)$$
$$MW = 1,504.33 \text{ lb/lbmol}$$

Assuming complete combustion, the stoichiometric equation for the combustion of the coal is written as follows:

$$C_{100}H_{85}S_{2.1}N_{1.5}O_{9.5} + 120.1\ O_2 \rightarrow 100\ CO_2 + 42.5\ H_2O + 2.1\ SO_2 + 1.5\ NO_2$$

The lbmol of coal per ton of coal is now calculated, noting that there are 2,000 lb per ton:

(2,000 lb/ton)/(1,504.33 lb/lbmol) = 1.33 lbmol/ton of coal

The amount of products generated in lbmol per ton of coal combusted can now be calculated as follows:

CO_2: $(1.33)(100) = 133.0$ lbmol
H_2O: $(1.33)(42.5) = 56.52$ lbmol
SO_2: $(1.33)(2.1) = 2.79$ lbmol
NO_2: $(1.33)(1.5) = 2.0$ lbmol

Hence, the amount of products generated in lb per ton of coal combusted can be calculated as follows:

CO_2: $(133.0 \text{ lbmol})(44 \text{ lb/lbmol}) = 5{,}852$ lb
H_2O: $(56.52 \text{ lbmol})(18 \text{ lb/lbmol}) = 1{,}017$ lb
SO_2: $(2.79 \text{ lbmol})(64.06 \text{ lb/lbmol}) = 179$ lb
NO_2: $(2.0 \text{ lbmol})(46 \text{ lb/lbmol}) = 92.0$ lb

11. SUMMARY

1. In addition to waste reduction, *energy conservation* and *accident and emergency management* also play a significant role in pollution prevention.

2. *Health risk assessments* provide an orderly, explicit, and consistent way to deal with scientific issues in evaluating whether a hazard exists and what the magnitude of the hazard may be. This evaluation typically involves large uncertainties because the available scientific data are limited, and the mechanisms for adverse health impacts or environmental damage are only imperfectly understood.

3. It is now increasingly clear that some treatment technologies, while solving one pollution problem, have created others. Most contaminants, particularly toxic compounds, present problems in more than one medium. Since nature does not recognize neat, jurisdictional compartments, these same contaminants are often transferred across media. In order to deal with the multimedia nature of environmental contamination, a *life cycle analysis* approach should be taken.

4. LCA is a procedure to identify and evaluate "cradle-to-grave" natural resource requirements and environmental releases associated with processes, products, packaging, and services. This includes identifying and quantifying energy and materials used and wastes released to the environment, assessing their environmental impact, and evaluating opportunities for improvement. LCA concepts can be particularly useful in ensuring that identified pollution prevention opportunities are not causing unwanted secondary impacts by shifting burdens to other places within the life cycle of a product or process.

5. Sustainable development requires conservation of resources, minimizing depletion of nonrenewable resources, and using sustainable practices for managing renewable resources when meeting the needs of the present so that the ability of future generations to meet their own needs is not compromised.

6. The *ISO 14000* standards are a voluntary series of guidelines designed to address environmental management. They focus on management, rather than specific performance based on the belief that correct management will lead to better performance. ISO 14000 asks industry leaders to accept their responsibility for the environment, and consider this in corporate deci-

258 Chapter 11

sion making. It also encourages companies to be proactive, and incorporate pollution prevention ideals into their environmental management activities.

7. There are numerous other interrelated pollution prevention components including: *Small Quantity Generators*, domestic activities, ethical considerations, and *environmental justice* issues. Pertinent aspects of these topics related to pollution prevention were discussed in the text.

12. PROBLEMS

1. Briefly describe the difference between environmental ethics and engineering ethics.
2. Discuss pollution prevention measures that can be instituted at home and at the office.
3. Refer to Illustrative Example 10.4. If the coal in question contains 1.3 lb of ash per ton of coal, determine the mass of ash released to the atmosphere per 10,000 kW it generates. Assume that 1% of the ash "flies," i.e., 99% is either captured or discharged from the boiler bottoms.
4. The New Jersey Power Company, Theo. Inc., has determined that for every 12,000 kW it generates, it must burn approximately 1 ton of coal. The coal used at this particular facility is a form of bituminous coal with the approximate chemical formula $C_{100}H_{82}S_{1.2}N_{1.0}O_{10.0}$. The company is embarking on a new public relations project to demonstrate to consumers the impact of wasted energy on the environment. As a Theo. engineer, you have been asked to calculate the amount of CO_2 and SO_2 discharged into the atmosphere for every kW of energy produced.
5. A large laboratory contains a reactor which behaves as a continuous stirred tank reactor (CSTR), where the concentration of the tank is spatially uniform. Outline the derivation of the equation that can be used to describe this CSTR. Suppose the reactor seal raptures and an amount of HC is emitted instantaneously. Discuss how this may constitute a health hazard and what pollution prevention measures could be implemented to decrease or eliminate them.
6. Provide examples of the difficulties experienced by minorities regarding health and/or environmental contamination problems.

13. REFERENCES

1. U.S. EPA, *Climate Wise Progress Report*, Office of Policy, Washington, D.C., EPA 231-R-98-015, 1998.
2. Michigan Manufacturing Technology Center, *Manufacturing Efficiency Decision Support, MEDS*, Ann Arbor, Michigan, 1999. Internet URL: http://meds.mmtc.org.
3. *Mark's Standard Handbook for Mechanical Engineers*, McGraw-Hill Book Company, New York, New York, 1987.
4. Thurman, A. and Mehta, D.P., *Handbook of Energy Engineering*, The Fairmont Press, Lilburn, Gerogia, 1992.
5. National Insulation Association, Internet URL: http://www.insulation.org.

6. Sprague, B., *Manufacturing Assessment Planner (MAP) Toolkit*, Michigan Manufacturing Technology Center (MMTC) and CAMP, Inc., Ann Arbor, Michigan, 1999.
7. U.S. EPA, *EPA Green Lights ProjectKalc for Windows, Version 3.00, User's Manual*, Office of Air and Radiation (6202J), Washington, D.C., 1996.
8. Theodore, M. and Theodore, L., *Major Environmental Issues Facing the 21st Century*, Prentice-Hall, Upper Saddle River, New Jersey, 1995.
9. Paustenbach, D., *The Risk Assessment of Environmental Management and Technology*, John Wiley and Sons, New York, New York, 1990.
10. Holmes, G, Singh, R., and Theodore, L., *Handbook of Environmental Management and Technology*, Wiley-Interscience, New York, New York, 1992.
11. Rodricks, J. and Tardiff, R, *Assessment and Management of Chemical Risk,*, American Chemical Society, Washington, D.C., 1984.
12. U.S. EPA, *Federal Facility Pollution Prevention Project Analysis: A Primer for Applying Life Cycle and Total Cost Assessment Concepts*, Office of Enforcement and Compliance Assurance, Planning, Prevention, and Compliance Division, Washington, D.C., 1995.
13. ERM, *Pollution Prevention Quarterly*, Miami, Florida, Winter, 1999.
14. U.S. EPA, *ISO 14000 Resource Directory*, Office of Research and Development, Washington, D.C., EPA/625/R-97/003, 1997.
15. Welch, T., *Moving Beyond Environmental Compliance, A Handbook for Integrating Pollution Prevention with ISO 14000*, Lewis Publishers, Boca Raton, Florida, 1998.
16. Abt Associates, *National Small Quantity Generator Hazardous Waste Generator Survey*, Cambridge, Massachusetts, 1985.
17. U.S. EPA, *Environmental Progress and Challenges*, EPA Update, Office of Enforcement and Compliance Assurance, Planning, Prevention, and Compliance Division, Washington, D.C., 1988.
18. Martin, M. W. and Schinzinger, R., *Ethics in Engineering*, McGraw-Hill, New York, New York, 1989.
19. U.S. EPA, *Environmental Equity: Reducing Risk for All Communities*, Office of Policy, Washington, D.C., 1992.

CHAPTER 12

Pollution Prevention Opportunity Assessment

A *pollution prevention opportunity assessment (PPOA)* is defined as a systematic, planned procedure with the objective of identifying ways to reduce or eliminate waste, preferably at the source. Generally, the assessment is preceded by careful planning and organization to set overall pollution prevention goals. Next, the actual assessment procedure begins with a thorough review of a plant's operations and waste streams, and the selection of specific areas of the plant to assess. Once an area is selected as a possible "minimization" area, various options with the potential for reducing waste generation can be developed and screened. The technical and economic feasibility of each option is evaluated and, finally, the most cost effective and technically feasible options are implemented. Initiating a successful program must begin with a secure commitment from top management, allocation of adequate funding and technical expertise, appropriate organization, and a good understanding of the goals of the assessment and planning required to make it effective. To be successful, pollution prevention must become an integral part of a company's operations. The PPOA offers opportunities to reduce operating costs, reduce potential liability, and improve the environment, while also improving regulatory compliance. Much of the material presented in this chapter has been drawn from EPA's original "waste minimization opportunity assessment" procedure, developed primarily for hazardous wastes.[1] This procedure has been modified and adapted to apply to what is defined above as a pollution prevention opportunity assessment.

Figure 12.1 depicts the pollution prevention opportunity assessment procedure. As shown, the assessment procedure can be divided into four phases which include the following:

1. *Planning and organization phase*, during which management buy-in and commitment are obtained, overall assessment goals are identified, and an assessment team is assembled.
2. *Assessment phase*, consisting of the collection of data and identification and screening of potential pollution prevention options.
3. *Feasibility analysis phase*, consisting of the technical and economic evaluation of each option.
4. *Implementation phase*, where recommended options are put into place, their performance is monitored, and feedback is provided for the next round of pollution prevention assessment activities at the facility.

Each stage depicted in Fig. 12.1 is discussed in detail below.

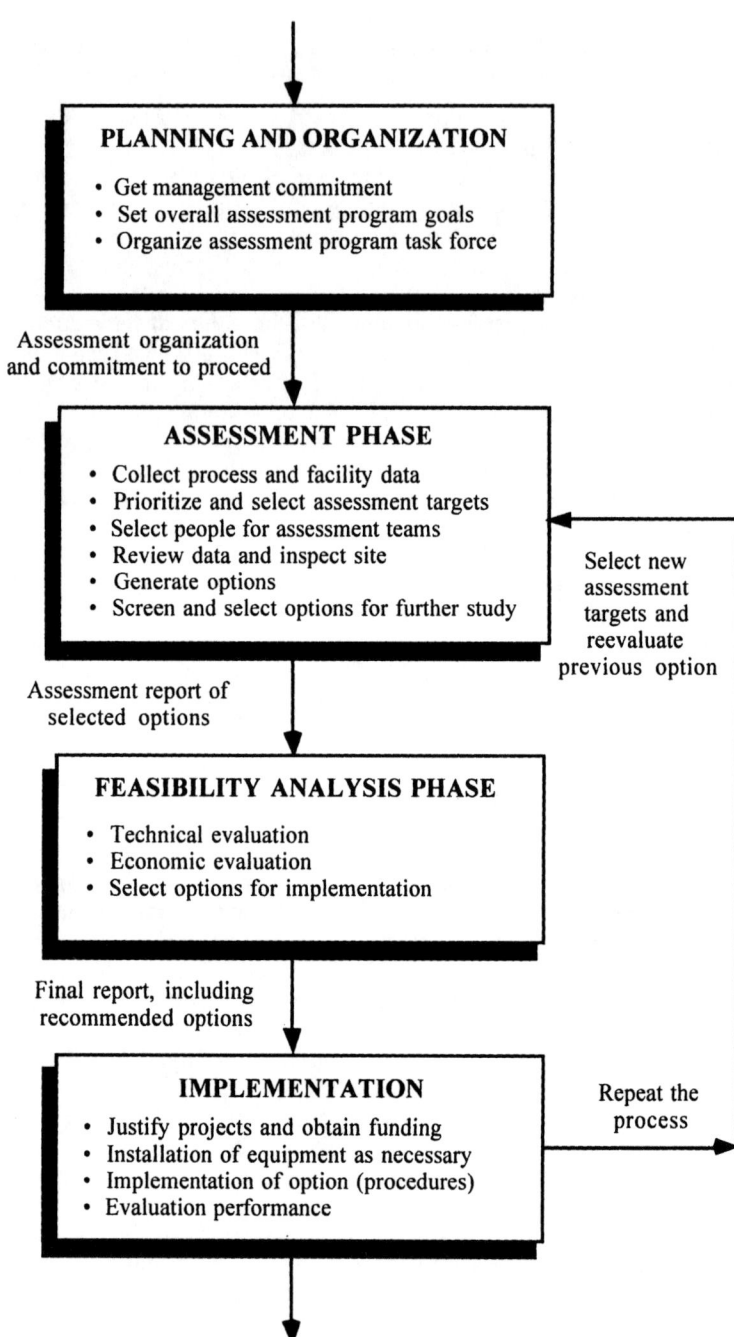

Figure 12.1. Pollution prevention opportunity assessment procedure.[1]

1. PLANNING AND ORGANIZATION

This section deals with aspects that are critically important to the ultimate success of any pollution prevention program. First, a commitment from top management to improve the environmental performance of the facility and reduce waste generation must be secured. To do this, the benefits, specifically economic ones, gained from pollution prevention must be clearly demonstrated. Top management must be made aware that the necessary personnel and financial resources must be committed for the program if it is to be a success. Support at the top management level only, however, will not be sufficient in maintaining a successful pollution prevention program. Personnel at every level in the company must be dedicated to the common goal of pollution prevention for it to succeed. Initial contact with employees includes solicitation of their views of how the assessment should be performed and which areas of the facility are likely candidates for pollution prevention. Gaining further employee support can be sought through incentives, bonuses, and awards for pollution prevention achievements. Employees are directly involved in the production of waste and can be instrumental in the overall pollution prevention program if they can be properly motivated. Recognition and rewards can increase employee cooperation and participation as shown in the WRAP program detailed in Chapter 10.

Once a company has committed to minimizing the generation of waste, a *task force* must be established to conduct the actual assessment and convey its findings to management. A team leader should be selected and serves as the motivator of the project. This individual needs to be an effective communicator with the rest of the team. This task force should be comprised of individuals from all groups within the company which have an interest in the outcome of the program. The size of the group should reflect the size of the company and the complexity of the processes being evaluated. Members of different departments within the company bring varying perspectives and suggestions to the program. The task force should be separated into different teams concentrating on different areas of the plant with which they have the most knowledge. Some example task forces, often referred to as audits teams in industry, are provided below.

Metal Finishing Department in a Large Aerospace Corporation
 Metal finishing department manager (Audit Team Leader)
 Process engineer responsible for metal finishing processes
 Facilities engineer responsible for metal finishing department
 Wastewater treatment department supervisor
 Staff environmental engineer

Small Pesticide Formulator
 Production manager (Audit Team Leader)
 Environmental manager
 Maintenance supervisor
 Pesticide industry consultant

Cyanide Plating Operation at a Government Defense Facility
Internal audit team
    Environmental coordinator (Audit Team Leader)
    Environmental engineer
    Electroplating facility engineering supervisor
    Metallurgist
    Materials science group chemist
External audit team
    Chemical engineers (2) (Audit Team Leader)
    Environmental engineering consultant
    Plating chemistry consultant

Large Offset Printing Facility
Internal audit team
    Plant manager (Audit Team Leader)
    Photoprocessing supervisor
    Printing supervisor
External audit team
    Chemical engineers (2) (Audit Team Leader)
    Environmental scientist
    Printing industry technical consultant

The top priority of the task force is to select the specific pollution prevention goals to be achieved as an outcome of the opportunity assessment. The scope of goals can be limited to one certain process or can be applied to an entire department. Quantifiable goals should be established (a commitment to reduce 10% of the overall waste generation per year is an example). Goals are not unchangeable and should be reviewed periodically. As the ultimate scope of the program is more clearly defined, goals should be reevaluated and updated or revised. As the assessment activities progress, the goals may be altered accordingly. Pollution prevention is an ongoing process and goals need to reflect this.

The task force must also be ready to recognize any potential barriers to implementation of the pollution prevention program. Bringing this type of new project into an existing operating company will often generate conflicts. Barriers are commonly caused by fear that the program will damage product quality, or that a great loss in production will be experienced while new process equipment is installed. The task force must be prepared to respond to these concerns in order to gain participation by the company as a whole. This general subject is treated in more detail in the last section of this chapter.

2. ASSESSMENT PHASE

The ultimate purpose of the *assessment phase* is to develop a comprehensive list of potential pollution prevention options. To accomplish this, a detailed analysis of the plant must be performed.

Collecting data from every process in the plant is the first step in performing the assessment. All waste streams must be identified and characterized.

Selecting the principal waste streams or waste producing operations within a facility provides the task force with the main focus for the assessment. The criteria used for the selection of principal waste streams should include the composition, quantity, and toxicity of wastes, method and cost of disposal, compliance status, and potential for minimization. Preliminary information should be acquired from (hazardous) waste manifests, biennial reports, environmental audits, TRI reports, and National Pollution Discharge Elimination System (NPDES) monitoring reports. Routine sampling can also help to provide data on certain chemicals within a particular waste stream. The best means for compiling information on all waste streams is the use of flow diagrams and, then, material balances. The material balance includes information on materials entering and leaving a process. Material balances can be used to quantify losses or emissions and provide essential data to estimate the size and cost of additional equipment, data to evaluate economic performance, and a baseline for tracking the progress of waste minimization efforts. Simply, the material balance is represented by the mass conservation principle (see Chapter 3):

$$\text{mass in} - \text{mass out} + \text{mass generated} = \text{mass accumulated} \qquad (12.1)$$

Material balances should be prepared individually for specific components entering and leaving a process and, to be most accurate, balances should be developed for individual units or processes, rather than for a large area of the facility. An overall material balance for the facility can be constructed from the individual material balances from its component units and processes. Although characterizing waste streams through material balances can require substantial effort, it is the best way to fully understand the fate of materials and makeup of wastes within a facility. Tracking waste flows and compositions should be performed periodically to ensure the accuracy of decisions made from these data. In addition, variable flows can be distinguished from constant, continuous flows, thereby providing a more accurate representation of process operations and waste generation rates.

Once all waste streams have been identified, priorities should be assigned to those of greater importance, which should be concentrated on first. Ideally, assessments should be performed on all waste streams. However, this is not always practical in larger plants with many waste streams considering the availability of limited resources. Waste streams of lesser importance can be assessed as time and budgets permit. As indicated above, prioritization of waste streams should take into consideration:

1. Compliance with current and future regulations
2. Costs of waste disposal and treatment
3. Potential safety and environmental liability
4. Quantity of waste and its hazardous makeup
5. Potential for minimization
6. Available budget for the PPOA

A practical approach in selecting a waste stream first is to choose one with a high probability for successful reduction of wastes (the "low hanging fruit" as they are sometimes called). Successful implementation of a pollution prevention opportunity assessment will help secure commitment for further pollution prevention efforts.

Once a specific area of the facility or process is selected, the task force should conduct a site inspection, starting with the point where raw materials enter and continuing to the point where waste leaves the process. Although collecting data is important, the assessment task force must be familiar with the actual operations at the facility from a hands-on perspective. The purpose of the inspection is to gain a clear understanding of the nature and causes of waste generation in the process. To ensure performance of a proper inspection, it is useful and practical to have prepared an inspection agenda in advance, to schedule the inspection to coincide with a particular operation, and to obtain permission to interview key personnel involved with all aspects of the running of the operation. The team should identify all suspected areas of waste generation. These include the production processes, maintenance operations, storage areas, finished products, and work-in-progress. Preliminary conclusions should result from this investigation. Additional data and/or site visits will generally be needed to supplement these preliminary conclusions.

With the nature and origins of waste generation understood, the assessment team can then begin to generate a set of pollution prevention options for further consideration. This process of identification should follow the hierarchy discussed in Chapter 10 to reflect environmental desirability in which source reduction options are looked to first, followed by closed, then open-loop recycling, treatment, and finally ultimate disposal. The list of options should be as large as possible, including those currently used in the facility. The identification process is a creative and analytical one by which members of the task force use their educational knowledge as well as their practical working experience to generate possible waste minimization options. Options can also be generated by examining technical literature and through discussions with equipment vendors or suppliers of process raw materials, plant personnel, trade association technical experts and other technical resources (manuals, guides, web resources, etc.) and regulatory agencies.

Each of the options identified by the assessment team must undergo a preliminary, qualitative evaluation so that a short-list of options with realistic potential for waste minimization and reduction of costs are clearly defined and screened. The screening procedure should be designed to eliminate options that appear marginal or impractical. One method for determining the most attractive option, after the screening process is complete, is the *weighted-sum method*. A set of criteria is developed and each criterion is assigned a specific numeric value related to its relative importance or weighting determined by the priorities of the facility and its assessment team members. Then each option is rated against each criterion. The final weighted rating is obtained by summing the product of the corresponding weight and rating for each criterion. The results of the screening process will promote certain options to be further evaluated in detail for their technical and economic feasibility, while identifying others that should not be considered for further evaluation. Example criteria of relevance for a pollution prevention opportunity assessment are shown below.

1. Reduction in waste quantity and/or toxicity
2. Main economic, compliance, safety, and liability benefits gained from implementation
3. Whether the process uses a proven technology
4. Cost-effectiveness
5. Time required for implementation

3. FEASIBILITY ANALYSIS PHASE

Once the assessment phase has identified potential pollution prevention opportunities, the *feasibility phase* is used to determine whether those options are technically and economically practical, through a detailed evaluation and analysis of each. The level of analysis is dependent on the complexity of the pollution prevention project under consideration. Simple options, such as *preventive maintenance*, would not require as detailed an evaluation as would an input *material substitution* option that may result in changes to product specification and process equipment. The detailed evaluation of options should begin with source reduction alternatives and end with ultimate disposal options if waste generation cannot be prevented.

The *technical evaluation* determines whether the potential option, once in place, will really work as intended. Performing a technical evaluation requires comprehensive knowledge of pollution prevention techniques, vendors, relevant manufacturing processes, and the resources and limitations of the facility and its staff. This evaluation may involve bench-scale or pilot-scale treatability testing. Some vendors will install equipment on a trial basis. Typical considerations for technical criteria evaluation include:

1. Technical reliability.
2. System safety.
3. Maintenance of product quality.
4. Space requirements.
5. Compatibility of proposed equipment with existing systems.
6. Downtime necessary for installation.
7. Special expertise requirements.
8. Additional labor, equipment and utility requirements.

All affected groups in the facility should be encouraged to review and comment on the technical evaluations that are conducted. If an option results in changes in production, effects on product quality must be determined. All options that are shown to be impractical after a technical evaluation are dropped from further consideration.

The *economic evaluation* is conducted using standard measures of profitability, return on investment, and net present value (see Chapter 9). Each firm uses individual economic procedures and criteria for selecting projects for implementation. The costs should be broken down into capital costs and operating costs. Capital costs include not only the fixed capital costs for designing, purchasing, and installing equipment, but also costs for working capital, permitting, training, start-up, and financing charges. Certain source reduction options will not need environmental permitting in order to be implemented. Some options, such as procedural or materials changes, will not have any capital costs. Pollution prevention projects generally need to show a savings in operating costs to be economically viable. Operating costs and savings associated with pollution prevention include reductions in waste treatment, storage, and disposal costs; raw material costs savings; insurance and liability savings; increased costs or savings associated with product quality; decreased or increased use of utilities; potential benefits of increased market-share from "greening" of the company; and increased or decreased revenues from changes in production of marketable byproducts. With capital and operating costs known, profitability can be determined. Those options requiring no capital investment should be implemented as soon as possible. In addition to the profitability margin, compliance with environmental regulations should be carefully considered, since violations may result in shut-down time and criminal penalties.

All results of the evaluation are compiled into a final report that contains the recommendations for implementation of the options. This final report will probably serve as the basis for obtaining the necessary funding for implementation and should include discussions on all benefits to the company, including reduced liabilities and an improved public image generated from pollution prevention efforts.

4. IMPLEMENTATION

Once the options for reducing waste have been established in the final report, the project must be justified in order to obtain funding. Here, the commitment from management is critical to overcome resistance to change within the company. With funding secured, the selected option(s) is (are) ready to be implemented. The first step, a detailed design of the system, is followed by construction. After any equipment is installed or equipment modifications are made, the personnel involved can be trained and the operation started. Options need to be monitored to ensure their effectiveness and demonstrate that the project has achieved its goals.

Issues related to quantifying pollution prevention outcomes and results are becoming more and more important as formal pollution prevention programs have matured within the various state and federal regulatory agencies with a vested interest in pollution prevention. New regulations and policies have contributed to a growing urgency in the need for adequate measures of pollution prevention program effectiveness. As a result of the *Government Performance and Results Act*, the federal government is under increasing pressure to assess program effectiveness and eliminate federal programs that are not successful. Therefore, a quantitative gauge of the success of pollution prevention programs is critical to the long-term survival of these programs. In addition, the *National Environmental Performance Partnership System*, which allows EPA to grant more regulatory flexibility to states, imparts an increased responsibility on the part of states to demonstrate that they are still meeting environmental goals and objectives. Due to the challenges associated with determining overall statewide pollution prevention progress, many states have focused on measuring the success of specific state pollution prevention program components.

States, as well as independent research organizations, are determining the extent to which specific components of state pollution prevention programs are resulting in actual implementation of pollution prevention measures at facilities. Typical measurement methods, which can be used individually or in combination, include: analysis of records, reports, and plans; surveys or in-depth interviews (either broadly covering the universe of relevant facilities, or narrowly focused on recipients of specific services); focus groups; and case studies.

Studies conducted by New Jersey, Washington, Massachusetts, North Carolina, and Iowa to evaluate the effectiveness of facility planning and/or technical assistance showed mixed results.[2] Some companies implemented pollution prevention recommendations resulting from on-site technical assistance visits, but costs and quality concerns formed significant impediments.

Three methods of measuring pollution reductions exist: actual quantity change, adjusted quantity change, and materials accounting. These methods rely on data that are readily available to facilities, states, and EPA. The data used to calculate actual quantity change or adjusted quantity change can be obtained from information reported to EPA's Toxics Release Inventory or under RCRA, or simply by recording the quantities of waste generated before and after implementation of the option. The difference in volume, weight or toxicity of waste before and after implementation of the pollution prevention option, divided by the original volume, weight or toxicity of waste generated, represents the percentage reduction in waste generation. Since waste generation is directly dependent upon the production rate, the ratio of waste generation to production is also a convenient means of representing waste reduction resulting from a pollution prevention project.

Some states, such as New Jersey, also require facilities to submit materials accounting data, i.e., process-level details of material balances in the facility. Material balances are particularly useful when there are points within a facility where it is difficult or uneconomical to collect analytical data, and for quantifying fugitive emissions such as evaporative losses from a process.[3] Other innovative pollution prevention performance techniques are being used in various state programs. For example, under a Pollution Prevention Incentives for States (PPIS) grant, the *Indiana Pollution Prevention and Safe Materials Institute* devised a pollution prevention measurement method that incorporates hazard rankings for chemicals.[2] The increased emphasis on pollution prevention program performance should spur the development of better measurement techniques in the years to come.

However pollution prevention performance is quantified, it is important to realize that a pollution prevention program is an ongoing effort. Once the highest priority waste streams have been assessed and dealt with, the assessment task force should continue to identify new opportunities, assess different waste streams, and consider attractive options that were not pursued earlier. The ultimate goal should be to establish a continuous process to seek out and eliminate as much waste generation as possible within a facility.

5. POLLUTION PREVENTION INCENTIVES

As described earlier, the problems associated with the handling of waste streams have become a major concern for industry as a whole, and for the chemical industry specifically. Overall costs for managing waste, coupled with increasing regulatory and economic consequences, continue to escalate. There are real incentives for avoiding the production of toxic and hazardous wastes up front that will only increase in value as more regulations are passed and treatment capacity for these substances becomes more and more limited. These incentives usually result in monetary encouragements and consist of economic benefits, regulatory compliance, reduction in liability, and enhanced public image.

5.1. Economic Benefits

Reducing the amount of pollutant/waste produced initially will result in a reduction of the costs associated with the handling of that waste. The obvious costs resulting from the transportation, disposal, and treatment of wastes will all be lowered as the amount of created waste is reduced. Current trends indicate that the cost of waste disposal will continue to increase. In addition to reduced waste handling and disposal costs, income can be derived through the sale, reuse, or recycling of certain wastes. A number of not-so-obvious gains that are associated with the reduction of waste can also be realized such as: reduction in raw materials costs; reductions in health and safety, and insurance costs; and a reduction in reporting, manifesting, and permitting costs.

Regulatory agencies can influence companies to investigate pollution prevention techniques through offers of reduced fines and penalties. For example, many of the Supplemental Environmental Projects (SEPs) carried out by Regional EPA offices and state regulatory agencies are initiated to invest fines and penalties into pollution prevention programs for the violating companies. Fines and penalties are waived in these SEPs if violators use this money to carry out pollution prevention assessment and implement cost-effective waste reduction options that will bring them back into compliance. A pollution prevention program can also be used to bring a firm into compliance with regulations by structuring permit fees in a way that promotes pollution prevention activities. For instance, instead of basing a permit fee on the size of a company an agency might base the fee on the volume and/or toxicity of waste substances produced, an approach currently being considered by some states. Firms actively pursuing pollution prevention techniques would be rewarded by paying a lower fee.

5.2. Regulatory Compliance

Federal and state laws require all firms classified as hazardous waste generators or small quantity generators (SQGs) to implement a pollution prevention program to reduce the quantity of waste to the extent that it is economically feasible. As indicated in Chapter 11, SQGs are facilities that generate more than 100 kg/month but less than 1,000 kg/month of hazardous waste. A facility generating more than 1,000 kg/month of hazardous waste is classified as a hazardous waste generator. Generators are required to sign the following statement on all manifests to certify their pollution prevention efforts:

> Unless I am a small quantity generator who has been exempted by statute or regulation from the duty to make a waste minimization certification under Section 3002(b) of RCRA, I also certify that I have a program in place to reduce the volume and toxicity of waste generated to the degree I have determined to be economically practicable, and I have selected the method of treatment, storage, or disposal currently available to me which minimizes the present and future threat to human health and the environment.

Firms permitted as hazardous waste generators are generally required under RCRA to report sampling data on a regular basis. When hazardous wastes are minimized, the frequency of sampling and data reporting is also minimized, thereby saving the firm a considerable amount of money.

In addition, land disposal restrictions and treatment standards provide indirect incentives for reducing (hazardous) waste. Under these regulations, only wastes meeting specified treatment standards are allowed to be disposed of in landfills. The level of treatment is based on pollution prevention technologies and, therefore, those facilities already employing these reduction techniques are best suited to meet the standards. With these reduction methods already in place, a company can save on costs associated with waste management.

5.3. Reduction in Liability

Both short-term and long-term liabilities can be reduced through pollution prevention programs. Short-term liabilities associated with releases to the environment resulting in noncompliance with permits can be reduced through the overall reduction of waste generation. Short-term liabilities connected with personnel exposure and workplace safety will also be lessened. Long-term liabilities resulting from the on-site or off-site disposal of wastes can be reduced as well. Disposing of wastes at a permitted disposal facility does not end a firm's connection to or liability for that waste. If a disposal facility is shown to be releasing contaminants into the environment, not only are the owners/operators of that facility liable under CERCLA, but the generators who arranged for the disposal of wastes in that facility are also liable for undertaking remedial actions to clean up the contamination. These generators can be ordered to finance the investigation of the extent of contamination and the remedial action required for site clean up. These CERCLA actions have generally been very costly ($\approx$ \$30 million per Superfund site), making the potential liability costs for some facilities with high hazardous wastes significant.

5.4. Enhanced Public Image

The public has placed a great deal of emphasis on environmental issues. Recent election campaigns on all levels of government continue to make the environment a priority platform issue. Because of this level of concern and awareness, it is becoming increasingly important for companies to share information with the public. Under *SARA*, the *Emergency Planning and Community Right-to-Know Act (EPCRA)* includes mandatory reporting of releases to the environment and optional reporting of pollution prevention activities. *Risk Management Plans* required by the *Clean Air Act* require the publication of worst case scenario information in response to an uncontrolled release of hazardous material from production facilities into neighboring communities. Implementation of a hazardous material pollution prevention program provides a good community relations baseline for improving a firm's public image as a good corporate citizen.

6. IMPEDIMENTS TO POLLUTION PREVENTION

The previous section presented the advantages of developing and implementing a pollution prevention program. This section briefly reviews some of the impediments to the success of pollution prevention efforts. A "dirty dozen" list is provided below, followed by a short description (and some comments) on each of these identified deterrents.

1. Management apathy.
2. Lack of financial commitment.
3. Production concerns.
4. Research, development, and design concerns.
5. Failure to monitor program success.
6. Middle-management decisions.
7. Lack of consensus within the organization.
8. Confusion regarding regulations.
9. Confusion about the economic advantages.
10. Bureaucratic resistance to change.
11. Lack of awareness of pollution prevention advantages.
12. Failure to apply multimedia approaches.

1. *Management Apathy*. It is not uncommon for upper-level management in most companies, particularly large ones, to take an indifferent attitude toward pollution prevention. Since administrators are often not technically oriented, most have difficulty realizing the potentially enormous benefits that can be gained from a pollution prevention program.

2. *Lack of Financial Commitment*. It is no secret that many considerations, including resources, individuals, money, economic incentives, and so forth, are critical to the success of a pollution prevention program. These "resources" are often just not available within small companies or within large companies that may be experiencing a financial downturn.

3. *Production Concerns*. The classic song of the production supervisor and/or plant operator is, "I'm meeting deadlines and making money, so don't

rock the boat." Sending a member of the pollution prevention team out to a plant is somewhat akin (at times) to letting a bull loose in a china closet.

4. *Research, Development, and Design Concerns.* The somewhat simplistic mentality described in Item 3 can be extended to apply to new projects at the research, development, or design stages. A very common misconception is that any pollution prevention activity is going to delay a (new) project and cost money.

5. *Failure to Monitor Program Success.* Pollution prevention programs are often instituted with high hopes of success, but later abandoned because of a failure to monitor the program properly. Responsible individuals must continuously monitor and record both the successes and failures of the program.

6. *Middle-Management Decisions.* In most companies, it is usually middle management that is directly responsible to the top company administrator(s) for earnings. Unfortunately, it is often this group that views pollution prevention programs as added expense that will eat into profits.

7. *Lack of Consensus Within Organization.* As indicated in Item 1 above, management often has a poor understanding of the advantages of pollution prevention. In addition, the economic, environmental, liability, social, and other advantages that various organizations within a company ascribe to a program of this nature will differ widely.

8. *Confusion Regarding Regulations.* It may be difficult to believe, but many companies, particularly small ones, are not aware of the applicable regulations regarding their processes or operations. Any permit review process is almost certainly doomed to failure if the responsible individual or group is not cognizant of the pertinent regulations, or does not want to know.

9. *Confusion About Economic Advantages.* Most companies are simply not aware of the true costs associated with generating and treating wastes. However, treatment and/or disposal costs have increased dramatically in recent years, and liability concerns continue to mount. As indicated above, many companies do not realize that pollution prevention opportunities, in many instances, result in increased profits. This situation is prevalent today because, for many companies, particularly large ones, it is difficult to evaluate the economic advantages of pollution prevention. Their accounting systems are often so complex, and not properly integrated within the various divisions or organizations of a company, that it is nearly impossible to quantify the effects arising from a pollution prevention program.

10. *Bureaucratic Resistance to Change.* It is natural, particularly in large corporations and utilities, to resist any change to an existing process or method of operation. This reluctance to adapt to a changing (environmental, regulatory, liability, and so on) climate is commonplace in industry.

11. *Lack of Awareness of Pollution Prevention Advantages*. Middle-level managers and upper-level administrators are often unaware of both the pollution problem at their facility and the associated true cost of its mitigation. Thus, it is understandable why these individuals are not aware of the overall advantages of a pollution prevention program.

12. *Failure to Apply Multimedia Approaches.* Issues related to multimedia analysis of environmental problems have been discussed in Chapter 11. Unfortunately, an overall approach that examines a system or

process from a multimedia point of view is rarely found in industry today. There is much work to be done in this area, and many environmental improvements can be expected in the years to come when a truly multimedia framework is routinely applied to environmental problems.

As has been noted, many companies and individuals are unaware of the advantages of pollution prevention. Until this situation is changed, progress in the environmental arena will–as it has done in the past–come slowly.

7. ILLUSTRATIVE EXAMPLES

Example 7.1. Briefly describe each of the following general principles necessary for a successful pollution prevention program:
  a. Management commitment
  b. Communicating the program to the rest of the company
  c. Performing waste audits
  d. Cost/benefit analysis
  e. Implementation of pollution prevention program
  f. Follow-up

SOLUTION. A waste reduction plan must be specific to the industry. However, some general principles that must be incorporated for the success of the pollution prevention plan are discussed below.

a. *Management commitment.* Gaining the approval and support of top management is vital to the success of the waste reduction plan. It will be necessary to educate management about the pollution prevention program and its benefits through seminars and meetings. 3M successfully uses a 12-minute video on "Pollution Prevention Pays" to communicate the program to management and employees.

b. *Communicating the program to the rest of the company.* Middle-managers and employees with direct process line experience are in the best position to make suggestions as to where process improvements can be made, so communication with these employees is essential for program success. In addition, a monetary incentive and corporate recognition of employees for their practical pollution prevention ideas will also be very effective in getting them to "buy-into" the process.

c. *Performing waste audits.* The company must identify the processes, products, and waste streams in which hazardous chemicals are used. Mass balances of specific hazardous chemicals will help to identify waste reduction opportunities. Engineering interns could be very valuable in conducting such audits. Since managers and employees are often busy performing their assigned duties, an outside person may be able to focus on pollution prevention opportunities, cutting through some of the management and personnel barriers of the industry and may achieve significant progress.

d. *Cost/Benefit analysis.* Any change or modification in the process requires additional capital and operation and maintenance costs. A cost analysis must be included to help management make informed decisions. Factors including cost avoidance, enhanced productivity and decreased liability risks from the pollution prevention effort should be factored into the study. Federal and State agencies provide matching grants to small industries to implement pollution prevention programs, so this opportunity should be investigated and factored into the cost/benefit analysis of the proposed projects.

e. *Implementation of pollution prevention program.* Resistance to change by management and employees will still be an impediment for the implementation of the pollution prevention program. People known to resist new ideas and changes in the company must be included in the planning stages of the program. The CEO must be convinced of the merits of the program and must fully support the implementation of it.

f. *Follow-up.* Reduced energy costs, reduced raw materials, and reduced waste disposal fees must be tracked and communicated to the company personnel. The tracking information will be very useful in the filing of the company's waste manifest and biennial reports on its waste reduction efforts required under new regulations. In addition, this information will show employees and management that pollution prevention programs not only make sense environmentally, but economically as well.

Example 7.2. As described earlier, a portion of the Assessment Phase is used to select and to prioritize assessment targets. From the list of campus sites below, select the departments/units most likely to be sources of pollution in a typical small college. First, consider the types and volumes of pollutants likely to be used by the various departments. Then prioritize the choices from 1 to 5 with 1 being most likely to be the largest source of pollution, and 5 being the least likely. The departments and units include: English/History Department, Library, Biology Department, Dormitories, Physics Department.

SOLUTION. Specific answers will differ depending on the school examined and the research carried out in the various departments. Below are a set of typical answers for a small college setting. The following departments and physical plant units are likely to be the major generators of waste on campus. They are listed in order of pollution prevention priority.

1. *Biology Department* - In the labs, formaldehyde, mercury salts, solvents and a small amount of osmium tetroxide are generated as wastes.

2. *Dormitories* - Household hazardous waste could be generated from this source on campus. This waste could contain small quantities of cleaning solvents, paints, waste oils, as well as discarded electronic equipment, art supplies, etc., that could represent a special handling problem.

All other sources would not be expected to generate hazardous waste and would be considered low-priority locations. Some consideration should be given to the office waste generated by all departments however, as waste toner from copying machines, batteries, and fluorescent tubes represent potential hazardous waste from these small quantity sources.

Example 7.3. You have been tasked with evaluating as part of a pollution prevention opportunity assessment the relative merits of transportation of goods by trucks versus by railroad. Provide specific comments regarding environmental pollution production and energy utilization requirements of each mode of transportation. Identify the type of data needed to provide a quantitative analysis of this problem.

SOLUTION. Data needed for a quantitative analysis of this problem must be in the form of relative emission rates per unit mass transported. Emission data for loading and unloading activities should also be provided.

Merits of truck transportation
  1. Trucks can usually travel a more direct route for a delivery (fewer miles traveled) than trains. This is due to the fact that the infrastructure of roads and highways is much more extensive than that of the railroads. In many cases, a truck will carry an entire load to a single destination, taking the most direct route. Because the railroad is limited by its infrastructure, trains will most likely have to travel farther to reach the same destination.
  2. Truck engines produce fewer emissions than railroad engines. Beginning in 1991, the EPA set emission standards for all classes of heavy-duty over-the-road trucks. The emission standards require heavy-duty diesel engine manufacturers to decrease engine emissions of particulates (carbon soot), hydrocarbons, NO_x and CO. To meet these standards, engine manufacturers had to redesign many engine components and develop better combustion technologies. So far, the EPA has not set emission standards for railroad diesel engines. Railroad diesels are typically much larger than truck diesels and are built by different manufacturers. Because there are no emission standards for the railroad diesels, they continue to be built with less expensive, obsolete engine technology that produces higher emissions. To compare exhaust emissions on an equal basis, they need to be measured in grams per hour per horsepower produced, or tons of load delivered per kg of pollution per mile.

Merits of railroad transportation
  1. Unlike trucks, trains can couple several hundred cars together and pull them with only a few large diesel engines. Because of this, trains accelerate very slowly but are able to reach and maintain cruising speeds (analogous to a large, heavy car with an undersized engine). Because a train has a much lower power-to-weight ratio than a truck, it will get better fuel economy. To compare a train and truck on an equal basis, the fuel economy (mpg) should be calculated for both and divided by the total weight carried, resulting in mpg per ton values.
  2. Trains travel at steadier speeds than trucks. Fuel economy and emissions are optimized during steady-state operating conditions with an internal combustion engine. Trucks are exposed to much more stop-and-go driving than trains during which the engine is repeatedly accelerated and decelerated. Engines get poor fuel economy during acceleration because more fuel must be used to prevent them from "bucking" or "stalling." Likewise, emissions are much higher during acceleration due to the overfueling that takes place. With a diesel engine, one can actually see the emissions (particulates) bellowing from the exhaust during acceleration.

Another section could be added discussing the proper balance of trucking and railroad use. Some would say that we have gone too far from the optimum in the direction of trucks or buses and make too little use of trains. This discussion could include an analysis of the optimal mix of trucking and trains to deliver goods in a fashion that maximizes pollution prevention.

8. SUMMARY

1. A *pollution prevention opportunity assessment (PPOA)* is defined by EPA as a systematic, planned procedure with the objective of identifying ways to reduce or eliminate waste. The assessment procedure can be divided into the following four major phases:

Planning and organization phase, during which management buy-in and commitment are obtained, overall assessment goals are identified, and an assessment team is assembled.

Assessment phase, consisting of the collection of data and identification and screening of potential pollution prevention options.

Feasibility analysis phase, consisting of the technical and economic evaluation of each pollution prevention option.

Implementation phase, where recommended options are put into place, their performance is monitored, and feedback is provided for the next round of pollution prevention assessment activities.

2. Careful *planning and organization* must start prior to the assessment phase. First, a commitment from top management to reduce waste generation must be secured. To do this, the benefits, specifically economic ones gained from a pollution prevention program, must be clearly identified. Once a company has committed to minimizing the generation of waste, a *task force* must be established to conduct the actual assessment and convey its efforts to management. The top priority of the task force is to select the specific goals for pollution prevention. As the assessment progresses, the goals may be altered accordingly.

3. The ultimate purpose of the assessment phase is to develop a comprehensive list of potential pollution prevention options. Collecting data from every process in the plant is the first step in performing the assessment. Selecting the principal waste streams or waste-producing operations provides the task force with the main focus for the assessment. The best means for identifying information on all waste streams is through the use of flow diagrams and, then, material balances.

4. The feasibility analysis phase is used to determine whether identified options are technically and economically practical. A detailed evaluation and analysis of identified options should begin with source reduction alternatives. The technical evaluation should determine whether the potential option(s), once in place, will really work as intended. Performing a technical evaluation requires comprehensive knowledge of pollution prevention techniques, vendors, relevant manufacturing processes, and the resources and limitations of the facility and its staff. This evaluation may involve bench-scale or pilot-scale treatability testing. An economic evaluation is conducted using standard

measures of profitability, return on investment, and net present value (see Chapter 9). Costs should be broken down into capital costs and operating costs. Pollution prevention projects need to show a savings in operating costs to be economically competitive. With capital and operating costs known, profitability can be determined. Those options requiring no capital investment should be implemented as soon as possible.

5. Once the assessment and feasibility phases are complete, the selected option is ready to be implemented. The initial commitment from management is critical to overcome resistance to change from within the company. The amount of waste reduction needs to be measured. Actual or adjusted quantity change can be used for this pollution prevention performance measurement metric. These values can be determined most efficiently by recording the quantities (weight, volume, and/or toxicity) of waste generated before and after implementation of the option. Some states may also require some form of material balance as verification of pollution prevention effectiveness. Once the highest priority waste streams have been assessed and dealt with, the assessment task force should continue to identify new opportunities, assess other waste streams, and consider attractive options that were not pursued earlier. The ultimate goal should be to continually search out and eliminate as much waste generation as possible.

9. PROBLEMS

1. Provide a brief but overall description of a pollution prevention opportunity assessment.

2. Discuss why pollution prevention has been adopted by regulatory agencies, industries, and environmental groups. Also indicate their interpretation of the pollution prevention principle.

3. As indicated in Illustrative Problem 8.2, a portion of the assessment phase is used to select and prioritize assessment targets. From the list of campus sites below, consider the types and volumes of pollutants likely to be used by the various departments. Then prioritize them from 1 to 5, with 1 being the most likely to be the largest source of pollution, and 5 being the least likely. The departments and units include: the Cafeteria, Chemistry Department, Maintenance/Shop, Psychology Department, Art/Theater Department.

4. Briefly explain the following source/waste reduction techniques that are used in industry:
    a. Good housekeeping
    b. Material substitution
    c. Equipment design modifications
    d. Recycling
    e. Waste exchange
    f. Detoxification

5. List five advantages and disadvantages associated with a pollution prevention program.

6. Solvent vapor emissions from a coating operation are diluted by air to bring the solvent concentrations below the lower explosive limit (the

lowest concentration of the vapor in air that can be ignited). As a pollution prevention measure you are asked to consider nitrogen as an alternative to air to transport the solvent vapor. What are the advantages of a nitrogen ventilation system over one that uses air?

10. REFERENCES

1. U.S. EPA, *Waste Minimization Opportunity Assessment Manual*, Office of Research and Development, Hazardous Waste Engineering Research Laboratory, Cincinnati, Ohio, EPA/625/7-88/003, 1988.
2. U.S. EPA, *Pollution Prevention 1997: A National Progress Report*, Office of Pollution Prevention and Toxics, Washington, D.C., EPA/742/R-97/00, 1997.
3. Perry, R. and Green, D., *Perry's Chemical Engineering Handbook*, McGraw-Hill, New York, New York, 1998.

CHAPTER 13

Source Reduction

Contributing Author: Mary Wrieden

Once opportunities for waste reduction have been identified in a pollution prevention opportunity assessment, *source reduction* techniques should be implemented first. Source reduction involves the reduction of wastes at their source before they are generated, usually within a process, and is the most desirable option in the waste management/pollution prevention hierarchy. By avoiding the generation of waste, source reduction eliminates the costs and energy and materials utilization associated with the proper handling and disposal of the waste. A wide variety of facilities can adopt procedures to minimize the quantity of waste that they generate.

1. SOURCE REDUCTION OPTIONS

Schematics of various source reduction process options that can be used in a variety of industrial settings are provided in Figures 13.1 through 13.4. Details of these process options are discussed in the sections that follow.

Many source reduction options involve a change in procedural or organizational activities, rather than changes in technology. For this reason, these options tend to affect the managerial aspect of production and usually do not demand large capital and/or time investments. This makes implementation of many of these source reduction options affordable to companies of any size. Figure 13.5 provides a schematic of the four major source reduction schemes.[1] Note that Options 1 through 3 represent source control changes in a strict sense, while Option 4 represents end product changes directed at reducing waste both in production and during end use.

1.1. Procedural Changes

Procedural changes involve the management, organization, and personnel functions of a production process. Many of these measures are used in industry largely as efficiency improvements for waste reduction and good management practices. They often require very little capital cost and result in a high return on investment. Companies of any size can implement these practices in all areas of a plant. Evaluating plant procedures can often reveal source reduction opportunities that are relatively inexpensive and easy to implement. As shown in Fig. 13.5, procedural changes can include:

1. Loss prevention
2. Materials handling
3. Schedule improvements
4. Segregation
5. Personnel practices

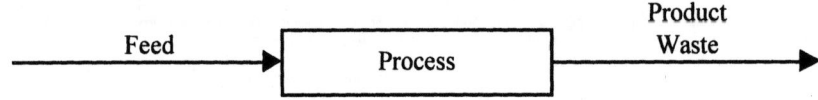

Figure 13.1. Typical process schematic.

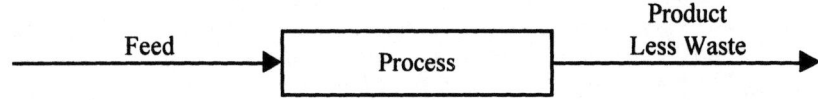

Figure 13.2. Typical process schematic with technology or process change(s).

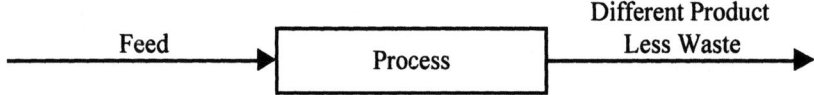

Figure 13.3. Typical process schematic with procedural change(s).

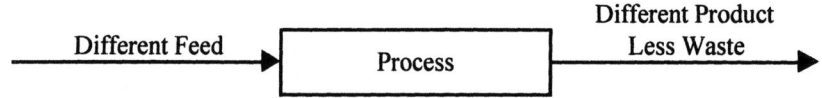

Figure 13.4. Typical process schematic with process change(s).

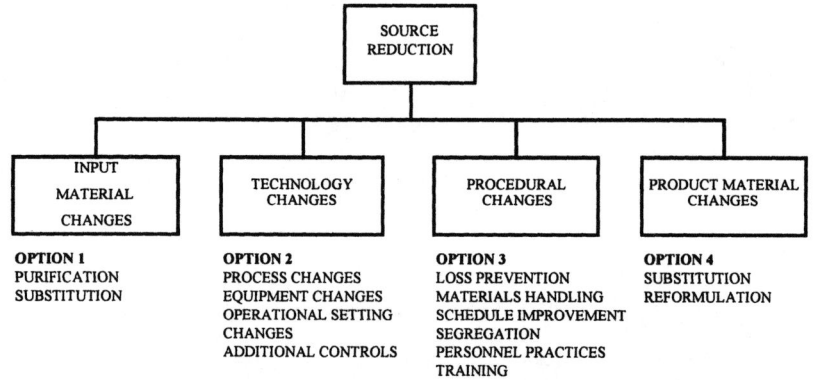

Figure 13.5. Source reduction options and alternatives.[1]

Loss prevention minimizes wastes by avoiding spills and leaks from equipment. Cleaning of chemical spills with typical absorbent/adsorbent materials results in the generation of additional waste, which in turn needs to be disposed of and may require treatment first. The most effective way to reduce the amount of waste generated by spills is to make precautionary modifications, ensuring that spills never occur in the first place. Several techniques can be designed into a plant and incorporated into its operations to reduce the likelihood of spills. These include:

1. Using properly designed storage tanks and containers according to manufacturer instruction and only for a specific purpose.
2. Installing overflow arms and automatic pump shutoffs.
3. Maintaining physical integrity of storage tanks and containers.
4. Installing spill containment dikes and curbing.
5. Implementing controlled and supervised loading, unloading, and transfer of all hazardous substances.
6. Installing a safe valving system, including interlocking devices that do not permit bypass by the operator.
7. Maintaining comprehensive records of all spills and leaks. Spills are most apt to occur during loading and unloading of materials. Therefore, it is important that safe practices be implemented and employees be educated in these practices.

Materials handling and inventory practices include programs to reduce loss of input materials resulting from mishandling, expired shelf life of time-sensitive materials, and proper storage conditions. All facilities transfer stored raw materials and wastes from different parts of a plant to another. The proper control over handling and transfer of these materials reduces the chances of spills. Obvious cost savings are the greatest benefit derived from handling materials correctly. Simple procedures, such as training employees in the operation of each type of transfer equipment, allowing adequate spacing for containers to be stored, stacking containers in a way that minimizes the chances of punctures and breaks, and labeling all containers to indicate the name and type of substance they contain, are easy ways to ensure proper materials handling. Computerized systems that link usage, shipping, storage, and raw material requests are the most efficient methods of inventory control and material tracking. Poor inventory control can result in the overstocking and subsequent disposal of expired materials. The economic loss associated with expired materials lies not just in the actual disposal costs that are incurred, but also in the cost of lost raw materials.

Schedule improvements in batch production runs can reduce the frequency of equipment and tank cleaning that can result in large amounts of cleaning solvent and reactant chemical waste. To reduce cleaning frequency, batch sizes should be maximized, or batches of one material should be followed by a batch containing a similar product, possibly eliminating cleaning between batches. For example, in the manufacturing of paint, scheduled batches should be sequential from light to dark to reduce the need for cleaning and to improve color uniformity.

Segregation of wastes reduces the volume of hazardous waste that must be specially handled by preventing the mixing of hazardous and nonhazardous waste. Because of RCRA's "mixture rule," any waste that is a combination of a listed hazardous and nonhazardous waste is classified in its entirety as a listed hazardous waste. This is true even if the mixture itself is characterized as nonhazardous. Thus, wastes are classified as hazardous if any component of the mixture is hazardous. By keeping these wastes separate, a facility can in one step significantly reduce the quantity of hazardous waste it must deal with. Separating nonhazardous from hazardous waste reduces the overall volume of hazardous waste requiring disposal and reduces the associated disposal costs. In addition, even the uncontaminated or undiluted hazardous wastes may be reusable in the production process or sent off-site for recovery. Waste segregation utilizes special storage or handling procedures to avoid mixing different waste streams resulting in cross-contamination. A commonly used segregation technique involves the collection and storage of wash water or solvents used to clean equipment for reuse in the production process. Consider a metal finishing facility as an example of an industrial application where segregation can yield economic benefits. Wastes containing different types of metals can be treated separately so that the metal values in the sludge can be recovered.

Personnel practices call on all employees to play vital roles in the pollution prevention program. Special attention should be given to employee training, incentives and bonuses, and other programs that encourage employees to strive for pollution prevention in the facility. Employees must first understand why waste management procedures are being used so that they are better able to make valuable suggestions regarding pollution prevention.

1.2. Technology Changes

Technology changes involve process and equipment modifications to reduce waste, primarily in a production setting. Technology changes can range from minor changes that can be implemented quickly and at low cost, to major changes involving replacement of processes at a very high cost. Since technology modifications usually require greater personnel and capital cost than procedural changes, they are generally investigated after all possible procedural changes have been instituted. As shown in Figure 13.5, categories of technology modifications include:

1. Process changes.
2. Equipment, piping, or layout changes.
3. Changes to operational settings.
4. Additional controls and/or automation.

Process changes often utilize innovative technology to develop a new process to achieve the same end product, while reducing the waste produced in the original process. A pollution prevention program should encompass process development activities. Research and development efforts should be encouraged in order to bring about new processes resulting in reduced energy demands and waste generation rates. Process redesign can include alteration of an existing process by addition of new unit operations or by implementation of new technology to replace out-of-date systems. For example, a metal manufacturer modified a process to use a two-stage abrasive cleaner and eliminated the need for a chemical cleaning bath.

Equipment changes can reduce waste generation by reducing equipment-related inefficiencies. The required capital involved in using more efficient equipment can be justified by higher productivity, reduced raw material costs, and reduced waste management costs. Modifications to certain types of equipment can require a detailed evaluation of process characteristics. In this case, equipment vendors should be consulted for information regarding the applicability of particular types of equipment for a given process and process application. Many equipment changes can be very simple and inexpensive to implement. Examples include installing better seals on equipment to eliminate leakage or simply putting drip pans under equipment to collect leaking material for reuse. Another minor modification is to increase agitation rates and altering temperatures to prevent the formation of fouling deposits resulting from crystallization, sedimentation, corrosion, and chemical reactions during formulating and blending procedures.

Changes to operational settings involve changes in the way Process equipment operates. including adjustments to the temperature, pressure, flow rate, and residence time parameters. These changes often represent the easiest and least expensive ways to reduce waste generation. Process equipment is designed to operate most efficiently at the optimum parameter settings. Less waste will be generated when equipment operates efficiently and, therefore, at optimum settings. Trial runs can be used to determine the actual optimum settings for specific pieces of process equipment. For example, a plating company can assess the flow rate of chromium in the plating bath to adjust chromium flow to the optimum setting, possibly reducing the chromium concentration used, thereby resulting in less chromium waste requiring treatment.

Additional controls and/or automation can result in improved monitoring and adjustment of operating parameters to ensure the highest level of process efficiency. Anything from simple steps involving one-stream set point controls to advanced statistical process control systems can be utilized depending on the need of the facility and the potential economic benefits additional controls would provide. Automation can reduce the likelihood of human errors resulting in spills and costly downtime. The resulting increase in efficiency owing to automation can increase product yields and could be highly cost effective for high value products.

1.3. Input Material Changes

Input material changes accomplish pollution prevention by reducing or eliminating waste materials that first enter the process through raw material impurities and/or undesirable reaction chemistry based on specific starting materials. Changes in materials can also be made to avoid the generation of wastes within production processes. Input material changes fall into the categories of material substitution and material purification. A substitute is selected that is either less hazardous, produces less pollutant, or results in lower waste generation rates than the original raw materials while still satisfying end-product specifications. Ideally, the best substitution is the replacement of a hazardous, impure, pollutant generating material with a pure, nonhazardous one, without damaging the quality of the product. Examples of industrial applications of input material changes include:

1. *Printing operations* - substitution of water-based inks for solvent-based inks.
2. *Furniture manufactures* - substitution of water-based or powder coatings for solvent-based coatings.
3. *Plating operations* - replacement of cyanide cadmium plating baths with non-cyanide baths.

1.4. Product Changes

Product changes are performed by the manufacturer of a product with the intent of reducing waste resulting from a product's production and use. *Product conservation* involves the way in which an end-product is used. For example, the manufacture of water-based paints instead of solvent-based paints involves no toxic solvents that make solvent-based paints hazardous. In addition, water-based paints greatly reduce volatile organic emissions to the atmosphere during their use.

Product reformulation involves manufacturing a product with a lower composition of hazardous substances or changing the composition so that no hazardous substances are present in the final product, or none are used in its manufacture. For example, a company can use a nonhazardous, water-based coating for finishing a product, or a nonhazardous solvent can be used in place of a chlorinated solvent in the cleaning preparation of a product for plating or coating. Other examples include the substitution of composite components for metal ones requiring plating in vehicle manufacturing, or modifying the packaging of computer components to utilize folded cardboard rather than expanded styrene beads and foam board for component shipping. Using less hazardous material within a process will reduce the overall amount of hazardous waste produced for a given product.

2. IMPEDIMENTS TO ACHIEVING SOURCE REDUCTION

Despite the numerous benefits attained with pollution prevention programs, certain misconceptions about possible drawbacks to pollution prevention efforts still exist. Concerns have primarily centered on maintaining product quality and the large capital expenditures people anticipate having to make for any pollution prevention project. Generally, product quality should not be compromised as a result of pollution prevention efforts, and in fact in many cases, waste generation is a symptom of poor product quality control and low process efficiency. Many pollution prevention projects have actually improved product quality because of improvements they have produced in the control and automation of the production process. Reusing or recycling waste material within a process sometimes creates fear that the overall quality will be jeopardized. This concern is valid only if the wastes are reused or recycled incorrectly or allowed to accumulate within a manufacturing process. Treatability studies involving lab screening, bench-scale, or pilot-scale demonstrations can be used to show the effectiveness of a pollution prevention technique prior to its actual implementation.

Pollution prevention technologies, involving the replacement of existing processes and product reformulations, can require substantial up-front capital costs. The expense of initiating and implementing a pollution prevention program presents an economic challenge to a company operating in a competitive market. Budget needs for the program must be weighed against other project needs, and against the potential benefits that can accrue from increased energy and materials use efficiency, reduced environmental and health and safety liability, and improved public relations for the company with the implementation of pollution prevention projects. The reader is referred to Chapter 9 and Perry's *Chemical Engineering Handbook*[2] for more detailed information regarding the economics associated with a pollution prevention program.

3. ILLUSTRATIVE EXAMPLES

Example 3.1. A surface coating facility must reduce VOC emissions in order to comply with new state regulations. Describe two source reduction options to achieve compliance for this facility.

SOLUTION. Two source reduction options that can be used to reduce VOC emissions at a surface coating facility are coating reformulation and a process change to improve the coating transfer efficiency. The advantage of coating reformulation is the reduction of disposal and storage requirements for hazardous wastes generated from paint residues, waste paint containers, etc. In addition, some industry-specific *national emission standards for hazardous air pollutants (NESHAPs)* specify compliance requirements in terms of coating characteristics or emission controls, making alternative coating formulations a cost-effective means of meeting environmental compliance requirements. The Wood Furniture Manufacturing Operations NESHAP regulation published in the December 7, 1995 edition of the Federal Register [60 FR 62930] and amended in June of 1997 [62 FR 30257; 62 FR 31361] is one such standard. Some disadvantages of reformulation may be additional pretreatment requirements, longer drying times, higher drying temperatures, and the need for a temperature-controlled storage area to produce the same final product that is currently being produced. A process change to improve transfer efficiency requires an investigation into application equipment and a performance evaluation with lab or pilot testing. An increase in transfer efficiency reduces coating product losses and subsequent air emissions from the facility.

Example 3.2. Chromium wastes have a long history of simply being "dumped" into the nearest sewer where they inevitably end up as a toxic pollutant in our environment. Since there is no satisfactory way to destroy chromium waste products, minimization of these wastes by source reduction is essential.

One such method of source reduction is for manufacturers to find a substitute for chromium(VI). When chromium(IV) is being used as an oxidizing agent, manufacturers have several choices. Common household bleach, NaOCl, is often an acceptable substitute for other oxidizing agents and generally generates no serious pollutants. Balance the equation below and calculate the volume in gallons of bleach required per year, on a 350 day per year basis, to produce 5,000 pounds of cyclohexanone per day. Household bleach contains 5.25% NaOCl by weight and has a density of 8.0 lb/gal.

$$C_6H_{12}O + NaOCl \rightarrow C_6H_{10}O + NaCl$$

SOLUTION: Balancing the equation yields:

$$C_6H_{12}O + NaOCl \rightarrow C_6H_{10}O + NaCl + H_2O$$

Calculate the number of moles of product per day as:

$$(5{,}000 \text{ lb } C_6H_{10}O/d)(1 \text{ lbmol } C_6H_{10}O/98 \text{ lb of } C_6H_{10}O)$$
$$= 51 \text{ lbmol } C_6H_{12}O/d$$

Calculate the number of moles of NaOCl per day as:

$$(51 \text{ lbmol } C_6H_{10}O/d)(1 \text{ lbmol NaOCl}/1 \text{ lbmol } C_6H_{10}O/d)$$
$$= 51 \text{ lbmol NaOCl}/d$$

Calculate the weight of the NaOCl per day as:

$$(51 \text{ lbmol NaOCl}/d)(74.4 \text{ lb of NaOCl}/1 \text{ lbmol NaOCl})$$
$$= 3{,}796 \text{ lb NaOCl}/d$$

Calculate the total weight of the bleach per day as:

$$(3{,}796 \text{ lb NaOCl}/d)(1 \text{ lb bleach}/0.0525 \text{ lb NaOCl})$$
$$= 72{,}303 \text{ lb bleach}/d$$

Calculate the total volume of the bleach per day as:

$$(72{,}303 \text{ lb bleach}/d)/(8 \text{ lb}/1 \text{ gal bleach}) = 9{,}038 \text{ gal bleach}/d$$

Calculate the volume of bleach per year as:

$$(350 \text{ d/yr})(9{,}038 \text{ gal bleach}/d) = 3{,}163{,}265 \text{ gal bleach/yr}$$

Example 3.3. The Noram Corporation's 100,000 square foot office center contains 1,253 standard four-lamp fluorescent fixtures consuming an average of 174 watts per fixture or about 43.5 watts per lamp. These lamps operate for an average of 16 hours/day, 6 days/week, thus accounting for a yearly operational time of 4,992 hours. Also, it is estimated that the local coal-fired electric power plant emits 0.0175 pounds of sulfur dioxide, 0.00824 pounds of nitrous oxides, and 2.25 pounds of carbon dioxide per kilowatt-hour generated.

Apply the most effective method of pollution prevention, i.e., source reduction, to effectively decrease the amount of pollutants resulting from the consumption of electricity in the building. It has been suggested that the number of lights be reduced by 20% and the watts per fixture be reduced to 106 watts. An effective guideline for properly retrofitting a new lighting system is to keep the lighting intensity to a maximum of 1.5 watts/square foot.

- a. Calculate the present watts per square foot, the new watts per square foot, and the number and new wattage of lamps to be installed.
- b. Determine the total present lighting load, the new lighting load, and the load reduction (in kW) as well as the present annual load, the new annual load, and the annual load reduction (in kW-hr).
- c. Calculate the present pollutant load, the new pollutant load, and the reduction in pollutant emissions that result from this source reduction effort.

SOLUTION.
a. Calculate the present total number of lights in the office as follows:

$$N = (1,253 \text{ units}) (4 \text{ lights/unit}) = 5,012 \text{ lights}$$

Calculate the number of lights that must be removed to retrofit the lighting system.

$$(5,012 \text{ lights}) (0.2) = 1,002 \text{ lights} = 250 \text{ units}$$

Calculate the present lighting load in W and kW as:

$$P = (1,253 \text{ units}) (174 \text{ W/unit}) = 218,000 \text{ W} = 218 \text{ kW}$$

Calculate the present watts per square foot:

$$(218,000 \text{ W})/(100,000 \text{ ft}^2) = 2.18 \text{ W/ft}^2$$

Calculate the new lighting load in W and kW as:

$$P = (1,253 \text{ units} - 250 \text{ units}) (106 \text{ W/unit}) = 106,000 \text{ W} = 106 \text{ kW}$$

Calculate the new watts per square foot:

$$(106,000 \text{ W})/(100,000 \text{ ft}^2) = 1.06 \text{ W/ft}^2$$

It should be noted that the present lighting intensity of 2.18 W/ft² is above the maximum recommended lighting intensity, while the proposed new lighting intensity falls below the maximum recommended number.

b. The total present lighting and new lighting loads were determined above to be:

$$\text{Present lighting load} = 218 \text{ kW}$$
$$\text{New lighting load} = 106 \text{ kW}$$

Calculate the present annual load in kW-h as:

$$P_a = (218 \text{ kW}) (16 \text{ h/d})(6 \text{ d/wk}) (52 \text{ wk/yr}) = 1.09 \times 10^6 \text{ kW-h}$$

Calculate the new annual load in kW-h as:

$$P_a = (106 \text{ kW}) (16 \text{ h/d})(6 \text{ d/wk}) (52 \text{ wk/yr}) = 529{,}000 \text{ kW-h}$$

Calculate the load reduction in kW as:

$$P_{red} = 218 \text{ kW} - 106 \text{ kW} = 112 \text{ kW}$$
$$P_{red} = (112 \text{ kW})/(218 \text{ kW}) (100) = 51.4\%$$

Calculate the annual load reduction in kW-h/yr as:

$$(112 \text{ kW}) (16 \text{ h/d}) (6 \text{ d/wk}) (52 \text{ wk/yr}) = 559{,}000 \text{ kW-h/yr}$$

c. Calculate the annual pollution contribution of SO_2, CO_2, and NO_x in lb/yr from the presenting lighting configuration as:

SO_2: $(0.0175 \text{ lb/kW-h}) (1.09 \times 10^6 \text{ kW-h}) = 19{,}075 \text{ lb/yr}$
CO_2: $(2.25 \text{ lb/kW-h}) (1.09 \times 10^6 \text{ kW-h}) = 2.45 \times 10^6 \text{ lb/yr}$
NO_x: $(0.00824 \text{ lb/kW-h}) (1.09 \times 10^6 \text{ kW-h}) = 8{,}982 \text{ lb/yr}$

Calculate the new annual pollution contribution of SO_2, CO_2, and NO_x in lb/yr as:

SO_2: $(0.0175 \text{ lb/kW-h}) (529{,}000 \text{ kW-h}) = 9{,}258 \text{ lb/yr}$
CO_2: $(2.25 \text{ lb/kW-h}) (529{,}000 \text{ kW-h}) = 1.19 \times 10^6 \text{ lb/yr}$
NO_x: $(0.00824 \text{ lb/kW-h}) (529{,}000 \text{ kW-h}) = 4{,}359 \text{ lb/yr}$

Calculate the pollution reduction achieved with the new lighting system as follows:

SO_2: $19{,}075 \text{ lb/yr} - 9{,}258 \text{ lb/yr} = 9{,}817 \text{ lb/yr}$
CO_2: $2.45 \times 10^6 \text{ lb/yr} - 1.19 \times 10^6 \text{ lb/yr} = 1.26 \times 10^6 \text{ lb/yr}$
NO_x: $8{,}982 \text{ lb/yr} - 4{,}359 \text{ lb/yr} = 4{,}623 \text{ lb/yr}$

As these results indicate, significant reductions in energy and pollutant emissions can be realized from source reduction activities such as the lighting upgrade suggested in this problem.

4. SUMMARY

1. Once opportunities for waste reduction have been identified in a pollution prevention opportunity assessment, source reduction techniques should be implemented first. *Source reduction* involves the reduction of wastes at their source, usually within a process, prior to waste generation, and is the most desirable option in the waste management/pollution prevention hierarchy. Source reduction can be broken down into options involving either product changes or various source control methods, including changes in procedure, technology, and input material.

2. *Procedural changes* involve the management, organizational, and personnel functions of production and include loss prevention, material handling, scheduling improvements, segregation, and personnel practices.

3. *Technology changes* involve process and equipment modifications to reduce waste, primarily in a production setting. *Technology changes* can range from minor changes that can be implemented quickly and at low cost, to major changes involving replacement of processes at a very high cost.

4. *Input material changes* accomplish pollution prevention by reducing or eliminating the waste materials that first enter the process. Changes in materials can also be made to avoid the generation of wastes within the production processes. Input material changes fall into the categories of *material substitution* and *material purification*.

5. *Product changes* are performed by the manufacturer of a product with the intent of reducing waste resulting from both a product's manufacture and its use. *Product conservation* implies changes that improve the hazardous waste production or emission rate of the product during its end use. *Product reformulation* involves manufacturing a product with a lower composition of waste substances or changing the composition so little or no wastes are generated during manufacturing or are present in the final product.

6. Despite the numerous benefits attained with a pollution prevention program, certain misconceptions about possible drawbacks to pollution prevention efforts still exist. Concerns have primarily centered on misconceptions of reduced product quality that result from pollution prevention efforts and of large capital expenditures that are required by all pollution prevention projects. Generally, product quality should not be compromised as a result of pollution prevention efforts, but in fact many times is improved through improved process control and optimization that is driven by pollution prevention efforts. In addition, pollution prevention opportunities are prioritized through a cost-benefit analysis, and many options, particularly source reduction ones, are inexpensive and highly effective waste reduction techniques. With more knowledge and experience with pollution prevention concepts these misconceptions will hopefully be eliminated.

5. PROBLEMS

1. Consider a degreasing operation in a metal finishing process. Give an example of process modifications that might be made to this part of the process that would represent source reduction.

2. Consider two types of diapers: commercially laundered cloth diapers and disposable super-absorbent diapers, which are known to need changing less often than cloth diapers.
   a. Test your intuition on the life cycle environmental impacts of the two types of diapers with respect to the following criteria: net energy requirements, atmospheric emissions, industrial and post consumer solid waste, and wastewater and water volume requirements.
   b. If available, study the report, *Energy and Environmental Profile Analysis of Children's Disposable and Cloth Diapers*, by Franklin Associates, Ltd., prepared for the American Paper Institutes Diaper Manufacturers Group, Prairie View, Kansas, 1990. Check your answers from Part a against this report.
3. The following table provides a weekly volume of solid wastes commonly generated in households.

Waste	Weight (lb)	Density (lb/ft^3)	Volume (ft^3)
Food	10	18.0	0.6
Paper	50	5.0	10.0
Cardboard	10	6.5	1.5
Plastics	2	4.0	0.5
Textiles	1	4.0	0.3
Leather	2	10.0	0.2
Garden Trimmings	10	6.5	1.5
Wood	5	15.0	0.3
Nonferrous Metals	2	10.0	0.2
Ferrous Metals	4	20.0	0.2
Dirt, Ash, Bricks, etc.	1	30.0	0.03

Note, it is assumed that glass and aluminum cans are recycled by the household.

Estimate the percent volume reduction that would be achieved in the amount of solid waste collected in the community if all households installed garbage compactors, which compacted the solid wastes to a density (ρ_{csw}) of 20 lb/ft^3. Note that garden trimmings, wood, ferrous metals, dirt, ash, and brick are usually not placed in trash compactors.

4. A suburban town council projects that there will be approximately 500 new homes built in and around the central town within the next 10 years. Currently, the town's wastewater treatment facility has a design capacity of 1.3 MGD (million gallons per day) of wastewater from homes (the current daily flow of wastewater is 1.0 MGD). It is projected that each new home will use 375 gpd of water, or a total of 0.1875 MGD for the 500 new homes. Unfortunately, the facility manager claims that the facility can handle no more than 0.16 MGD of additional wastewater before the end of 12 years (at which time a new 2.0 MGD facility will have been completed).

In an attempt to solve the above problem, the owner of a plumbing supply outlet in the town suggests that each new home be fitted with the company's new Airflush® 7 toilets. He claims the Airflush® 7 uses considerably less water per flush than conventional toilets. Projections also predict that each new home will have 2.5 toilets. If the average number of flushes per toilet is 5/day and the Airflush® 7 toilet saves 4.5 gallons per flush over the conventional toilet, will the plumber's proposition work? Note, wastewater discharges from other domestic sources are to be neglected in this analysis.

5. A power plant generates 2.76×10^5 megawatt hours (MW-hr) of electricity per year by burning 1.66×10^5 tons of coal containing 3.2% sulfur and 15.5% ash. The ratio of fly ash to the bottom ash is 0.65 and the plant's particulate collection efficiency is 85%. The Jones family illuminates each of the six rooms in their apartment with one 100 watt incandescent light per room. Each light is used for 5 hours each day. Assuming that electricity costs $0.136/kW-hr, determine the following based on the coal-fired plant emission information below.

GIVEN: The pollutant emission equations, which apply to coal-fired plants, have been adapted from Ronald White's *The Price of Power Update - Electric Utilities and the Environment*.

Particulates: $e = (a)(b)(c)[1 - (P/100)] \times 10^{-2}$
Sulfur Dioxide: $e = (1.90)(c)(s) \times 10^{-2}$
Nitrous Oxides: $e = (15)(c) \times 10^{-2}$

where e = emissions in tons/year; a = mass percent ash content of coal burned; b = ratio of fly ash to bottom ash; c = coal consumption in tons/year; P = percent particulate collection efficiency of the precipitator, baghouse and/or mechanical collector; and s = average annual mass percent coal sulfur content.

a. The reduction in the amount each of pollutant (SO_2, NO_x and particulates) achieved each year by utilizing 40 watt fluorescent bulbs instead of incandescent bulbs.
b. The monetary savings to the Jones family each year.

6. A nickel electroplating line uses a dip-rinse tank to remove excess plating metal from the parts. Currently, a single tank is used which requires R gal/hr of fresh rinse water to clean F parts/hr (see figure below). Assume the following equilibrium relation governs the cleaning:

$\lambda = f_1/r_1 =$ (ounces of metal residue/part)/(ounces of metal residue/gal batch)

Calculate the reduction in rinse water flow rate (a pollution prevention measure) if a two-stage counter-current rinse tank is used (as compared to the single-stage unit) and 99% of the residue must be removed. Assume the drag-out volume is negligible.

The flow diagram for a one stage operation is given as:

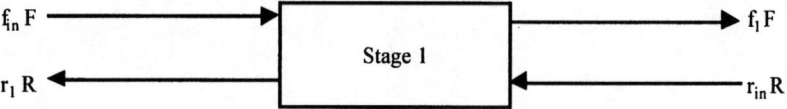

6. REFERENCES

1. U.S. EPA, *Waste Minimization, Issues and Options*, *Vol. 1*, Office of Solid Waste, Washington, D.C., EPA 530-SW-86-041, 1986.
2. Perry, R. and Green, D., *Perry's Chemical Engineering Handbook*, McGraw-Hill, New York, New York, 1998.

CHAPTER 14

Recycling

Contributing Author: Joseph Lanzillotti, Jr.

Recycling or reuse can take two forms: preconsumer and postconsumer applications. *Preconsumer* recycling involves raw materials, products, and by-products that have not reached a consumer for an intended end-use, but are typically reused within an original process. *Postconsumer* recycled materials are those that have served their intended end-use by a business, consumer, or institutional source and have been separated from municipal solid waste for the purpose of recycling. Although this chapter mainly addresses preconsumer recycling, postconsumer recycling concepts and practices can be found in the illustrative examples and problems at the end of this chapter.

Recycling techniques allow waste materials to be used for a beneficial purpose. A material is recycled if it is used for other than its originally intended purpose after being discarded, or if it is reused or reclaimed to serve its intended function once again. Recycling through use and/or reuse involves returning waste material to either the original process as a substitute for an input material, or to another process as a raw material. Recycling through reclamation is the processing of a waste for recovery of a valuable component or for regeneration of the waste for reuse. Recycling of wastes can provide a very cost-effective waste management option. This option can help eliminate waste disposal costs, reduce raw material costs, and provide income from saleable waste.

Recycling is the second most preferred option in the waste management/pollution prevention hierarchy and as such should be considered only when all feasible, cost-effective source reduction options have been investigated and implemented. Reducing the amount of waste generated at the source will often be more cost-effective than recycling, since waste often represents lost raw material or product which requires time, energy, and money to recover. It is important to note that recycling can increase a generator's risk or liability as a result of the associated handling and management of the waste materials involved. The measure of effectiveness in recycling is based upon the ability to separate any recoverable materials from other process waste that is not recoverable.

1. RECYCLING OPTIONS

Consider the basic process described in Figure 14.1. Four recycle options for this process are presented in Figures 14.2 through 14.5. These include Figure 14.2 which shows a basic process with a recycle stream placed into the feed, Figure 14.3 which indicates a recycle stream that feeds directly into the process, Figure 14.4 which incorporates additional treatment to further reduce waste, and Figure 14.5 that adds an energy recovery process to reduce waste to an even lower level, while at the same time producing energy.

296   Chapter 14

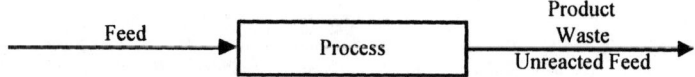

Figure 14.1. Typical process schematic.

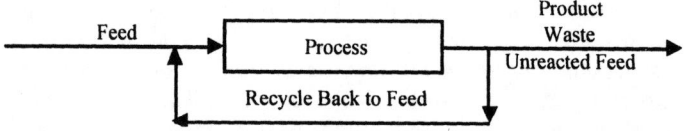

Figure 14.2. Basic process schematic with recycle back to feed.

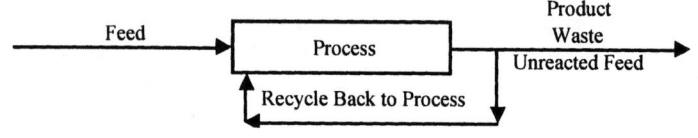

Figure 14.3. Basic process schematic with recycle back to process.

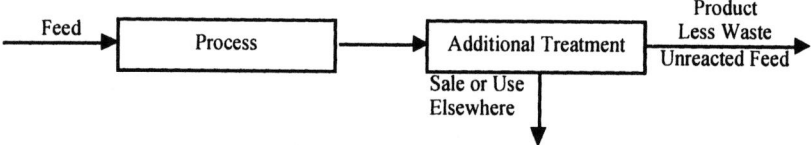

Figure 14.4. Basic process schematic with additional treatment for by-product recovery.

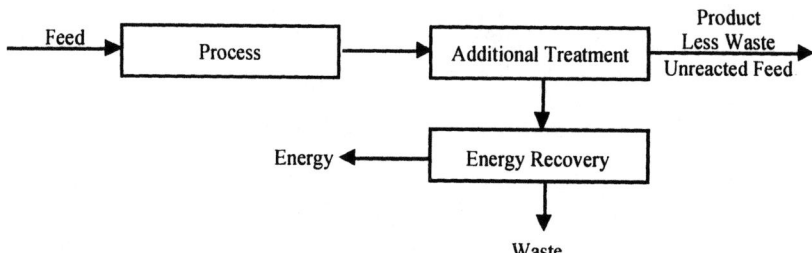

Figure 14.5. Basic process schematic with additional treatment and energy recovery.

In a generic sense, recycling options can be listed in the following order of preferability (for most systems):

1. Direct reuse on-site
2. Additional recovery on-site
3. Recovery off-site
4. Sale for reuse off-site
5. Energy recovery

These recycling options are summarized in the following sections.

1.1. Direct Reuse On-Site

Direct reuse on-site involves finding a beneficial purpose for a recovered waste in a different process. There are three factors that must be considered when determining the potential for waste on-site reuse. These are

1. The chemical composition of the waste and its effect on the process it is to be reused in.
2. Whether the economic value of the reused waste justifies any modification to a process in order to accommodate it.
3. The extent of availability and consistency of the waste to be reused.

1.2. Additional On-Site Recycling

Recycling alternatives can be accomplished either on-site or off-site and may depend on a company's staffing or economic constraints. On-site recycling alternatives directly result in less waste leaving a facility and in associated reductions requirements for reporting and manifesting the waste. The disadvantages of *on-site recycling* lie in the capital outlay for recycling equipment, the need for operator training, and additional operating costs recycling might incur. In some cases, the amount of waste generated does not warrant the cost for installation of in-plant recycling systems. In general, however, since on-site alternatives do not involve transportation of waste materials off-site, and the incurred liability therein, they are preferred over off-site recycling alternatives.

1.3. Recovery Off-Site

If insufficient amounts of waste are generated on-site to make an in-plant recovery system cost effective, or if the recovered material cannot be reused on-site, *off-site recycling* is an option. Some materials commonly reprocessed off-site include oils, solvents, electroplating sludges and process baths, scrap metal, and lead-acid batteries. The cost of off-site recycling is dependent upon the purity of the waste and the market for the recovered material.

1.4. Sale for Reuse Off-Site

As an alternative to both on-site and off-site recycling, the generator may transfer waste to another facility for use as a raw material in its manufacturing operations. Facilities receiving the waste either use it as is or subject it to a minimal amount of pretreatment prior to its use. Supply and demand is the key criterion for the success of waste transfer of this type, but off-site reuse becomes more feasible if a method exists within the company to identify facilities capable of utilizing the waste and marketing the waste to them. This need has purportedly been fulfilled by *waste exchanges*, which serve as brokers of wastes and/or clearinghouses for information on the availability of waste streams. Waste exchanges (discussed later in this chapter) can be either privately owned or government-funded organizations that facilitate waste transfer by identifying potential users and matching them up with suppliers. This exchange often proves to be economically advantageous to both firms involved, since the generator experiences a reduction in waste disposal costs and the purchasing firm experiences a reduction in raw material costs. Liability concerns remain with this option.

1.5. Energy Recovery

Recycling can also be achieved in the recovery of energy through the use of waste as a fuel supplement or fuel substitute. Waste may be processed in fossil-fuel-fired plants or in incinerators equipped with energy recovery systems. Note that processes with overall energy efficiencies of less than 60% are generally regarded strictly as incineration and not energy recovery. Usually, a variety of high-Btu wastes with different compositions are blended to produce a fuel with a desired specification.

1.6. Regulatory Considerations[1]

Under RCRA Subtitle C, Congress granted EPA the authority to regulate hazardous wastes. The principle objective of hazardous waste regulation is the protection of human health and the environment. RCRA regulation is also intended to encourage the conservation and recovery of valuable materials. The definition of solid waste under RCRA, which serves as the starting point for the hazardous waste management system, reflects EPA's effort to obtain the proper balance between these two underlying objectives.

According to RCRA regulations, a material must be defined as a solid waste before it can be considered a hazardous waste. Materials that are recycled are a special subset of the solid waste universe. When recycled, some materials may qualify for an exclusion from the definition of solid waste and fall out of RCRA regulation or be subject to less-stringent regulatory controls. Based on the material and the type of recycling, the generator of a recyclable solid waste must determine if it is subject to reduced requirements or full regulation.

RCRA Subtitle C §261.6 and Parts 266, 273, and 279 define management standards for different types of hazardous waste recycling. This range of management, from no regulation to full regulation, is based on the type of recycling activity involved and the hazards posed, and demonstrates EPA's intent to encourage recycling while still protecting human health and the environment.

Under the RCRA, EPA defines recycling broadly. Use constituting disposal, burning for energy recovery, and reclamation are all forms of recycling. The direct use or reuse of a secondary material is also a form of recycling using the RCRA definition. Section 261.2(e)(1) of RCRA does provides exclusions from the definition of solid waste, however, for materials that are used or reused in one of the following ways: used or reused as an ingredient, used or reused as a product substitute, or returned to the production process, and significant relief from RCRA regulatory requirements may be realized of a recycling program that is implemented for a formally disposed of solid or hazardous waste. Examples of the types of exemptions RCRA provides for in the recycling area include the following.

1. *Used as an Ingredient (Direct Reuse On-Site or Off-Site).* If a secondary material is directly used as an ingredient in a production process without first being reclaimed (e.g., carbon tetrachloride still bottoms used in producing tetrachloroethylene), then that material is not a solid waste and is exempt from RCRA regulation.
2. *Used as a Product Substitute (Direct Reuse On-Site or Off-Site).* If a secondary material is used as an effective substitute for a commercial product without first being reclaimed (e.g., hydrochloric acid by-product from chemical manufacturing used by the steel industry for pickling steel), it is exempt from the definition of solid waste and RCRA regulation.
3. *Returned to the Production Process (Direct Reuse On-Site).* When a material is returned to the original production process from which it was generated, it is not a solid waste and is RCRA exempt. When this exclusion was originally promulgated on January 4, 1985, it applied only to materials returned to a primary production process. The September 19, 1994, Federal Register extended this exemption, however, to include materials returned to secondary processes. This exclusion only applies if the material is used as a raw material or feedstock in the production process and if it is not reclaimed prior to its reintroduction into the system (e.g., emission control dust returned directly to a primary zinc smelting furnace). The material does not have to be returned to the exact unit, but may be returned to any unit associated with the production of a particular product. In the case where the original process to which the material is returned is a secondary process, the material must be managed such that there is no placement on the land.

Even if a waste that is being recycled is still considered a RCRA waste, in general, the actual recycling activity and treatment prior to recycling are not regulated. Thus, only storage of these materials prior to such recycling is regulated. If the waste is not stored prior to recycling, the recycler only needs to notify EPA of the activity and comply with the use of the manifest when receiving shipments of recyclable materials from off-site.

For all recycling activities, the premise is that legitimate reclamation or reuse is taking place. To encourage recycling, EPA subjects these activities to reduced regulation. Some facilities, however, may claim that they are "recycling" a material in order to avoid being subject to RCRA regulation, when in fact the activity is not legitimate recycling. Therefore, EPA has established guidelines for what constitutes legitimate recycling and has described activities it considers to be "sham recycling." Considerations include whether the secondary material is effective for the claimed use, the secondary material is used in excess of the amount necessary, and whether or not the facility has maintained records of the recycling transactions.

For all questions regarding regulatory requirements and exemptions related to waste recycling, the appropriate EPA Regional or authorized state personnel, or the RCRA Hotline (800/424-9346) should be contacted.

2. RECYCLING TECHNOLOGIES

There are numerous treatment technologies for recovering a beneficial component from a waste stream for reuse. The remainder of this chapter will discuss treatment methods applicable to the recovery of vapor-liquid phases, solid-liquid phases, liquid-liquid phases, and solute recovery. The reader is reminded that some material in this chapter overlaps with other parts of this book because of an effort to accommodate those readers who chose not to review Part 1. One may refer to *Chapter 4, Unit Operations*, and *Chapter 5, Plant Equipment*, for an introduction to the general subject of separation techniques. Gas-solid methods, including particulate separation techniques and gas adsorption, can be found in *Chapter 6, Ancillary Processes and Equipment*. Additional details, and in some cases extensive information, on treatment technologies not employed for recovery, recycle, or reuse can be found in *Chapter 7, Waste Treatment Processes and Equipment*.

2.1. Vapor-Liquid Separation

Common separation techniques involving liquids to be recycled include *distillation, evaporation*, and *gas absorption*. These separation techniques are briefly described below.

2.1.1. Distillation

Distillation is the most widely used liquid phase separation process for recovering organic components from hazardous waste product streams. It may be performed by either of two principal methods which rely on the differences in boiling points among various components in a waste stream. The first method involves producing a vapor, through boiling of a liquid mixture to be separated, and condensing the vapors without allowing any liquid to return to the still in contact with the vapors. The second method involves returning part of the condensate to the still in such a way that the returning liquid is brought into intimate contact with the vapors on their way to the condenser. Either method can be performed as a continuous distillation process (fractionation) or as a batch distillation process. Since distillation is a nondestructive process with no effluent problems, rising chemical prices and stricter emissions regulations are making distillation a more competitive and popular recovery process. Distillation cannot be used to separate thick wastes such as sludges or slurries. Capital, operating, and utility costs for distillation systems depend on the nature of the feed and product streams.

Continuous distillation may be used when the difference among vapor pressures of various components within the waste stream is not large or when a high degree of purity is sought. The vapor and liquid phases are brought into intimate contact through the use of trays or plates, stacked one above the other, enclosed in a cylindrical shell forming a column. The feed material to be separated is introduced into the column at one or more feed points. As the liquid runs down the column, flowing past each tray, the vapor rises through the column and contacts the liquid. Vapors rising to the top are cooled and condensed to a liquid. Some of this liquid is removed as distillate overhead product, and part is returned to the column as reflux. The lower-boiling point components concentrate in the vapor phase, where as the higher-boiling point components tend toward the liquid phase. The result is a vapor phase that becomes enriched in the lower-boiling point components as it phases upward and a liquid phase that becomes enriched in higher-boiling components as it flows downward. One application of this process is the recovery of solvents from carbon adsorption systems and paint wastes.

Batch distillation may be used when the difference in vapor pressures among components is large. This type of distillation involves separating a specific quantity of a liquid mixture into its components. In its simplest form, batch distillation consists of a heated vessel, a condenser, and one or more receiving tanks. The waste to be separated is charged into the vessel, or pot, and the liquid is brought to boiling. The vapors are condensed and collected in a receiving tank. A batch process is effective for separating materials containing high solids contents, i.e., tars, or resins, which might otherwise foul a continuous distillation unit.

2.1.2. Evaporation

Evaporation is commonly used with a variety of waste streams including slurries, sludges, streams containing suspended or dissolved solids, and streams containing nonvolatile, dissolved liquid. The objective is to concentrate a solution consisting of a nonvolatile solute and a volatile solvent. The process and equipment used in evaporation are similar to those used in distillation, except that the vapor is not collected and condensed unless organic compounds are present. Evaporation is carried out by vaporizing part of the solvent to produce a concentrated solution or thick liquor. Normally, the thick liquor is the valuable product. This approach is an energy-intensive process and equipment costs may be high. Heat transfer is the most important factor in evaporator design, since the heating surface represents the largest part of evaporator cost.

A major type of evaporation involves agitated *film evaporators* that employ a heating surface consisting of a large-diameter tube. The liquid waste stream is spread onto the walls of the tube through a rotating assembly of blades, allowing the solvent to be evaporated, condensed, and collected. High agitation and power intensities provided in these systems permit the handling of extremely viscous materials.

2.1.3. Gas Absorption

Gas absorption is a process by which one or more components of a gas mixture are dissolved in a liquid. The soluble vapor is absorbed from the gas mixture by a liquid in which the solute gas is more or less soluble. The equipment used for continuous contacting of the vapor and liquid can be a tower filled with solid packing material or one filled with plates. The vapor and liquid almost always flow countercurrent to each other. One application is the washing of ammonia from a mixture of ammonia and air using water as the absorbing liquid.

2.2. Solid-Liquid Separation

Common techniques involving the separation of solids and liquids include *filtration*, *centrifugation*, and *sedimentation*. These separation techniques are briefly described below.

2.2.1. Filtration

Filtration is a popular method of purifying a liquid by removing suspended solids. In the process, a liquid containing suspended solids is passed through a porous medium. The solids are trapped against the medium, and separation of solids from the liquid results. The porous medium may be a fibrous fabric (paper or cloth), a screen, or a bed of granular material. The filter may also be precoated with a filtration aid such as a ground cellulose or diatomaceous earth. For large solid particles, a thick barrier such as sand may be used. This is known as *granular media or bed filtration*. For smaller particles, a fine filter such a filter cloth may be used. Gravity, positive pressure, or a vacuum may be used to induce fluid flow through the filter medium. The filter is cleaned and the solids collected by passing a stream of liquid (usually water) in the direction opposite to the waste stream flow.

Filtration is most often a preliminary purification step to remove solids from a solvent waste mixture. A typical use is the removal of solids from tetrachloroethylene spent solvent wastes so that the solvent can then be reused. *Granular media filtration* is typically used after gravity separation processes for additional removal of suspended solids prior to other treatment steps. It is also used as a polishing step for treated waste, to reduce suspended solids and associated contaminants to low levels.

2.2.2. Centrifugation

Centrifugation is a well-established solid-liquid separation process used in commercial and municipal waste treatment facilities. The method is performed in a closed system and is therefore an excellent choice when treating volatile or toxic fluids. The liquid and solid phases, having dissimilar densities, are mechanically separated by centrifugal force. This is a technically and economically competitive process, commonly used as a preliminary step before other recycling processes. One application of centrifugation in the recycling context is the reduction of crystalline solids from spent dry cleaning solution before the solution goes through a further distillation step.

2.2.3. Sedimentation

Sedimentation, another process used to separate waste streams containing both liquid and solids, works best when the waste stream contains a low concentration of contaminated solids. This process is generally not used to treat non-aqueous waste streams or semisolid waste streams such as sludges or slurries. Sedimentation is a physical process in which suspended solids are settled out through the use of gravitational and inertial forces acting on the solids. Once the solids have settled out, the clarified liquid is decanted and moved to subsequent reuse or further processing steps.

2.3. Liquid-Liquid Separation

Two common liquid-liquid separation techniques are *liquid-liquid extraction* and *decantation*. These separation techniques are briefly described below.

2.3.1. Liquid-Liquid Extraction

Liquid-liquid extraction is used for the removal and recovery of organic solutes from aqueous and non-aqueous waste streams. The concentration of solutes in these streams range from a few hundred parts per million to a few percent. This process may remove most types of organic solutes, but it is not commonly used in treating pollutant wastes. As environmental regulations on industrial wastewaters are made more stringent, its usage is likely to increase. Liquid-liquid extraction techniques have been used in removing and recovering freons and chlorinated hydrocarbons from organic solvent streams.

Liquid-liquid extraction may be performed in a single stage or a multi-stage unit. In this process, the waste stream comes into contact with another liquid stream in which the waste stream solvent is immiscible. The organic solutes are soluble in the extracting solvent and are distributed between the two streams. Increased contact of the waste and solvent streams increases the recovery efficiency of the organic solutes by the liquid extractant. Recovery of the solvent from the product stream is carried out by distillation or by simple *decantation* if the solvent and product streams are immiscible. Recovery of the solute from the solvent must usually be carried out by distillation. The recovered solute may be treated and disposed of, reused, or resold, while the recovered solvent can be reused for further extraction of the product stream. Capital investment in this type of process depends on the particular waste stream to be processed.

2.3.2. Decantation

Decantation is used in the separation of two immiscible liquids of differing densities. The liquid waste is fed into one end of a tank, then allowed to flow slowly through the vessel, where it separates into two distinct layers that are then withdrawn. The rate of separation of one liquid phase from another is an important variable and will dictate the size of the separator. The two types of decanters used are the *continuous gravity decanter* and the *centrifugal decanter*. In cases where the difference in densities of the two liquids is small, gravitational force may not be sufficient for separation, and centrifugal force is required. The rate of separation depends on the different densities involved, the viscosity of the continuous liquid phase, the size of the droplets of the dispersed liquid, and, in centrifugal decanters, the speed of rotation of the decanter.

2.4. Solute Recovery

A variety of physical and chemical processes are available for the removal of organic and inorganic solutes from aqueous or organic liquid phases. Distillation of organic liquids containing recoverable organic solutes from product streams or from liquid-liquid extractor solvent streams was described above. Additional solute processes that are briefly discussed below include: *membrane separation, ion exchange,* and *precipitation.*

2.4.1. Membrane Separation

Membrane separation methods involve both *reverse osmosis* and *ultrafiltration* techniques employing selective membranes.

Reverse osmosis is used for removing dissolved organic and inorganic components from aqueous waste streams by using a semi-permeable membrane that allows only the passage of the solvent molecules, not the dissolved organic and inorganic material. A sufficient pressure gradient is applied to the concentrated solution to overcome the osmotic pressure across the membrane. This applied pressure produces a net flow through the membrane toward the dilute phase, causing separation of the solvent and solute. This allows the concentration of solute (impurities) to be built up in a circulating system on one side of the membrane, while relatively pure solvent is transported through the membrane. Ions and small molecular weight compounds in true solution can be separated from an aqueous solvent by this technique.

The basic components of a reverse osmosis unit are the membrane, a membrane support structure, a containing vessel, and a high-pressure pump. The semi-permeable membrane can be flat or tubular, but regardless of its shape, it acts like a filter because of the pressure-difference driving force. Any components that will damage the membrane must be removed prior to initiating the process.

Reverse osmosis has been most widely used for the desalination of seawater, but is also beneficial to manufacturers of thermally and chemically unstable products such as biologicals, drugs, and food products, in which traditional purification and separation techniques often lead to product loss or flavor deterioration. An industrial application for reverse osmosis is in metal machining operations, where an oil-water emulsion is used to lubricate and cool tools. This emulsion ends up containing a concentration of the metals. A reverse osmosis unit can be utilized to separate this emulsion, yielding a water that can be discharged and an oil concentrate that can be further refined and reused.

Similarly, *ultrafiltration* uses a selective membrane to remove solutes or colloids from a pressurized waste stream, which is passed through the membrane. The membrane retains the larger particles as the liquid phase and smaller particles pass through it. This process can remove particles as small as 10^{-3} to 10^{-2} μm. A very common application for ultrafiltration lies in the electrophoretic paint industry. Here, ultrafiltration is used to process the paint by retaining the polymer resins and pigment solids while allowing the inorganic salts, water, and solvent to permeate through the membrane. The retained species are then returned to the electropaint tank. The permeate is used to rinse the freshly painted components as they emerge from the paint and to recover the drag-out excess paint. For both systems, capital investment and operating costs depend primarily on the composition of the waste stream being separated.

2.4.2. Ion Exchange

Ion exchange is generally used for the removal of dilute concentrations of metallic cations and anions, inorganic anions, organic acids, and organic amines from aqueous waste streams. This approach is a two-stage chemical process in which a solid material, the ion exchanger, collects specific ions after coming into contact with the aqueous waste stream. The ions in the waste stream are removed from the aqueous phase by electrostatic exchange with relatively harmless ions that are held by electrostatic forces to charged functional groups on the surfaces of a solid ion exchange resins. The process is generally accomplished by sending the aqueous waste stream through one or more fixed beds of exchangers. The ion exchange bed is regenerated by then exposing it to a concentrated aqueous regenerating solution that exchanges the resin ions for the ions recovered from the waste stream.

Fixed-bed and *countercurrent systems* are the most widely used ion exchange systems. The continuous countercurrent systems are most suitable for high flows. One very common application of ion exchange is the recovery of hexavalent chromium from plating wastes. The ion exchange unit selectively removes the chromium, and the purified stream can be recycled back to the plating bath. Another application of ion exchange is the removal of copper crystals that form in dipping solutions in the manufacture of brass parts. The dipping solution can be continuously fed through an ion exchange unit for removal of copper, and can then be recycled to the bath. Once the ion exchange resin becomes saturated with copper, the copper ions are removed during bed regeneration as a concentrated copper solution that can then be sold as scrap metal.

2.4.3. Precipitation

Precipitation involves the alteration of the ionic equilibrium to produce insoluble precipitates. Chemical precipitation is combined with solids separation processes, such as filtration, in order to remove the precipitate from the liquid once it forms. The process is sometimes preceded by chemical reduction of the metal ions to forms that can be more easily precipitated. The chemical equilibrium can be affected in different ways to change the solubility of particular species of interest. Precipitation can be induced by the addition of the following materials.

1. *Alkaline agents*, such as lime or caustic soda, added to waste streams to raise the pH. The solubilities of most metal ions decrease as the pH increases, precipitating out of solution as hydroxide species.
2. *Soluble sulfides*, such as hydrogen or sodium sulfide, and insoluble sulfides, such as ferrous sulfide, are used for precipitation of heavy metals. Sodium bisulfide is commonly used for precipitating chromium from aqueous solutions.
3. *Sulfates*, including zinc sulfate or ferrous sulfate, are used for precipitation of cyanide complexes.
4. *Carbonates*, especially calcium carbonate, can be used directly for precipitation of metals. In addition, hydroxides can be converted into carbonates with carbon dioxide and easily filtered out.

Precipitation is an effective and reliable treatment process. Energy requirements are low compared with those of other processes. However, the resulting sludge is generally subject to RCRA regulations and may be considered hazardous if metal concentrations are high enough. Consideration should be given to the disposition of this sludge if recycling via precipitation is being considered. As discussed above, if some form of on-site or off-site reuse of the material can be accomplished as an ingredient or product substitute, then regulatory compliance issues would be greatly simplified for this material.

3. MATERIAL AND WASTE EXCHANGES

The waste from one industry is often an acceptable raw material for another. In response to this common interest, *material and waste exchanges* have been established. Material and waste exchanges are organized to enable industrial process wastes, by-products, surpluses, or materials that do not meet specifications to be transferred from one company to another company where they can be used as a process input. Because many of these materials are typically of low or negative value, it usually does not pay to transport them long distances. As a result, these exchanges are generally regional ventures servicing a wide variety of industries. Many of these exchanges have been operating in the U.S. since the early 1970s. Initially, a major stumbling block to widespread adoption of this method of waste recycling in this country has been the lack of an intermediary or an information bank to help match supply and demand. Although more waste exchanges have been set up in this country in recent years, they have not been fully accepted by industry because of liability concerns, particularly for wastes that can be considered hazardous under RCRA.

Waste exchanges are usually non-profit operations; however, some waste exchanges charge for access to their database services or for copies of their catalogs. The cost to run waste exchanges is typically in the range of $1 to $2/ton of waste exchanged.[2]

There are approximately 50 active waste exchanges operating in the U.S., and can be accessed through a variety of technology transfer and information channels. An Internet site containing updated information on material and waste exchanges located across the U.S. and in Canada can be found at the following web location:

http://www.enviroworld.com/Resources/matexchs.html#anchor1633682

A partial list of waste exchanges in operation in the U.S. is provided below in Table 14.1.[2,3]

4. ILLUSTRATIVE EXAMPLES

Example 4.1. Consider a degreasing operation in a metal finishing process. Give two examples of process modifications that might be made to this part of the process that represents recycling, and waste disposal.

SOLUTION. One process modification that could be made to a degreasing operation in a metal finishing process that would represent recycling is filtering and reusing the solvents for non-critical cleaning jobs.

Another process modification that could be made to a degreasing operation in a metal finishing process that represents waste disposal is disposing still bottoms from a solvent distillation process by shipping them to an off-site landfill.

Example 4.2. As a means of raising one's awareness to the pervasiveness of waste within our society and the ease with which significant changes can be made by each individual, complete the following exercise:
1. Make a list of your activities over the last 24 hours (or 48 hours, or 1 week, etc.) in which you personally wasted energy resources. A list of five items should be easily achievable. (Hint: Try not to limit yourself to only the choices you currently have, but think in terms of the best of all possible worlds, e.g., assume mass transit is available to you., etc.)
2. Write a phrase or brief sentence to indicate how you could have been more environmentally conscious (assuming other options were available).

SOLUTION. There is, of course, no single or correct answer to this question; it is an open-ended question. Below are some examples of some suggested answers for common wasteful practices around the home.

Table 14.1 Selected Waste Exchanges Operating in the United States.[2,3]

California Integrated Waste
Management Board
CALMAX
8800 Cal Center Drive
Sacramento, CA 95826
(916) 255-2369 phone
(800) 553-2962 fax

California Waste Exchange
Department of Health Services
Toxic Substances Control Division
Alternative Technology Section
714/744 P Street
Sacramento, CA 94234-7320
(916) 324-1807 phone
(916) 327-4494 fax

Great Lakes Waste Exchange
400 Ann Street, N.W. Suite 201-A
Grand Rapids, MI 49504-2054
(616) 363-3262 phone

Pacific Materials Exchange (PME)
S. 3707 Godfrey Boulevard
Spokane, WA 99204
(509) 623-4244 phone
(509) 623-4276 fax

Industrial Metals Exchange (IMEX)
Seattle-King County Environmental
Health
172 20th Avenue
Seattle, WA 98122
(206) 296-4633 phone
(206) 296-0188 fax

Industrial Material Exchanges Service
(IMES)
2200 Churchill Road, #24
Springfield, IL 62794-0276
(217) 782-0450 phone
(217) 524-4193 fax

Industrial Waste Information Exchange
New Jersey Chamber of Commerce
5 Commerce Street
Newark, NJ 07102
(201) 623-7070 phone
(201) 623-8739 fax

Iowa State University Waste
Exchange Project
205 Engineering Annex
IMSE
Ames, IA 50011
(515) 294-4056 phone
(515) 294-1682 fax

Montana Industrial Waste Exchange
Montana Chamber of Commerce
2030 11th Avenue
Helena, MT 59601
(406) 442-2405 phone
(406) 442-2409 fax

Hudson Valley Materials Exchange
and Buy Recycled Consortium
P.O. Box 550
New Paltz, NY 12561

Northeast Industrial Waste
Exchange (NIWE)
90 Presidential Plaza, Suite 122
Syracuse, NY 13210
(315) 422-6572 phone
(315) 422-9051 fax

Resource Exchange Network for
Eliminating Waste (RENEW)
Texas Water Commission
P.O. Box 13087
Austin, TX 78711-3087
(512) 463-7773 phone
(512) 463-8317 fax

Southeast Waste Exchange (SEWE)
Urban Institute
Department of Civil Engineering
University of North Carolina
Charlotte, NC 28223
(704) 547-2307 phone
(704) 547-2767 fax

Wasteful Practice	Solution
Took a long shower	Take shorter showers, use a water saver device on your shower head
Left the lights and/or electrical appliances on in an empty room	Make sure you turn off lights and/or appliances before leaving the room
Kept water running while brushing teeth	Turn the water off or let it run less while brushing
Washed only a few articles of clothing instead of a full load	Wash a full load of clothes
Left the air conditioning running all day while not at home	Turn the air conditioning off when not at home, or set it to a warmer temperature

Example 4.3. Consider the use of pollution prevention measures to improve a company's method of handling paper waste. At present all of the company's paper waste is combined with the rest of their waste stream and is hauled off to landfills. They are seeking a more environmentally sound method of dealing with the waste paper produced by their 1,850 employees who, on average, each produce 1.5 lb of paper waste per day. Approximately 0.5 lb per each 1.5 lb of paper waste contain recyclable paper. The average number of days per year that the employees are at work is 250.

Apply source reduction and recycling strategies to this situation assuming that 75% of the employees will be willing to actively participate in a source reduction and recycling program. Assume an average decrease in waste paper generation of 0.3 lb/d per participating person. Discuss the programs usefulness and calculate the changes in waste production rates after a recycling program is implemented.

SOLUTION. First calculate the amount of waste paper that is initially being generated by the company.

Waste Paper Generated = (1,850 people) (1.5 lb/d) (250 d/yr) = 693,750 lb/yr

Next, calculate the annual waste paper generated after implementing both reduction and recycling measures.

Waste Paper Generated = (0.75) (1,850 people) (1.5 - 0.3 - 0.5 lb/d)(250 d/yr)
+ (0.25) (1,850 people) (1.5 lb/d) (250 d/yr) = 416,250 lb/yr

Finally, calculate the percent reduction in waste paper generated.

Percent Reduction = (693,750 - 416,250 lb/yr)/(693,750 lb/yr)
= 0.40 or 40% reduction

The company can reduce its waste paper generated by 40% by implementing these reduction and recycling measures.

5. SUMMARY

1. *Recycling* techniques allow waste materials to be used for beneficial purposes. A material is recycled if it is used, reused, or reclaimed. Recycling is the second most preferred option in the waste management/pollution prevention hierarchy and as such should be considered only when all feasible, cost-effective source reduction options have been investigated and implemented. Recycling options can be listed in the following order of preferability: *direct reuse on-site, additional recovery on-site, recovery off-site, sale for reuse off-site,* and *energy recovery.*
2. Reuse involves finding a beneficial purpose for a recovered waste.
3. Recycling alternatives can be accomplished either *on-site* or *off-site* and may depend on a company's staffing or economic constraints. On-site recycling alternatives directly result in less waste leaving a facility and an associated reduction in the reporting and manifesting for the waste.
4. If an insufficient amount of waste is generated on-site to make an in-plant recovery system cost-effective, or the recovered material cannot be reused on-site, *off-site recovery* is preferable.
5. As an alternative to both on-site and off-site recycling, a generator may transfer waste to another facility for use as a raw material in its manufacturing operations.
6. Recycling can also be achieved by the recovery of energy through the use of waste as a fuel supplement or fuel substitute. Waste may be processed in fossil-fuel-fired plants or in incinerators equipped with an energy recovery system.
7. *Regulatory issues* arise when a RCRA solid or hazardous waste is being considered for recycling. Exemptions from the RCRA definition of solid waste are available for waste materials recycled in the following ways: if the waste is *used as an Ingredient (Direct Reuse On-Site or Off-Site)* in a production process without first being reclaimed; if the waste is *used as a Product Substitute (Direct Reuse On-Site or Off-Site)* for a commercial product without first being reclaimed; or if a waste is *returned to the Production Process (Direct Reuse On-Site)* from which it was generated. These considerations emphasize the desirability of direct reuse options for waste recycling over recycling on- or off-site or energy recovery options.
8. Recycling methods include numerous treatment technologies for recovering a beneficial component from a waste stream for reuse. Vapor-liquid separation techniques involving liquids to be recycled include *distillation, evaporation,* or *gas adsorption.* Solid-liquid separation techniques include *filtration, centrifugation,* and *sedimentation.* Liquid-liquid separation involves *liquid-liquid extraction* and *decantation.* Solute recovery techniques include *distillation, membrane separation* processes, involving *reverse osmosis* and *ultrafiltration, ion exchange,* and *precipitation.*
9. *Material and waste exchanges* are organized to enable industrial process wastes, by-products, surpluses, or materials that do not meet

specifications to be transferred from one company to another company where they can be used as a process input. These exchanges are generally regional ventures servicing a wide variety of industries because many of these materials are typically of low or negative value and cannot be economically transported long distances. Waste exchanges are usually non-profit operations, however, some waste exchanges charge for access to their database services or for copies of their catalogs. The cost to run waste exchanges are typically in the range of $1 to $2/ton of waste exchanged.

6. PROBLEMS

1. From the list of campus sites below, select the three most likely waste/pollution sources that may be reduced or eliminated by applying recycling principles.

        English/History Department   Cafeteria
        Library   Chemistry Department
        Biology Department   Maintenance Shop
        Dormitories   Psychology Department
        Physics Department   Art/Theater

2. The use of solvent recovery has increased due to the increased utilization of industrial solvents. To prevent pollution and minimize waste, industry has to choose between on-site recovery and off-site recovery for pollution prevention purposes. Briefly discuss the advantages and disadvantages of both.

3. Glass food and beverage containers are excellent candidates for recycling as these can be reused many times. Suggest containers that should not be recycled and why they should not be recycled. Also discuss the limitations that must be considered when recycling plastics.

4. Assume that the average water usage of a community is approximately 119 gallons/person/d. After implementing water saving practices, the average water usage drops to 73 gallons/person/d. Estimate the reduction in pollution loading for biochemical oxygen demand (BOD) and suspended solids (SS) possible from a 10 million gallon per day wastewater treatment plant for which 60% of the flow is residential wastewater. The plant effluent quality is generally 30 mg/L BOD and 25 mg/L SS. Express the answers in lb/day. Do you think the pollution reduction is significant? Support your answer. Do you think this approach is practical? Why or why not?

5. Calculate the annual cost for the use of rechargeable nickel-cadmium AA batteries as opposed to the annual cost for conventional AA batteries. They are to be used in a portable radio, which uses four such batteries. The rechargeable batteries are recharged every month with a solar battery charger, which is not designed to get wet, and which requires 2 to 4 d to recharge the batteries. What is the pay back time if any? The payback time is defined as the time required to realize (recover) the principal of an initial investment.

GIVEN: Conventional Batteries - Number Required = 4, Cost (each) = $0.89, Rotation Frequency = monthly, Lifetime = 1 month with average intensity of use. Rechargeable Batteries - Number Required = 8, Cost (each) = $2.75, Rotation Frequency = monthly, Lifetime = 5 yr (1,000 charges).[4]

6. Perchloroethylene is utilized in a degreasing operation and is lost from the process via evaporation from a degreasing tank. This degreasing process has an emission factor (estimated emission rate/unit measure of production) of 0.78 lb PCE released/lb PCE entering the degreasing operation. The PCE entering the degreaser is made up of recycled PCE from a solvent recovery operation plus fresh PCE make-up. The solvent recovery system is 75% efficient, with the 25% reject going off-site for disposal. (Adopted from EPA[5]).
   a. Draw a flow diagram for the process.
   b. Develop a mass balance around the degreaser.
   c. Develop a mass balance around the solvent recovery system.
   d. Develop a mass balance around the entire system.
   e. Determine the mass of PCE emitted/lb of fresh PCE utilized.
   f. What is the impact of the emission factor in the degreasing operation on the flow rates within the solvent recovery unit? Please quantify.

7. REFERENCES

1. U.S. EPA, *RCRA, Superfund & EPCRA Hotline Training Module Introduction to: Definition of Solid Waste and Hazardous Waste Recycling (40 CFR Parts 261.2 and 261.9) Updated July 1997*, Office of Solid Waste and Emergency Response (5305W), Washington, D.C., EPA 530-R-97-051, 1997.
2. Higgins, T.E., ed., *Pollution Prevention Handbook*, CRC Press, Boca Raton, Florida, 1995.
3. Allen, D. T. and Rosselot, K. S., *Pollution Prevention for Chemical Processes*, John Wiley & Sons, Inc., New York, New York, 1997.
4. Real Goods, *Real Good News*, Spring Issue, pp. 10-11, 1992.
5. U.S. EPA, *Eliminating Releases and Waste Treatment Efficiencies for the Toxic Chemical Release Inventory Form*, Office of Pesticides and Toxic Substances, Washington, D.C., 560/4-88-002, 1988.

PART 3
POLLUTION PREVENTION APPLICATIONS

Today's industry is faced with the major technological challenge of identifying ways to manage wastes effectively. As described earlier, technologies designed to treat and dispose of wastes are usually no longer the optimal strategy for handling these wastes for two major reasons. First, the potential liabilities associated with handling and disposing of wastes have increased significantly and are expected to increase even more in the future. Second, restrictions placed on land disposal of wastes have caused a considerable increase in waste disposal costs. The economic impact of these changes is causing various industries to explore alternatives to conventional treatment and disposal techniques.

Technologies are now being developed to enable industries to recover valuable materials from their waste streams and to reduce or eliminate (hazardous) waste generation. These waste reduction technologies are more accessible to larger industrial facilities with sufficient capital and staff who can identify and implement waste reduction plans than they are to small manufacturers. Most smaller facilities do not have the in-house expertise and/or capital to develop such technologies themselves. Although the materials in the following chapters in Part 3 apply to all facilities, they are primarily focused on the smaller facilities.

Part 3 of the text focuses primarily on the first component of the waste management/pollution prevention hierarchy, namely source reduction. However, specific resource recovery and recycling techniques that are applicable to a specific industry are discussed in each chapter. Readers are referred to Parts 1 and 2 of the text for a more detailed discussion of recycling and treatment techniques.

Each chapter in Part 3 consists of four sections:

1. Process description.
2. Waste description.
3. Pollution prevention options available to the specific industry.
4. Case studies

A general description of the industry is also included in the beginning of each chapter. The pollution prevention option section of each chapter is further broken down into the following sections:

1. Process in modifications.
2. Resource recovery and recycling.
3. Good housekeeping and operating practices.

The reader should note that although many actual pollution prevention techniques are specific not only to particular industries but also to particular production processes, the approaches to waste reduction and the methods for implementing pollution prevention practices are similar and can be generalized among industries and plants. Likewise, many of the advantages and disadvantages of pollution prevention techniques may be common among many industrial applications.

The industries reviewed in Part 3 (with chapter number) are as follows:

Commercial printing industry with extensive case study (15)
Metal finishing industry (16)
Electronics industry (17)
Drug manufacturing industry (18)
Paint manufacturing industry with extensive case study (19)
Pesticides formulating industry (20)
Paper and pulp industry (21)

CHAPTER 15

Commercial Printing

The printing and publishing industry consists of establishments engaged in printing by one or more of the common printing processes, such as letterpress, lithography, gravure, or screen. The industry also includes establishments that perform services for the printing trade, such as bookbinding, typesetting, engraving, photoengraving, and electrotyping. Printing establishments are scattered all over the U.S., the largest concentrations being in California and New York. Those two states share approximately 13% and 8% of the industry, respectively. Illinois, Texas, Florida, New Jersey, Pennsylvania, Ohio, and Michigan each share about 5% of the total. The industry is characterized by a large number of small establishments. As of 1987, there were approximately 62,000 facilities located in the United States.[1,2,3]

Ink and substrates are the primary raw materials used in printing. Since different printing processes and substrates may require different final-product qualities, many different types of inks are used. Some inks may contain toxic heavy metals and solvents. The substrate can be any material on which ink is impressed, such as paper, plastic, wood, or metal. Other chemicals/materials that are used include photoprocessing and platemaking chemicals, fountain solutions, cleaning solvents, and lubricating oils. Waste characteristics of these materials/chemicals are discussed in "Waste Description" later in this chapter.

The five most common printing processes (in the order of their market share) are lithography, gravure, flexography, letterpress, and screen printing. The details of the printing process are discussed in "Process Description."

Among other pollution prevention options, source reduction and recycling are those most effective for the printing industry. On-site waste treatment techniques are generally not economical nor practical for most printers, mainly because of the small size of the typical establishment. Source reduction and recycling options available to the printing industry are discussed under "Pollution Prevention Options."

1. PROCESS DESCRIPTION

As of 1988, the distribution of the printing processes for printing, publishing, and packaging (excluding copying and duplicating) was: lithography 44%, gravure 18%, letterpress 18%, flexography 14%, and screen printing and other processes 6%.[4]

Letterpress is the most versatile and the oldest method of printing. Letterpress is printed by the relief method. Printing is done from cast metal type or plates on which the image or printing areas are raised above the nonprinting areas. Ink rollers touch only the top surface of the raised areas. The nonprinting areas are lower and do not receive ink. There are four types of presses for letterpress: platen, flat-bed cylinder, rotary, and belt. On platen and flat-bed cylinder presses, the type or plates are mounted on a flat surface or bed. Printing is done on sheets of paper on sheet-fed presses or rolls of paper on web-fed presses. Sheet-fed presses are usually used for general printing, books, catalogs, and packaging. Web-fed presses are used for newspapers and magazines.

Flexography is a form of rotary web letterpress that uses flexible rubber plates and fast-drying solvents or water-based inks. The rubber plates are mounted to the printing cylinder. Products printed by the flexographic process range from decorated toilet tissue to polyethylene and other plastic films, since almost anything that can go through a web press can be printed by flexography.

Gravure is a type of intaglio printing. It uses a depressed (or sunken) surface for the image. The image areas consist of cells or wells etched into a copper cylinder or wraparound plate. The printing area is the cylinder or plate surface. The plate cylinder is rotated in an ink bath, and the excess ink is wiped off the surface by a flexible steel "doctor blade." The remaining ink in the thousands of sunken cells forms the image as the paper passes between the plate cylinder and the impression cylinder (Figure 15.1). Gravure presses are manufactured to print sheets of paper (sheet-fed gravure) or rolls of paper (web-fed gravure); however, the web-fed gravure is more popular.

Offset lithography uses the planographic method. The image and the nonprinting areas are on the same plane of a thin metal plate, and the areas are distinguished by chemicals. Lithography is based on the principle that grease and water do not mix. The ink is offset first from the plate to a rubber blanket and then from the blanket to the paper. The printing areas in the plate are made ink receptive and water repellent, and the nonprinting areas are made ink repellent and water receptive. The plate is mounted on the plate cylinder, which rotates and comes into contact with rollers that are wet by a dampening solution (or water) and rollers are wet by ink, in succession. The ink wets the image areas, which are then transferred to the intermediate blanket cylinder. The image is printed to the paper as the paper passes between the blanket cylinder and the impression cylinder (Figure 15.2). The major advantage of transferring the image from the plate to a blanket before transferring to the paper (offsetting) is that the soft rubber surface of the blanket creates a clearer impression on a wide variety of paper surfaces and other substrate materials.

Screen printing (or *silk screen*) employs a porous screen of fine silk, nylon, or stainless steel mounted on a frame. Nonprinting areas are protected by producing stencils on the screen manually or photochemically. Printing is done on the paper by applying ink to the screen, then spreading and forcing the ink through the fine openings with a rubber squeegee. Versatility is the major advantage of screen printing since any surface (e.g., wood, glass, metal, plastic, fabric, etc.) can be printed.

Because lithography is the predominant printing process, the following process descriptions will focus on lithography. A typical commercial offset lithographic printing operation is illustrated in Figure 15.3. Printing begins with the preparation of artwork (or copy) which is photographed to produce

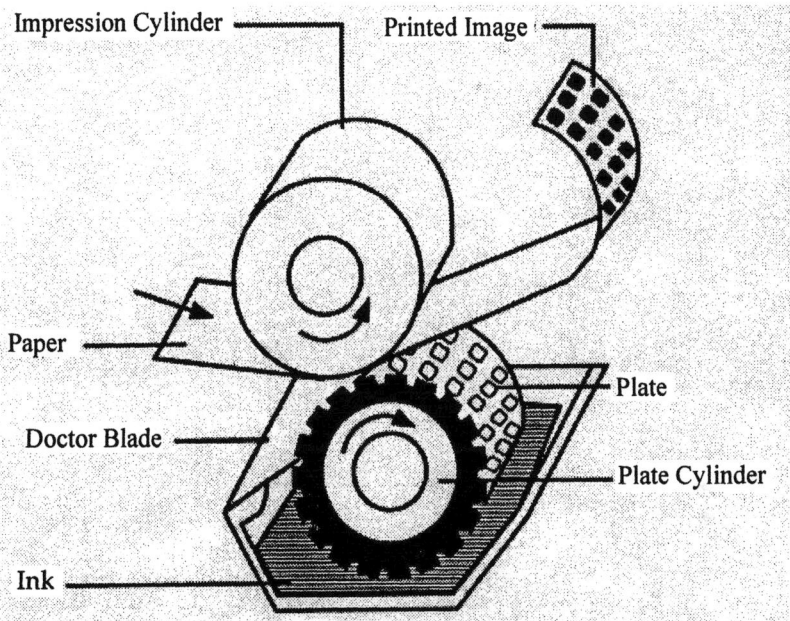

Figure 15.1. The gravure printing process.

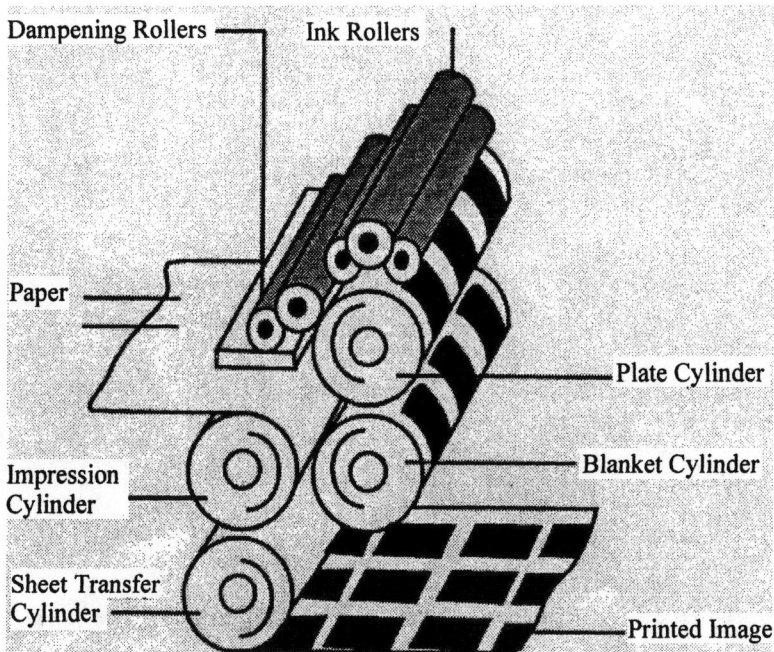

Figure 15.2. The offset lithography printing process.

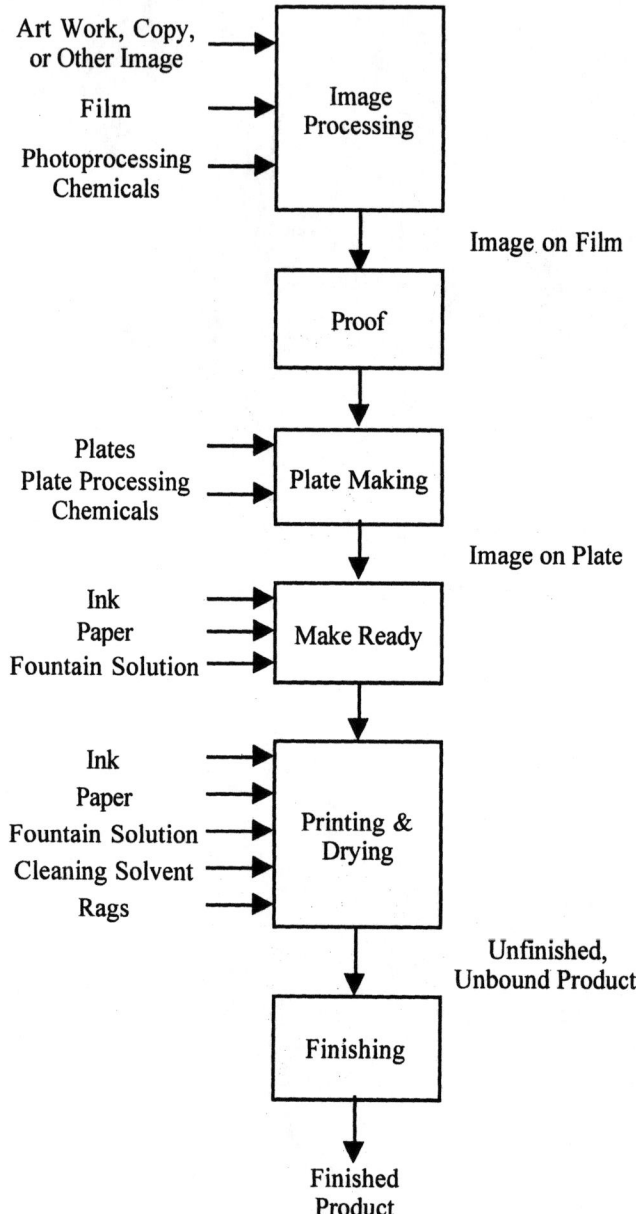

Figure 15.3. Commercial offset lithographic printing operation.

an image. A proof is made, and it is used to compare with the printed product and make adjustments to the press. The photographic image is then transferred to a plate. In the platemaking step, the image areas of the plate are made receptive to the ink. During the printing step, ink is applied to the plate and the image is transferred to a rubber blanket, and then to the substrate. The substrate is then cut, folded, and bound to produce the final product.

The four main steps of the printing process are: image processing; platemaking; printing; and finishing. Each of these steps is described in detail below.

1.1. Image Processing

The printing industry employs graphic arts photography in the reproduction of artwork (or copy). Materials similar to those in other fields of photography are used. The materials include paper, plastic film, or glass base covered with a light-sensitive coating (photographic emulsion). The photographic emulsion is usually made of silver halide salts (e.g., silver chloride, silver bromide, and silver iodide) in gelatin. Most photographic films are made of polyester.[5]

After the photographic emulsion is exposed to light, oxidation and reduction reactions develop the emulsion. The exposed film or plate is immersed in a developer, where the silver halide in the emulsion is converted to metallic silver. Developers typically contain benzene derivatives, including pyrogallol, hydroquinone, catechol, p-phenylene diamine, p-aminophenol, metol, amidol, and pyramidal. The two most common developing agents are hydroquinone and metol. The developer solution also contains an accelerator, a preservative, and a restrainer. The accelerator is an alkaline compound such as sodium hydroxide, sodium bicarbonate, or sodium tetraborate (borax), which increases the activity of the developer by neutralizing the acid formed during the development process. The preservative, usually sodium sulfite, reduces oxidation damage to the developing agent. The restrainer, typically potassium bromide, reduces the formation of "fog" on the images.

The developing action is then stopped by immersing the film in a fixing bath containing "hypo" (sodium thiosulfate), ammonium sulfate, or sodium hyposulfite. These compounds convert the silver halides to soluble complexes, preventing them from turning to metallic silver.[5] In addition to hypo, the fixing bath contains acetic acid, sodium sulfite, potassium alum, and boric acid. Potassium alum, which prevents excess swelling of the photo emulsion gelatin, is not stable in a neutral or basic solution, and the precipitation of aluminum hydroxide will result unless an acid (e.g., acetic acid) is added to keep the pH low. Hypo will decompose in an acid solution, forming elementary sulfur as a precipitate. Therefore, sodium sulfite is added to react with the sulfur precipitate and form more sodium thiosulfate. Boric acid is added to buffer the fixing solution and limit pH changes, preventing precipitation of aluminum salts (from potassium alum).

Each time photographic film or paper is immersed in the fixing bath, a small amount of silver enters the bath from the photographic emulsion.

Above a certain silver concentration, insoluble compounds that are formed cannot be removed from the photographic emulsion. The bath, therefore, should be recycled before this concentration is reached. The critical silver concentration for fixing baths is 2 g/L. Use of ammonium thiosulfate doubles the maximum allowable concentration of silver.[5]

After a negative (or positive) is fixed, some of the fixing bath chemicals remain in the gelatin emulsion. One of the chemicals remaining is hypo, which can react with the silver or complex silver salts to form yellowish-brown silver sulfide, lowering the quality of the image. To prevent the formation of silver sulfide, fixing chemicals are washed from the emulsion. Films are washed in water, which dissolves the hypo from the emulsion until an equilibrium concentration is reached between the hypo in the water and that in the emulsion, at which point the water is changed. Films can also be washed in running water. Wash water at about 80°F increases the efficiency of hypo removal.

A proof is produced after the image-processing step to check whether all the elements fit, whether the color is right, and how the job will look when it is printed.

1.2. Plate Processing

The printing process revolves around the intermediate image carrier (i.e., plate or cylinder), which accepts ink from a roller and transfers the image to the rubber blanket. The blanket, in turn, transfers the image to the paper. The type of image carrier determines the type of ink and press to be used, the number of impressions that can be printed, the speed with which they are printed, and the characteristics of the image. The image carriers are made by the following methods:

- Manual
- Mechanical
- Electrostatic
- Photomechanical

Manually made images are seldom used now, except commercially in screen printing. Manually made image carriers consist of hard set composition, woodcuts, linoleum blocks, copper plate or steel-die engravings.

Mechanically produced image carriers are used mainly for relief printing. The plates can be made by either hot metal machine composition or duplicate printing plates. Intaglio (i.e., gravure cylinder) printing uses mechanically made plates.

Electrostatic plates are popular in offset duplicating where electrophotographic cameras convert original images to lithographic plates. Electrostatically produced plates are also used in newspaper printing.

Photomechanical platemaking is the most common method of platemaking. It uses a light-sensitive coating, the physical properties of which change after exposure to light. The exposed coating areas become insoluble in water or other solutions (e.g., diazo solutions). The unexposed areas dissolve, leaving the exposed areas as an image.

Natural organic compounds such as asphalt, shellac, and gum arabic have been used as photomechanical coatings. New types of coatings include polyvinyl alcohol, diazo compounds, and synthetic photopolymer resins. The two most common coatings are diazo compounds and photopolymer resins. Diazo coatings have the advantage of not being affected by changes in temperature and humidity. Photopolymer coatings are inert and abrasion resistant, allowing for longer press runs than diazo-coated plates. The photopolymer coatings also have a low sensitivity to temperature and humidity changes.

In lithographic printing, both the image and nonimage areas are on the same plane. *Lithographic plates* have image areas that are ink receptive and water receptive. This is achieved by a chemical change via a photochemical process on the plate surface. Most of the printing industry uses presensitized plates, to which light-sensitive coatings have already been applied by the manufacturer. Three methods are employed to make lithographic plates: surface, deep-etch, and bimetal.

Surface plates have a light-sensitive coating which becomes ink receptive when exposed. Most surface plates are made from negatives. Generally, the light-sensitive coatings contain a diazo compound. Other organic compounds such as azide compounds, hydrazine derivatives, quinone diazides, and quinone esters are also used as light-sensitive materials. Aluminum or anodized aluminum is used extensively as surface lithographic plates.

Deep-etch plates are made from positives. When the plate is exposed, the coating in the exposed nonimage areas hardens and unexposed areas remain soft. Then the developing solution washes the image area away and the stencil remains. The developing solution is typically calcium, zinc, or magnesium chloride combined with a weak acid. The image area is next plated with copper and/or coated with lacquer, making it ink receptive. The lacquer is a combination of polyvinyl chloride, polyvinyl acetate, and a small percentage of malic acid. The nonimage areas are treated with a desensitizing etch and finally gummed with gum arabic solution. The wastewater from the deep-etch plate processing is acidic and contains copper or other heavy metal compounds. Because of its heavy metal toxicity and high cost, the deep-etch process is becoming obsolete.

Bimetal plates can be exposed through a negative or positive. The lifetime of bimetal plates is long, since the image and nonimage areas are established by two different metals (capable of printing several million impressions). Bimetal plates are also becoming obsolete owing to water pollution problems associated with the toxicity of heavy metals.

Gravure printing cylinders begin with steel cylinders plated with copper. The surface is either mechanically engraved with a diamond stylus or chemically etched with ferric chloride. The negative image is transferred (as a resist) to the cylinder before etching. The resist protects the nonimage areas of the cylinder from the etchant. After etching, the resist may be stripped off. Then the cylinder is proofed and chrome plated.

The developing process of *letterpress and flexography relief plates* (plates with raised images) is similar to that of lithographic plates, with an extra step that involves etching the nonimage areas with an acid solution. The relief plates are typically made of zinc, magnesium, and copper, and the plates contain light-sensitive coatings (e.g., polyvinyl alcohol). The plates are then etched with nitric acid. Baths used for developing the relief plates eventually build up a high concentration of heavy metals. These solutions need to be treated before being discharged to most sewer systems.

1.3. Printing

Once the plates are prepared and all the adjustments are made (make-ready), the actual printing begins. The printing operations are generally the same for each of the major processes, with the exception of screen printing. Sheet-fed presses and web presses are the two most common types of presses. The plates are attached to the plate cylinder of the press. Because they are made of thin flat aluminum sheets, lithographic plates can be wrapped around and attached to the plate cylinder. Generally, all presses print from a plate cylinder (rotating) rather than a flat plate.

As the plate cylinder rotates, a water-based dampening solution, followed by an oil-based ink, is applied to the plate's image area. The inked image area repels the water-based solution and the nonimage area repels the ink. As the cylinder is rotated, the inked image is transferred to a blanket and then onto a substrate.

In gravure printing, the cylinder is placed in the press and partially immersed in an ink bath or fountain. Solvent may be added to the ink to maintain the proper level and appropriate viscosity of the bath. As the cylinder is rotated, ink coats the entire surface. A metal wiper (doctor blade) is then pressed against the surface of the cylinder and removes the ink from the nonetched (nonimage) areas. The substrate is then pressed against the rotating cylinder and the ink is transferred. After printing, the substrate may pass through a drying step, depending on the type of ink used. Lithography can use heat-set and nonheat-set inks. In heat-set lithography, the substrate is passed through a tunnel or floater dryer, which employs hot air and/or direct flame. Gravure printing uses inks that dry by solvent evaporation.

After printing is completed, the substrate goes through "finishing," which involves final trimming, folding, collating, binding, and laminating. These finishing operations are typically accomplished by an outside company.

2. WASTE DESCRIPTION

The principal wastes associated with printing operations are summarized in Table 15.1.[5] The wastes generated by gravure cylinder making are not included in Table 15.1, but are similar to the other metal processing operations described in Chapter 16.

Paper is the major waste associated with the printing industry; almost 98% of the total waste generated is spoiled paper and paper wrap. Waste paper comes from rejected print runs, scraps from the starts and ends of runs, and overruns (excess number of copies made to ensure that there are enough acceptable copies). Most paper is recycled, incinerated, or disposed of as trash.

Table 15.1. Wastes from Commercial Printing Processes.[5]

Waste Description	Process Origin	Composition
Trash	Image processing,	Empty containers, packages, used proof film, outdated materials
	Platemaking	Damaged plates, developed film, outdated materials
	Printing	Ink containers (if not recycled), used blankets, used plates, unacceptable printings, paper wrappings
Wastewater	Image processing	Photographic chemicals, silver (if not recovered)
	Platemaking	Acids, alkali, solvents, plate coatings (may contain dyes, photopolymers, binders, resins, pigments, organic acids), developers (may contain isopropanol, gum arabic, lacquers, caustics), and rinse water
	Printing	Spent fountain solutions (may contain chromium)
Recovered Silver	Image Processing, platemaking	
Empty ink containers	Printing	
Paper (recycled)	Makeready	Inked and clean sheets
	Printing	Inked sheets
Equipment-cleaning wastes	Printing, proof	Lubricating oils, waste ink, cleanup solvent (halogenated and nonhalogenated), rags
Air emissions	Makeready, printing	Solvent from heat-set inks, isopropyl alcohol (fountain solution), and cleaning solvents

Spent photoprocessing chemicals are generally biodegradable with high BOD (biochemical oxygen demand), and it is generally necessary to treat the waste before discharging to sanitary sewers. For larger printing companies, it may be economical and necessary to recover silver from the spent solutions.

Platemaking wastes (e.g., acids and bases used to clean or develop the plates) must be either sent to a wastewater treatment facility or drummed for disposal. Platemaking wastes are minimal for those facilities that use presensitized plates. Fountain solutions used in lithography contain gum arabic, phosphoric acid, defoamers, and fungicides. Isopropyl alcohol (IPA) is usually added to reduce the surface tension of the solution, making it adhere better to the nonimage areas of the plate cylinder. Most of the IPA evaporates with water, and the other chemicals remain on the paper. Some printing chemical manufacturers offer low volatility fountain solutions that do not use IPA or other volatile compounds.

Equipment-cleaning wastes include spent lubricants, waste inks, cleanup solvent(s), and rags. Waste ink is the ink removed from the ink fountain at the end of a run, or contaminated ink. Although most of the ink used by a printing company ends up on the paper (or other substrates), other ink losses include spills and ink printed on waste paper. Most waste inks are either incinerated (if considered hazardous) or discarded with trash.

Cleanup solvents are used to clean the press. The rubber blankets are cleaned once or twice per 8-hour shift to minimize the imperfections resulting from dirt or dried inks. When lower quality paper is used, cleaning is required more frequently. The cleaning solvents include methanol, toluene, naphtha, trichloroethane, methylene chloride, and specially formulated blanket washes.

Inks may contain solvents (e.g., xylene, ketones, alcohols, etc.), depending on the type of printing process and substrate. For example, gravure printing inks contain solvents. Inks used for offset lithography are

- Sheet-fed inks that dry by oxidative polymerization.
- Heat-set inks that dry by evaporation of aliphatic ink oils.
- Non-heat-set web inks that dry by absorption of the ink on the substrate.

No significant amount of VOC (volatile organic compounds) is emitted from sheet-fed inks or nonheat-set web inks. For heat-set inks, the printed web passes through a dryer where ink oils are evaporated. The resulting VOC emissions can be controlled by catalytic or thermal incineration, or by condenser systems. For the VOC emissions from gravure printing, carbon adsorption is the most commonly applied control method.

3. POLLUTION PREVENTION OPTIONS

Pollution prevention options available to the commercial printing industry are source reduction through process modifications, recycling and resource recovery, and good housekeeping and operating practices. Process modifications include input material changes and product reformulation. Source reduction techniques also include proper equipment cleaning techniques, proper storage and handling of materials, and employee training. Most good operating and housekeeping practices can be implemented with little cost and can reduce the costs of input materials, waste treatment, and disposal.

Advanced technology and government regulations have led to rapid process changes in the printing industry. Metal etching and metal plating operations are being replaced by platemaking processes that do not produce hazardous wastewater discharges. The use of presensitized plates in offset lithography generates very little or no hazardous wastes. Computerized imaging and proof systems can greatly improve the productivity and reduce hazardous waste generation from printing operations.

Major applications of recycling and resource recovery in the printing industry are recycling/reuse of waste inks, recovery of silver from photoprocessing wastewater and used film, and sale of used plates and waste paper to off-site recyclers.

The pollution prevention options available to the printing industry are described in the following subsections. The process modification options are categorized and discussed for each process step.

3.1. Process Modifications

3.1.1. Image Processing

The major waste stream associated with image processing is wastewater, which contains photographic chemicals and silver removed from film. The use of computerized electronic prepress systems for typesetting and copy preparation is a recent advance in image-processing steps. The electronic scanner scans the image fed by text, photos, and graphics, and the copy is edited on a computer display monitor rather than on paper. This reduces the number of films and the amount of developing chemicals and paper used. The high initial cost, however, may prohibit smaller printing operations from using electronic prepress systems.

The wastes from photoprocessing that uses silver films may be considered hazardous, depending on the silver concentration. Photographic materials that do not contain silver are available, but they are slower to develop than silver halide films. Diazo and vesicular films have been used for many years. Vesicular films have a honeycomb-like cross section and are coated with a thermoplastic resin and a light-sensitive diazonium salt. Recently, photopolymer and electrostatic films have been used. Photopolymer films contain carbon black as a substitute for silver, and the films are processed in a weak basic solution that needs to be neutralized before disposal. Electrostatic films are nonsilver films that can be developed at a speed comparable to that of silver films. An electrostatic charge makes the film light-sensitive, and a liquid toner brings out the image after the film is exposed to light. Electrostatic films also have high resolution.

Waste from photographic processing can be reduced by extending the life of fixing baths. Techniques for extending bath life include:

1. Addition of ammonium thiosulfate, which increases the maximum allowable silver concentration in the bath from 2 to 4 g/L.
2. Use of an acid stop bath prior to the fixing bath.
3. Addition of acetic acid to keep the pH low.

Close monitoring of the process bath and optimizing the bath conditions will minimize the use of bath chemicals.

Squeegees can be used to wipe excess liquid from the film and paper in nonautomated processing systems. This can reduce the chemical carryover from one process bath to the next by as much as 50%. Minimizing chemical contamination of process baths increases recyclability and bath life, and reduces the amount of replenisher chemicals required. However, squeegees must be used only after the film image has hardened because squeegees can damage the film image if it has not fully hardened.

During photographic processing, films are commonly washed with water, using parallel tank systems to remove hypo from the emulsions. In a parallel system, fresh water enters each wash tank and effluent leaves each wash tank. The removal efficiency of hypo can be increased by employing a countercurrent washing system. In a countercurrent rinsing system, the water from previous rinsing is used in the initial film washing stage and fresh water enters the system only at the final rinse stage. The countercurrent rinse system decreases the amount of wastewater generated and also increases the rinse efficiency. The disadvantages of the countercurrent system are greater space requirement and higher initial equipment cost.

3.1.2. Plate Processing

Recent advances in plate-processing techniques have reduced the quantity and/or toxicity of hazardous wastes and improved worker safety. In gravure printing, metal etching and metal plating operations involve chemical compounds that are generally considered hazardous. Waste solutions from metal etching or metal plating usually require treatment before being discharged to a municipal sewer. The same is true for all wastewater used in plate rinsing operations. The use of multiple countercurrent rinse tanks can reduce the amount of wastewater generated. The toxicity of wastewater from plating can be reduced by minimizing the drag-out from the plating tanks. The drag-out can be reduced by:

1. Installing a draining rack.
2. Using draining boards to collect the drag-out and returning it to the plating tank.
3. Raising the plate tank temperature to reduce the viscosity and surface tension of the solution.

The printer should consider replacing metal etching or plating processes with presensitized lithographic plates, plastic or photopolymer plates, or hot metal plates, which do not generate hazardous wastes. The wastes generated by presensitized lithographic plates are wastewater from developing and finishing baths and used plates. Consumption of chemicals can be reduced by frequently monitoring the bath for pH, temperature, and solution strength, thereby extending the bath life. Automatic plate processors may also be used since they are designed to maintain the optimum bath conditions.

Nonhazardous developers and finishers are also available. For example, some developers and finishers have a flash point of 213°F and are therefore considered nonflammable. Presensitized plates that are processed only with water are also available.

3.1.3. Makeready

Paper is the largest raw material item and is the most expensive component of this printing operation. The printed paper produced in makeready is frequently the largest waste a printer generates, but it is nonhazardous. The amount of paper waste is determined by the efficiency of the press adjustments needed to achieve the desired print quality (e.g., proper ink density and accurate registration).

With proper use, the automated press adjustment devices can speed up the makeready step and save paper and ink. Examples of these devices are automated plate benders, automated plate scanners, automatic ink density setting systems, computerized registration and ink/water ratio sensors. It is important, however, that the cost of these items be considered against the degree of quality improvement and the extent of waste reduction they provide.

3.1.4. Printing and Finishing

The major wastes associated with printing and finishing are scrap paper, waste ink, and cleaning solvents. The solvent waste stream consists of waste ink, ink solvents, lubricating oil, and cleaning solvents. Adopting a standard ink sequence can reduce the amount of waste ink and waste cleaning solvents. If a standard ink sequence is employed, ink rotation is not changed with a job and it is not necessary to clean out fountains in order to change ink rotation.

A web break detector can note tears in the web. If tears are not detected, the broken web begins to wrap around the rollers and force them out of the bearings. Although web break detectors are primarily used to avoid severe damage to the presses, they also reduce paper and ink wastes by preventing press damage. An automatic ink level controller can also be used to maintain the desired ink level in the fountain and to optimize process conditions.

Water-based inks may be used in place of inks that contain oils. Applications for water-based inks are flexographic printing on paper and gravure. Although water-based ink reduces emissions that result from evaporation of ink oils, it is more difficult to dry and makes equipment cleaning more difficult.

Another option is the use of UV (ultraviolet) inks. UV inks consist of one or more monomers and a photosynthesizer that selectively absorbs energy. UV inks do not contain any solvents, and the inks are not "cured" until they are exposed to UV light. Therefore, UV inks can remain in the ink fountains (and plates) for longer periods of time, reducing cleanup frequency. UV inks are particularly recommended for letterpress and lithography. Although UV inks reduce the amount of wastes generated, they cost 75 to 100% more than the conventional heat-set inks and some of the chemicals in these inks are toxic. In addition, conventional commercial paper recycling procedures cannot de-ink papers printed by UV inks.

Automatic blanket cleaners can also be used to increase process efficiency, thereby reducing the amount of waste generated. An automatic blanket cleaner consists of a control box, a solvent metering box for each process unit, and a cloth (rag) handling unit.

Less toxic and less flammable blanket washes are now available, replacing cleaning solvents that contain benzene, carbon tetrachloride, and trichloroethylene. However, these new blanket washes have a lower cleaning efficiency.

To reduce the amount of cleaning solvent, ink fountains should be cleaned only when a different color ink is used or when the ink may dry out between runs. Aerosol sprays are available to spray onto the ink fountains to prevent overnight drying and to eliminate the need for cleaning the fountains at the end of the day. This reduces the amount of waste ink generated and the amount of cleaning solvents used.

Although rollers are cleaned with solvents and roller wash-up blades, several factors affect the efficiency of cleaning the rollers. They are

- Condition of the rollers
- Condition of the blade
- Angle of the blade against the roller
- Press speed during washup

Alternative printing technologies should also be considered. An example is electrostatic screen printing, also known as pressureless printing. In electrostatic screen printing, a thin and flexible printing element with a finely screened opening is used to define the image to be printed. An electric field is established between the image element and the surface to be printed. Finely divided "electroscopic" ink particles, metered through the image openings, are attracted to the printing surface and held by electrostatic force until they are fixed by heat or chemicals.

3.2. Recycling and Resource Recovery

Photoprocessing waste chemicals basically consist of developer, fixer, and rinse water. Minimizing cross-contamination of process baths is the key to successful recycling of these chemicals. The technologies available for reuse of developer and fixer are ozone oxidation, electrolysis, and ion exchange.

Silver is present in most photographic film and paper, and it is also present in the wastewater. Various economical methods of recovering silver are available, including metallic replacement, chemical precipitation, and electrolyte recovery. The most popular method of silver recovery is electrolyte deposition. In an electrolytic recovery unit, a direct-current low voltage is created between a carbon anode and stainless steel cathode, with metallic silver depositing onto the cathode. Once the silver is removed, the fixing bath may be reused by mixing the desilvered solution with fresh solution.

In metallic replacement recovery, the spent fixing bath is pumped into a cartridge containing steel wool, in which an oxidation-reduction reaction occurs. The iron in the wool is replaced by silver through the oxidation-reduction reaction, and the silver settles to the bottom of the cartridge as sludge.

Most waste inks can also be recycled. One recycling technique relies on blending waste inks of different colors together to make black ink. Small amounts of certain colors or black toner may be needed to obtain an acceptable black color. Off-site ink recycling (by ink manufacturers or large printers) may be more economical for smaller printers. The waste ink is reformulated into black ink and sold back to the printer.

Cleaning solvents can be collected and reused. In some cases, used solvents with a particular ink color can be used to formulate new inks of the same color.

The used oil (from lubricating the printing presses) should be collected and turned over to a recycler. The recycler can either refine the oil into new lubricating oil, use it as fuel grade oil, or use it for blending into asphalt.

Paper use and the disposal of waste paper is a critical concern for the printer. Segregating and recycling the paper according to its grade is the key to effective recycling. Inked paper is of one grade and is recycled separately. Unprinted white paper should also be sent separately for recycling.

3.3. Good Housekeeping and Operating Practices

In addition to the pollution prevention options that are categorized as process modifications and material/product substitutions, good housekeeping and good operating practices play a major role in waste minimization in the commercial printing industry. Good housekeeping and operating practices should include:

1. Proper material handling and storage.
2. Material tracking and inventory control.
3. Good procedural measures, including good documentation and scheduling.
4. Good personnel practices, including management initiatives and employee training.
5. Loss prevention practices, including spill prevention, preventive maintenance, and emergency plans.

Proper handling and storage of materials deserve special attention since some printers waste up to 25 percent of the raw materials as a result of improper storage.[6] Many photoprocessing and plate-developing chemicals are sensitive to temperature and light. Photosensitive film and paper storage areas should be designed for proper storage of these chemicals, following the recommended storage conditions specified by the manufacturer.

Paper waste can also be reduced by proper handling and storage of rolls or packages of paper. Paper should be stored in a location with proper temperature and humidity to minimize the absorption of moisture. It is generally recommended that paper be conditioned to the temperature and humidity of the pressroom for a day before printing.

4. CASE STUDY

The following pollution prevention case study of a commercial printing plant is not one of an existing or real plant, but is constructed from many different case studies of existing plants in order to illustrate the many possible pollution prevention options available to the printing industry.

An overview of the plant operation is discussed first, followed by a description of the process. The potential pollution prevention options are discussed, along with an economic analysis of some of the options.

Plant A is a medium-size commercial printer located in a large metropolitan area. The company employs 70 people, 55 of whom are involved in production, while the rest are administrative staff. Plant A handles a wide range of commercial printing, including advertising inserts, pamphlets, brochures, circulars, and product labels.

4.1. Process and Waste Description

Plant A performs its own photoprocessing. Approximately 8 gallons of developer and 5 gallons of fixer are used each month for film processing. The developer and fixer are diluted with water to a 20% solution before use.

For platemaking, presensitized aluminum plates are used. The average monthly usage of plate processing developer and fixer is 10 gallons for each. The developer and fixer are diluted with water to a 20% solution before use.

The spent developers and fixers are sent to the sewer. The processing baths are dumped, and fresh new baths are made up every 6 to 8 weeks. These dumped baths also go to the sewer (under the jurisdiction of the local sanitation district) since the manufacturer of the developers and fixers claims these chemicals are nonhazardous and biodegradable. The silver concentration in the spent fixing baths is below the limit set by the sanitation district, although the value is very close to the limit.

The printing operation at Plant A includes five web-fed press lines and one sheet-fed line. Four of the web press lines use four color printing units, and the fifth uses a six color printing unit. Non-heatset inks are used for web-fed press lines, and heatset inks are used for the sheet-fed line. The heatset inks are dried by passing the paper through a tunnel dryer, which uses warm air to dry the ink. Approximately 1,400 pounds of non-heatset inks in 20 colors are used every month. Fresh black inks cost from $1.50 to $3.50 per pound, and colored inks cost from $3.50 to $7.50 per pound.

Approximately 25 gallons of "fountain soup A" is used to make fountain solution for the web presses. Fountain soup A is a concentrate that includes phosphoric acid, gum arabic, a defoamer, and a fungicide. The concentrate contains high boiling point surfactants that help wet the nonimage areas of the plate. The fountain solution is made up by adding 1 to 4 ounces of concentrate to each gallon of water.

A different fountain soup ("fountain soup B") is used for the sheet-fed unit because fountain soup A was found to not work efficiently in the sheet-fed unit. Fountain soup B requires the addition of 5 to 15% of isopropyl alcohol. The average usage of isopropyl alcohol is 12 gallons per month.

The fountain solutions are never drained from the printing unit reservoirs. They either end up on the paper or are lost through evaporation. The components that evaporate are isopropyl alcohol and water.

Approximately 250,000 pounds of paper are used per month. Eight to 9% of the web-fed paper ends up as waste, with 2% as wrapper slab and core waste, and 6 to 7% as other waste. Of the sheet-fed paper, approximately 2 to 6% ends up as waste.

Ink containers, along with photoprocessing and plate processing chemical containers, are scraped out when empty and thrown in the trash. Worn rubber blankets from the process are also thrown in the trash. Eighteen blankets are replaced each month on the web presses, and one blanket per month is replaced on the sheet-fed unit. Cleanout of the blankets is accomplished by specially formulated blanket wash solvents. Most equipment cleanout is done using methanol.

4.2. Pollution Prevention via Process Modification

Modern image-processing technology, including computerized prepress systems for copy preparation, can be used to minimize the waste generated during the image-processing step. However, computerized image-processing units require high capital cost, and they are not likely to be economical for a medium-size plant like Plant A.

Automatic processors may be used for photoprocessing and plate processing. Fresh developer and fixer are added automatically, and an equivalent amount of bath solution is simultaneously removed by overflow to prevent the buildup of impurities. This improves process efficiency by maintaining optimum operating conditions for the baths. Automatic processors are less expensive than computerized image-processing units, and some medium-size printing plants have used them successfully.[5]

The life of the processing baths can be extended by the following techniques:

1. Addition of ammonium thiosulfate to increase the maximum allowable concentration of silver in the fixing bath
2. Use of an acid stop bath prior to the fixing bath
3. Addition of acetic acid to keep the pH low

However, the addition of ammonium thiosulfate may also increase the level of silver in the waste fixing bath above the level allowed by the local sanitation district. Silver recovery techniques should then be considered if they are economically feasible. Since the maximum silver concentration with ammonium thiosulfate is approximately 4 g/L (the maximum silver concentration is approximately 2 g/L without ammonium thiosulfate, and ammonium thiosulfate can double the silver concentration), the following calculation can be performed to estimate the amount of recoverable silver in the waste streams.

Recoverable silver = (4 g/L fixer solution) (5 gal fixer solution/gal fixer) (5 gal fixer/month) (12 months/yr) (3.785 L/gal) = 4,542 g/yr

An automated press adjustment device can be used for larger printers (if economically feasible) to speed up the makeready step and save paper and ink. A web break detector can be used to detect tears in the web-fed units to reduce paper and ink wastes.

Replacing heatset inks used in the sheet-fed unit with non-heatset inks or UV inks should be considered to reduce the VOC emissions from the dryer. The use of non-heatset or UV inks will also lower the energy input to the plant by eliminating the use of the dryer.

Countercurrent rinse systems should be considered in rinsing operations to minimize the amount of wastewater generated. Automatic blanket cleaners can be used to increase cleaning efficiency, although a better method is dry cleaning with rags.

4.3. Pollution Prevention via Resource Recovery and Recycling

The silver in the waste fixing bath and wastewater should be recovered if economically feasible. It appears that silver recovery is not economically attractive at the current silver concentration, especially since the level in the wastewater is below the level specified by the local sanitation district. Some large printers have successfully recovered silver from their waste streams. One printer recovers about 3 lb of silver per month.[5]

Waste inks may be sent to the manufacturer where they are reformulated into black inks, and Plant A can purchase the reformulated black inks. For example, one plant sends its waste inks to the supplier and the supplier adds 50 to 100% of fresh black ink to the waste inks to obtain an acceptable black color. The plant then purchases the reformulated black ink at $3.00/lb. Depending on the cost of fresh black ink available to Plant A (the cost ranges from $1.50 to $3.50/lb), this is an excellent recycling option. Even if Plant A has to pay top dollar for the fresh black ink (i.e., $3.50/lb), this option is still attractive since the cost of disposing of the waste inks will be eliminated.

Another option for recycling waste inks is on-site recycling. The waste inks can be reformulated on-site by purchasing a small ink recycler which can blend the waste inks with fresh inks. One manufacturer (KMI Marketing, Inc.) sells a small recycler that can blend 60 lb of waste ink with up to 120 lb of fresh ink to produce 180 lb of reformulated black ink. (The reformulated ink has to be filtered before use). Depending on the capital cost of the recycler and payback periods, this is also an excellent option. A simple economic analysis comparing off-site and on-site ink recycling is shown in Table 15.2. (The reader is referred to Chapter 9 for more detailed information on economic analyses.) The assumptions made for the analysis are as follows:

1. The capital cost of the ink recycler is $6,000.
2. The buy-back price of reformulated paint is $3.00/lb.
3. Fresh ink requirement for on-site recycling is 200% of the waste ink amount.

Table 15.2 Economic Comparison of On-Site versus Off-Site Ink Recycling.

	On-Site	Off-Site
Capital cost ($)	6,000	0
Operating cost ($/month)		
Waste ink (250 lb)	0	0
Fresh ink for blending (@ $2.00/lb)	1,000	500
Buy back reformulated ink (@ $3.00/lb)	0	750
Buy additional fresh ink (@ $2.00/lb)	0	500
Total operating cost	1,000	1,750

Saving in operating cost with on-site recycling = $750/month
Payback period = $6000/$750 per month = 8 months

Note: The increase in maintenance and operating costs resulting from on-site ink recycling is assumed to be negligible.

4. Fresh ink requirement for off-site recycling is 100% of the waste ink amount.
5. Plant A generates 250 lb of waste ink per month.
6. Cost of buying additional fresh black ink is $2.00/lb.

Based on the results of the economic analysis, it appears that on-site recycling is a more attractive option than off-site processing. Since the benefit of the on-site recoloring is mainly proportional to the amount of waste ink generated, on-site waste ink recycling will be even more attractive to larger printers.

Solvent wastes can also be recycled by on-site or off-site distillation. An example of an on-site distillation unit is a Cardinal distillation unit.[5] The Cardinal distillation unit boils the solvent out of a 55-gallon drum of waste solvents and condenses the clean solvent vapor into another drum for reuse. One printer employed the Cardinal distillation unit at an approximate cost of $16,000 and recovered 85% of the waste solvent, resulting in a savings of $15,000 per year in fresh solvent costs.[5]

Used and spoiled films can be collected and sold to a recycler, and used aluminum plates can be sold to an aluminum recycler. All waste paper can be collected and sold to a paper recycler.

4.4. Pollution Prevention via Better Housekeeping and Operating Practices

A dry cleaning method (using rags) can be employed to minimize the solvent wastes generated by blanket and roller washes. One printer accomplishes cleanup with a rag moistened with blanket wash. Almost all of the blanket wash ends up on the rags, and waste solvent generation is minimized. A local industrial laundry cleans the rags for the printer at a cost of $40/week for about 2,000 rags.[5]

Proper handling and storage of photoprocessing and plate developing chemicals can significantly reduce the amount of raw material waste because these chemicals are easily spoiled by exposure to light and high temperature. In addition to the specific pollution prevention options outlined above, a common sense approach to improve housekeeping and operating practices also plays an important role. Refer to the guidelines on better housekeeping and operating practices presented earlier in this chapter for more details.

5. SUMMARY

This chapter describes pollution prevention options available to the commercial printing industry, along with descriptions of the processes and wastes generated, and a detailed case study at a commercial printing plant. Among other pollution prevention options, source reduction and recycling are the most effective methods for the printing industry. On-site waste treatment techniques are generally not economical or practical for most printers, mainly because of the small size of typical establishments. The case study did show the economic attractiveness of on-site recycling of waste ink, however.

1. Printing operations generally include four process steps: image processing, plate processing, printing, and finishing. The five most common printing processes (in the order of their market share) are lithography, gravure, letter press, flexography, and screen printing.

2. Paper is the major waste associated with the printing industry. Other wastes include spent photoprocessing chemicals, platemaking wastes (acids, bases, solvents, resins, photopolymers, pigments, developers, and rinse water), and equipment cleanup wastes (waste ink, lubricating oils, and solvents).

3. The general trend in the printing industry is to replace the metal etching and metal plating operations with platemaking processes that do not produce hazardous wastewater discharges. The use of presensitized plates in offset lithography generates very little or no hazardous wastes. Computerized imaging and proof systems can greatly improve the productivity and reduce hazardous waste generation of commercial printing facilities.

4. Major applications of recycling and resource recovery in the printing industry are recycling/reuse of waste inks, recovery of silver from photoprocessing wastewater and used film, and the sale of used plates and waste paper to off-site recyclers.

5. Waste from photographic processing can be reduced by extending the life of fixing baths. Techniques for extending bath life include:
   - Addition of ammonium thiosulfate, which increases the maximum allowable concentration of silver in the bath.
   - Use of an acid stop bath prior to the fixing bath.
   - Addition of acetic acid to keep the pH low.

6. The toxicity of wastewater from plating can be reduced by minimizing the drag-out from the plating tanks. The drag-out can be reduced by:

- Installing a draining rack
- Using draining boards to collect the drag-out and returning it to the plating tank
- Raising the plate tank temperature to reduce the viscosity and surface tension of the solution

7. Wastes generated by the printing and finishing operations can be reduced by:
   - Using a web break detector
   - Using an automatic ink level controller
   - Using water-based inks or UV inks
   - Using an automatic blanket cleaner
   - Using alternative printing technologies (e.g., electrostatic screen printing)

8. Minimizing cross-contamination of process baths is the key to the successful recycling of photoprocessing chemical wastes. The technologies available for reuse of developer and fixer are ozone oxidation, electrolysis, and ion exchange.

9. Various economical methods of recovering silver are available, including metallic replacement, chemical precipitation, and electrolyte recovery. The most popular method of silver recovery is electrolyte deposition.

10. Most waste inks can also be recycled. One recycling technique relies on blending waste inks of different colors together to make black ink.

11. Good housekeeping and good operating practices play a major role in pollution prevention in the commercial printing industry. Good housekeeping and operating practices should include:
    - Proper material handling and storage
    - Material tracking and inventory control
    - Good procedural measures, including good documentation and scheduling
    - Good personnel practices, including management initiatives and employee training
    - Loss prevention practices, including spit prevention, preventive maintenance, and emergency plans

12. The case study presented for Plant A, a medium-size commercial printer located in a large metropolitan area, indicated that:
    - Computerized image processing units require high capital cost, and they are not likely to be economical for a medium-size plant like Plant A.
    - Automatic processors may be used for photoprocessing and plate processing. Fresh developer and fixer are added automatically, and an equivalent amount of bath solution is simultaneously removed by overflow to prevent the buildup of impurities.
    - The life of processing baths can be extended using the techniques listed above.
    - Replacing the heatset inks used in the sheet-fed unit with non-heatset inks or UV inks should be considered to reduce the VOC emissions from the dryer. The use of non-heatset or UV inks will

also lower the energy input to the plant by eliminating the use of the dryer.

Based on an economic analysis, on-site recycling of waste inks should be considered for Plant A, providing a payback of only 8 months for the $6000 capital investment made by the plant for an on-site ink recycling system.

6. REFERENCES

1. U.S. Department of Commerce, *Commercial Printing and Manifold Business Firms, 1987 Census of Manufacturers,* Bureau of Census, MC87-1-27B, Washington, D.C., 1987.
2. U.S. Department of Commerce, *Greeting Cards, Bookbinding, Printing Trade Services, 1987 Census of Manufacturers,* Bureau of Census, MC87-1-27B, Washington, D.C., 1987.
3. U.S. Department of Commerce, *Books, and Miscellaneous Publishing, 1987 Census of Manufacturers,* Bureau of Census, MC87-1-27B, Washington, D.C., 1987.
4. Bruno, M. H., *Pocket Pal - A Graphic Arts Production Handbook*, 13th edition, International Paper Company, New York, New York, 1988.
5. Jacobs Engineering Group, Inc., *Waste Audit Study of the Commercial Printing Industry,* Prepared for the California Department of Health Services, Alternative Technology and Policy Development Section, Pasadena, California, 1989.
6. Campbell, M. E. and Glenn, W. M., *Profit from Pollution Prevention: A Guide to Industrial Waste Reduction and Recycling*, Pollution Probe Foundation, Toronto, Canada, 1982.

CHAPTER 16

Metal Finishing Industries

The metal finishing industry uses a variety of physical, chemical, and electrochemical processes to clean, etch, and plate metallic and nonmetallic substrates. Chemical and electrochemical processes contribute more to the generation of hazardous waste than physical processes such as blasting, grinding, and polishing. The chemical and electrochemical processes are performed in numerous chemical baths, which are followed by a rinsing operation. The most common wastes generated by the metal finishing industry are rinse water effluent and spent process bath chemicals. Typically, rinse water effluent and spent process bath chemicals are treated on-site before being discharged to a local publicly owned treatment works (POTW).

1. PROCESS DESCRIPTION

The metal finishing industry employs a variety of process baths, depending on the services required. Typical metal finishing processes include plating, anodizing, stripping, etching, degreasing, cleaning, tooling, and other finishing steps. Examples of the plating baths are nickel, tin, zinc, copper, cadmium, and gold baths. Most metal finishing establishments receive workpieces from their clients and perform metal finishing processes in accordance with the clients' specifications. The process steps that generate wastes include plating, anodizing, etching, stripping, and rinsing steps. Wastes include spent baths, cleaning baths, and contaminated rinse water.[1,2]

Although the metal finishing industry utilizes a variety of processing steps, this description of the metal finishing industry will be limited to only a general outline of common processes that are used. As examples, zinc plating operation steps are shown in Figure 16.1 and anodizing operation steps are shown in Figure 16.2. Generally, workpieces are processed in various plating, etching, and anodizing baths, dragged out, and carried to degreasing and rinsing baths in preparation of final products.

2. WASTE DESCRIPTION

Metal finishing industry wastes are generated from spent plating baths, spent process baths, rinsing/cleaning baths, and degreasing operations. Common hazardous wastes generated by the metal finishing industry are summarized in Table 16.1.

3. POLLUTION PREVENTION OPTIONS

For the metal finishing industry, available pollution prevention technologies include extending the life of chemical process baths, reducing the volume of wastewater generated, and considering alternative, less hazardous metal finishing techniques.

338 Chapter 16

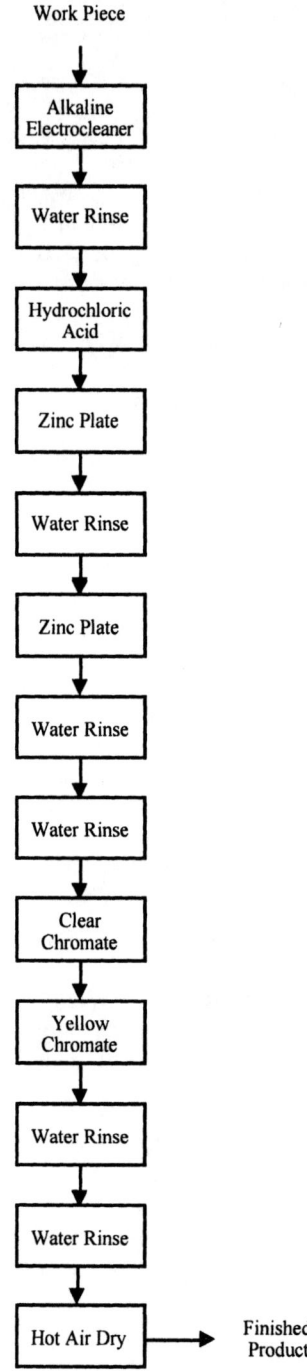

Figure 16.1. Typical zinc plating process.

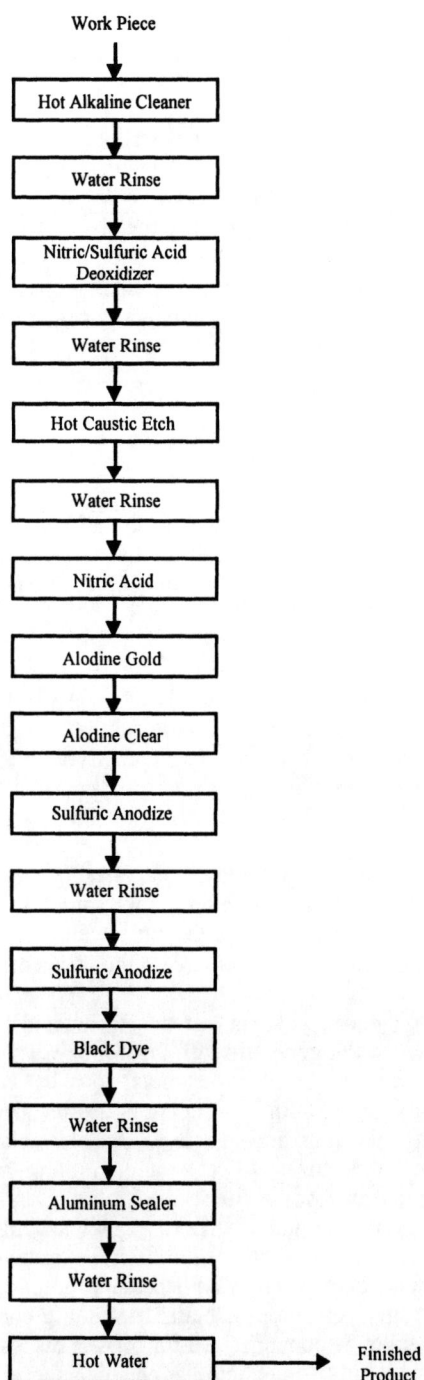

Figure 16.2. Typical metal anodizing operation.

Table 16.1. Wastes Generated by the Metal Finishing Industry.[1]

Operation	Wastes
Plating	Contaminated or spent electroplating/electroless baths.
Etching	Contaminated or spent etchants and cleaners.
Cleaning	Nitric acid, sulfuric acid, hydrochloric acid, and hydrofluoric acid used to strip metals from work piece racks and parts.
Degreasing	Contaminated or spent solvents.
Miscellaneous	Absorbants, filters, empty containers, aisle grates, and abrasive blast residues.

3.1. Process Modifications

Since pollution prevention in the metal finishing industry usually involves extending the bath life and reducing the amount of wastewater generated, this section is divided into two subsections: *process baths* and *rinse systems*.

3.1.1. Process Baths

Source/waste reduction at the process-bath level can be established by material substitution, bath life extension, and/or drag-out reduction. Material substitution options include modifying the chemistry of the process baths or replacing the chemicals used for a given process. Because process bath chemistries vary widely from plant to plant, these options are described below only in general terms.

Deionized water can be used to replace tap water for process bath makeup and rinsing operations. Natural contaminants found in tap water (e.g., carbonates, phosphates, etc.) can reduce the rinse water efficiency, minimize the potential for drag-out recovery, and increase the frequency of changing the process bath.

The use of a nonchelated process instead of chelated chemicals for the process baths can reduce the generation of hazardous wastes. Chelators are used in chemical process baths to allow metal ions to remain in solution beyond their normal solubility limit. Chelators are usually found in baths used for metal etching, cleaning, and selective electroless plating. In general, mild chelators such as phosphates, silicates, and ammonia are used for most cleaning and etching processes. Electroless plating baths use stronger chelating compounds, such as organic acids (e.g., maleic acid, oxalic acid, and citric acid). Chelating compounds can also inhibit the precipitation of metals, and additional treatment chemicals may be necessary to treat the waste stream. In addition, many of the spent process baths containing chelators cannot be treated on-site and need to be containerized for off-site disposal. To minimize the problems associated with using chelating chemicals, process baths can be filtered to remove metals. However, the use of filtering techniques is not very feasible for electroless plating since the chelators play a significant role in the plating process.

Converting to noncyanide process baths can simplify wastewater treatment, saving the cost of treatment chemicals and reducing sludge generation. Replacement chemistries are available for most cyanide containing process baths, with the exception of copper strike baths used for copper plating.

Many chlorinated and nonchlorinated solvents, which are used for degreasing operations prior to processing workpieces, can be replaced with alkaline cleaners. Hot alkaline cleaning baths used for degreasing can be treated on-site and discharged to a POTW.

Maintenance of the process bath is important in extending its life and, subsequently, in reducing the frequency of the bath replacement. Filtration systems can be used to remove solids that build up in the process baths (cleaning baths and etching baths). Continuous filtration of a bath can remove the solid contaminants and allow the bath to be used longer.

The life of a cleaning bath can be extended by proper and regular bath maintenance (i.e., regular replenishing, pH measurement, metal content measurement, etc.). The life of a plating bath can also be extended through bath treatment that removes metal contaminants. Copper is a common metal contaminant that builds up in plating baths. Copper can be removed from a bath by "*dummying*." The dummying process is based on the principle that the copper can be plated at a low electrical current. When the copper content of a process bath becomes too high, an electrolytic panel is placed in the bath and a low current (1 to 2 amperes per square foot) is then run through the system. At this low current, only the copper in the bath will be plated out on the panel and other bath additives will remain unaffected by the current.[1]

Process chemical loss as a result of drag-out is the most significant source of chemicals present in the wastewater. The factors that contribute to drag-out are the workpiece size and shape, bath viscosity, bath concentration, surface tension, and temperature. The available techniques of reducing process chemical drag-out are summarized in the following guidelines:

1. Maintain the minimum bath chemical concentration acceptable within operating range.
2. Maximize bath operating temperature to lower solution viscosity.
3. Reduce surface tension by using wetting agents in the process bath.
4. Maintain proper racking orientations to achieve the best possible drainage.
5. Withdraw workpieces at slow rates and allow for sufficient drainage time.
6. Use a spray rinse above the process tank.
7. Use drainage boards between process baths and rinse tanks to collect and return drippage to the process baths.
8. Use drag-out tanks to recover process chemicals for reuse in the process baths.

Keeping the chemical concentration in the process bath at the lowest acceptable level can reduce drag-out losses; the lower the concentration, the lower the viscosity of the solution. At an elevated temperature, the viscosity of the process solution is also lowered, allowing the chemical solution to drain faster from the workpiece and subsequently reduce the volume of drag-out loss.

The volume of drag-out loss can be reduced by adding wetting agents to a process bath. However, wetting agents can sometimes create foaming problems in the process baths. In addition, some process bath chemistries may not tolerate the use of wetting agents.

The amount of drag-out loss is also affected by the position of a workpiece on the rack. Workpieces should be oriented so as to allow the chemical solution to drain properly from the workpieces and not be trapped in grooves or cavities.

The drag-out loss can be reduced by removing the workpiece at a slow rate since the film on the surface of the workpiece is thinner at a lower removal rate. However, it is difficult to control the rate of removal if work pieces are removed manually. Drag-out losses can be recovered by using spray rinses, drain boards, and drag-out tanks. For example, the evaporative loss in a heated bath can be replenished by adding the drag-out solution from the drag-out tank. Care must be taken when using drag-out tanks because contaminated drag-out solution can also contaminate the process baths.

3.1.2. Rinse Systems

The greatest amount of wastes generated by the metal finishing industry comes from the treatment of wastewater resulting from the rinsing operations required after the plating, stripping, and cleaning processes. The use of most wastewater treatment chemicals depends on the volume of wastewater generated. Therefore, it is important that rinse water usage be reduced in order to minimize waste generation in the metal finishing industry. Rinse water usage may be reduced by improving the rinsing efficiency and/or by controlling the water flow rate.

Rinsing efficiency may be improved by providing high turbulence between the work piece and the rinse water, sufficient contact time between the work piece and the rinse water, and a sufficient volume of water during contact time. However, the last strategy causes the use of significantly more rinse water than might otherwise be required.

To improve the turbulence between the work piece and the rinse water, spray rinsing or rinse water agitation systems may be used. Spray rinsing uses between 15 to 25% of the volume of the water that a dip rinse system uses. However, it is not applicable to all metal finishers because the spray rinse may not reach all parts of certain workplaces. Spray rinse systems may be used along with dip rinse systems (immersion rinse systems) as a first rinse step after the workpieces are removed from the process tank. This removes most of the drag-out before the work piece is submerged in the dip rinse tank. Spray rinses can also be installed above the heated process tanks with rinse water volume less than or equal to the water loss resulting from evaporation.

This allows the drag-out and rinse water to drain directly to the process bath, thus replenishing the bath solution.

Agitation between the work piece and the rinse water can be achieved by using forced air or forced water systems. In these systems, air or water is pumped into the immersion rinse water tank. Air agitation is thought to provide the best type of turbulence for removing process bath chemicals. An air sparger at the bottom of the rinse water tank is frequently used for forced air systems.

The use of multiple-stage rinse tanks can increase contact time between the workpiece and the rinse water, thus improving the rinse efficiency. A countercurrent multiple-stage rinse tank system can also reduce the amount of water usage (by 90% when compared with a conventional single-stage rinse system).[3] In a multistage countercurrent rinse system, the rinse water flow moves in an opposite direction to the workpiece flow. The disadvantages of multistage countercurrent rinse systems are that more process steps, additional equipment, and greater work space are required.

Rinse water usage can also be reduced by controlling the flow rate. Flow restrictors can be used to limit the volume of the rinse water and maintain a constant flow of fresh water into the system. A conductivity or pH meter can also be used to control the flow of fresh water into the rinse system by monitoring the level of dissolved solids or pH in the rinse water. When the concentration or pH reaches the preset maximum level, the probe can activate a valve which opens to allow an additional flow of fresh water into the rinse water system.

3.2. Recycling and Resource Recovery

This section describes the recycling and resource recovery technologies available to the metal finishing industry. To reuse or recover a waste stream for another process and to recycle rinse water, the waste stream must be separated from other wastes that may interfere with the reuse or recovery process. Therefore, implementation of recycling and resource recovery techniques will generally require process piping modifications and additional storage tanks to provide the necessary segregation of materials.

After the rinse water becomes too contaminated for the original process, it may be useful for other rinsing operations in which the purity of the water is not crucial to the process. For example, the effluent from a rinse tank following an acid cleaning bath can be reused as influent water to a rinse tank following a basic cleaning bath. Not only is water usage reduced by 50% (assuming both rinse tanks require the same water flow rates), but at the same time rinsing efficiency is improved, owing to the neutralization reaction that reduces the concentration of alkaline chemicals. Neutralization reactions can also reduce the viscosity of the alkaline drag-out film.[4]

Acid cleaning rinse water effluent can be used as rinse water for workpieces that have gone through a mild acid etch process. Rinse water from a final rinse, which is less contaminated than other rinse waters, can be used as influent for other rinsing operations that do not require high rinse efficiencies.

Alkaline or acidic spent process chemical baths can be used to adjust pH during waste treatment. However, they should not be used for final pH adjustment since they often contain high concentrations of metals.

Because of the increased regulatory requirements on handling and disposal of wastes containing metals, and the subsequent increase in treatment costs, it has become more economical to recover metals and metal salts from process baths. Metal recovery may be used in two ways: recovered elemental metals can be sold to a metal reclaimer, and recovered metal salts can be recycled back to the process baths. The available techniques for recovering metals and metal salts include:

- Evaporation
- Reverse osmosis
- Ion exchange
- Electrolysis
- Electrodialysis

These techniques are generally used to recover metals and metal salts from rinse water effluent. A short description of each is provided below; the reader is referred to Chapters 7 and 14 for more details. The applications of each technique for different metals are summarized in Table 16.2.

Evaporation is frequently used to recover a variety of plating bath chemicals. Water is boiled off from the contaminated rinse water to allow the concentrated chemicals to be returned to the process bath. Evaporation is typically performed under vacuum to minimize the thermal degradation of the chemicals. Because it is an energy-intensive technology, evaporation is economically feasible only when the reduced amount of rinse water (e.g., multistage countercurrent systems) is involved.

Table 16.2 Applications of Recovery Technologies[4]

	Evaporation	Reverse Osmosis	Ion Exchange	Electrolysis	Electrodialysis
Chromium (decorative)	X		X		
Nickel	X	X	X		X
Electroless nickel			X		
Cadmium	X			X	X
Zinc	X	X		X	X
Copper	X	X		X	X
Tin	X			X	X
Silver	X		X	X	X

Reverse osmosis is a membrane separation system driven by pressure. It uses a semipermeable membrane that allows the passage of the purified water while not permitting the passage of other chemicals. The chemicals can be returned to the process bath, and the water can be returned to the rinse system. It should be noted that the membranes should be able to withstand a wide range of pH and exposure to long-term high pressure conditions.

Ion exchange can be also used to recover drag-out from a dilute rinse solution. The rinse solution is passed through a bed (one or more) of ion exchange resins which selectively remove cations and anions. The metals are recovered from the resin by cleaning the resin with an acid or alkaline solution. The treated water is returned to the rinse system. Ion exchange systems can be used effectively on dilute rinse water systems and are less sensitive than reverse osmosis systems.

Electrolysis is used only to recover the metallic components of rinse water. A cathode and an anode are placed in the rinse water, and, as the current passes from the anode to the cathode, the metallic ions deposit on the cathode, generating a solid metallic slab that is recoverable.

Electrodialysis is employed to concentrate the ionic components in rinse water solutions. The rinse water solution is passed through a series of alternately placed cation and anion permeable membranes, with an anode and cathode placed on the opposite sides of the membranes. The anode and cathode create an electropotential across the membranes, which forces the ions in the rinse solution to migrate across the membranes.

These recovery technologies can be used to recycle rinse water in a closed-loop or open-loop system. In a *closed-loop system*, the treated effluent is returned to the rinse system. In an *open-loop system*, the treated effluent is reused in the rinse system, but the final rinse is accomplished with fresh water. Examples of closed- and open-loop systems are illustrated in Fig. 16.3 and Fig. 16.4, respectively.

Solvent wastes generated from degreasing operations of the metal finishing industry are also significant and can be recycled on-site through distillation. The residue from the distillation is handled as a hazardous sludge.

3.3. Good Housekeeping and Operating Practices

Improved housekeeping and operating practices are directed at reducing waste generation from many sources. Although it is difficult to quantify the contribution of each source to the total wastes generated in a plant, regular inspection and maintenance schedules, as well as proper controlling and handling of raw materials, can reduce the generation of wastes.

Frequent inspection can identify leaks in piping systems, storage tanks, and process tanks. The inspection list should also include checks for defective racks, air sparging system(s), automated flow systems, and operator's procedures (e.g., draining time, rinse method, etc.). Maintenance schedules should be developed to ensure that the process baths and rinse systems are operating at their optimal efficiencies.

Proper purchasing and handling of raw materials can also reduce waste generation significantly. Strict procedures should be developed to ensure that the chemicals are mixed properly for process baths, minimizing spills and assuring the proper concentration (the lowest concentration acceptable to min-

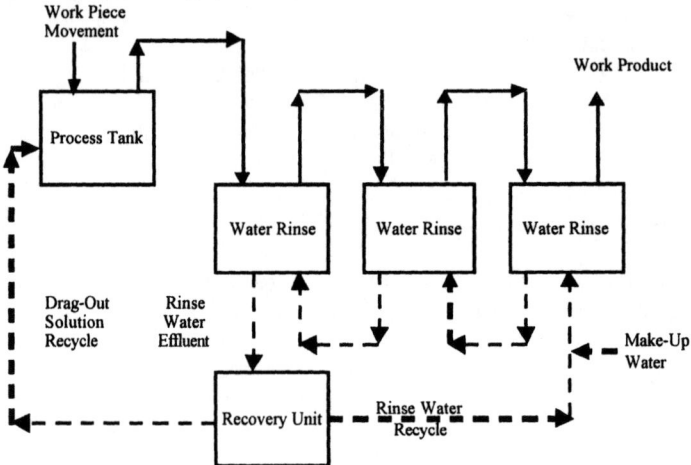

Figure 16.3. Closed-loop water recovery system.

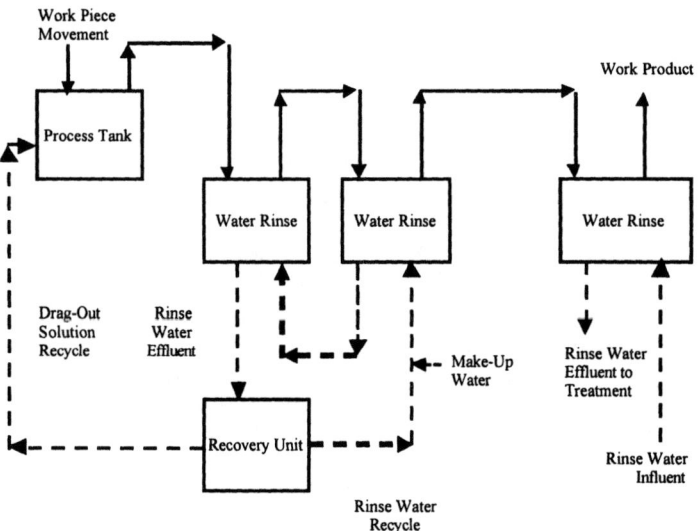

Figure 16.4. Open-loop rinse water recovery system.

imize the drag-out). A limited number of personnel should be designated to handle and mix the chemicals, to improve the consistency of the solution concentration and to minimize waste.

4. CASE STUDY

4.1. Problem Description

A facility manufactures stationary power tools, requiring both metal finishing and painting operations. Large amounts of solvent waste, paint sludge waste, and wastewater were being generated.

4.2. Pollution Prevention Techniques Used

A waterborne electrostatic paint system was installed, replacing the conventional organic solvent-based electrostatic paint system. A computerized robotic electroplating system was installed to improve process efficiency. An ultrafiltration system was also installed for the wastewater streams, to recover oil and process chemicals. In addition, a comprehensive chemical waste management program, along with an incentive program for new pollution prevention ideas, were introduced.

4.3. Benefits

The new paint system reduced hazardous waste disposal costs from $10,000 to $300/yr. The water-based system also allows 99.5% recovery and reuse of paint. This has led to a raw material cost reduction of $600,000/yr. The ultrafiltration system reduced the load to the wastewater treatment system and recovered $11,000 in raw materials. The automated metal electroplating system increased annual productivity by $200,000 and decreased the system downtime by half from 8% to 4%. The new automated system also decreased chemical consumption, water use, and the volume of plating wastes.

5. SUMMARY

This chapter describes pollution prevention options available to the metal finishing industry, along with descriptions of the processes and wastes generated. For the metal finishing industry, the available pollution prevention technologies mainly involve extending the life of chemical process baths and reducing the volume of wastewater generated.

1   The metal finishing industry uses a variety of physical, chemical, and electrochemical processes to clean, etch, and plate metallic and nonmetallic substrates. Chemical and electrochemical processes contribute more to waste generation than do physical processes such as blasting, grinding, and polishing. Chemical and electrochemical processes are performed in numerous chemical baths, which are followed by a rinsing operation.

2.   The most common wastes generated by the metal finishing industry are rinse water effluent and spent process bath chemicals.

3.   Waste reduction at the process-bath level can be established by material substitution, extension of bath life, and/or drag-out reduction.

4.   The material substitution option includes the use of deionized water, use of a nonchelated process, and conversion to noncyanide process baths.

348 Chapter 16

5. Process bath maintenance is important in extending the life of a process bath and subsequently reducing the frequency of bath replacement. Filtration systems can be used to remove solids that build up in process baths (cleaning baths and etching baths).

6. The life of a cleaning bath can be extended by proper and regular bath maintenance (i.e., regular replenishment, pH measurement, metal content measurement, etc.).

7. The life of a plating bath can be extended through bath treatment that removes metal contaminants. Copper, a common metal contaminant in plating baths, can be removed from a bath by "dummying."

8. The available techniques for reducing process chemical drag-out are summarized in the following guidelines:
- Maintain the minimum bath chemical concentration acceptable within operating range.
- Maximize the bath operating temperature to lower solution viscosity.
- Reduce surface tension by using wetting agents in the process bath.
- Maintain proper racking orientations to achieve the best possible drainage.
- Withdraw work pieces at slower rates and allow for sufficient drainage time.
- Use a spray rinse above the process tank.
- Use drainage boards between process baths and rinse tanks to collect drippage and return to the process baths.
- Use drag-out tanks to recover process chemicals for reuse in the process baths.

9. The rinsing efficiency in rinsing baths may be improved by providing enough turbulence between the workpiece and the rinse water, sufficient contact time between the workpiece and the rinse water, and sufficient volume of water during contact time.

10. In the metal finishing industry, it has become more economical to recover metals and metal salts from process baths. Metal recovery may be used in two ways: recovered elemental metals can be sold to a metal reclaimer, and recovered metal salts can be recycled back to the process baths. The available techniques of recovering metals and metal salts are:
- Evaporation
- Reverse osmosis
- Ion exchange
- Electrolysis
- Electrodialysis

11. Frequent inspection can identify leaks in piping systems, storage tanks, and process tanks. Maintenance schedules should be developed to ensure that process baths and rinse systems are operating at their optimal efficiencies.

12. Strict procedures should be developed to ensure that chemicals are mixed properly for the process baths, minimizing spills and assuring the proper concentration (the lowest concentration acceptable to minimize the

drag-out). A limited number of personnel should be designated to handle and mix the chemicals, to improve the consistency of the solution concentration and to minimize waste.

6. REFERENCES

1. PRC Environmental Management, Inc., *Waste Audit Study - Metal Finishing Industry*, Prepared for the U.S. Environmental Protection Agency and California Department of Health Service, Alternative Technology Section, 1988.
2. U.S. EPA, *Profile of the Fabricated Metal Products Industry*, Office of Enforcement and Compliance Assurance, Washington, D.C., EPA/310-R-95-007, 1995.
3. Couture, S. D., *Source Reduction in the Printed Circuit Industry*, Proceedings of the Second Annual Hazardous Materials Management Conference, Philadelphia, Pennsylvania, June 5 to 7, 1984.
4. U.S. EPA, *Guide to Cleaner Technologies, Alternative Metal Finishes*, Office of Research and Development, Washington, D.C., EPA/625/R-94/007, 1994.

CHAPTER 17

Electronics Industry

This chapter describes the pollution prevention options available to the electronics industry, specifically the microelectronics industry. The microelectronics industry manufactures high quality electronic components (e.g., computer microchips). Fabrication steps required to produce microchips generate a substantial amount of waste solvents and wastes containing heavy metals.

Microelectronics is a revolutionary technology which has expanded greatly in the last two decades, becoming of major economic and technological importance to the world. The development of semiconductors and the subsequent development of the transistor and the integrated circuit were the most important factors in the evolution of microelectronics. The electrical conductivity of semiconductors can be enhanced enormously with small changes in chemical composition. Unlike other electrical devices that depend on the flow of electrical charges through a medium (vacuum or gas), semiconductor devices use the flow of current through a solid. As the primary semiconductor material, silicon has dominated the development of integrated circuit technology. In recent years, the principal focus of semiconductor development has been on valence III-V compounds (e.g., gallium arsenide) in addition to silicon.

1. PROCESS DESCRIPTION[1]

The manufacture of silicon ingots and wafers, from which microchips are produced, involves a growth process in a high temperature furnace at controlled pressure. Several hundred individual chips can be made from a single wafer. Typically, the valence III-V compound crystals (e.g., gallium arsenide) grown by this ingot growth process are used as a substrate material on which thin layers of the same or other III-V compounds are grown. Such crystal growth, in which the type of substrate determines the crystallinity and the orientation of the growth layer, is called epitaxial growth. Many different epitaxial growth processes are used to produce valence III-V compound devices. The major process steps involved in manufacturing microchips are described below.

1.1. Ingot Growth

Polycrystalline III-V compounds are typically grown by reacting the vapor of valence V compounds (e.g., arsenic) with valence III metal (e.g., gallium) at an elevated temperature in sealed quartz ampules (shown in Figure 17.1 for gallium arsenide crystals). Single crystal III-V compounds are grown by reacting elemental forms of valence III and V compounds at an

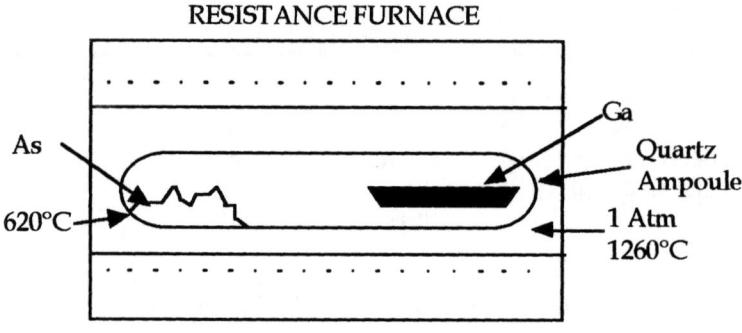

Figure 17.1. Gallium arsenide ingot growth.

elevated temperature. One example of a single crystal growth process for GaAs (gallium arsenide) is the liquid encapsulated Czochralski ingot growth system, in which the single crystal GaAs is slowly pulled out of the melted bulk GaAs (shown in Figure 17.2 for GaAs).[2] The single crystal ingots must be sandblasted and cleaned to remove surface oxides and contaminants. Generally, the crystals are "doped" with small amounts of dopants during growth to increase electrical conductivity. Typical dopants are silicon, tellurium, chromium, arsenic, phosphorus, and boron.

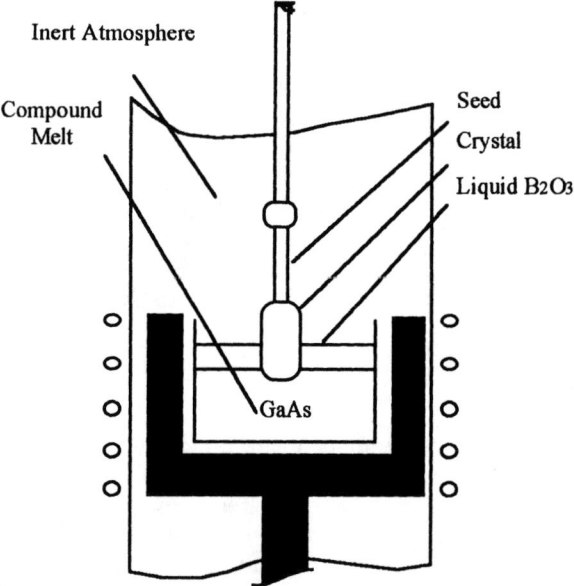

Figure 17.2. Liquid encapsulated Czochralski ingot growth system.

1.2. Ingot Sandblasting and Cleaning

Single crystal and polycrystalline defects must be sandblasted and cleaned to remove the oxides formed on the surface and other contaminants. Sandblasting is done in a bead blaster (glove box type), using either calcined alumina or silicon carbide media. After the sandblasting, wet cleaning is performed in chemical baths with alcohol (isopropyl alcohol or methanol) rinses.

1.3. Ingot Cropping

The ends or tails of the single crystal ingots are removed, using a water-lubricated single blade diamond saw with coolants added to the water.

1.4. Wafer Slicing

The ingots are sliced into individual wafers by use of multiple internal diameter diamond-blade saws. The slicing is done while the ingots are wetted with lubricants. Sliced silicon wafers are then stored in plastic reservoirs containing water or methanol. The thickness of the sliced wafers depends on the initial diameter of the ingot and the intended usage of the wafers. Gallium arsenide ingots are usually wax mounted onto a graphite beam and sliced into the individual wafers by use of mechanical diamond-blade saws.

1.5. Wafer Washing

Sliced silicon wafers are then transferred from plastic reservoirs (containing methanol or water) to wash tanks. The washes are accomplished with soap solution, weak caustic (10% NAOH) solution, and the final rinsing is done with water. Gallium arsenide wafers which were wax mounted are removed from the graphite beam in a vapor degreaser (typical solvent used is 1,1,1-trichloroethane). The wafers are then cleaned with sequential rinsing in wet chemical baths containing methanol, DI (deionized) water, and an acid mixture of sulfuric acid, hydrogen peroxide, and water.[3]

1.6. Wafer Lapping

Lapping is performed to smooth the roughness of the outer surface of the sliced wafers. Silicon wafers are lapped by machines that apply a specific pressure and rotational speed to smooth the surface. While the wafers are being lapped, a lapping solution is fed onto the surface. GaAs wafers are wax mounted to a lapper by use of a hot plate and lapped with lapping solution being fed onto the surface. After the lapping is completed, the wax is removed by use of a hot plate and the wafers are rinsed in a soap solution and wiped dry.

1.7. Wafer Etching

After the lapping and rinsing, silicon wafers are chemically etched. The purpose of the etching is to remove damages on the wafer surface and to reduce the wafer thickness. The wafers are held by fluorinated plastic cassette holders during etching. Etching solutions may contain nitric acid, hydrofluoric acid, or acetic acid.

For GaAs wafers, many different wet chemical acid solutions are used as etchants, typically sulfuric acid, hydrofluoric acid, hydrochloric acid, and phosphoric acid. Ammonium hydroxide is also used as a caustic etch.[2]

1.8. Polishing

After etching, silicon wafers are mounted to a metal carrier plate, which is in turn attached to a polishing machine. First, "stock removal" is done to provide the initial polish and to decrease the thickness. Then the finishing polish is performed to yield a mirrorlike surface. The polishing solution contains sodium hydroxide, colloidal silica, and water. For polishing GaAs wafers, a polishing solution containing sodium bicarbonate, chlorine, colloidal silicon slurry, and water is used.

After the polishing, the particles and residues on the surface are removed by chemical solution baths. These baths may contain 25% hydrofluoric acid, soap, oxidizers, or other cleaning agents.

1.9. Epitaxial Growth

Single crystal GaAs wafers can be used as substrates for the growth of thin layers of the same or other Ill-V compounds to obtain the desired electronic and optical properties (epitaxial growth). For epitaxial growth, the wafers are degreased and polished as described above and undergo a preepitaxial etch and cleaning step. The preepitaxial etch involves a sequential wet chemical dipping using sulfuric acid, hydrogen peroxide, and water; DI water rinse; and isopropyl alcohol rinse and dry. This etch cycle is repeated after each growth cycle (each layer of growth) for the wafers and also for the quartz reactor in which the epitaxial growth is performed. After growth is completed, another lapping is done to remove the unwanted deposited materials from the back side of the wafer.

2. WASTE DESCRIPTION

The major source of waste generated by the microelectronics industry is the production of solid valence III-V compound wastes from ingot grinding, cropping, wafer slicing, lapping, polishing, and back-lapping. Some of the solid wastes are present in the coolants used for grinding/slicing machines, and some are present in the noncoolant side (e.g., etch bath, etc.). The amount of solid waste generated is a function of the weight of III-V compounds lost per ingot during grinding or cropping, average wafer weight loss during lapping, and throughput coolant volume, among other variables. A waste sludge cake resulting from the treatment system of a GaAs wafer manufacturing facility is considered extremely hazardous because of its arsenic contents.

Other wastes generated by the microelectronics industry include etching bath chemicals such as nitric acid, sulfuric acid, hydrofluoric acid, phosphoric acid, and ammonium hydroxide. The waste polishing solution may include sodium bicarbonate, chlorine, colloidal silicon slurry, and sodium hydroxide. Waste organic solvents generated by the industry include methanol, acetone, trichloroethane, and isopropyl alcohol from the wafer-washing process and reactor/equipment cleaning.

3. POLLUTION PREVENTION OPTIONS[4]

Pollution prevention options in the electronics industry are source reduction through process modifications, recycling and resource recovery, and good housekeeping and operating practices. Owing to the ultrahigh purity requirements for wafer fabrication, the recycling and resource recovery option is generally not the best choice for this industry.

3.1. Process Modifications

A well-controlled ingot growth process can minimize the amount of defects formed during crystal growth, thus minimizing the need to sandblast and crop the ingots. Automated control systems can be used to reduce the generation of off-specification batches or product. Computerized process control systems have already been used successfully and have proven to be highly economical. Such systems can, for example, allow better control of slicing operations (yielding thinner and smoother slices), thus reducing the degree of lapping required.

Waste methanol or water from wafer-washing operations may be reduced by reducing the drag-out from the reservoirs before it is transferred to the wash tanks. The techniques that can reduce drag-out in the metal finishing industry are discussed in more detail in Chapter 16. Although the exact processing steps may be different, the same basic approaches may be used in the electronics industry.

Etching bath life may be extended by installing a filtration system to remove solids buildup and by regular monitoring. Extended bath life will result in a reduction in the quantity of etching chemical wastes.

The rinse efficiency in wash tanks may be improved by using the following techniques:

1. Provide enough turbulence between the wafer and rinse solution by using a spray rinse system or a forced air or water system.
2. Use multiple-stage countercurrent wash tanks in place of discrete wash tanks.
3. Control the rinse solution flow to reduce solution usage.

Efficient cleaning methods can also minimize solvent consumption. When solvent-based cleaning is used, the "flow-over" method should be selected over the "fill and empty" method, which consumes far more solvents.

Aqueous cleaning solutions are often viable substitutes for organic solvents. The disadvantages of using an aqueous cleaning solution are reduced cleaning efficiency, possible incompatibility between the cleaning solution and the equipment material of construction, and the presence of moisture on equipment parts, which may require air or heat drying. Trichloroethane, frequently used as a degreasing solvent, can also be replaced by an aqueous degreasing solution.

3.2. Recycling and Resource Recovery

The ultrahigh purity requirements for wafer fabrication make recycling and reuse of the solvents difficult in the microelectronics industry. However, waste solvents can be recycled and used for the steps in which ultrahigh purity is not required. Examples of those steps are the wafer-washing step before wafer lapping, and the washing step after etching and before polishing.

Slurry of III-V compounds from the noncoolant side can be filtered and recycled off-site. For example, recycled GaAs may be used in liquid encapsulated Czochralski ingot growth systems to grow single crystal GaAs. Slurry in the coolant can also be filtered, and the coolant may be recycled and reused in steps in which ultrahigh purity is not required.

3.3. Good Housekeeping and Operating Practices

As in other industries, good housekeeping and operating practices are the easiest and often cheapest means of reducing wastes in the electronics industry. Good housekeeping and operating practices include:

1. Proper material handling and storage.
2. Material tracking and inventory control.
3. Good procedural measures, including proper documentation and scheduling.
4. Good personnel practices, including management initiatives and employee training.
5. Loss prevention practices, including spill prevention, preventive maintenance, and emergency plans.

Special attention should be paid to keeping work areas clean in order to minimize product contamination during ingot growth, polishing, and epitaxial growth.

4. CASE STUDIES

4.1. Case Study 1 - Electronic Circuit Manufacturing[5]

4.1.1. Problem Description

In an electronic circuit manufacturing plant, flexible electronic circuits are made from copper sheeting. Before sheeting can be used, it has to be cleaned. Cleaning had been accomplished by spraying with ammonium persulfate, phosphoric acid, and sulfuric acid. This cleaning operation created a hazardous waste stream that required special handling and disposal.

4.1.2. Pollution Prevention Techniques Used

Equipment for cleaning by chemical spraying was replaced by a specially designed machine with rotating brushes that scrubbed the copper sheet with pumice.

4.1.3. Benefits

Use of the fine, abrasive pumice material resulted in a slurry that was not hazardous and could be disposed of in a sanitary landfill. Although the new cleaning machine cost $59,000, savings of $15,000 in raw material, disposal, and labor costs were achieved in the first year. This process change also resulted in the elimination of 40,000 pounds of hazardous liquid wastes a year.

4.2. Case Study 2 - Manufacture of Printed Circuit Boards[6]

4.2.1. Problem Description

A facility was generating several hazardous waste streams that contained high concentrations of heavy metals. The company's goal was to reduce its reliance on hazardous waste landfills.

4.2.2. Pollution Prevention Techniques Used

A management initiative to develop markets for untreated wastes resulted in the identification of markets for the following waste streams:

1. Spent cupric chloride etchant
2. Spent plating baths (acid copper bath, palladium catalyst bath, and spent sulfuric acid from the etch-back process on the electroless copper line)
3. Copper sulfate crystals from regeneration of the sulfuric peroxide bath
4. Spent ammonia etchant
5. Spent nickel-plating bath
6. Still bottoms from the 1,1,1-trichloroethane recovery process

These wastes were shipped to buyers at no cost or for profit for the facility. The total amount of waste was approximately 90,000 lb/yr.

4.2.3. Benefits

The wastes shipped to buyers yielded the company $9,000/yr in disposal cost savings and revenue.

5. SUMMARY

This chapter describes pollution prevention options available to the electronics industry, along with descriptions of the processes and the wastes generated. Pollution prevention options in the electronics industry are source reduction through process modification(s), recycling and resource recovery, and good housekeeping and operating practices. Owing to the ultrahigh purity requirements for wafer fabrication, the recycling and resource recovery option is generally not the best choice for this industry.

1. The manufacture of silicon ingots and wafers, from which microchips are produced, involves a growth process in a high temperature furnace with controlled pressure. The major process steps are: ingot growth, ingot sandblasting and cleaning, ingot cropping, wafer slicing, wafer washing, wafer lapping, wafer etching, and wafer polishing.

2. The major source of waste generated by the microelectronics industry is the production of solid valence III-V compound wastes from ingot grinding, cropping, wafer slicing, lapping, polishing, and backlapping. Other wastes generated by the microelectronics industry include etching bath chemicals such as nitric acid, sulfuric acid, hydrofluoric acid, phosphoric acid, and ammonium hydroxide.

3. A well-controlled ingot growth process can minimize the amount of defects formed during crystal growth, thus minimizing the need to sandblast and crop the ingots. Automated control systems can be used to reduce the generation of off-specification batches or product.

4. Waste methanol or water from the wafer-washing operation may be reduced by reducing the drag-out from the reservoirs before it is transferred to the wash tanks.

5. Etching bath life may be extended by installing a filtration system to remove solids buildup and by regular monitoring.

6. The rinse efficiency in wash tanks may be improved by using the following techniques.
- Provide enough turbulence between the wafer and rinse solution by using a spray rinse system or a forced air or water system.
- Use multiple-stage countercurrent wash tanks in place of discrete wash tanks.
- Control the rinse solution flow to reduce solution usage.

7. Good housekeeping and operating practices include:
- Proper material handling and storage.
- Material tracking and inventory control.

- Good procedural measures, including proper documentation and scheduling.
- Good personnel practices, including management initiatives and employee training.
- Loss prevention practices, including spill prevention, preventive maintenance, and emergency plans.

8. Special attention should be paid in keeping work areas clean, in order to minimize product contamination during ingot growth, polishing, and epitaxial growth.

6. REFERENCES

1. U.S. EPA, *Profile of the Electronics and Computer Industry*, Office of Enforcement and Compliance Assurance, Washington, D.C., EPA/310-R-95-002, 1995.
2. Envirosphere Company, *The Reduction of Arsenic Wastes in the Electronics Industry*, Prepared for the State of California, Department of Health Services, Toxic Substances Control Division, Alternative Technology and Policy Development Section, Santa Ana, California, 1987.
3. Envirosphere Company, *The Reduction of Solvent Wastes in the Electronics Industry*, Prepared for the State of California, Department of Health Services, Toxic Substances Control Division, Alternative Technology and Policy Development Section, Santa Ana, California, 1988.
4. U.S. EPA, *Guides to Pollution Prevention, The Printed Circuit Board Manufacturing Industry*, Risk Reduction Engineering Laboratory, Cincinnati, Ohio, EPA/625/7-90/007, 1990.
5. 3M, *A Compendium of 3P Success Stories*, Environmental Engineering and Pollution Control Department, St. Paul, Minnesota, 1989.
6. Institute for Local Self-Reliance, *Proven Profits from Pollution Prevention: Case Studies in Resource Conservation and Waste Reduction*, 1986.

CHAPTER 18

Drug Manufacturing and Processing Industry

The drug manufacturing and processing industry can be categorized into five areas:

1. Fermentation
2. Chemical synthesis
3. Natural extraction.
4. Formulation and packaging
5. Research and development

Drugs are manufactured and processed in batch, semi-batch, and continuous operations. Batch operations account for 87% of all types of drug manufacturing.[1,2] Owing to the diversity of operations and products, drug manufacturing processes are described only in general terms for the first four areas in this chapter.

1. PROCESS DESCRIPTION

1.1. Fermentation

Batch fermentation processes are typically used for producing steroids and antibiotics. The fermentation process consists of three steps: inoculum and seed preparation, fermentation, and product recovery and purification.

Inoculum preparation begins with a population of a microbial strain. A few cells from this culture are grown into a dense suspension through a series of test tubes and flasks. The cells are then transferred to a seed tank for further growth. The seed tank is essentially a fermenter designed for maximum growth.

The cells from the seed tank are charged to a sterilized fermenter, and sterilized nutrients are added to the fermenter to begin the fermentation process. During fermentation the fermenter contents are agitated and aerated by a sparger, maintaining an optimum air flow rate and temperature. After the cells are matured, the fermentation broth is filtered to remove the solid remains of the inoculum microorganisms (i.e., mycelia). The product is then recovered from the resulting filter beer through various recovery processes, for example, solvent extraction, precipitation, adsorption, or others (see Chapters 7 and 14 for details of these recovery processes). An example of a fermentation process is illustrated in Figure 18.1.

1.2. Chemical Synthesis

The majority of drugs today are produced by chemical synthesis. Typically, one or more batch reactors (depending on the number of chemical steps) are utilized in series to make the final product. The product recovery (or isolation) steps may involve separation and/or purification. The intermediates can also be isolated in order to avoid carrying impurities to the next chemical reaction. (Impurities may interfere with the next reaction, reducing the yield). The types of chemical reactions and the isolation processes used are quite diverse, depending on the chemical structure of the desired final product.

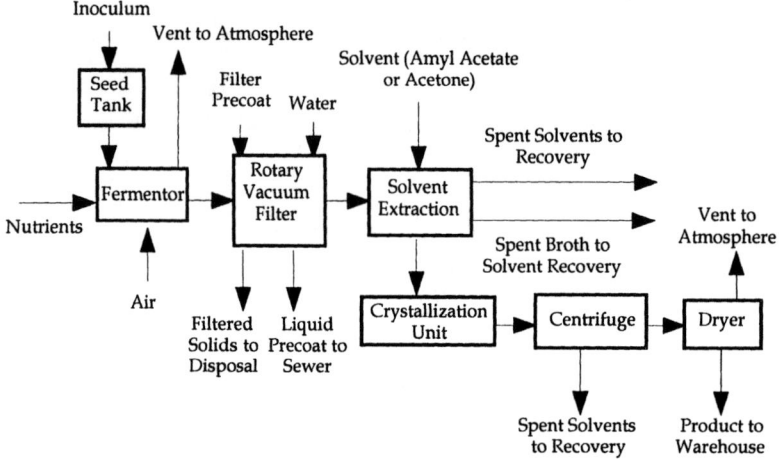

Figure 18.1. Typical fermentation process flowsheet.

An example of a chemical synthesis process is illustrated through a process flow diagram of product X (see Figure 18.2). Potassium permanganate (product X precursor) and water are mixed in a 3,000-gallon reactor, and manganese dioxide is formed. Manganese dioxide is removed from the solution by a rotary drum filter coated with a filter aid (celite). The filter cake containing manganese dioxide and celite is deposited into trash bins for disposal. The filtrate is neutralized with sulfuric acid and sent to an evaporator to remove water. The product solution is then sent to an 800-gallon Pfaudler reactor where a final pH adjustment is made with additional sulfuric acid. The mixture is cooled, potassium sulfate (product of neutralization) crystallizes, and the potassium sulfate crystals are filtered by centrifuge. Butyl acetate is then added to the filtrate, and the resulting mixture is azeotropically dried (by azeotropic distillation) until all the water is removed. The mixture is again filtered to remove additional potassium sulfate salt, which comes out of solution as more water is removed. When the filtrate is cooled, the product crystallizes out and the product crystals are isolated by filtering (centrifuge). The product is then dried in a tumble dryer and packaged.

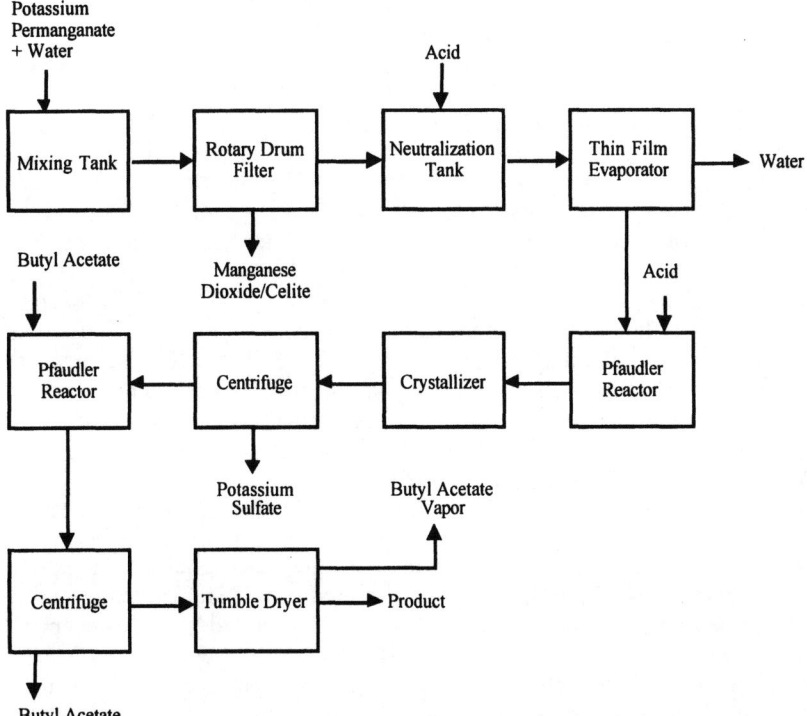

Figure 18.2. Example chemical synthesis process flowsheet.

1.3. Natural Extraction

Drugs manufactured by natural extraction include allergy relief medicines, insulin, and morphine. These drugs are extracted from natural sources (e.g., roots, leaves, animal glands, fungi, etc.) typically because chemical synthesis of these compounds is difficult owing to their large molecule size and complex molecular structure. During each extraction (or process) step, the volume of the intermediates diminishes greatly and the final purification can occur on volumes less than 1% of the initial volume. Therefore, conventional batch or continuous processes are not usually suitable for natural extraction.

1.4. Formulation

The active ingredient produced by either fermentation, chemical synthesis, or natural extraction is then formulated to tablets, capsules, liquids, creams, and ointments. The active ingredient is blended to the right dosage by adding fillers or binders. Fillers or binders used for tablet formulation are starch and sugar. The blend is then compressed via wet granulation, direct compress, or slugging.

In *wet granulation*, the active ingredient and filler in powder form are blended and wetted with a binder solution to form coarse granules. These granules are dried, mixed with lubricants (e.g., magnesium stearate), and compressed into tablets. Direct compression is done by using a die that holds a measured amount of material and a punch that compresses the tablet. Slugging is used for drugs that are not sufficiently stable for wet granulation formulation or cannot be compressed directly. The slugging process blends and compresses relatively large size tablets (20 to 30 grams) which can then be ground and screened to a desired mesh size and recompressed into the final tablets.

After tablets, capsules are the most widely used oral dosage form for solid drugs. Hard gelatin capsules are prepared by depositing a gelatin film on the capsule (using a solution of gelatin and cooling). The capsules are dried, trimmed, and the separate sections of the capsule are filled and joined.

Liquids are prepared for injection and oral use in aqueous solutions, syrups, suspensions, and tinctures. The liquids are prepared by mixing the active ingredient with a solvent and preservatives (to prevent mold and bacterial growth), and filtering at high pressure. Suspensions and emulsions are prepared by using colloid mills and homogenizers, respectively. Although liquid solutions prepared for oral or topical use do not require sterilization, liquid solutions for ophthalmic use must be sterilized. The liquid dosages that are injected into the fluid systems of the body (e.g., intradermal, hypodermic, intramuscular, intravenous, etc.) also need to be sterilized by dry or moist heat under pressure. Products that are not stable at high temperatures can be sterilized by using bacteria-retaining filters.

Creams and ointments are prepared for topical use, usually by blending the active ingredient with petroleum derivative petrolatum and passing the mixture through a colloid or roller mill.

2. WASTE DESCRIPTION

2.1. Fermentation

Fermentation processes generate large volumes of spent fermentation medium. Filtration processes, which remove the solid remains of microorganisms (mycelia), yield large quantities of solids as spent filter cake. The filter cake contains mycelia, filter aids, and small amounts of residual product. After product recovery, the spent filtrate is discharged as wastewater. Other sources of wastewater are equipment-cleaning operations and fermenter vent gas scrubbing liquids. Wastewaters from fermentation processes are typically high in BOD (biological oxygen demand), COD (chemical oxygen demand), and TSS (total suspended solids), with a pH range of 4 to 8.

2.2. Chemical Synthesis

Wastes from chemical synthesis are complex and diverse as a result of the variety of reactions and operations employed during the synthesis process. *Chemical synthesis* processes generate acids, bases, solvents, cyanides, and metal wastes among others. Solid wastes are generated as waste filter cake, which may include various inorganic salts (e.g., sodium sulfate, ammonium sulfate, etc.), organic by-products, and metal complexes. Typically, spent solvents are recycled on-site within the process (via extraction and distillation). Still bottom wastes are generated as a result of recovery and recycling operations. Chemical synthesis processes also generate aqueous waste streams, which include mother liquors from crystallization operations, aqueous wash layers from extractions, contaminated scrubber water, and equipment-cleaning wastes. Wastewaters from chemical synthesis processes typically have high BOD, COD, and TSS with the pH ranging from 1 to 11.

2.3. Natural Extraction

Wastes resulting from natural extraction processes include raw materials such as leaves and roots, spent solvents, and wastewater. Wastewaters from natural extraction processes are typically low in BOD, COD, and TSS levels, with the pH ranging from 6 to 8.

2.4. Formulation

Wastes generated in formulation processes include wastes resulting from cleaning, sterilization, spills, and rejected products. During blending and/or tableting operations, dusts are generated, which are typically recycled back to the formulation process. During tablet coating and capsule manufacturing, there are air emissions from solvent handling. The primary source of wastewater is equipment cleanouts. Wastewaters typically contain inorganic salts, sugars, syrups, and residual amounts of the active ingredient. Wastewaters are usually low in BOD, COD, and TSS, with near neutral pH.

3. POLLUTION PREVENTION OPTIONS[1,3]

Pollution prevention options in the drug manufacturing and processing industry are source reduction through process modifications, recycling and resource recovery, and good housekeeping and operating practices. Process modifications include product reformulation, material substitution, and process step modifications. The following sections describe some of the pollution prevention options available to the drug manufacturing and processing industry.

3.1. Process Modifications

Product reformulation involves the substitution of one or more raw materials, which may reduce the volume and/or toxicity of wastes generated. However, product reformulation is difficult for the pharmaceutical industry because of the time required for product redevelopment and testing. Testing is required to ensure that the reformulation has the same medicinal effect as the original drug. The time and the testing required for FDA approval of the reformulated drug can be long and drawn out. Therefore, product reformulation is usually not an economic or practical option for the drug manufacturing and processing industry.

Material substitution is the replacement of a hazardous material in manufacturing or processing with a nonhazardous material. Material substitution may also reduce the volume and toxicity of the wastes generated, but without changing the product itself. For example, organic solvents used during tablet-coating processes can be substituted with water-based solvents, thus reducing the volatile organic compound emissions to the atmosphere. Another example is the use of less hazardous organic solvents (e.g., ethyl acetate, isopropyl acetate, toluene, etc.) in place of more toxic (or more rigidly controlled by EPA) solvents, such as methylene chloride and benzene. In fact, many drug companies have specific guidelines prohibiting the use of certain solvents (e.g., methylene chloride and benzene) even at the research and development stage. Another approach in material substitution includes the use of aqueous-based cleaning solutions (e.g., soap and water) in place of solvent-based solutions. The current trend in the industry is to replace chlorinated solvents with nonchlorinated solvents wherever possible.

Generation of wastes can be reduced through *process modification* or modernization. Process modifications involve changes in process operating conditions and/or process equipment. Process modernization includes installing updated control systems or increasing control levels. Another form of modernization is increasing the degree of process automation.

Incomplete chemical reactions can result in increased formation of byproducts, thus increasing the amount of wastes generated. Possible causes of incomplete reaction are inadequate feed flow control and/or reactor temperature control. Installing updated control systems can improve reactor and/or reaction efficiency and reduce by-product formation. An increased degree of automation can also reduce wastes generated by preventing operator errors.

Other process considerations for waste minimization are outlined as follows:

1. Install internal recycling systems for solvents and cooling water.
2. Select new or improved catalysts to improve reaction conversion.
3. Consider using a continuous process in place of batch process.
4. Modify tank and vessel dimensions to minimize the wetting loss (e.g., use of conical vessels).
5. Optimize crystallization conditions to reduce the amount of product loss to the mother liquor and cake wash.
6. Optimize crystallization conditions to obtain the desired crystal size, thus minimizing the need for recrystallization (crystal size too small to filter) or milling (crystal size too large).
7. Consider new reactor-filter-dryer systems that can handle more than one operation (e.g., Rosenmund filter-dryer).
8. Utilize available computer programs/expert systems that can evaluate the entire process for efficiency and characterize the waste streams, thus determining the major source of waste generation and optimizing the process early.

One example of pollution prevention via process modification is the use of *supported liquid membranes (SLM)* for liquid solution separation during reaction, workup, and/or isolation sequences. The SLM are thin layers of pure liquid or liquid solutions immobilized in microporous inert supports. They have the extraordinary capacity of removing solutes from a feed solution to a strip solution. For example, an aqueous solution containing a solute is fed into an SLM and the solute is extracted into a contained or supported organic solvent, which in turn is stripped to the aqueous strip solution. The organic solvent is continuously renewed, and this specific feature allows significant reduction in solvent usage.

Other advantages of SLM are the high separation factor per stage and low capital/operating costs. The major disadvantage is the instability of the SLM, which prevents a long-term operation. In an attempt to overcome SLM's shortcomings, the hollow fiber contained liquid membrane (HFCLM) technique was developed.[4] The HFCLM uses microporous hollow fibers to contain liquid membranes in a permeator (see Figure 18.3). In theory, the solvent contained in the membrane could be used indefinitely, thus reducing the amount of solvent used for extractions. However, depending on the type of process, the solvent may have to be replaced as the performance starts to degrade because of the accumulation of impurities.

Although significant waste reduction can be achieved through process modifications, there are problems associated with this option for the drug manufacturing industry. Extensive process changes can be very costly owing to the installation of new equipment and downtime. New processes must also be tested and certified by the FDA to ensure that the resulting product is acceptable. An important parameter to consider is the impurity profile of the drug manufactured by the new process, because an identical or better impurity profile (i.e., fewer impurities) is essential for the validation of clinical testing. In addition, the existing management may be reluctant to change a working (successful) process. Such resistance is particularly likely when suggesting adopting continuous processes in place of existing batch processes.

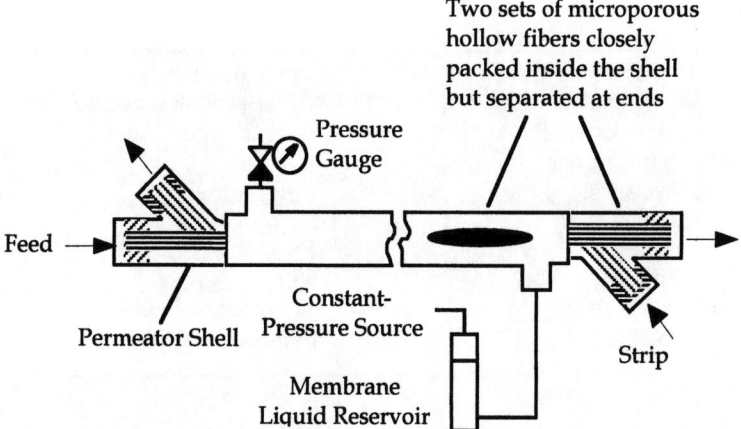

Figure 18.3. Hollow fiber contained liquid membranes.

Equipment-cleaning efficiency can be improved significantly by employing high-pressure spray nozzles (as a preliminary cleaning method) to clean reactor walls, rather than conventional solvent or caustic solution boil-outs. The use of high-pressure nozzles can also reduce solvent usage appreciably.

3.2. Recycling and Resource Recovery

Recycling and resource recovery include direct reuse of the waste generated, i.e., reclamation for a separate use or direct reuse within the process via separation of impurities. However, the high degree of quality control required in the pharmaceutical industry limits the options available for reuse within these processes.

Fermentation wastes may be used as supplements to animal feeds, soil conditioners, and fertilizers. For example, mycellium filter cake and other by-products of the production of antibiotics by fermentation are widely sold as protein supplements to animal feed, as long as the antibiotic activity in the filter cake does not exceed 2 g/ton of cake and less than 3 lb of cake is used per ton of feed.[2]

Inorganic salts such as ammonium sulfate and sodium sulfate (typical by-products of some chemical synthesis processes) can be reclaimed. Ammonium sulfate can be used as raw material for fertilizer, and sodium sulfate can be dried for sale to the glass industry.

Recycling is a particularly good option for spent solvents. Spent solvents can be used for equipment cleaning. They can also be purified for reuse in the process for reaction mediums, extraction mediums, and coating mediums. The types of recycling methods available are discussed in Chapter 14 of this text. Commonly recycled solvents in the drug manufacturing and processing industry are summarized in Table 18.1.

Table 18.1. Commonly Recycled Solvents in the Pharmaceutical Industry.[2]

Tetrachloroethylene	Trichloroethylene
Methylene chloride	1, 1, I-Trichloroethane
Carbon tetrachloride	Chlorofluoroethanes
Chlorobenzene	Acetone
Ethyl acetate	Ethylbenzene
Benzene	Butanol
Xylene	Cyclohexane
Methanol	Nitrobenzene
Toluene	Methyl ethyl ketone
Isobutanol	Pyridine

The following steps can improve solvent waste recyclability.

1. Segregate chlorinated solvents from nonchlorinated solvents, aliphatic from aromatic solvents, and aqueous from flammable solvents.
2. Minimize the solid concentration in solvent wastes.
3. Label all solvent wastes properly, including the composition and generation source.

It should also be noted that the decision to recycle on-site or off-site depends on the capital cost, operating costs, waste volume, and in-house expertise required.

3.3. Good Housekeeping and Operating Practices

Good housekeeping and operating practices can reduce the amount of wastes generated and material losses. Such practices may be categorized into three components: plant management, raw materials handling, and waste management.

Good operating and housekeeping practices in plant management include the following:

1. Management's awareness of environmental responsibilities.
2. Closer supervision of plant personnel to increase production efficiency and reduce waste generation.
3. Effective employee training for safe equipment operation, material handling, and spill cleanup.
4. Improved documentation of process steps, including start-up, shutdown, emergency, special, and normal operating procedures.
5. Effective scheduling to reduce waste generation.

Raw material handling procedures should be updated to improve materials tracking and inventory control. Utilization of a computer to monitor inventories can reduce overstocking. Proper labeling and use of material safety data sheets (MSDS) facilitate material handling. Spill and leak prevention programs are also important to pollution prevention. A spill and leak prevention program should start with properly designed storage tanks and process vessels (including level indicators and alarms), and include installing secondary containment, setting up procedures that prevent by-pass of the interlock systems, and keeping a regular maintenance schedule that includes checking the integrity of containers and testing alarms. In addition, a proper maintenance program can prevent the generation of (hazardous) waste by equipment failure. Preventive maintenance programs should include cleaning, making minor adjustments, lubrication, testing, calibrating, and replacing of minor parts.

Once wastes are generated, it is essential that waste tracking and inventory control are established. There should also be frequent environmental and waste audits (see Chapter 12). The waste streams can be segregated (hazardous from nonhazardous, liquid from solid, etc.) to reduce waste volume, to simplify disposal, and to facilitate recycling and recovery.

4. CASE STUDY[5]

4.1. Problem Description

A facility engaged in the coating of medicine tablets found that solvent emissions had the potential to exceed the air pollution limits set by an upcoming state regulation.

4.2. Pollution Prevention Techniques Used

A water-based coating was developed to replace the solvent-based tablet coating. Different spraying equipment was installed. Spray guns, located inside the drum, spray the aqueous mixture onto the tablets, which are tumbling through the drum. The spray system was changed from an airless to an air system, permitting quick and precise application of the coating (too much coating can result in soggy tablets). The heating capacity of the drying process had to be increased also, since water does not evaporate as quickly as solvents. These changes were approved by the U.S. Food and Drug Administration.

4.3. Benefits

Solvent emissions were eliminated (air pollution reduced 24 tons annually). The changes required $60,000 in capital costs, but eliminated the need to spend $180,000 for pollution control equipment. Solvent cost was eliminated, thus saving $15,000 a year in solvent costs alone.

5. SUMMARY

This chapter describes pollution prevention options available to the drug manufacturing industry, along with descriptions of the processes and wastes generated. Pollution prevention options in drug manufacturing and processing include *source reduction* through *process modification(s)*, *recycling* and *resource recovery*, and *good housekeeping and operating practices*. Process modifications include *product reformulation*, *material substitution*, and *process step modifications*.

    1. Activities in the drug manufacturing and processing industry can be categorized into five areas: fermentation, chemical synthesis, natural extraction, formulation and packaging, and research and development. Drugs are manufactured and processed in batch, semi-batch, and continuous operations.

    2. Fermentation processes generate large volumes of spent fermentation medium as spent filter cake. The filter cake contains mycelia, filter aids, and small amounts of residual product.

3. Chemical synthesis processes generate acids, bases, solvents, cyanides, and metal wastes, among others. Solid wastes are generated as waste filter cake, which may include various inorganic salts (e.g., sodium sulfate, ammonium sulfate, etc.), organic by-products, and metal complexes. Chemical synthesis processes also generate aqueous waste streams that include mother liquors from crystallization operations, aqueous wash layers from extractions, contaminated scrubber water, and equipment-cleaning wastes.

4. Wastes from natural extraction processes include raw materials such as leaves and roots, spent solvents, and wastewater.

5. Wastes generated by formulation processes include wastes from cleaning, sterilization, spills, and rejected products. The primary source of the wastewater is equipment cleanouts. Wastewaters typically contain inorganic salts, sugars, syrups, and residual amounts of the active ingredient.

6. Product reformulation is difficult for the pharmaceutical industry because of the time required for product redevelopment and testing (required by FDA).

7. Installing updated control systems can improve reactor/reaction efficiency and reduce by-product formation.

8. Process modification options in the drug manufacturing industry are suggested as follows:
   - Install internal recycling systems for solvents and cooling water.
   - Select new or improved catalysts to improve reaction conversion.
   - Consider using a continuous process in place of a batch process.
   - Modify tank and vessel dimensions to minimize the wetting loss (e.g., use of conical vessels).
   - Optimize crystallization conditions to reduce the amount of product loss to the mother liquor and cake wash.
   - Optimize crystallization conditions to obtain the desired crystal size, thus minimizing the need for recrystallization (crystal size too small to filter) or milling (crystal size too large).
   - Consider new reactor-filter-dryer systems that can handle more than one operation (e.g., Rosenmund filter-dryer).
   - Utilize available computer programs/expert systems that can evaluate the entire process for efficiency and characterize the waste streams, thus determining the major source of waste generation and optimizing the process early.

9. The high degree of purity required in the pharmaceutical industry limits the resource recovery and recycle options available. Recycling is a good option for spent solvents. Solvent waste recyclability can be improved by:
   - Segregating chlorinated solvents from nonchlorinated solvents, aliphatic from aromatic solvents, and aqueous from flammable solvents
   - Minimizing solid concentration in solvent wastes
   - Labeling all solvent wastes properly, including the composition and generation source

10. Good operating and housekeeping practices may be categorized into three areas: plant management, raw materials handling, and waste management. Good operating and housekeeping practices in plant management include the following:
- Management's awareness of environmental responsibilities
- Closer supervision of plant personnel to increase production efficiency and reduce waste generation
- Effective employee training for safe equipment operation, material handling, and spill cleanup
- Improved documentation of process steps, including start-up, shut-down, emergency, special, and normal operating procedures
- Effective scheduling to reduce the waste generated

6. REFERENCES

1. U.S. EPA, *Profile of the Pharmaceutical Manufacturing Industry*, Office of Enforcement and Compliance Assurance, Washington, D.C., EPA/310-R-97-005, 1997.
2. ICF Technology Inc., *Waste Audit Study-Drug Manufacturing and Processing Industry*, Prepared for the State of California, Department of Health Services, Toxic Substance Control Division, Alternative Technology and Policy Development Section, Universal City, California, 1989.
3. U.S. EPA, *Guides to Pollution Prevention, The Pharmaceutical Industry*, Office of Research and Development, Washington, D.C., EPA/625/7-91/017, 1991.
4. Segupta, A., Basu, R., and Sirkar, R. R., *Separation of Solutes from Aqueous Solutions by Contained Liquid Membranes*, AIChE Journal 34(10): 1698-1708.
5. 3M, *A Compendium of 3P Success Stories*, Environmental Engineering and Pollution Control Department, St. Paul, Minnesota, 1989.

CHAPTER 19

Paint Manufacturing Industry

The paint manufacturing industry consists of establishments engaged in the manufacture of paints, varnishes, lacquers, enamels, paint and varnish removers, and allied paint products. The paint manufacturing industry described in this chapter does not include the manufacturers of pigments (organic or inorganic), resins, printing inks, adhesives, sealants, or art materials. There are approximately 1,375 establishments that are categorized as paint manufacturing facilities nationwide.[1]

The paint manufacturing industry generates large quantities of both hazardous and nonhazardous wastes. Therefore, reducing waste is a high priority for this manufacturing sector. Paint, coating, and ink manufacturers represent approximately 44 percent of the solvent market.[2] The amount of solvent wastes (or solvent containing wastes) disposed of by the paint manufacturing industry is the highest volume of solvent wastes generated in most states.[3]

Most small paint manufacturing facilities produce paint in 10- to 500-gallon batches. For an average paint plant located in the U.S., 60% of its total annual production is solvent-based paints, 35% is water-based paints, and 5% is other allied products. For architectural coating, water-based paint is mostly used (more than 70%), and solvent-based paints are used generally for product coatings and special purpose coatings.

1. PROCESS DESCRIPTION

The paint manufacturing industry's two main products are solvent-based paints and water-based paints. At a typical plant, both types of paints are produced. The steps involved in manufacturing paint are illustrated in the block diagram in Figure 19.1.

The production of solvent-based paint begins with mixing of resins, dry pigment, and pigment extenders, along with solvents and plasticizers, in a high-speed mechanical mixer. After the mixing operation, the batch is transferred to a mill for additional grinding and mixing. The type of mill employed is a function of the types of pigments being handled. Examples of the types of mills used are roll mills, sand mills, pin mills (e.g., Alpine mill), hammer mills (e.g., Fitz mill, Mlkro-Pulverizer), and sieve mills. Then the paint base or concentrate is transferred to an agitated tank where tints, thinners, and additional resins are added and mixed.

After reaching the proper consistency (viscosity and/or concentration), the batch is filtered via a cartridge filter to remove any nondispersed pigment and the paint is transferred to a loading hopper. From the hopper, the paint is poured into cans, labeled, and moved to storage or to the warehouse.

The water-based paint manufacturing process is similar to the solvent-based paint manufacturing process except for the use of water in place of sol-

374 Chapter 19

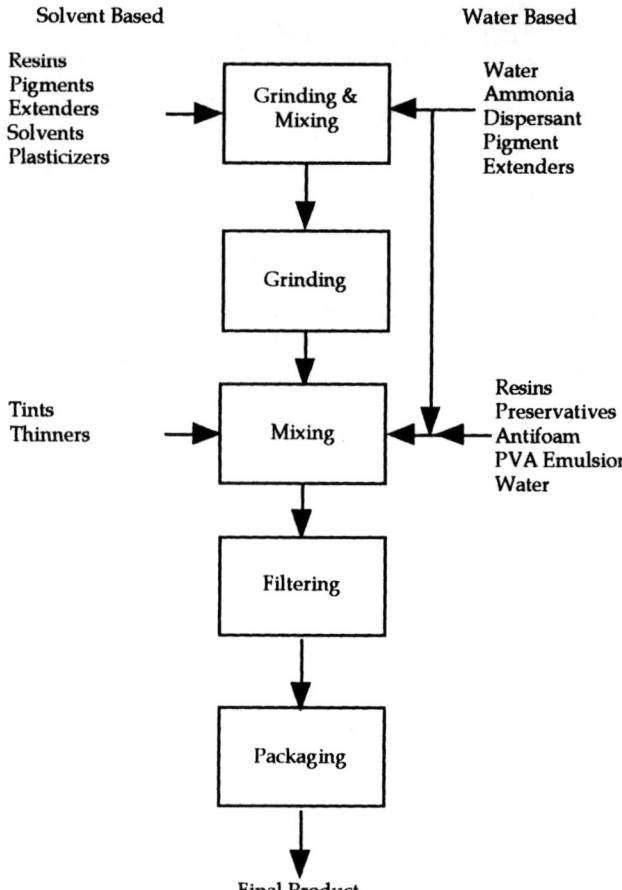

Figure 19.1. Block diagram of typical paint manufacturing steps.

vent and the sequence of addition of materials. Manufacturing of water-based paints begins by mixing water, ammonia, and a dispersant in a high-speed mixer. Dry pigment and pigment extenders are also added and mixed into the mixture. After mixing, the batch is milled and transferred to an agitated tank where resin, plasticizers, preservatives, anti-foaming agents, polyvinyl acetate emulsion, and water as thinner are added (in the order of addition). After the mixing operation, the water-based paint undergoes the same procedure as used in the solvent-based paint manufacturing process. At many paint manufacturing facilities, all dispersion operations are accomplished in a single high-speed mixer rather than through grinding and mixing operations in a mill and then in an agitated tank.

2. WASTE DESCRIPTION

The major raw materials used to manufacture paints are resins, solvents, drying oils, pigments, and extenders. Types of resins include alkyd, acrylic, vinyl, and others. The pigments include titanium dioxide and other inorganic (e.g., iron oxide, zinc oxide, zinc dust, aluminum paste, lead compounds, and chrome compounds) and organic pigments. Calcium carbonate is the most common extender used, along with talc and clay.

The major wastes generated by the paint manufacturing industry are empty raw material packages (containing trace amounts of raw materials), off-specification paint, equipment-cleaning wastes, and spills. Empty raw material packages are generated during unloading of materials to high-speed mixers or mixing tanks. Off-specification paints are generated during small-scale color matching production. Waste solvents are generated from equipment cleaning. Even after distillation of waste solvents for reuse, a residual paint sludge remains. The paint sludge contains solvents and residual toxic metals such as mercury, lead, and chromium. Waste rinse water is generated from equipment cleaning with water and/or caustic solutions. Wastes containing undispersed pigments are contained in waste filter cartridges.

3. POLLUTION PREVENTION OPTIONS

The wastes generated by the paint manufacturing industry include equipment cleaning wastes, spills and off-specification paints, leftover pigment in packages, pigment dust from air pollution control equipment (usually a baghouse), waste filter cartridges, and obsolete products and rejects. Equipment cleaning generates most of the waste associated with paint manufacturing.

Equipment-cleaning wastes can be minimized by employing more efficient cleaning methods. Other process modifications may be used to minimize wastes from off-specification paints and obsolete products and rejects. Better housekeeping and operating practices should be applied to all waste streams.

3.1. Process Modifications

Because equipment cleaning is the largest source of waste in the paint manufacturing industry, reduction in the frequency of equipment cleaning (e.g., mixing tanks) is the most effective approach to pollution prevention. Methods that can reduce the frequency of cleaning include:

1. *Use of rubber wipers* to reduce the amount of paint left on the walls of a mix tank. This operation can be done either manually or mechanically. Many new mixers are equipped with automatic wall scrapers.
2. *Use of Teflon-lined tanks* to reduce adhesion and to improve drainage.

3. *Use of a plastic or foam "pig"* to clean pipes. Plastic or foam pigs (slugs) can be used to clean paint from pipes. The pig is forced through the pipe from the mixing tank to the product hopper. This increases the yield and reduces the frequency of pipe cleaning required. Inert gas is typically used to propel the pig, and the equipment (launcher and catcher) must be designed carefully to prevent spills, sprays, and potential operational hazards.[1]

Equipment used for producing water-based paints is cleaned with either a water or caustic solution. Equipment used for producing solvent-based paints is cleaned with solvents or a caustic solution. However, caustic solutions are more frequently used for water-based paint equipment than for solvent-based paint equipment. Therefore, the amount of waste rinse water generated is a function of the ratio of the water-based paints manufactured to the solvent-based paints in a given facility. The amount of wastewater or waste caustic solution generated may be minimized by employing the following methods:

1. *Use of high-pressure spray heads.* After the tank walls are scraped, high-pressure spray hoses can be used in place of regular hoses to clean water-based paint tanks. High-pressure wash systems can reduce water use by 80 to 90 percent.[4] For tanks used in making solvent-based paints, a built-in high-pressure cleaning system may be employed.
2. *Use of a countercurrent rinsing sequence.* For facilities with additional space, a countercurrent rinsing operation may be employed to increase the rinsing efficiency and to minimize the amount of wastewater generated.
3. *Encouragement of dry cleaning methods.* Equipment may be cleaned by hand with rags rather than by using hoses.

Pigment waste from bags and packages may be reduced by using water-soluble bags. In water-based paint manufacturing, bags containing toxic pigments can be dissolved or mixed in with the paint. This method is frequently used for mercury compounds and other paint fungicides. However, soluble bags cannot be used for producing high quality, smooth finish paints since the bag material can affect the paint quality. Dissolved bag material can also increase the load on the filters.

Paste pigments, which are wetted or mixed with resins, may be used in place of dry powder pigments that cause particulate or dust emissions. Bag filters or metal mesh filters may be used instead of cartridge filters. Metal mesh filters are available in micron sizes and can be cleaned and reused indefinitely.

Spills and rejects can be reduced by increased automation. Obsolete products (resulting from changing customer demands, newer and better products, and expired shelf life) and rejects can also be blended into new batches of paint.

3.2. Recycling and Resource Recovery

Waste solvents generated by equipment cleaning can be collected and used in the next compatible batch as part of the formulation. For example, wash solvent from each solvent-based paint batch can be collected separately and stored. When the same type of paint is being produced later, the waste solvent from the previous batch is used in place of virgin solvent. Waste solvents can also be collected and recycled by distillation, either on-site or off-site. Because of the current costs of solvent disposal, on-site distillation can be economically justified for as little as 8 gallons of solvent wastes generated per day.[1]

3.3. Good Housekeeping and Operating Practices

Scheduling paint production for long runs or cycles from light to dark colors can significantly reduce the amount of solvent wastes generated, in that it minimizes the frequency of equipment cleaning. An immediate cleaning after use can reduce the amount of paint drying in a tank and thus reduce cleaning frequency. Continuous cleaning operations without any "wait time" – which allows paint to dry – is essential in reducing solvent wastes and/or caustic solution wastes.

The use of dry cleanup methods should be maximized through operator training and by closing floor drains to reduce the amount of wastewater generated. Other effective ways to reduce the amount of wastewater are the use of volume-limiting hose nozzles, use of recycled water for clean-ups, and actively involved supervision.

The most effective way of reducing wastes associated with bags and packages is to segregate hazardous materials from nonhazardous materials. For example, empty bags and packages containing residual amounts of hazardous materials should be stored in plastic bags to eliminate dusting, which can cause contamination of nonhazardous materials.

By installing a separate dedicated baghouse for each production step, the collected pigment dust or resin dust can be recycled to the process step. However, the cost of an additional baghouse installation should be considered against the benefit of recycling the dusts.

4. CASE STUDY[5]

This pollution prevention case study of a paint manufacturing plant, like that presented in Chapter 15, is not based on an existing or real plant. The case study of this facility, Plant B, was also constructed from many different case studies of existing plants in order to illustrate many possible pollution prevention options available to the paint manufacturing industry.

An overview of the plant operation is given first, followed by a description of the processes it uses to manufacture paint. Potential pollution prevention options are discussed, along with an economic analysis for some of the options.

Plant B produces a wide variety of industrial coatings. Approximately 70% of the coatings produced are solvent based; the rest are water based. The solvent-based paints include pigmented nontints, pigmented tints, lacquer thinners, unpigmented paints (clears), and stains. The water-based paints are emulsion paints (i.e., latexes). The total annual production rate for Plant B is 2 million gallons.

4.1. Process and Waste Description

4.1.1. Solvent-Based Paints

The block flow diagram of the solvent-based paint manufacturing process in Plant B is shown in Fig. 19.2. Production begins with dispersing the pigments in a roll mill or in sand mills. The sand mills are either horizontal or vertical and use sand/glass/steel beads to disperse the pigments in a small quantity of solvent-resin mixture. Typically, the dispersion is carried out in batches of 30 or 55 gallons. After passing through the mill, the mixture of pigments and solvent-resin is collected in another container or an intermediate storage tank and sent to let-down tanks. Another container is used to allow multiple passes through the mill since it is sometimes necessary to pass the mixture through the mill up to three times to obtain the required degree of dispersion. The containers are cleaned after each pass.

The let-down step consists of filling the mixing tank with the primary dispersions (from the dispersion step), solvents, plasticizers, and other additives. Mixing tanks are either portable or stationary. The solvents are pumped into the tanks using a solvent metering system, and the contents are then mixed. When the tank contents reach the proper viscosity, color, and gloss, mixing is stopped and the contents are filtered and dispensed into product containers. Filtration is accomplished by using cartridge filters. Batch sizes are 55, 100, 200, 300, or 500 gallons for portable tanks, and 1000, 1500, 3000 gallons for stationary tanks. Stationary tanks are usually dedicated to one product and no cleaning is required between batches.

4.1.2. Water-Based Paints

The process block flow diagram of the water-based paint manufacturing process in Plant B is shown in Fig. 19.3. As for solvent-based paints, the first step in water-based paint production is the dispersion of pigments. The pigments in emulsion or slurry form, along with water, resins, and additives, are added directly to a mill in the primary dispersion step. The dispersed material from the mill is then pumped directly to the let-down tanks. In the let-down step, the dispersed pigments from the milling operation are mixed with additional water, resins, and additives. Mixing tanks are either portable or stationary. The additives consist of bactericides, fungicides, surfactants, defoamers, or extenders. The bactericides and fungicides used are mercury based. Solvents such as diethylene glycol and propylene glycol are added to water-based paints to extend the drying time and act as an antifreeze in cold climates.

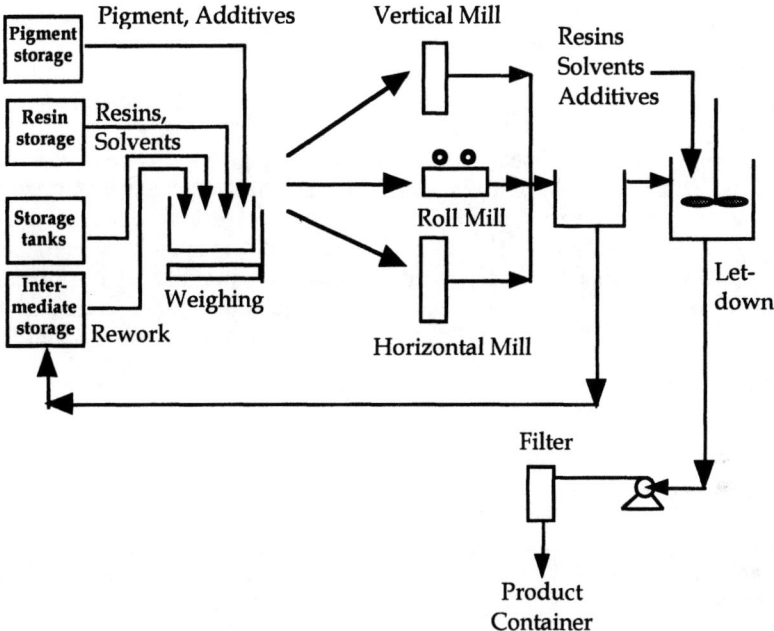

Figure 19.2. Block flow diagram of the solvent-based paint process in Plant B.

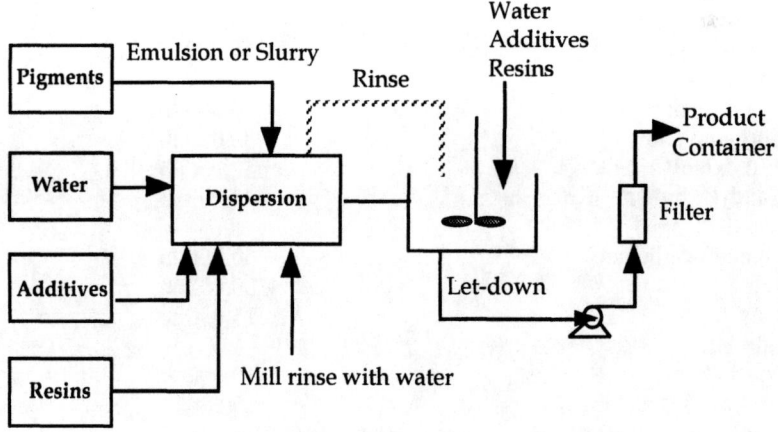

Figure 19.3. Block flow diagram of the water-based paint process in Plant B.

The stationary let-down tanks have a capacity greater than 400 gallons, whereas portable tanks have a 50- to 400-gallon capacity. About 30 percent of the batches are processed using portable tanks. When the properties of the batch meet the specifications (e.g., viscosity, color, gloss, etc.), mixing is stopped and the mixing tank contents are pumped through cartridge filters to the filling unit.

The raw materials used in Plant B include pigments, resins, various solvents, extenders, and additives. Solvents used in the solvent-based paint process include methanol, methyl ethyl ketone (MEK), toluene, lacquer thinner, and mineral spirits. These solvents are purchased in bulk or in drums.

Powdered pigments are delivered in plastic or paper bags, and slurried pigments in drums. Slurried pigments are used predominantly for water-based paint formulations. A list of raw materials and their uses are summarized in Table 19.1.

Table 19.1. Raw Materials Used in Plant B.

Raw Material	Annual Consumption (Gallons)
Solvents	
Methanol	40,000
Methyl ethyl ketone	150,000
Toluene	350,000
Lacquer thinner (blend)	170,000
Mineral spirits	100,000
Isobutyl isobutyrate	35,000
Resins	
Alkyds	60,000
Acrylics	30,000
Vinyl-acrylics	15,000
Pigments	
Titanium dioxide	360,000 lb
Chrome yellow	30,000 lb
Red oxide	52,000 lb
Vandyke brown	45,000 lb
Extenders	
Calcium carbonate	50,000 lb
Talc	130,000 lb
Clay	25,000 lb
Additives	
Plasticizers	15,000
Bactericides and fungicides	5,000
Viscosity modifiers	2,000
Drying oils	20,000

The principal waste streams generated by Plant B include the following:

- Equipment-cleaning wastes
- Obsolete stock
- Returns from customer
- Off-specification products
- Spills
- Spent cartridge filters
- Empty bags and packages

Equipment (such as mills and mixing tanks) is cleaned after each batch to prevent cross-contamination. Unusable containers (e.g., drums and pails) are cleaned before sending them to off-site reclamation. Mills are cleaned with a solvent or water that will be used in the formulation of the next batch. Portable tanks are cleaned with a commercially available caustic cleaning solution. The cleaning solution is recirculated, and the blow down (purge) is sent to wastewater treatment (flocculation and pH adjustment). Stationary tanks are usually dedicated to a single product formulation and frequent cleaning is not done. If necessary, however, they are cleaned with the solvent (or water) used in formulation. The spent solvent is drummed and sent to an off-site recycler. Filling units are cleaned with solvents (predominantly MEK), and the waste solvents are drummed and sent to an off-site recycler. Equipment cleaning generates two waste streams: spent solvent from solvent rinsing operations and paint sludge from caustic cleaning.

Spills are usually cleaned by "dry" methods. Sawdust or sand is sprinkled on the spill, then drummed for disposal in a landfill. The spill area is then mopped with a thinner.

Spent filter cartridges are disposed of in a landfill. Empty pigment bags and packages are considered hazardous and are sent to a landfill for disposal.

During the most recent year of operation, about 200 tons of wastes (or 700 55-gallon drums) were sent to a landfill.

4.2. Pollution Prevention via Process Modifications

Rather than collecting the dispersed mixture and storing it in an intermediate storage or other container, the process can be modified to send the dispersed mixture directly to the let-down tanks. This modification will eliminate the use of an extra container and the wastes associated with cleaning the container.

The installation of more efficient mills that would not require multiple passing of the batch should be considered. This will eliminate the use of an extra container as well as the waste associated with cleaning the container. Another option is the installation of a direct recirculation loop on the existing setup. Additional pumps can be installed to pump the batch back to the mill inlet directly, thus avoiding the use of an extra tank.

Since used filter cartridges have to be sent to a landfill for disposal, the use of bag filters in place of cartridge filters should be considered. Although bag filters are more expensive, spent bag filters contain much less paint than spent cartridges and can be reused several times. Unreusable bag filters can easily be washed with solvent or water and dried before their disposal as nonhazardous waste. Wash solvent or water can be recycled along with other solvent wastes either on-site or off-site. Another option is the use of wire screens. Wire screens can be reused almost indefinitely when washed with a solvent or water.

Installing high-pressure nozzles to improve efficiency in cleaning the tanks should also be considered. The improved cleaning efficiency will result in a reduction in the generation of the equipment cleaning wastes.

Manual cleaning with spatulas can be done to remove clingage from portable tanks before cleaning with solvents or caustic solutions. The removed clingage can be recycled back to the process if the same paint is being formulated, or it can be sent to an off-site recycler.

Plant B uses lead and chromate pigments for making special primers. These can be replaced with less hazardous pigments. For instance, chrome yellow pigment can be replaced with organic pigments or yellow iron oxide. However, yellow oxide does not yield as bright a yellow color as chromate yellow, and customer acceptance may be a problem.

4.3. Pollution Prevention via Resource Recovery and Recycling

Flush solvent or water from mill cleaning can be mixed with the batch in the let-down step, eliminating the mill cleaning wastes. Filling unit cleaning solvent or water can also be reused in the let-down step for the same product or a similar product.

When customers return unused paint, it can be reformulated into other products. Off-specification products can also be reworked into other products. Since off-specification products are often generated as a result of the lack of good quality control, quality control should be tightened.

The increase in operating expense for reworking customer returns and off-specification products can be easily offset by the increase in revenues derived from the sale of reworked products. However, the avoided disposal costs can be more significant. At a disposal cost of $170 per drum, a total savings of $119,000 per year can be realized if all of the 700 drums that are sent to the landfill are reworked.

Cleanup solvents can be reused several times for rinsing tanks, thus minimizing the amount of solvents used for cleaning. When the rinse solvent is considered too dirty for direct reuse, it can be distilled on-site. The solvent recovered by distillation is recycled to the dispersion/let-down process or cleaning operation.

The solvent wastes typically sent to an off-site recycler can be recycled on-site by a distillation unit. For feasible on-site recycling, the following conditions must be met:

1. The distillation unit must meet the technical requirements for recovering the solvents.
2. The economics of on-site recovery must be favorable.

3. The unit must be proven to be an environmentally safer option (long- and short-term) than the currently used off-site recycling operation.

A simple economic analysis of an on-site distillation unit is presented in the example below. The assumptions made for the analysis are as follows:

1. The total amount of wastes sent to the off-site recycler is 250 tons (or 50,000 gallons at a density of 10 lb/gal).
2. The distillation unit has a capacity of 120 gallons, and the total capital cost is $51,000.
3. The installation cost of the unit is $4000.
4. The distillation unit has an 85% solvent-recovery efficiency.
5. The on-site distillation residue is to be incinerated at a cost of $200/ton.
6. The estimated operating/utility cost of the distillation unit is $35,000/yr.
7. The off-site recycling cost is $150/ton.
8. The off-site recycler charges extra for the disposal of distillation residues at $170/ton of wastes.

An economic feasibility study is summarized in Table 19.2. The rule of thumb is that if the payback period is less than 3 years, it is economically feasible. Therefore, the on-site distillation unit with a payback period of 1.5 years is economically feasible, based on the results of this economic analysis.

4.4. Pollution Prevention via Better Housekeeping and Operating Practices

Batch scheduling should be done in the order of light to dark paints to minimize the need for equipment cleaning. An effort should be made to have dedicated let-down tanks. This is especially effective in the production of white paints, thus minimizing the intermediate washing of the tanks.

In addition, when large batches of paints are formulated, the laboratory scale formulations should be repeated two or three times to ensure that the formulation is correct. This practice can lower the probability of a large volume batch being spoiled.

Computerized inventory control can improve raw material tracking and help identify raw material losses at an early stage. Computerized waste documentation and control systems can also help track the wastes in the process and identify the major sources of waste generation.

Cleaning the tanks immediately after use prevents scaling as a result of paint drying. Proper coordination between production and cleaning is essential to reduce scaling, which increases solvent or water usage by making cleaning more difficult. Dry cleaning methods for spills should be discouraged because spent solid adsorbent wastes cannot be reworked or recycled and need to be disposed of off-site.

Table 19.2. Economics of On-Site Distillation.

Distillation Unit Cost	
Capital cost of the unit	$51,000
Installation cost	4000
Total Installed Cost	$55,000
Current Annual Disposal Cost	
Recycling cost @ $150/ton and (@ 250 ton/yr of wastes)	$37,500
Surcharge for distillation residue (@ $170/ton of wastes)	42,500
Total Disposal Cost	$80,000
Annual Cost with On-Site Recycling	
Disposal cost of distillation residue (@ $200/ton of residue*)	$7500
Operating/utility cost	35,000
Total cost with on-site recycling	$42,500
Annual savings by on-site recycling = $80,000 - $42,500	
= $37,500/yr	
Payback period = $55,000/$37,500/yr	
= 1.5 years	

*The amount of distillation residue was assumed to be 15% of the total waste, since the unit has 85% efficiency.

Segregation of water-based and solvent-based waste improves the overall recycling efficiency of these wastes. Segregating the bags and packages that contain hazardous materials (e.g., lead, chromate, etc.) from those that do not contain hazardous materials can reduce the amount of wastes sent off-site for disposal.

5. SUMMARY

This chapter describes pollution prevention options available to the paint manufacturing industry, along with descriptions of the processes and wastes generated. Because equipment cleaning is the largest source of waste in the paint manufacturing industry, reduction in the frequency of equipment cleaning (e.g., mixing tanks) is the most effective approach to pollution prevention.

1. Most small paint manufacturing facilities produce paint in 10- to 500-gallon batches. For an average paint plant located in the U.S., 60% of its total annual production is solvent-based paints, 35% is water-based paints, and 5% is other allied products.

2. The production of solvent- and water-based paints involves: mixing of resins, dry pigment, plasticizers, and pigment extenders, along with

solvents or water, in a high-speed mechanical mixer, additional grinding and mixing, mixing of the paint base with tints, thinners, and additional resins, filtering, and packaging.

3. The major wastes generated by the paint manufacturing industry are empty raw material packages (containing trace amounts of raw materials), off-specification paint, equipment cleaning wastes, and spills.

4. Methods that can reduce the frequency of cleaning include:
- Use of rubber wipers to reduce the amount of paint left on the walls of a mix tank
- Use of Teflon-lined tanks to reduce adhesion and to improve drainage
- Use of a plastic or foam "pig" to clean pipes

5. The amount of wastewater or waste caustic solution generated may be minimized by employing the following methods:
- Use of high-pressure spray heads
- Use of a countercurrent rinsing sequence
- Encouragement of dry cleaning methods

6. Waste solvents from equipment cleaning can be collected and used in the next compatible batch as part of the formulation. Waste solvents can also be collected and recycled by distillation, either on-site or off-site. Because of the current costs of solvent disposal, on-site distillation can be economically justified for as little as 8 gallons of solvent wastes generated per day.

7. Scheduling paint production for long runs or cycles from light to dark colors can significantly reduce the amount of solvent wastes generated, in that it minimizes the frequency of equipment cleaning.

8. The facility, Plant B, evaluated in the case study produces a wide variety of industrial coatings with approximately 70% solvent-based, and the balance water based. The solvent-based paints include pigmented nontints, pigmented tints, lacquer thinners, unpigmented paints, and stains. The water-based paints are emulsion paints.

9. The dispersed paint mixture can be sent directly to the let-down tanks rather than collected and stored in an intermediate storage or other container.

10. The installation of more efficient mills that would not require multiple passing of the batch should be considered. This will eliminate the use of an extra container as well as the waste associated with cleaning the container.

11. Using bag filters in place of cartridge filters should be considered, since used filter cartridges have to be sent to a landfill for disposal.

12. Installing high-pressure nozzles to improve the cleaning efficiency of the tanks should also be considered, as well as manual cleaning with spatulas to remove clingage from portable tanks before cleaning with solvents or caustic solutions.

13. Pigments that contain lead and chromate can be replaced with less hazardous pigments.

14. Flush solvent or water from mill cleaning can be mixed with the batch in the let-down step, thus eliminating the mill cleaning wastes.

15. Customer returns and off-specification products can be reformulated into other products.

16. Cleanup solvents can be reused several times for rinsing tanks, thus minimizing the amount of solvents used for cleaning. When the rinse solvent is considered too dirty for direct reuse, it can be distilled on-site along with the solvent wastes typically sent to an off-site recycler. The on-site distillation unit is economically feasible for Plant B with a payback period of only 1.5 years.

17. Batch scheduling should be done in the order of light to dark paints to minimize the need for equipment cleaning.

18. When large batches of paints are formulated, the laboratory scale formulations should be repeated two or three times to ensure that the formulation is correct.

19. Computerized inventory control can improve raw material tracking and help identify raw material losses at an early stage.

20. Cleaning tanks immediately after use prevents scaling as a result of paint drying.

21. Segregation of water-based and solvent-based waste improves the overall recycling efficiency of these wastes. In addition, segregating the bags and packages that contain hazardous materials from those that do not contain hazardous materials can also reduce the amount of these wastes sent off-site for disposal.

6. REFERENCES

1. U.S. EPA, *Guides to Pollution Prevention: The Paint Manufacturing Industry*, Risk Reduction Engineering Laboratory, Cincinnati, Ohio, EPA/625/7-90/005, 1990.
2. Pace Company Consultants and Engineers, Inc., *Solvent Recovery in the United States 1980-1990*, Houston, Texas, 1990.
3. U.S. EPA, *Seminar Publication, Solvent Waste Reduction Alternatives*, Center for Environmental Research Information, Cincinnati, Ohio, EPA/625/4-89/021, 1989.
4. U.S. EPA, *Development Document for Proposed Effluent Limitation Guidelines, New Source Performance Standards, and Pretreatment Standards for the Paint-Formulating Point Source Category*, Office of Water and Waste Management, Washington, D.C., 4, 1979.
5. 3M, *A Compendium of 3P Success Stories*, Environmental Engineering and Pollution Control Department, St. Paul, Minnesota, 1989.

CHAPTER 20

Pesticide Formulating Industry

The pesticide formulating industry includes facilities that formulate and prepare agricultural pest control chemicals and pesticides, including insecticides, herbicides, and fungicides. These products are formulated from pesticide concentrates manufactured elsewhere and are sold to farmers in ready-to-use form.

Most pesticide formulating facilities are located near agricultural areas, and approximately 59% of these facilities are located in 10 states. The pesticide industry comprises about 330 establishments nationwide.[1]

There are two major steps in the production of pesticides for agricultural use. The first step is the *manufacturing* of the pesticide concentrate from basic chemical raw materials, which include petrochemicals, inorganic acids, and other chemicals. The second major step is the *formulation* and preparation of the pesticide in final ready-to-use form. Discussions in this chapter are focused mainly on the second step rather than the first, since the processes employed to manufacture the concentrated pesticides are too diverse to be covered in this text.

The agricultural chemicals industry produces pesticides and other agricultural chemicals (e.g., soil conditioners, etc.). In the U.S., over 600 different pesticides are produced. Most pesticides can be classified as either insecticides, herbicides, or fungicides, with many minor classifications. The major classes of pesticides and their approximate annual production rates are listed in Table 20.1.

Approximately 75% of all insecticides and herbicides and 66% of all pesticides are used for agricultural purposes; the balance are used in homes, gardens, and commercial and industrial properties.[2] The majority of pesticides are used on a few major crops. Presently, corn, cotton, and apples receive 67% of all insecticides used in agriculture.

Corn and soybeans receive 60% of the herbicides used, while 80% of the fungicides are applied to fruits and vegetables.[3]

1. PROCESS DESCRIPTION

There are three types of pesticide formulations: *solvent based, water based*, and *solid based.* In solvent-based formulations, the solvent serves as the carrier solution for the active pesticide ingredient. A solvent-water emulsion may also be used as the carrier liquid. Typical solvents used are light aromatics (e.g., xylene), chlorinated organic solvents (e.g., 1,1,1-trichloroethane), and mineral spirits. For water-based formulations, water serves as the carrier liquid for the active pesticide ingredients. The solvent-based and water-based formulations are applied directly as a liquid or sprayed as an aerosol.

Table 20.1. Pesticide Production in the United States.[4]

Product	Quantity Produced (tons/yr)
Insecticidal formulations	
Inorganic compounds	54,300
Organic compounds	206,750
Chlorinated hydrocarbons	18,900
Carbamates	78,400
Organophosphates	73,150
Biological (botanical, bacterial)	11,250
Other organics	25,050
Herbicide formulations	
Inorganic compounds*	N/A
Organic compounds	541,750
Phenoxy	101,400
Metal organic	9450
Triazine	97,250
Urea, amide, benzoic, other organics	333,150
Fungicide formulations	
Inorganic compounds*	N/A
Organic compounds	56,250
Other pesticidal formulations	
Fumigants	17,450
Defoliants and desiccants	3500
All others*	N/A

*Data not available.

Solid pesticide formulations are prepared by blending solid active ingredients with inert solids (e.g., clay, sand, etc.). Some solid pesticide formulations are prepared by adsorbing liquid active ingredients with solid carrier materials. Common solid formulations are dusts, wettable powders, granules, treated seed, bait pellets, and cubes.

Block diagrams of the steps involved in formulating liquid-based (solvent and water) pesticides and dry pesticides are shown in Figures 20.1 and 20.2, respectively.

The pesticide formulating process generally consists of blending operations in which the active ingredients are mixed with the inert ingredients (carriers). Milling and coating operations are also employed for granule and treated-seed productions to reduce the particle size.

Blending is accomplished with conventional blenders, which consist of tanks with mixers (for liquid formulations), or blending mills (for solid formulations). Additional equipment includes storage tanks, rotary kilns for curing solid formulations, pumps, hoppers, and conveyers. Mixing tank capacities range from less than 100 gallons to several thousand gallons.

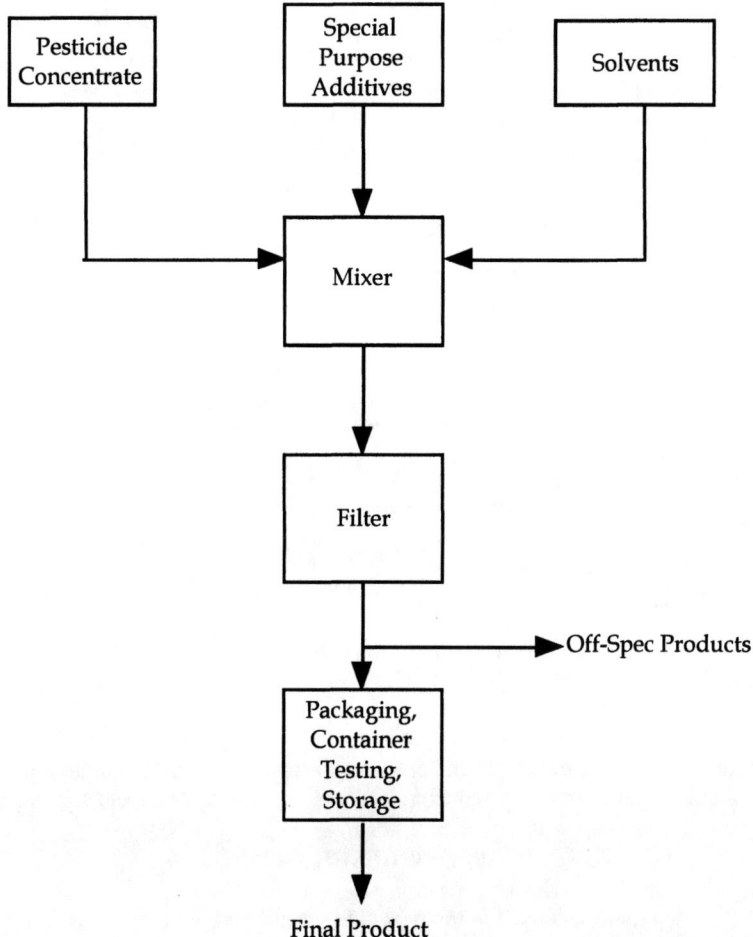

Figure 20.1. Block diagram for liquid pesticide formulation.

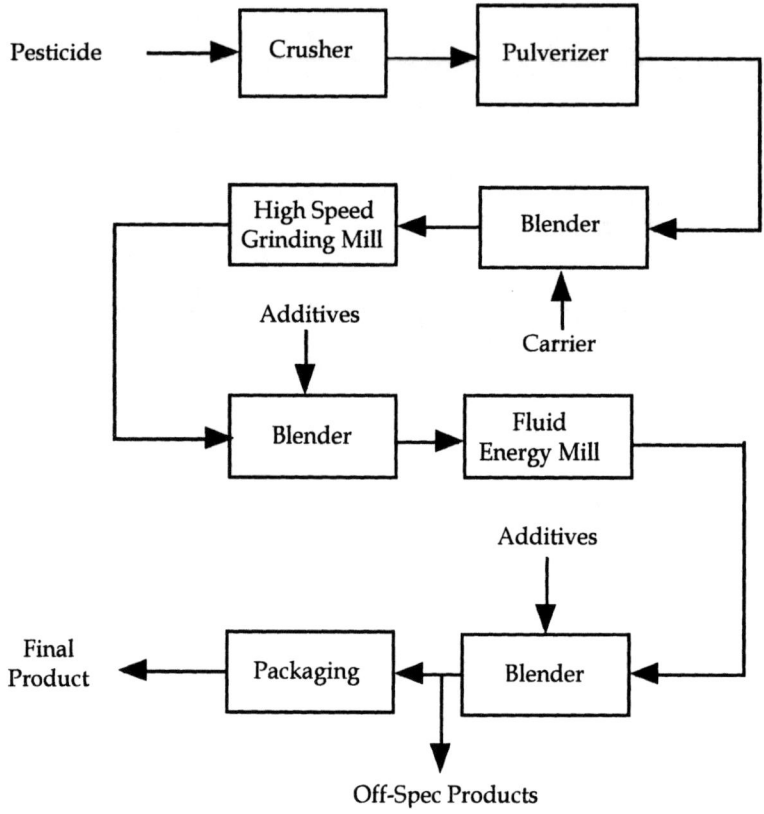

Figure 20.2. Block diagram for solid pesticide formulation.

Final packaging involves pouring liquid formulations into product containers (e.g., 55-gallon drums or glass bottles). For solid formulations, the product is generally gravity fed from hoppers into drums or paper bags.

2. WASTE DESCRIPTION

The raw materials used in the pesticide formulating industry are the active ingredients (organic and inorganic pesticides) and formulation and preparation materials. Formulation and preparation materials for dry formulations include organic flours, sulfur, silicon oxide, lime, gypsum, talc, pyrophyllite, bentonites, kaolins, attapulgite, and volcanic ash. For solvents in liquid formulations, xylenes, kerosenes, methyl isobutyl ketone, amyl acetate, and other chlorinated solvents are used. Carbon dioxide and nitrogen are used as propellants.[1]

Decontamination (or cleaning) of liquid pesticide mixing and storage tanks generates pesticide-contaminated solvents or water, depending on whether a solvent-based or water-based formulation is performed in a specific tank. Cleaning of the blending equipment also generates pesticide-contaminated wastes. Cleaning is typically performed by using high-pressure water hoses equipped with spray nozzles, using portable steam generators, or pumping a fixed volume of solvent through the formulating equipment. High-pressure water nozzles are also used to clean active ingredient containers.

Solid dust wastes are generated from the decontamination of solid-based pesticide blending mills. These solids contain mostly diluents or carriers (typically, clay for dust mills and sand for granule mills) with trace amounts of the active ingredients. Many dust or granule mills are equipped with vacuum systems to minimize fugitive emissions of dusts.

Wastes containing pesticides are also generated from the residual raw materials in the containers during the unloading of these materials to blend tanks or blending mills. Off-specification products, laboratory analysis wastes, and spills are also sources of pesticide-containing wastes.

Another source of wastewater is the hot water bath used to check the leakage of aerosol cans. Aerosol cans are immersed in a hot water bath, where bubbles are observed if the can has not been sealed properly. Leakage checks of aerosol cans are required by the U.S. Department of Transportation.

3. POLLUTION PREVENTION OPTIONS

Waste streams generated by the pesticide formulating industry include equipment cleaning wastes, off-specification products, spills and area washdowns, empty bags and drums, and water used for aerosol leak testing. Source reduction through product substitution is not a feasible option for the pesticide formulating industry because of the high level of effort and cost associated with registering a new pesticide with EPA. Such registration requires the development and submission of a health and ecological risk data set to EPA for review.

3.1. Process Modifications

One of the most concentrated wastes produced at a pesticide formulating facility results from the cleaning of the process equipment. Squeegees and wiper blades may be used to remove the residual material remaining in the mix tank and subsequently reduce the level of cleaning solution (solvent or water) contamination. Mixers with automatic wall scrapers are also available commercially.

High-pressure spray nozzles can be used in place of standard rinsing hoses to reduce the amount of wastewater generated. Other low volume, high efficiency systems include *water knives* and *portable steam cleaners*. Steam cleaning can be used in place of the "boil-out" cleaning procedure whereby a tank is filled with water and heated to boiling and refluxing to provide the cleaning.

Pigs can be employed to clean pipes between the mixing tank and the hopper. "Pigs" are propelled pipe inserts that push ahead any pesticide formulation left clinging to the pipe wall. Inert gas is used to propel the pig.

Dedicated vacuum systems can be used to clean spilled powders in facilities producing dry formulations. If a dedicated vacuum system is available, rather than being tied into the facility's main dust collection system, the collected spill can be reused or recovered. Dedicated dust collector systems may be employed to collect dust streams from different production lines so that the collected pesticide dust can be reused.

The manual opening and emptying of pesticide dust containers generates dust that must be collected. Use of an enclosed cut-in hopper can allow bags to be opened and emptied while minimizing the release of dust.

The volume of off-specification products can be minimized by employing strict *quality control* and process *automation*. Even though the formulation of pesticides is a relatively simple process, improved process automation and control will yield high quality products consistently, avoiding the generation of off-specification products resulting from operator errors.

3.2. Recycling and Resource Recovery

When cleaning is performed by passing a dry, inert material (sand or clay) through a system for dry formulation, this inert material can be saved and used in the next production run of the same product.

For liquid pesticide formulating facilities, cleaning is normally done by rinsing the equipment with the process solvent, followed by rinsing with water. Waste solvents can be saved and reused in the next batch of a similar or same pesticide formulation. The waste rinse water can be stored and reused as a first water rinse or in an initial cleanup of spills. In addition, off-specification products can be reformulated to an acceptable quality.

3.3. Good Housekeeping and Operating Practices

In addition to standard *good housekeeping and operating practices* such as waste stream segregation, employee training, better documentation, better material handling and storage, and material tracking/inventory control, production runs can be scheduled to maximize efficiency. Production runs of a given formulation should be scheduled together to reduce the need for equipment cleaning between batches.

Another major source of wastes, pesticide ingredient containers, can be cleaned for reuse or nonhazardous waste disposal. The containers can also be returned to the pesticide supplier for refilling with the same material. However, very few suppliers accept used drums from formulators. Therefore, formulators should consider the feasibility of receiving raw materials in returnable bulk containers such as the "tote bins" used by the paint manufacturing industry.

4. CASE STUDY

This case study[1] summarizes the results of a plant waste audit inspection and pollution prevention opportunity assessment during which sources of hazardous waste handling and disposal and waste minimization opportunities were reviewed. This facility is mainly engaged in the formulation and distribution of fertilizer, but also formulates a small quantity of liquid pesticides. The pesticides formulated are primarily insecticides consisting of malathion and diazinon in a xylene carrier solution. On an average annual basis, approximately 400 gallons of pesticide are formulated.

4.1. Process and Waste Description

The mixing of the active ingredients with the xylene is performed in a 40-gallon stainless steel mixing tank equipped with a portable mixing propeller. The active ingredients and the xylene are received at the plant in 55-gal drums, and are added to the mixing tank by pumping. Following mixing, the mixing tank is raised by a forklift, and the formulated product is drained through a spigot on the side of the tank into a 55-gal drum. The drum is then transported to the packaging area where the pesticide is pumped into 8-ounce bottles for distribution.

The only sources of hazardous waste generation at this plant relating to the pesticide formulating operation are rags used to wipe down the 40-gal mixing tank and small spills following formulation of each batch. An estimated 10 to 20 lb of pesticide-contaminated rags are generated annually. The rags are disposed of in the dumpster and ultimately in a local sanitary landfill.

4.2. Audit Findings

The pesticide-contaminated rags generated by cleaning the mixing tank and small infrequent spills are possibly classified as a hazardous waste, depending primarily on the quantity of pesticide in the rags. Sampling and analysis of the rags to determine the content of pesticide and carrier solvent in the rags would be required to determine if their content exceeds the hazardous waste criteria set forth by the state in which the plant resides. If the rags are determined to be a classified hazardous waste, they should not be disposed of in a sanitary landfill and steps should be taken to reduce the quantity of this material generated during pesticide formulation.

4.3. Pollution Prevention via Better Housekeeping and Operating Practices

One alternative is to collect the rags in a 55-gal drum for ultimate disposal at an approved hazardous waste disposal facility. If the rags are not accumulated longer than 90 days prior to disposal, then a hazardous waste storage facility permit will not be required. However, the drum must be stored and transported in accordance with the requirements for generators of hazardous waste. If the plant generates less than 100 kg of waste during any calendar month, the 90-day accumulation time limit identified above would begin when the amount of pesticide-contaminated rags reached 100 kg. At a maximum generation rate of approximately 20 lb or 9 kg of rags/yr, the rags could be accumulated up to 11 years before the small quantity exclusion limit of 100 kg would be exceeded. Assuming that up to 220 lb of contaminated rags could be placed in a 55-gal drum, only one drum would be required to store the rags over the 11-year period. However, because of the future landfill disposal ban of hazardous wastes and the potential risk associated with possible spontaneous combustion of the rags due to solvent carrier solution content, disposal of the rags on a more frequent basis would be prudent. Additionally, the above 100 kg limitation is reduced to 1 kg for waste classified as extremely hazardous. The current disposal cost for one drum is approximately $200.

As the costs for hazardous waste disposal increase, an alternative method of cleaning the pesticide mixing tank that generates a reusable product may be considered. An example of such an alternative is the use of a small pressure washer to apply carrier solution and a squeegee to clean the tank, with collection of the resulting spent solution for reuse in subsequent formulations. This alternative also includes the placement of a shallow stainless steel pan underneath the mixing tank to catch any small spills which may occur during formulation or transfer operations. The pan should be slightly elevated and equipped with a spigot, so spills could be collected for reuse in pesticide formulation. The capital and additional operating costs to implement this alternative are estimated to be approximately $200 and $150/yr, respectively. These costs estimates are based on the following assumptions: capital cost of catchment pan and portable pressure washer = $200, additional operational requirement of 10 labor hours per year for tank cleaning, labor rate of $15/hr.

5. SUMMARY

This chapter describes pollution prevention options available to the pesticide formulating industry, along with descriptions of the processes and wastes generated. Source reduction through product substitution is not a feasible option for the pesticide formulating industry because of the high level of effort and cost associated with registering a new pesticide with EPA.

1. There are three types of pesticide formulations: solvent based, water based, and solid based.
2. The waste streams generated by the pesticide formulating industry include equipment cleaning wastes, off-specification products, spills and area washdowns, empty bags and drums, and water used for aerosol leak testing.
3. Squeegees and wiper blades may be used to remove the residual material remaining in the mix tank and subsequently reduce the level of cleaning solution (solvent or water) contamination.
4. High-pressure spray nozzles can be used in place of standard rinsing hoses to reduce the amount of wastewater generated, while steam cleaning can be used in place of the "boil-out" cleaning procedure.
5. "Pigs" can be employed to clean pipes between the mixing tank and the hopper.
6. Dedicated vacuum systems can be used to clean spilled powders for facilities producing dry formulations.
7. Use of an enclosed cut-in hopper can allow bags to be opened and emptied while minimizing the release of dust.

8. The volume of off-specification products can be minimized by employing strict quality control and process automation.

9. Waste solvents can be saved and reused in the next batch of a similar or same pesticide formulation.

10. Waste rinse water can be stored and reused as a first water rinse or for an initial cleanup of spills.

11. Off-specification products can be reformulated to an acceptable quality.

6. REFERENCES

1. U.S. EPA, *Guides to Pollution Prevention: The Pesticide Formulating Industry*, Risk Reduction Engineering Laboratory, Cincinnati, Ohio, EPA/625/7-90/004, 1990.
2. Dillon, A. P., ed., *Pesticide Disposal and Detoxification Processes and Techniques*, Noyes Data Corp., Princeton, New Jersey, 1981.
3. Dahlston, D. L., Pesticides in an Era of IPM, *Environment* 25(10):45-54, 1990.
4. U.S. Department of Commerce, *1982 Census of Manufacturers*, Bureau of Census, Washington, D.C., 1985.

CHAPTER 21

Pulp and Paper Industry

The paper and pulp industry ranks third in terms of fresh water withdrawal, after the primary metals and the chemical industries. It is believed that the paper and pulp industry will eventually be the largest manufacturing user of water in the U.S.[1] It ranks fifth among the major industries in its contribution to the water pollution problem.[2] Therefore, in this chapter the pollution prevention options available to the paper and pulp industry will focus on reducing wastewater generation.

The principal waste parameters of concern to the paper and pulp industry are biochemical oxygen demand (BOD), pH, and total suspended solids (TSS) in all categories of manufacturing. For groundwood, chemimechanical, and thermochemical operations, the heavy metal ion zinc is of concern. For certain types of pulp manufacture (e.g., Kraft, soda, and neutral sulfite semichemical pulping), the color of effluent is also of concern.[3]

The dominant pulping process currently used in the world is the Kraft process, largely because it can produce a strong pulp from a wide variety of species. Seventy-six percent of the pulp produced in the United States in 1982 was made by the Kraft process. The process also has an excellent chemical recovery system, which is important in the paper and pulp industry because of the high cost of chemicals.

1. PROCESS DESCRIPTION

A block diagram of an idealized paper mill process that uses bleached Kraft pulp is shown in Figure 21.1; typical effluent loadings are also included. This type of mill normally produces 500 to 1000 tons of product per day. The processes are divided into five separate areas:

1. Wood yards
2. Pulping
3. Recovery
4. Bleaching
5. Paper manufacturing

The largest volume of discharge comes from the paper mill, and almost as much from the bleaching plant. Therefore, volume reduction programs should be focused on these two sources. The highest BOD loads are produced by the bleaching plant, with the pulp mill second and the paper mill a close third. The paper mill produces the highest TSS. Pollution prevention plans should therefore be aimed primarily at the paper mill and the bleaching plant.

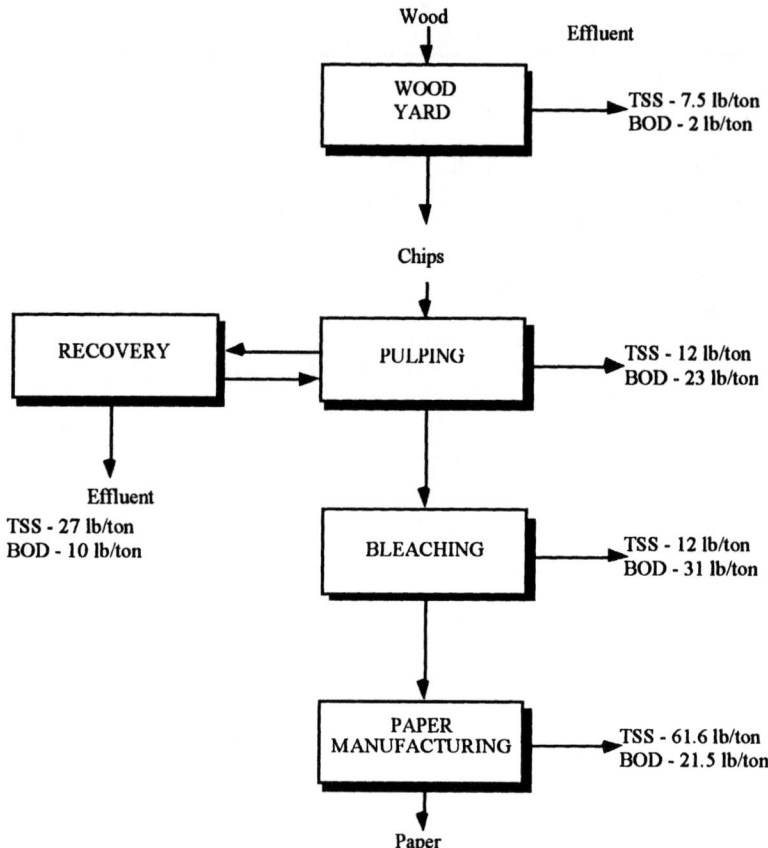

Figure 21.1. Representative bleached Kraft mill processes and pollutant loads.

A variety of pulping processes are employed in the industry; they include Kraft, sulfite, neutral sulfite semichemical (NSSC), chemimechanical, and groundwood processes. Each process is different, producing wastes of varying natures. In general, the amount of BOD_5 (5-day biochemical oxygen demand) discharged is inversely proportional to the pulp yield. The higher the process yield, the less BOD_5/ton of production, since the major source of BOD is the escape of spent liquor, which contains all of the organic matter dissolved during the pulping process. The lower the yield, the more materials present in the liquor, and the lower the amount of pulp produced from a given amount of wood.

Pulp is essentially ground-up wood. A wide range of processes are employed to produce mechanical pulp. These include stone groundwood, chemigroundwood, refiner groundwood, thermochemical, and chemimechanical techniques. The groundwood and thermomechanical processes are those that produce the least amount of BOD (i.e., highest yield), while the chemical processes (e.g., calcium-based sulfite process) produce the greatest amount of BOD. The TSS released by various processes are a function of the efficiency of the equipment used to remove solids. The groundwood pulp process is used mainly in the manufacture of newsprint, toweling, tissue, wallpaper, and coated specialty papers. A block process diagram of a stone groundwood pulp mill is shown in Figure 21.2. In this process, the rejects from the screening operation are passed to a reject refiner and returned to the process.

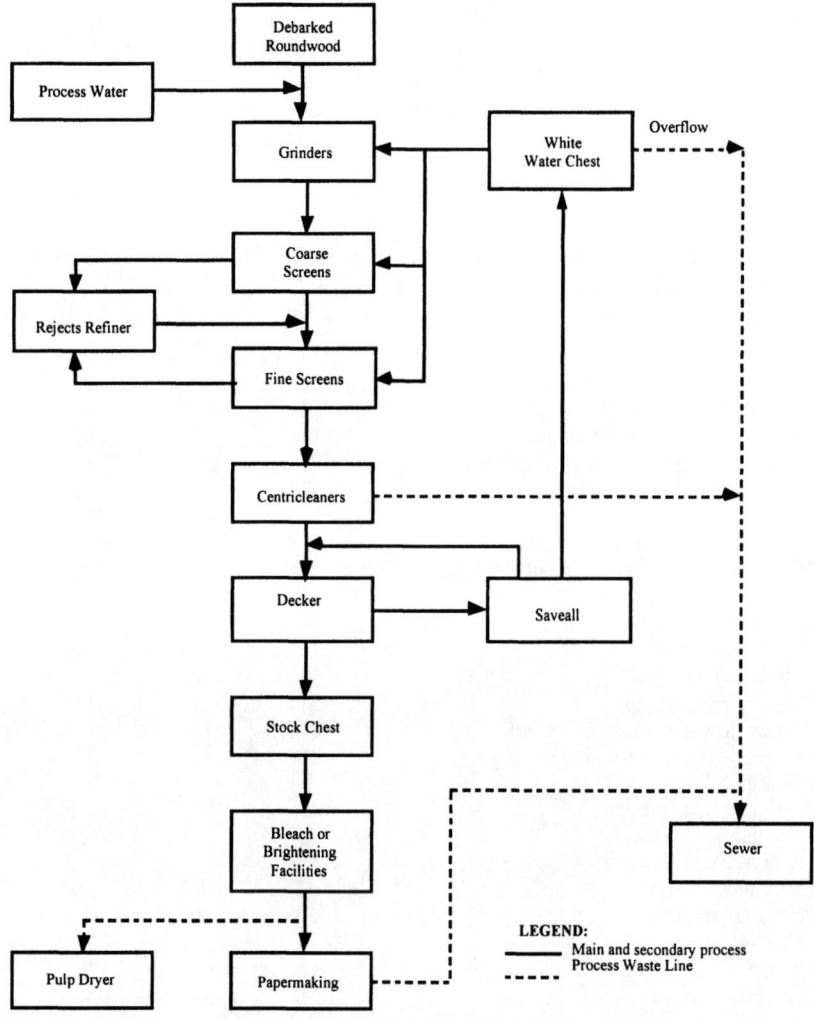

Figure 21.2. Stone groundwood pulp mill.

In the paper-making process, water is used to transport the stock and additives through the refining, cleaning, and paper-forming steps of the operation. There is no general trend in BOD load/ton of product for different types of paper-manufacturing processes. However, the lower the quality specifications for the product and the less sensitive it is to manufacture, the more internal process water reuse is possible, thus reducing the volume of the wastewater generated.

2. WASTE DESCRIPTION

The types of wastes from the paper and pulp industry are widely varied, depending on the types of manufacturing process employed. The major sources of wastes are unused (or residual amounts in containers) or spilled raw materials (e.g., bleaching chemicals, additives, pulping chemicals, etc.), overflows, wash-ups, and the pollutants present in wastewater. The raw wastewater loadings for different types of manufacturing are summarized in Table 21.1.

Table 21.1 Raw Wastewater Loadings for Different Pulping Processes.[3]

	Flow (kL/kkg)	BOD_5 (kg/kkg)	TSS (kg/kkg)
Bleached Kraft: dissolving pulp	241	55.0	113
Bleached Kraft: market pulp	171	40.0	71.5
Bleached Kraft: paperboard, coarse and tissue	151	37.9	70.5
Bleached Kraft: fine papers	133	32.4	82.0
Soda	144	43.3	142.5
Groundwood: chemimechanical pulp	113	95.5	52.0
Groundwood: thermomechanical pulp	99	28.0	48.5
Groundwood: fine papers	91	16.9	52.0
Groundwood: coarse papers, molded products, and newspapers	99	17.4	70.0
Sulfite: paper	220	126.5	89.5
Sulfite: market pulp	244	123	32.6
Sulfite: low alpha dissolving pulp	251	134	92.5
Sulfite: high alpha dissolving pulp	247	243.5	92.5
Deinking	104	82.5	178.5
Nonintegrated fine paper	63	10.8	30.8
Nonintegrated tissue	96	11.6	34.1
Nonintegrated tissue from waste paper	94	13.0	110.5
Unbleached Kraft	53	16.9	21.9
NSSC-ammonia	35	33.5	17
NSSC-sodium	43	25.2	12.3
Kraft-NSSC	58	19.4	12.3
Paperboard waste	30	11.2	--

3. POLLUTION PREVENTION OPTIONS

As mentioned earlier, the pollution prevention options available to the paper and pulp industry will be focused on the reducing wastewater generation rates. Among other pollution prevention options, process modification plays the most important role, and as such, it is emphasized in this section.

3.1. Process Modifications

3.1.1. Pulping Process Modifications

Pulp washer system efficiency is an important factor in reducing the amount of wastewater produced during pulp and paper production. The number of washers required depends on whether the pulping is carried out on a batch or continuous basis. Generally, continuous digesters employ *diffusion washing* as a final step. This final diffusion washer may be equivalent to one to three vacuum drum washers, the type commonly used in most mills. For batch processing, four or five stages of drum washers are common in a modern mill. A diffusion washer consists of a series of concentric baffles that move with the pulp as it flows through the washer body and return to their original position at the end of a wash cycle. The washing solution is forced hydraulically through the pulp across the baffles, providing a wash cycle at least 50 times as long as in a drum washer. The pulp is kept at high consistency during the wash cycle and is not exposed to air, thus eliminating odor problems and the need to use defoamers. Therefore, diffusion washer systems should be employed to maximize overall washing efficiency.

The operating conditions of a *drum washer* are also important in maintaining high efficiency in these washing units. The pulp mat that is formed should be uniform, and the showers must spray evenly across the surface. The optimum temperature of the spray water is 60°C to 70°C. Although the high water temperature decreases the liquor viscosity (the lower the viscosity, the higher the efficiency), a temperature that is too high increases the vapor pressure, interfering with the washer vacuum. The wastewater volume may also be reduced by using high-pressure, low-volume showers.

There are alternatives to drum washers and diffusion washers. One example is a horizontal, enclosed *belt washer*. This washer is operated in a pressure hood under positive pressure. Another example is the Impco Compaction Baffle Filter.[4] It is a slightly smaller unit than a drum washer and operates under pressure. Because it is fully enclosed, it is free of odor and foam problems. It operates at high feed consistency, and the volume of the liquor (or wastewater) handled is reduced to approximately 25% of the volume required by a drum washer.

Another general pollution prevention option applicable to the pump and paper sector is to use closed cycle mill processes. An example of a closed cycle bleached Kraft pulp mill is shown in Figure 21.3. This system is completely closed, and water is added only to the bleached pulp decker or to the last dioxide stage washer of the bleach plant. The bleach plant is countercurrent, and a major portion of the filtrate from this plant is recycled to the stock washers, after which it flows to the black liquor evaporators and then to the recovery furnace. The evaporator condensate is steam stripped and is used as a major water source at various points in the pulp mill. A white liquor evaporator is used to separate NaCl since the inlet stream to the white liquor evaporator contains a large amount of NaCl owing to the recycling of the bleach plant liquors to the recovery furnace.

3.1.2. Bleaching Process Modifications

The most commonly used chemicals for bleaching pulps are chlorine, sodium hypochlorite, calcium hypochlorite, chlorine dioxide, peroxide, and oxygen. Bleaching is usually performed in a number of stages to preserve the strength of the pulp by avoiding extremely strong chemical treatments. All stages of bleaching are typically carried out at greater than 10% consistency, with the exception of the chlorination stage which is usually carried out at 3% consistency. Therefore, a considerable amount of process water must be added and extracted before and after the chlorination stage. Thus, the chlorination stage plays an important role in wastewater pollution prevention efforts. Approaches to minimize the volume of wastewater are summarized in the following guidelines:[2]

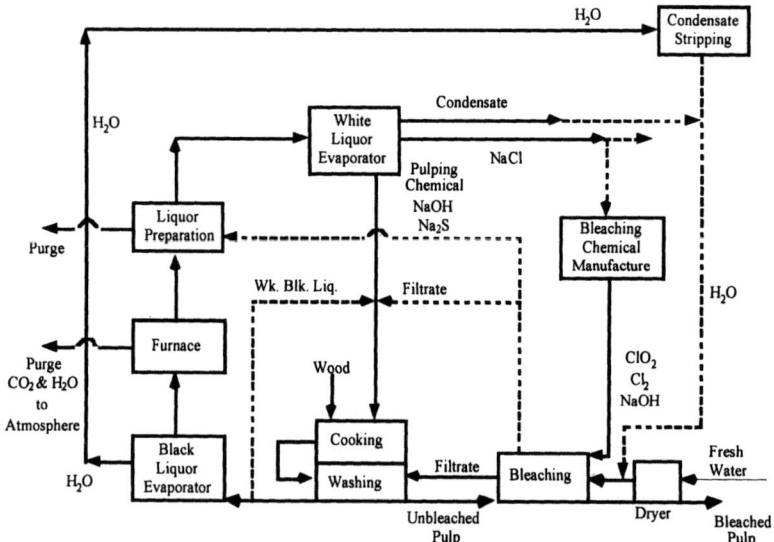

Figure 21.3. Process flow schematic of closed-cycle mill.

1. Replace the chlorination stage with an oxygen or ozone stage.
2. Recycle the chlorination stage process water.
3. Use water from the countercurrent washing system in the chlorination stage.
4. Perform high consistency gas phase chlorination.

Although using water directly from the countercurrent wash system produces the least volume of effluent, serious corrosion problems can occur if washers or piping in contact with chlorine dioxide filtrates are not made of corrosion-resistant materials.

3.2. Recycling and Resource Recovery

When paper-making stock is run on a paper machine wire, a certain amount of solids are passed through the wire, together with most of the water used for suspending the fiber. The water removed from the stock on the wire is known as "*white water*." White water can be recycled to reduce the volume of effluent from a paper mill.

Typically, white water contains fiber debris, small fibers, soluble matter, and a high percentage of nonfibrous suspended matter such as pigments, starches, and dyestuffs. The total amount of solid matter in the white water varies widely from mill to mill, depending on the quality of paper and the types of mill process employed. The richest white water is obtained from the trays under the first part of the wire. This rich white water can generally be returned directly to the mixing box, refiner, or other points as makeup water. A less rich white water is obtained from the pit under the wire. In some mills, the pit is divided into sections to collect water of different richness, with the richest white water used directly as makeup and any excess or less rich white water sent to saveralls.

The design of a closed water system with a white water return should follow the general guidelines listed below:

1. Use rich white water from the wire pit and couch pit in as many applications as possible.
2. Make maximum use of saveall clarified clear water. This water should be used for all machine showers that can tolerate water of this quality (e.g., wire showers, wire knockoff showers, headbox showers, and breast roll showers).
3. Consider better handling of the large-volume seal water used on water ring vacuum pumps.
4. Reduce the amount of fresh water used in felt showers.
5. Reduce the amount of fresh water used in gland seals and cooling water.

The large volume of seal water used on water ring vacuum pumps may be reduced by installing a separate recirculation system for the vacuum pumps. The use of saveall clarified water is not a practical option, unless the pH of the system is kept sufficiently low, because of pump corrosion problems. Use of vacuum pump seal water can be eliminated entirely by replacing the water ring pumps with high-speed exhausters or lobe-type pumps.[5]

The volume of water used in felt showers can be reduced by the use of high-pressure, low-volume showers. If each low-pressure cleaning shower in a mill is replaced by a high-pressure intermittent cleaning shower, an approximate 90% reduction in water volume can result.[2]

Many of the gland seals can be replaced by mechanical seals. The cooling water can be handled by a clear water sewer for the mill, in which all uncontaminated cooling water is collected and discharged untreated.

There are some potential operating problems that arise from recycling of the white water. The higher level of suspended solids can cause plugging problems in wire showers, headbox showers, felt filling, and lines. Suspended solids can also cause deposit formations that may be biological and/or non-biological in nature. Dissolved and colloidal matter presents a more serious problem than suspended solids. After long periods of running under conditions of low fresh water use (or of high recycling), problems of scale and corrosion are observed. With very high degrees of recirculation, the concentration of dissolved matter increases to a point where the chemistry of the paper-making system may be affected and difficulties with foam and sizing can occur.[2]

3.3. Good Housekeeping and Operating Practices

To minimize sewer losses and spills from pulping processes, it is recommended that a segregated system be developed to treat each sewer as a separate stream. Sewer systems can be classified in four categories:

1. Low suspended solids
2. High suspended solids
3. Strong wastes
4. Sanitary sewers

Low suspended solids streams are cooling water, bleaching effluent, and evaporator condensates that do not require clarification. High suspended solids streams are decker filtrate, paper machine effluent, and wood yard effluent. Strong waste sewers collect floor drains, overflows, spills, and strong condensates. The strong waste streams should be contained and reused if possible.

Waste generation resulting from spills can be minimized by following the guidelines listed below:[2]

1. Prevent spills by improving process controls.
2. Redirect overflow pipes to trenches leading to "contaminated" surge vessels for ultimate use.
3. Minimize overflows or spills by installing level controls in process and storage tanks.
4. Install redundant key pumps and other equipment to avoid losses caused by equipment failure and routine maintenance.

5. Install monitoring systems to alert operating personnel of potential and actual spills so that corrective action can be initiated promptly.
6. Provide a storage lagoon before the biological treatment system to accept long-term shock loads.
7. Train personnel to prevent spills whenever possible and to take immediate corrective actions when they occur.

4. CASE STUDY

4.1. Process and Waste Description

In silver reader paper manufacturing, several coats of solvent-based resins and a layer that contains silver are applied to the paper; the coated paper is passed through a series of high-temperature ovens for drying. Product quality depends very strongly on the drying temperature. In one particular plant, temperature control became a problem during the machine start-up phase. The amount of off-specification product generated was significant.

4.2. Pollution Prevention via Better Housekeeping and Operating Practices

The oven temperature at start-up was raised higher than normal. This extra heat simulated the normal operating condition, compensating for cooling that occurred during the downtime. The amount and duration of the temperature boost were a function of the duration of the downtime. In addition, a computer program/application was developed to control start-up temperature.

4.3. Benefits

The product wastes from the start-up phase were eliminated, with an annual savings of $533,200 in silver, paper, solvents, and labor. The capital cost involved in the project was $16,000. Annual solid waste generation was reduced by 137 tons, while air pollution was lowered by 53 tons.

5. SUMMARY

This chapter describes pollution prevention options available to the paper and pulp industry, along with descriptions of the processes and wastes generated. The paper and pulp industry ranks fifth among the major industries in its contribution to water pollution. Therefore, pollution prevention efforts should be focused on reducing wastewater generation in this industrial sector.

1. Paper and pulping processes are divided into five separate areas: wood yards, pulping, recovery, bleaching, and paper manufacturing.
2. A variety of pulping processes are employed in this industry; they include Kraft, sulfite, neutral sulfite semichemical (NSSC), chemimechanical, and groundwood processes.

3. The types of wastes from the paper and pulp industry are widely varied, depending on the types of manufacturing process employed. The major sources of wastes are unused or spilled raw materials (e.g., bleaching chemicals, additives, pulping chemicals, etc.), overflows, wash-ups, and the pollutants present in wastewater.

4. Diffusion pulp wash systems can be employed in place of drum washers to maximize overall washing efficiency.

5. Maintaining the temperature of the spray water at 60°C to 70°C can improve the rinse efficiency.

6. Employing a closed cycle mill process can minimize wastewater generation.

7. The volume of wastewater generated in the bleaching process can be minimized by
- Using rich white water from the wire pit and couch pit in as many applications as possible.
- Making maximum use of saveall clarified clear water.
- Considering better handling of the large volume of seal water used on water ring vacuum pumps.
- Reducing the amount of fresh water used in felt showers.
- Reducing the amount of fresh water used in gland seals and cooling water.

8. Employing a white water recycling system can reduce the volume of effluent from a paper mill.

9. The development of a segregated sewer system can minimize sewer losses and spills from the pulping process. Sewer systems can be classified into four categories: low suspended solids, high suspended solids, strong wastes, and sanitary sewers.

10. Waste generation caused by spills can be minimized by following the guidelines listed below:
- Prevent spills by improving process controls.
- Redirect the overflow pipes to trenches leading to "contaminated" surge vessels for ultimate use.
- Minimize overflows or spills by installing level controls in process and storage tanks.
- Install redundant key pumps and other equipment to avoid losses caused by equipment failure and routine maintenance, and install monitoring systems to alert operating personnel of potential and actual spills so that corrective action can be initiated promptly.
- Provide a storage lagoon before the biological treatment system to accept long-term shock loads.
- Train personnel to prevent spills whenever possible and to take immediate corrective actions when they occur.

6. REFERENCES

1. Yulke, S. G., O'Rouke, J. T., and Rudd, D. F., *Fundamental Concepts. Strategy of Pollution Control*, John Wiley & Sons, New York, New York, 1977.

2. Springer, A. M., *Industrial Environmental Control - Pulp and Paper Industry*, John Wiley & Sons, New York, New York, 1986.
3. U.S. EPA, *Development document for interim final and proposed effluent limitations guidelines and proposed new source performance standards for the bleached Kraft, groundwood, sulfite, soda, de-ink, and non-integrated paper mills segment of the pulp, paper, and paperboard mill point source category*, Office of Water and Wastewater Programs, Washington, D.C., EPA 440/1-76/047-A, 1976.
4. Sanjana, B.H. and Shockford, L.D., *Westvaco Starts Up New Brown Stock Washer*, Proceedings of the TAPPI Pulping Conference, Toronto, Ontario, pp. 75-61, 1982.
5. Coates, J.G., *Journal of TAPPI*, 51(8):95A, 1968.

APPENDIX A

Units, Conversion Factors and Mathematical Symbols

Contributing Author: Kerwin L. Somoza

This Appendix contains a brief history of the Metric System and what led to the development and adoption of the International System of Units (SI) in the United States. It includes a discussion of the defined seven base units and derived units which, when combined, form the "coherent system of SI units." Also included is a Glossary of Units used throughout the book as well as some selected Conversion Factors.

The Metric System

The need for a single worldwide coordinated measurement system was recognized more than 300 years ago. In 1670 Gabriel Mouton, Vicar of St. Paul in Lyons, proposed a comprehensive decimal measurement system based on the length of 1 minute of an arc of a great circle of the earth. In 1671 Jean Picard, a French astronomer, proposed the length of a pendulum beating seconds as the unit of length. (Such a pendulum would have been fairly easily reproducible, thus facilitating the widespread distribution of uniform standards.) Other proposals were made, but over a century elapsed before any action was taken.

In 1790, in the midst of the French Revolution, the National Assembly of France requested the French Academy of Sciences to "deduce an invariable standard for all the measures and weights." The Commission appointed by the Academy created a system that was, at once, simple and scientific. The unit of length was to be a portion of the earth's circumference. Measures for capacity (volume) and mass (weight) were to be derived from the unit of length, thus relating the basic units of the system to each other and to nature. Furthermore, the larger and smaller versions of each unit were to be created by multiplying or dividing the basic units by 10 and its multiples. This feature provided a great convenience to users of the system, by eliminating the need for such calculating and dividing by 16 (to convert ounces to pounds) or by 12 (to convert inches to feet). Similar calculations in the metric system could be performed simply by shifting the decimal point. Thus the metric system is a *base-10 or* decimal system.

The Commission assigned the name metre (which we now spell meter) to the unit of length. This name was derived from the Greek word metron meaning "a measure." The physical standard representing the meter was to be constructed so that it would equal one ten-millionth of the distance from the north pole to the equator along the meridian of the earth running near Dunkirk in France and Barcelona in Spain.

The metric unit of mass, called the *gram,* was defined as the mass of one cubic centimeter (a cube that is 1/100 of a meter on each side) of water at its temperature of maximum density. The cubic decimeter (a cube 1/10 of a meter on each side) was chosen as the unit of fluid capacity. This measure was given the name liter.

Although the metric system was not accepted with enthusiasm at first, adoption by other nations occurred steadily after France made its use compulsory in 1840. The standardized character and decimal features of the metric system made it well suited to scientific and engineering work. Consequently, it is not surprising that the rapid spread of the system coincided with an age of rapid technological development.

By the late 1860s, even better metric standards were needed to keep pace with scientific advances. In 1875, an international treaty, the "Treaty of the Meter," set up well-defined metric standards for length and mass, and established permanent machinery to recommend and adopt further refinements in the metric system. This treaty, known as the *Metric Convention*, was signed by 17 countries, including the United States.

As a result of the Treaty, metric standards were constructed and distributed to each nation that ratified the Convention. Since 1893, the internationally agreed to metric standards have served as the fundamental weights and measures standards of the United States.

By 1900 a total of 35 nations - including the major nations of continental Europe and most of South America - had officially accepted the metric system. Today, with the exception of the United States and a few small countries, the entire world is using predominantly the metric system or is committed to such use. As of December 31, 1997, the convention includes members from 48 countries.[1]

The International Bureau of Weights and Measures located at Sevres, France, serves as a permanent secretariat for the Metric Convention, coordinating the exchange of information about the use and refinement of the metric system. As measurement science develops more precise and easily reproducible ways of defining the measurement units, the General Conference on Weights and Measures – the diplomatic organization made up of adherents to the Convention – meets periodically to ratify improvements in the system. and the standards.

The International System of Units (SI)

In 1960, the General Conference on Weights and Measures adopted an extensive revision and simplification of the metric system. The name *Le Systeme International d'Unites* (International System of Units), with the international abbreviation SI, was adopted for this modernized metric system. Further improvements in and additions to the SI have been made by the General Conference since its inception in 1875.

In 1995 the General Conference on Weights and Measures decided to eliminate the class of supplementary units as a separate class in the SI.[2] Thus the SI now consists of only two classes of units, base units and derived units which, when combined, form what is called "the coherent system of SI units."[3]

Base units are seven base quantities assumed to be dimensionally independent upon which the SI is founded. The seven base units in the SI are the *meter* (length), *kilogram* (mass), *second* (time), *ampere* (electric current), *kelvin* (temperature), *mole* (amount of substance) and *candela* (luminous intensity). All are commonly used and are defined below.

Derived units are units which may be expressed in terms of the base units by means of a mathematical relationship such as multiplication and division. Some derived units have been given special names and symbols for convenience. These names and symbols may themselves be used to express other derived units.[4]

Seven Base Units[5]

Length - meter (m)

The *meter* (common international spelling, metre) is defined as the length of the path traveled by light in vacuum during a time interval of 1/299 792 458 of a second.

Mass - kilogram (kg)

The standard for the unit of mass, the *kilogram*, is a cylinder of platinum-iridium alloy kept by the International Bureau of Weights and Measures in Paris. A duplicate in the custody of the National Institute of Standards and Technology (NIST) in Maryland serves as the mass standard for the United States. This is the only base unit still defined by an artifact but discussions are currently underway to develop an accurate reproducible replacement.

Time - second (s)

The *second* is defined as the duration of 9 192 631 770 periods of the radiation corresponding to the transition between the two hyperfine levels of the ground state of the cesium-133 atom.

Electric current - ampere (A)

The *ampere* is defined as that constant current which, if maintained in two straight parallel conductors of infinite length, of negligible circular cross-section, and placed one meter apart in vacuum, would produce between these conductors a force equal to 2×10^{-7} newton per meter of length.

Temperature - kelvin (K)

The *kelvin* is defined as the fraction 1/273.16 of the thermodynamic temperature of the triple point of water.

Amount of substance - mole (mol)

The *mole* is defined as the amount of substance of a system which contains as many elementary entities as there are atoms in 0.012 kilogram of carbon-12. When the mole is used, the elementary entities must be specified and may be atoms, molecules, ions, electrons, other particles, or specified groups of such particles.

Luminous intensity - candela (cd)

The candela is defined as the luminous intensity, in a given direction, of a source that emits monochromatic radiation of frequency 540×10^{12} hertz and that has a radiant intensity in that direction of 1/683 watt per steradian.

Derived Units

Derived units include the m^2 (area), m^3 (volume), m/s (velocity), and the m/s^2 (acceleration). Derived units which have been given special names include the *newton* (force), the *joule* (energy), the *watt* (power), the *pascal* (pressure), and the *hertz* (frequency). There are also a number of electrical units such as the *Coulomb* (charge), *volt* (potential), *farad* (capacitance), *ohm* (resistance), *henry* (inductance), and *weber* (magnetic flux). The *radian* (plane angle) and *steradian* (solid angle) which were formerly in the category of supplementary units are now included in the category of derived units.

One of the major advantages of the SI is that larger and smaller units are given in powers of ten. In the SI a further simplification is introduced by recommending only those units with multipliers of 10^3. Thus for lengths in engineering, the *micrometer* (previously micron), millimeter and kilometer are recommended, and the *centimeter* is generally avoided.

The SI in the United States

In the United States, by passage of the Kassen Act in 1866, it was made "lawful throughout the United States of America to employ the weights and measures of the metric system in all contracts, dealings, or court proceedings."[4] In spite of this, the United States has not established a national policy of committing itself and taking steps to facilitate conversion to the metric system. This has led to a competitive disadvantage when dealing in international markets since world trade is increasingly geared toward the metric system.

In 1968 Congress authorized a three-year study of the nation's system of measurement with the passage of the Metric Study Act of 1968. The study emphasized the feasibility of adopting the SI in the United States. The study included testimony from consumers groups, business organizations, labor groups, manufacturers, and state and local officials.

In 1971 the Secretary of Commerce, in transmitting to Congress the results of the study, recommended that the U.S. change to predominant use of the metric system through a coordinated national program with a target time period of 10 years.

This led to the passage of the Metric Conversion Act in 1975. The purpose of this Act was to coordinate and plan the increasing use of the metric system in the United States, although it did not include a specific time period.

The Act also established the U.S. Metric Board. The Board was charged with "devising and carrying out a broad program of planning, coordination and public education consistent with other national policy and interests, with the aim of implementing the policy set forth in the Act."

In 1981, the Board reported to Congress that it lacked the necessary clear Congressional mandate to bring about a national conversion to the predominant use of the metric system. As a result, the Board was ultimately disestablished in 1982. The demise of the Board led to doubts about the United States' commitment to metrification. Public and private transition also slowed during this period.

In 1988, the Congress, recognizing the urgency of the conformance with international trade standards, amended the Metric Conversion Act of 1975 with the passage of Omnibus Trade and Competitiveness Act of 1988. The Act included new strong incentives for industrial metrification in the United States.

The Act changed the name of the National Bureau of Standards (NBS), which was formed in 1902, to the National Institute of Standards and Technology (NIST) and gave the NIST the added task of assisting industry to increase its competitiveness in global markets.

The Omnibus Trade and Competitiveness Act of 1988 made the metric system the preferred system of weights and measures for the United States in trade and commerce. It also required that all federal agencies use the metric system in their business related activities to the extent economically feasible by the end of fiscal year 1992.

The future of metrification in the United States will require the support of the federal government, private industry and the public, if it is to be fully implemented as the preferred system in trade and commerce. Conversion to the metric system is in the best interest of the Nation, particularly in view of the increasing importance of technology in American life and international trade.

Glossary of Units

Ao, A	angstrom unit of length
abs	absolute
acfm	actual cubic feet per minute
amb	ambient
app. MW, M	apparent molecular weight
atm	atmospheric
at. wt.	atomic weight
b.p.	boiling point
bbl	barrel
Btu	British thermal unit
cal	calorie
cg	centigram

Glossary of Units (continued)

cm	centimeter
cgs system	centimeter-gram-second system
conc.	concentrated, concentration
cc, cm^3	cubic centimeter
cu ft, ft^3	cubic feet
cfh	cubic feet per hour
cfm	cubic feet per minute
cfs	cubic feet per second
m^3, M^3	cubic meter
°	degree
°C	degree Celsius
°F	degree Fahrenheit
°R	degree Reamur, degree Rankine
ft	foot
ft-lb	foot pound
fpm	feet per minute
fps	feet per second
fps system	foot-pound-second system
f.p.	freezing point
gr	grain
g, gm	gram
h	hour
in.	inch
kcal	kilocalorie
kg	kilogram
km	kilometer
liq	liquid
L	liter
log	logarithm (common)
ln	logarithm (natural)
m.p.	melting point
m, M	meter
µm	micrometer (micron)
mks system	meter-kilogram-second system
mph	miles per hour
mg	milligram
ml	milliliter
mm	millimeter
mµ	millimicron
min	minute
mol wt, MW, M	molecular weight
oz	ounce
ppb	parts per billion
pphm	part per hundred million
ppm	parts per million
lb	pound

Glossary of Units (continued)

psi	pounds per square inch
psia	pounds per square inch absolute
psig	pounds per square inch gage
rpm	revolutions per minute
s	second
sp. gr.	specific gravity
sp ht	specific heat
Sp wt	specific weight
sq	square
scf	standard cubic foot
scfm	standard cubic feet per minute
STP	standard temperature and pressure
temp	temperature
wt	weight

Appendix A

Conversion Factors

LENGTH

TO CONVERT FROM	TO	MULTIPLY BY
m	cm	100
m	mm	1000
m	microns (μm)	10^6
m	angstroms (A)	10^{10}
m	in	39.37
m	ft	3.281
m	mi	6.214×10^{-4}
ft	in	12
ft	m	0.3048
ft	cm	30.48
ft	mi	1.894×10^{-4}

MASS

TO CONVERT FROM	TO	MULTIPLY BY
kg	g	1000
kg	lb	2.205
kg	oz	35.27
kg	ton	1.102×10^{-3}
kg	grains	1.543×10^4
lb	oz	16
lb	ton	5×10^{-4}
lb	g	453.6
lb	kg	0.4536
lb	grains	7000

TIME

TO CONVERT FROM	TO	MULTIPLY BY
s	min	0.01667
s	h	2.78×10^{-4}
s	day	1.157×10^{-5}
s	week	1.653×10^{-6}
s	yr	3.171×10^{-8}

FORCE

TO CONVERT FROM	TO	MULTIPLY BY
N	kg-m/s^2	1
N	dynes	10^5
N	g-cm/s^2	10^5
N	lb$_f$	0.2248
N	lb-ft/s^2	7.233
lb$_f$	N	4.448
lb$_f$	dynes	4.448 x 10^5
lb$_f$	g-cm/s^2	4.448 x 10^5
lb$_f$	lb-ft/s^2	32.17

PRESSURE

TO CONVERT FROM	TO	MULTIPLY BY
atm	N/m^2	1.013 x 10^5
atm	kPa	101.3
atm	bars	1.013
atm	dynes/cm^2	1.013 x 10^6
atm	lb$_f$/in^2 (psi)	14.696
atm	mm Hg @ 0 °C (torr)	760
atm	in Hg @ 0 °C	29.92
atm	ft H$_2$O @ 4 °C	33.9
atm	in H$_2$O @ 4 °C	406.8
psi	atm	6.804 x 10^{-2}
psi	mm Hg @ 0 °C (torr)	51.71
psi	in H$_2$O @ 4 °C	27.70
in H$_2$O @ 4 °C	atm	2.458 x 10^{-3}
in H$_2$O @ 4 °C	psi	0.0361
in H$_2$O @ 4 °C	mm Hg @ 0 °C (torr)	1.868

VOLUME

TO CONVERT FROM	TO	MULTIPLY BY
m^3	L	1000
m^3	cm^3 (cc, mL)	10^6
m^3	ft^3	35.31
m^3	gal (US)	264.2
m^3	qt	1057
ft^3	in^3	1728
ft^3	gal (US)	7.48
ft^3	m^3	0.02832
ft^3	L	28.32

ENERGY

TO CONVERT FROM	TO	MULTIPLY BY
J	N-m	1
J	erg	10^7
J	dyne-cm	10^7
J	kW-h	2.778×10^{-7}
J	cal	0.2390
J	ft-lb$_f$	0.7376
J	Btu	9.486×10^{-4}
cal	J	4.186
cal	Btu	3.969×10^{-3}
cal	ft-lb$_f$	3.088
Btu	ft-lb$_f$	778
Btu	hp-h	3.929×10^{-4}
Btu	cal	252.0
Btu	kW-h	2.928×10^{-4}
ft-lb$_f$	cal	0.3240
ft-lb$_f$	J	1.356
ft-lb$_f$	Btu	1.285×10^{-3}

POWER

TO CONVERT FROM	TO	MULTIPLY BY
W	J/s	1
W	cal/s	0.2390
W	ft-lb$_f$/s	0.7376
W	kW	10^{-3}
kW	Btu/s	0.9486
kW	hp	1.341
hp	ft-lb$_f$/s	550
hp	kw	0.7457
hp	cal/s	178.2
hp	Btu/s	0.7074

HEAT CAPACITY

TO CONVERT FROM	TO	MULTIPLY BY
cal/g°C	Btu/lb-°F	1
cal/g-°C	kcal/kg-°C	1
cal/g-°C	cal/gmol-°C	Molecular Weight
cal/gmol-°C	Btu/lbmol-°F	1
J/g-°C	Btu/lb-°F	0.2389
Btu/lb-°F	cal/g-°C	1
Btu/lb-°F	J/g-°C	4.186
Btu/lb-°F	Btu/lbmol-°F	Molecular Weight

CONCENTRATION

TO CONVERT FROM	TO	MULTIPLY BY
$\mu g/m^3$	lb/ft^3	6.243×10^{-11}
$\mu g/m^3$	lb/gal	8.346×10^{-12}
$\mu g/m^3$	gr/ft^3	4.370×10^{-7}
gr/ft^3	$\mu g/m^3$	2.288×10^6
gr/ft^3	g/m^3	2.288
lb/ft^3	$\mu g/m^3$	1.602×10^{10}
lb/ft^3	$\mu g/L$	1.602×10^7
lb/ft^3	lb/gal	7.48

VISCOSITY

TO CONVERT FROM	TO	MULTIPLY BY
P (poise)	g/cm-s	1
P	cP (centipoise)	100
P	kg/m-h	360
P	lb/ft-s	6.720×10^{-2}
P	lb/ft-h	241.9
P	lb/m-s	5.6×10^{-3}
lb/ft-s	P	14.88
lb/ft-s	g/cm-s	14.88
lb/ft-s	kg/m-h	5.357×10^3
lb/ft-s	lb/ft-h	3600

References

1. Bureau International des Poids et Mesures (BIPM), *Le Systeme International d'Unites (SI), The International System of Units (SI)*, 7th ed., Sevres, France, (English version), 1998.
2. Taylor, B. N., *The International System of Units (SI)*, National Institute of Standards and Technology, Gaithersburg, Maryland, NIST Special Publication No. 330, 1991.
3. Taylor, B. N., *Guide for the Use of the International System of Units (SI)*, National Institute of Standards and Technology, Gaithersburg, Maryland, NIST Special Publication No. 811, 1995.
4. Carver, G. P., *A Metric for Success*, U.S. Department of Commerce, National Institute of Standards and Technology, Gaithersburg, Maryland, 1994.

APPENDIX B

State Pollution Prevention Programs
Contributing Author: Roberto Diaz

This appendix provides information and details on state activities regarding pollution prevention programs. All information is presented in the following format (where available):

State: State pollution prevention web site address
Contact
e-mail address (E)
Postal address
Telephone number (T)
Fax number (F)

Following this information, details on the state's pollution prevention programs are provided. If no information for a particular state is listed, then it was not available at the time of the preparation of this text.

Alabama:

Gary L. Ellis, Chief
E: gle@adem.state.al.us
Alabama Department of Environmental Management
P.O. Box 301463
Montgomery, AL 36130-1463
T: (334) 213-4303
Pollution Prevention Programs: Alabama Department of Environmental Management stresses pollution prevention in the form of awards, on-site assessments, P2 fact sheets, training, presentations, workshops, and seminars. Waste Reduction and Technology Transfer (WRATT) Foundation: provides free, confidential, non-regulatory, non-binding waste reduction opportunity assessments to businesses or other entities upon request.

Alaska: www.state.ak.us/local/akpages/ENV.CONSERV/dsps/compasst/ptnrshp.htm

David Wigglesworth
E: DWiggles@envircon.state.ak.us
Alaska Department of Environment Conservation
Pollution Prevention Office
555 Cordova Street
Anchorage, Alaska 99501-2617
T: (907) 269-7582
F: (907) 269-7600

Green Star: program which promotes the benefits of pollution prevention and energy conservation and recognizes businesses, schools, and other organizations which implement a series of waste reduction, recycling, reuse, and resource conservation activities.

Alaska Materials Exchange: an information clearinghouse to help Alaska businesses reuse materials and find alternatives to throwing valuable materials into local landfills.

Oil & Gas Pollution Prevention Committee: works to prevent pollution by conducting oily water recycling evaluations, publishing a P2 opportunity training guide, recycling lead-acid batteries, documenting pollution prevention success stories, and preparing product substitution guide for cleaning solvents.

Interagency Pollution Prevention (IPP) Initiative: a cooperative effort between several state and federal agencies whose objectives are to: improve communication between agencies; increase focus on environmental results and prevention; share resources for pollution prevention projects; improve efficiency and limit duplication of efforts; develop a unified mission/approach to pollution prevention; and incorporate pollution prevention into agency programs.

Pitstop: a voluntary program to help automotive service shops reduce, recycle and properly manage solid and hazardous waste.

Federal Facility Partnership: promotes a series of projects that help prevent pollution through training programs for a majority of state agencies.

Recycling Market Development: program which concentrates on pollution prevention by increasing in-state markets for recycled materials including the production of medium density fiberboard and the use of crushed glass.

Arizona: www.adeq.state.az.us/waste/p2/index.htm

Sandra Eberhardt, Pollution Prevention Unit Manager
E: eberhardt.sandra@ev.state.az.us
3033 North Central Avenue
Pollution Prevention Unit (7th Floor)
Phoenix, AZ 85012-2809

Pollution Prevention Partnership: a voluntary partnership between state government and industry which works together to expediate reducing hazardous waste in Arizona. Efforts include sharing information on successful reduction efforts and identifying and resolving impediments to pollution prevention.

P2 Publications/Fact Sheets/Newsletters: library providing pollution prevention techniques and strategies for the following industries: automotive, painting, fiberglass fabrication, printing, and wood furniture manufacturing.

Arkansas: www.adeq.state.ar.us/envpres/main.htm

John Giese, Chief
E: cox@adeq.state.ar.us
Environmental Preservation
P.O. Box 8913
Little Rock, AR 72209
T: (501) 682-0018
F: (501) 682-0010

Environmental Preservation Division: reviews a wide variety of project types within the state which have not yet been developed. This allows the pollution prevention concept to be considered and incorporated into many of these new projects while still in the planning stages and can prevent problems in later years.

Small Business Assistance Program: provides technical advice and support services to establish pollution prevention strategies. Also, provides free, limited engineering assistance to evaluate voluntary compliance and pollution prevention technologies.

California: www.dtsc.ca.gov/txpollpr.htm

E: techdev@cwo.com
California Environmental Protection Agency
Department of Toxic Substances Control
Office of Pollution Prevention and Technology Development
P.O. Box 806
Sacramento, CA 95812-0806
T: (916) 322-3670
F: (916) 327-4494

Incinerable Hazardous Waste Minimization Project (IHWMP): reduces the volume of incinerable waste being shipped off-site to California facilities for disposal, treatment, or incineration.

California Waste Exchange: attempts to bring together a waste generator with businesses that can use the waste as a product material.

Pollution Prevention Programs: stresses pollution prevention through publications, case studies, training, presentations, and workshops.

Colorado: www.sni.net/light/p3/

Parry Burnap
Colorado Department of Public Health and Environment
Pollution Prevention Unit
4300 Cherry Creek Drive South
Denver, CO 80222
T: (303) 692-2975
F: (303) 782-4969

Pollution Prevention Program: provides free, confidential on-site assessments, telephone consultations, industry-specific fact sheets and case studies, training programs and technical workshops, a resource library, presentations to trade and industrial organizations, program development and support for local governments and tribes, and grants for entities involved in providing P2 educational and outreach activities and technical assistance.

Pollution Prevention Forum: encourages and facilitates open communication among any and all individuals and companies interested in or involved with pollution prevention.

Pollution Prevention Partnership: a voluntary alliance of business, government, and public interest groups organized to develop and promote pollution prevention and waste minimization in industries.

Connecticut: dep.state.ct.us/deao/ca/assist.htm

Kim Trella, Program Coordination Supervisor
E: kim.trella@po.state.ct.us
State of Connecticut
Department of Environmental Protection
79 Elm Street
Hartford, CT 06106-5127
T: (860) 424-3234
F: (860) 424-4081

Compliance Assistance and Pollution Prevention Activities: consists of several activities including training, inspections, initiatives, and educational programs which prevent pollution in such fields such as automotive services, analytical laboratories, metal finishers, printers/publishers, government/ institutions, furniture refinishing, utility boilers, agricultural applications, machine shops, pesticide applications, and construction/excavation.

Hartford Neighborhood Environmental Project: addresses environmental concerns and identifies pollution prevention opportunities in African-American and Hispanic neighborhoods.

Pollution Prevention for Printers: consists of a training package focusing on pollution prevention opportunities and implementation in the print shop.

Integrated Pest Management (IPM): provides training to commercial lawn care companies to discourage the use of routine applications of pesticides.

Pollution Prevention Awards: includes the Governor's Awards for Environmental and Economic Progress which recognizes outstanding businesses and individuals for their advancements in pollution prevention.

Storm Water Pollution Prevention: developed storm water general permits which contain requirements for controlling and minimizing pollutants picked up by storm water.

Quinnipiac River Project: provides technical assistance to the public and local industry to prevent pollution in the Quinnipiac Rivershed.

Aquifer Protection Program: creates regulations which restrict land use in areas that feed ground water to the supply wells. Regulated businesses within the aquifer protection areas will be required to implement a series of P2 measures.

Emission Reduction Credit Trading Program: provides sources of emissions with an incentive to undertake early emission reductions and to experiment with innovative emission reduction strategies.

Delaware: www.dnrec.state.de.us/ppguide.htm

T: (302) 739-3822

Pollution Prevention Guides: pollution prevention fact sheets for various types of businesses including: auto repair shops, building construction and demolition, dry cleaning industry, food processors, metal fabricators, printing industry, and small chemical manufacturing operations.

P2-Education Program: provides education to children (K-8) on the advantages of pollution prevention.

Florida: www2.dep.state.fl.us/waste/programs/p2/index.htm

Julie Abcarian, Environmental Manager
E: abcarian_j@dep.state.fl.us
Pollution Prevention Program
Florida Department of Environmental Protection
2600 Blair Stone Road
Tallahassee, FL 32399-2400
T: (850) 921-9227

Technical Assistance: On-site technical assistance is one of the established services that the Pollution Prevention Program provides. The purpose of the visit is to help company staff look at inefficiency leading to pollution as a controllable factor in their business. The main goal during the visits is to provide the company with the ability to solve their future environmental problems in a more cost effective manner than traditional end of the pipe technology. This is done by working through different options for a current process or plant. The Retired Engineers segment of the program often lends a hand and valuable industrial experience.

Resource Center: The Pollution Prevention Resource Center (PPRC) provides information about pollution prevention techniques to anyone in Florida. The collection is focused on industrial processes. Fact sheets and success stories are available on-line. Most of the information is in printed form and may be requested free of charge within Florida. Videos are available for lending on various subjects.

County Training: provides compliance assistance staff with additional tools to help businesses and the environment benefit from pollution prevention options.

Georgia: www.ganet.org/dnr/p2ad/

Janice Hatcher, Information Manager
E: p2ad@ix.netcom.com
Pollution Prevention Assistance Division
Suite 450
7 Martin Luther King, Jr. Dr.
Atlanta, GA 30334-9004
T: (404) 651-5120 or 1-800-685-2443
F: (404) 651-5130

Technical Assistance: is offered to businesses, industries, and local and state government agencies in response to phone inquiries, electronic communications, mail, workshops, seminars, and on-site visits. Pollution prevention services include an information center with P2 publications and videos, on-site assessments, employee training sessions, seminars, presentations and workshops. Programs cover such areas as the manufacturing sector (toxic and hazardous waste, solid waste), the commercial and institutional sector, the agricultural sector, and the public sector (household hazardous waste, automotive fluids, radon).

Georgia Environmental Partnership (GEP): is comprised of the Pollution Prevention Assistance Division (P2AD), the Georgia Institute of Technology's Economic Development Institute (EDI), and the University of Georgia's Department of Biological and Agricultural Engineering (BAE). The GEP was formed as a mechanism to provide more coordinated environmental technical assistance to Georgia companies.

BOD/COD Project: cooperative program sponsored by the P2AD and Dalton Utilities on a project with the goal of eliminating or delaying the need to adopt industrial wastewater pre-treatment requirements for Dalton area industrial dischargers. A strategy of promoting voluntary water conservation and reduction of Biochemical Oxygen Demand (BOD) and Chemical Oxygen Demand (COD) in industrial effluents being discharged by Dalton area manufacturers is taken.

National Industrial Competitiveness through Energy, Environment, and Economics (NICE3) Grant Program: funding is provided to state and industry partnerships for the development and commercial deployment of innovative technologies which use less energy and generate less waste. Past research was applied to the carpet industry and the pulp & paper industry.

Farm*A*Syst: an interagency that provides Georgia's farmers with information and a voluntary means to become environmentally proactive in managing their farms.

Neighborhood Environmental Partnership: promotes the adoption of P2 practices by industries in the neighborhoods. It also provides a forum for both community residents and surrounding industry to discuss environmental and other issues of community concern and to provide a vehicle to address them.

Recognition Program: recognizes outstanding businesses and individuals for their advancements in the field of pollution prevention. Included in these awards are the Pollution Prevention Partners award which recognizes and rewards continuous pollution prevention performance and environmental improvement among Georgia industries and the Governor's Award for Pollution Prevention which recognizes special and significant pollution prevention projects from companies and organizations in Georgia.

Other Programs: Georgia also supports several other pollution prevention programs in other areas such as toxic waste reduction, solid waste reduction, household hazardous waste prevention, radon awareness, and agricultural fields such as animal waste management, bioconversion and by-product utilization, and horticulture.

Hawaii: www.state.hi.us/doh/eh/eiswmo01.htm

Office of Solid Waste Management
919 Ala Moana Blvd.
Honolulu, HI 96814
T: (808) 586-4240

Hawaii's Waste Minimization Project: consists of pollution prevention tips, case studies, and publications for the following industries: autobody painting, dry cleaning, laboratories, parts cleaning, photoprocessing, and printing.

Hawaii's Materials Exchange Program (HIMEX): an educational program encouraging residents and businesses to reuse, rather than discard unwanted items.

Idaho: www.state.id.us/deq/ptwo.htm#PPP

John A. Bernardo, Pollution Prevention Coordinator
E: jbernard@deq.state.id.us
Pollution Prevention Program
Department of Environmental Quality
1410 N. Hilton
Boise, ID 83706
T: (208) 373-0502
F: (208) 373-0169

Idaho Pollution Prevention Programs: consists of case studies, P2 recognition programs, site assistance visits and programs which provide pollution prevention techniques and strategies for the following industries: auto service and repair, agriculture, metal finishing, printing, etc.

Illinois: www.epa.state.il.us/p2/index.html

Illinois EPA - Office of Pollution Prevention #34
1021 North Grand Avenue East
P.O. Box 19276
Springfield, IL 62794-9276
T: (217) 782-8700
F: (217) 557-2125

Partners in Pollution Prevention(PIPP): partnership between Illinois EPA and businesses designed to encourage industrial facilities and others to voluntarily adopt pollution prevention programs. Benefits include expedited permits, internship priority, access to technical information, and leadership recognition including an awards program.

Illinois Graduate Internship P2 Program: a summer internship program which places college graduate students with industrial facilities to help them evaluate and implement pollution prevention projects.

Regulatory Integration: stresses the fact that the best way for businesses to stay in compliance is to not generate pollution in the first place. Accomplishes the above by providing environmental management training and through the formation of a regulatory integration team which helps develop strategies, projects, and measures that will help achieve environmental goals.

Greater Chicago Pollution Prevention Project: effort among federal, state, local environmental agencies, and non-governmental organizations to promote pollution prevention activities in the greater Chicago area.

The Great Printers Project (GPP): established to incorporate pollution prevention as the standard business practice for the lithographic printing industry. Such practices include using less hazardous inks and fountain solutions, and installing energy-efficient equipment. This program is also implemented in Michigan, Minnesota, and Wisconsin.

Common Sense Initiative (CSI) - Metal Finishing Sector: U.S. EPA program which stimulates the development of more cost-effective environmental technologies and processes that deal with the generation of waste and pollution at the source instead of after the fact. Illinois is supporting this program with an emphasis on the metal finishing sector.

Community Outreach: provides techniques to help companies and organizations meaningfully involve the public in their environmental programs and projects.

Indiana: www.ai.org/idem/oppta/

John Chavez, Pollution Prevention Branch Chief
Indiana Department of Environmental Management
The Office of Pollution Prevention and Technical Assistance
100 North Senate Avenue
P.O. Box 6015
Indianapolis, IN 46206-6015
T: (317) 232-8172 or 1-800-451-6027 ext. 2-8172 (in Indiana only)

Indiana's Material Exchange (IMEX): The Indiana Materials Exchange facilitates recycling and reuse of industrial and commercial waste. This is accomplished by maintaining and distributing listings of materials available and materials wanted.

Indiana 5-Star Environmental Recognition Program for Drycleaners: recognizes and ranks Indiana's dry cleaners based upon their pollution prevention efforts.

Mercury Thermostat Reduction and Recycling Pledge Program: The Mercury Thermostat Reduction and Recycling Pledge Program is the beginning of several other initiatives to voluntarily reduce the amount of mercury-containing devices which may be found in homes.

Business and Community Recognition Programs: The Office of Pollution Prevention and Technical Assistance (OPPTA) offers a multiplicity of award recognition programs commending a broad spectrum of businesses in their efforts statewide. These recognition programs serve to recognize those businesses that have worked diligently to excel at uniting both economic and environmental objectives successfully in their respective industries, while serving as models to their competitors.

Compost Program: stresses pollution prevention by encouraging the public to reuse and recycle of agricultural waste.

Other Programs: other facets of Indiana's Pollution Prevention Office consist of case studies, fact sheets, newsletters, training, grants, awards, workshops, and on-site assessments.

Iowa: www.iwrc.org/programs.html

Brian Tormey, Manager
E: btormey@max.state.ia.us
Waste Management Assistance Division
Iowa Department of Natural Resources
502 E. 9th St.
Des Moines, IA 50319-0034
T: (515) 281-8927
F: (515) 281-8895

Waste Reduction Assistance Program (WRAP): provides free, non-regulatory, on-site pollution prevention and waste reduction technical assistance to Iowa businesses and public institutions with 100 or more employees.

Painting and Coating Compliance Enhancement: helps small businesses to implement pollution prevention techniques and maintain compliance with environmental regulations. Included in the program are training programs and installing computer applications which help with record keeping.

Small Business Pollution Prevention Center: conducts applied research, education and training in small business pollution prevention. Specifically, it targets the automotive and lithographic printing industries.

Pollution Prevention Assessment and Implementation of New Tools and Techniques: program which helps small businesses with painting pollution prevention. The two step process consists of site visit in which a staff member will gain familiarity with its processes and then a report which will help to identify potential pollution prevention opportunities.

Spray Technique Analysis and Research: improves and optimizes the efficiency of manual spray coating operations as well as reducing material consumption and cost, reduced over-spray and emissions (specifically VOC's), reduced health risks, and improving finish quality.

<u>Iowa Waste Exchange</u>: a free state program that helps businesses retain the value of waste materials and diverts these materials from disposal facilities.

<u>On-site Reviews</u>: a free, non-regulatory and confidential environmental assistance for small companies in Iowa. The process consists of a waste reduction professional touring the facility and learning about the company's operations and current waste management practices. Within a month, a report is sent evaluating waste production areas and identifying pollution prevention opportunities.

Kansas: www.ink.org/public/kdhe

Janet E. Neff, Public Advocate
Planning and Prevention Section
Division of Environment
Forbes Field, Building 283
Topeka, KS 66620-0001
T: (785) 296-0669 or 1-800-357-6087
F: (785) 296-3266

<u>Kansas State Pollution Prevention Program</u>: the Kansas Department of Health and Environment (KDHE) provides pollution prevention assistance in partnership with Kansas State University (KSU) Engineering Extension. KSU is responsible for technical assistance and provides information on equipment and processes which eliminate, reduce, or control emissions. KSU also provides confidential on-site assessments at no costs.

<u>Small Business Environmental Assistance Program (SBEAP)</u>: provides guidance in compliance and technical matters to businesses that otherwise would not have access to such help because of financial constraints. It does so by offering newsletters, manuals, pamphlets, fact sheets, workshops and on-site assessments.
SEE ALSO: www.sbeap.niar.twsu.edu

Kentucky: www.state.ky.us/agencies/nrepc/waste/programs/p2/ppdivisi.htm

Division of Waste Management
Director's Office
14 Reilly Road
Frankfort, KY 40601
T: (502) 564-6716

<u>Pollution Prevention Programs</u>: several pollution prevention programs which focus on hazardous waste generators, hazardous waste treatment, storage, and disposal facilities, and underground storage tanks. Other programs consist of seminars, recycling activities, and used oil recycling.
SEE ALSO: www.kppc.org/

Kentucky Pollution Prevention Center
420 Academic Building
University of Louisville
Louisville, KY 40292
T: (502) 852-0965
F: (502) 852-0964

On-site Technical Assistance: free, non-regulatory, and confidential service which assists businesses by studying their processes and determining where pollution prevention strategies can be applied.

Kentucky Materials Exchange (KIME): an information clearinghouse to help Kentucky businesses reuse materials and find alternatives to throwing valuable materials into local landfills.

Pollution Prevention Training: KPPC offers pollution prevention training through free workshops, seminars, satellite teleconferences, manuals, interactive CD-Roms, and the Internet.

Wood Waste Alliance: promotes pollution prevention by attempting to find cost-effective ways of making products from wood waste.

Louisiana: www.deq.state.la.us/osec/latap.htm

Gary Johnson, Pollution Prevention Coordinator
E: osec@deq.state.la.us
Office of the Secretary
Technical Program Support
P.O. Box 82263
Baton Rouge, LA 70884-2263
T: (225) 765-0720
F: (225) 765-0742

Louisiana Environmental Leadership Pollution Prevention Program: a program which encourages Louisiana facilities to voluntarily reduce their waste and emissions beyond the level required by environmental regulations.

Technology Transfer: provides books, articles, and videos dealing with pollution prevention / waste minimization; computer searches for information not available in library; electronic bulletin board service to EPA P2 databases; conferences and workshops, and newsletters.

Technical Assistance: provides waste assessments to selected industries through reviews of company data, on-site waste surveys, evaluation of data collected, confidential written reports, references to applicability, and suggestions for pollution prevention/waste minimization.

Technical Bulletins: provides pollution prevention information on used oil management, reducing volatile emissions in the fiber reinforced plastic industry, and general P2 techniques for the automotive, wood products, painting, and printing industries.

Govenor's Award: recognizes outstanding people and business for their achievements in the field of pollution prevention.

Maine: www.state.me.us/dep/p2home.htm

Ron Dyer, Director
E: ron.e.dyer@state.me.us
Office of Pollution Prevention
Department of Environmental Protection
#17 State House Station
Augusta, ME 04333
T: (207) 287-4152

<u>Environmental Leader Program</u>: a voluntary certification program for gas stations in which they are recognized for being in 100% compliance with all environmental requirements. The public is encouraged to fulfill their automotive needs at these gas stations.

<u>Small Business Technical Assistance Program</u>: assists small businesses achieve compliance and reduce waste and pollution through the use of education, outreach, on-site assistance, and pollution prevention techniques.

<u>Small Business Compliance Incentives Policy</u>: provides an opportunity for small businesses to not be penalized for environmental violations given that they are discovered through this voluntary program and corrected by a given date.

<u>Other Programs</u>: provide case studies, publications, and pollution prevention tips and strategies.

Maryland: www.mde.state.md.us/permit/p2prog.html

Laura Armstrong, Pollution Prevention Coordinator
E: larmstrong@mde.state.md
Maryland Department of the Environment
2500 Broening Highway
Baltimore, MD 21224
T: (410) 631-4119 or 1-800-633-6101 x4119

<u>Businesses for the Bay</u>: a program consisting of a voluntary team of industries, commercial establishments, and small business within the Chesapeake Bay watershed which are committed to implementing pollution prevention in daily operations and reducing the chemical releases to the Chesapeake Bay.

<u>Pollution Prevention Solid Waste Program</u>: The Pollution Prevention Solid Waste Program provides educational programs, publications, and technical assistance on ways to reduce and recycle paper, cardboard, plastic, and other types of solid waste. The program provides assistance to businesses, municipalities, counties, schools, colleges, and other institutions interested in starting recycling or composting programs.

Massachusetts: www.state.ma.us/dep/bwp/dhm/tura/turhome.htm

Cynthia Chaves
E: cynthia.chaves@state.ma.us
Toxics Use Reduction: stresses pollution prevention through the reduction of toxics use.

Michigan: www.deq.state.mi.us/ead/p2sect/

Environmental Assistance Division
E: deq-ead-env-assist@state.mi.us
Michigan Department of Environmental Quality
P.O. Box 30457
Lansing, MI 48909-7957
T: 1-800-662-9278
F: (517) 335-4729
Agriculture: consists of programs which try to reduce or eliminate the generation potential sources of environmental contamination.
Auto Project: voluntary effort by Chrysler, Ford, and General Motors to promote pollution prevention throughout their business operations, products, and practices.
Clean Corporate Citizen: allows regulated sources that have demonstrated environmental stewardship and a strong environmental ethic through their operations in Michigan to be recognized as Clean Corporate Citizens.
Clean Air Assistance Program: helps small businesses understand their regulatory obligations and assists them in achieving compliance through phone consultations, publications, and workshops.
The Great Printers Project: established to incorporate pollution prevention as the standard business practice for the lithographic printing industry.
Health Care: program which promotes pollution prevention in hospitals which are estimated to generate in excess of one percent of all solid waste.
Lake Superior P2 Initiative: address the need for pollution prevention to be integral part of the "Binational Program to Restore and Protect The Lake Superior Basin" and to develop a comprehensive P2 strategy that would address the nine toxic substances of concern in Lake Superior.
Michigan Business P2 Partnership: voluntary pollution prevention program open to all Michigan businesses which would like to initiate or expand their P2 practices and receive recognition for their efforts.
Mercury (Hg) Pollution Prevention: program comprising of representation from industry, trade associations, environmental, government, and academic groups focusing upon mercury waste in seven sectors including: general public, healthcare, dental, automotive, electrical users/manufacturers, chemical users/manufacturers, and utilities.
Pulp & Paper: voluntary environmental initiative open to all forest products companies in Michigan who want to reduce the amount of waste generated through pollution prevention techniques.

Recycling: pollution prevention program which emphasizes the recycling of as much waste as possible. The program includes a material exchange program and a recycled materials market directory.

MDEQ Regulatory Integration of Pollution Prevention: established on the premises that the regulatory staff, through their day-to-day contacts with businesses, industries and municipalities, were in the best position of all agency staff to promote pollution prevention concepts on a widespread basis.

"Frontiers in Pollution Prevention": concentrates on pollution prevention research for the Michigan Great Lakes Protection program.

Retired Engineer Technical Assistance Program (RETAP): provides free, confidential, and non-regulatory on-site waste reduction assessments for Michigan businesses and institutions provided by retired engineers.

Turfgrass Pollution Prevention Assistance: designed advance environmental stewardship of the turfgrass industry, prevent pollution, and recognize environmental achievements.

Minnesota: www.pca.state.mn.us/programs/p2_p.html

Raymond Bissonnette
E: raymond.bissonnette@pca.state.mn.us
T: (612) 297-8588
F: (651) 297-8676

Metro District Planning: coordinates, supports, and provides training for P2 work in the District, assist in agency-wide initiatives, and administer Pollution Prevention Incentives for States grants from USEPA.

Small Business Ombudsman and Assistance Programs: offers compliance, P2 assistance, and loans to small businesses to upgrade process equipment.

Remediation Program: provides technical assistance and administrative or legal assurances to those seeking to investigate or clean up contaminated property.

Environmental Auditing Program: encourages businesses and governments to conduct environmental audits of facilities and correct problems they may discover.

Clean Air Act: companies may use P2 techniques to qualify for a non-expiring and flexible alternative permit option, to gain quicker consideration of a permit application, or to avoid or minimize the regulatory impact of new maximum achievable control standards for priority hazardous air pollutants.

Environmental Review: companies or individuals seeking to build or expand operations with potential to effect the environment can reduce the project's impact by designing in P2 technologies or techniques.

Supplemental Environmental Projects (SEPs): on a case-by-case basis, an investment in P2 techniques with long-term environmental benefit can be used in SEPs and other enforcement options to offset penalties.

Clean Water Program: the entire Point Source program is developing strategies to create new and stronger partnerships, including ones which will enable staff to be more upfront and proactive by incorporating P2 opportunities.

Environmental Regulatory Innovations Act: provides incentives or rewards, this 1996 statute gives the Minnesota Pollution Control Agency the authority to waive Minnesota statutes or rules if such relief will result in improved environmental performance or reduced administrative costs.

The Great Printers Project: established to incorporate pollution prevention as the standard business practice for the lithographic printing industry.

Common Sense Initiative (CSI) - Metal Finishing Sector: U.S. EPA program which stimulates the development of more cost-effective environmental technologies and processes that deal with the generation of waste and pollution at the source instead of after the fact; Minnesota is supporting this program with an emphasis on the metal finishing sector.

Pollution Prevention Programs: other programs which help to promote pollution prevention concepts are the materials exchange program, the recycling program, and general training programs.

Mississippi: www.deq.state.ms.us/domino/eroweb.nsf/ (Pollution Prevention links)

Jim Hardage, Hazardous Waste Coordinator
E: Jim_Hardage@deq.state.ms.us
Pollution Prevention Division
Mississippi Department of Environmental Quality
P.O. Box 10385
Jackson, MS 39289-0385
T: (601) 961-5321

Pollution Prevention Solid Waste Program: provides educational programs, publications, and technical assistance on ways to reduce and recycle paper, cardboard, plastic, and other types of solid waste. The program provides assistance to businesses, municipalities, counties, schools, colleges, and other institutions interested in starting recycling or composting programs.

Missouri: www.dnr.state.mo.us/deq/tap/polprev.htm

Becky Shannon
E: tap@mail.dnr.state.mo.us
Missouri Department of Natural Resources
Division of Environmental Quality
Technical Assistance Program
P.O. Box 176
Jefferson City, MO 65102
T: (573) 526-6627 or 1-800-361-4827

Pollution Prevention Program: promotes pollution prevention by helping to provide answers on how to deal with household hazardous waste; helping businesses learn how to deal with their small quantity hazardous waste generation; providing information that can help in waste minimization efforts; helping groups with P2 coordination and promotional material; and by providing a wide variety of information in its pollution prevention library.

Montana: www.montana.edu/wwwated

Michael P. Vogel, Director
E: acxmv@montana.edu
Montana State University Extension Service
Taylor Hall, P.O. Box 173580
Bozeman, MT 59717-3580
T: (888) 678-6822
DEQ (Dept. of Environmental Quality) Pollution Prevention Bureau: oversees several programs which provide businesses and general public with pollution prevention tips, pollution prevention strategies, compliance assistance, a recycling program, etc.
EcoStar Awards Program: recognizes businesses that go out of their way to create a safer work environment, reduce waste, reuse, recycle, and conserve water and energy.
Montana Material Exchange: program created to prevent pollution by creating a database of materials that can be reused instead of being sent to disposal facilities.
Library: information on pollution prevention is provided including newsletters, publications, and videos.

Nebraska: no information available

Nevada: www.scs.unr.edu/nsbdc/bep.htm

Kevin Dick, Manager
University of Nevada, Reno
College of Business Administration
Business of Environmental Program
Mail Stop 032
Reno, NV 89557-0100
T: (702) 784-1717
F: (702) 784-4337
Business Environmental Program: works with the Nevada Division of Environmental Protection to provide confidential environmental management assistance to businesses throughout Nevada. It can provide assistance by phone, by distributing informational materials, and by conducting on-site assessments. It also stresses pollution prevention through case studies, newsletters, fact sheets, and free training seminars.

New Hampshire: www.state.nh.us/des/nhppp/nh01000.htm

E: nhppp@des.state.nh.us
New Hampshire Pollution Prevention Program
Department of Environmental Services
Waste Management Division
6 Hazen Drive
Concord, NH 03301
T: (603) 271-2902 or 1-800-273-9469 (In New Hampshire only)
F: (603) 271-2456

On-Site Assistance: reviews waste generating processes and makes advisory waste reduction recommendations.

Technical Assistance: technical staff provides research information relating to process modifications, solid and hazardous waste management, regulatory questions, recycling, and air & waste issues.

Information Clearinghouse: provides information on new technologies, pollution prevention products and vendors, fact sheets, case studies, and success stories.

Pollution Prevention Internship Program: teams chemical engineering students with interested companies to work on pollution prevention projects during the summer.

WasteCap of New Hampshire: a program of the Business and Industry Association of New Hampshire specializing in solid waste reduction assistance for businesses, providing material exchange, business recycling consortia, site visits, hotline technical assistance, and waste reduction recognition programs.

New Hampshire Small Business Development Center: offers free and confidential business counseling and environmental assessments and referrals. These services are aimed at identifying reduction and resource conservation in order to enhance profitability through the promotion of a cleaner environment.

Small Business Environmental Assistance Alliance: helps small businesses become more profitable while maintaining or enhancing environmental quality. It provides free and confidential one-on-one counseling to small business.

Outreach Programs: the New Hampshire Pollution Prevention Program sponsors and co-sponsors various workshops, training sessions and conferences throughout the state which encourage both regulatory and economic incentives to prevent pollution and to minimize the generation of pollutants and other wastes.

New Jersey:

Melinda M. Dower
E: mdower@dep.state.nj.us
Office of Pollution Prevention & Permit Coordination
401 E. State St.
P.O. Box 423, 3rd Floor
Trenton, NJ 08625-0423
T: (609) 292-1122

Technical Assistance: provided by the statewide Technical Assistance Program (NJTAP) for small and medium-sized businesses; technical assistance comes in the form of free, non-regulatory, and confidential assistance by way of literature reviews and on-site assessments of a particular facility's operations.

New Mexico: no information available

New York: www.dec.state.ny.us/website/ppu/

John Iannotti, Director
New York State Department of Environmental Conservation
Pollution Prevention Unit
50 Wolf Road
Room 202
Albany, NY 12233-8010
T: (518) 457-7267
F: (518) 457-2570

Governor's Awards for Pollution Prevention: Annual awards for outstanding achievements in pollution prevention.

Comparative Risk Project: Using pollution prevention to reduce risk to New York State residents and the environment by comparing different environmental problems so that there may be a better understanding of which problems are posing the highest risks to people and the environment.

Facility Reporting: Toxic Release Inventory (a report of toxic emissions in New York State) and State agency audit reports.

Pollution Prevention Initiatives: New pollution programs in New York State, including geographic- and sector-based pollution prevention initiatives, Great Lakes Mercury program, ISO 14000, environmental indicators, and other projects to encourage and monitor pollution prevention.

Pollution Prevention Clearinghouse: The clearinghouse has computerized databases of over 15,000 references of articles and publications on pollution prevention and a library of selected hard copy documents.

Great Lakes and Lake Champlain Projects: evaluates and recommends improvements to the environmental conditions in the Great Lakes and Lake Champlain areas.

North Carolina: www.p2pays.org/

Gary E. Hunt, Director
E: nowaste@p2pays.org
N.C. Division of Pollution Prevention and Environmental Assistance
P.O. Box 29569
Raleigh, NC 27626-9569
T: (919) 715-6500 or 1-800-763-0136
F: (919) 716-6794

Small Business Ombudsman: assists small businesses with air quality and other regulatory requirements and encourages environmental compliance and stewardship.

Recycling Business Assistance Center: promotes environmentally sound economic development through reuse and remanufacture of recyclable materials. It is accomplished by providing technical assistance, and business development assistance, promoting the need to develop, transfer, and apply technologies that will increase the use of recycled materials, promoting partnerships among government and industry to stimulate and facilitate the recovery and use of secondary materials, and promoting waste prevention and reuse within both the public and private sectors.

Waste Reduction Resource Center: provides multimedia waste reduction support to the states in EPA Regions III (Alabama, Florida, Georgia, Kentucky, Mississippi, North Carolina, South Carolina, and Tennessee) and IV (Delaware, Maryland, Pennsylvania, Virginia, West Virginia, and Washington, D.C.). It provides a clearinghouse of journal articles, case studies, technical reports, factsheets, books, and videotapes.

Pollution Prevention Programs: N.C. Division of Pollution Prevention and Environmental Assistance (DPPEA) stresses pollution prevention through technical assistance, workshops, training, case studies, newsletter, fact sheets, permit assistance and financial assistance in the form of grants. Programs include an awards program, an information clearinghouse, a Small Business Environmental Assistance program, and a Materials Assistance for Recyclable Materials program.

North Dakota:

Jeffrey L. Burgess
E: jburgess@state.nd.us
North Dakota Department of Health
Environmental Health Section
P.O. Box 5520
Bismark, ND 58502
T: (701) 328-5153

Pollution Prevention Programs: offer pollution prevention advice and suggestions during routine compliance work.

Ohio: www.epa.ohio.gov/opp/oppmain.html

Michael W. Kelley, Chief
E: michael.kelley@epa.state.oh.us
Office of Pollution Prevention
Ohio EPA
P.O. Box 1049
Columbus, OH 43216-1049
T: (614) 644-3469

Ohio Prevention First: voluntary pollution prevention initiative that targeted the top 100 companies that report the most releases to the environment and helped them develop a comprehensive P2 plan.

Pollution Prevention Technical Assistance: provides technical assistance to over 8,300 companies, organizations and/or individuals in the form of on-site visits, free publications, presentations and training events.

EnviroPrint initiative: coordinated by the Printing Industry of Ohio (PIO) with the support of the Ohio Office of Pollution Prevention (OPP) in an attempt to implement P2 techniques in the printing industry. This initiative developed a self help guide to environmentally sound printing operations which was designed to help printers chart their own way through the complicated maze of environmental requirements, while emphasizing P2.

Pollution Prevention Loan Program: established in 1994, this is a low-interest loan program for P2 to be jointly administered by Ohio EPA and the Ohio Department of Development to businesses and projects which will result in a large reduction of pounds of pollution.

Award Programs: consists of programs including the Govenor's Award and the Ohio EPA Recognition of Superior Pollution Prevention (Or Reduction) Accomplishments which recognize outstanding achievements in the field of pollution prevention.

Pollution Prevention Education and Training Grants: provides funds to educate small and mid-sized businesses about the benefits of P2.

Lake Erie Basin Pollution Prevention Activities: consists of a series of projects to identify P2 opportunities for Ohio businesses in the Lake Erie Basin. These activities have been used to support technical assistance and regulatory integration efforts.

Common Sense Initiative (CSI): CSI is an industry-by-industry initiative which is looking for ways to transform the current process of environmental regulation into a comprehensive system for strengthened environmental protection. Ohio is supporting this program in the iron and steel sector.

Waste Minimization Measurement Pilot Project: the Ohio EPA Waste Minimization Measurement Pilot Project (Pilot Project) was a joint U.S. EPA/Ohio EPA pilot project to determine effective measures in waste minimization and P2.

Automotive Pollution Prevention Project: a voluntary partnership between Chrysler, Ford, General Motors and the Great Lakes States whose purposes are to identify Great Lakes persistent toxic substances (GLPTS) and reduce their generation and release; to advance P2 within the auto industry and its supplier base; to reduce releases of GLPTS beyond regulatory requirements; and to address regulatory barriers that inhibit P2.

Tri-State Geographic Initiative: works to create an improved environment for the area where Ohio, Kentucky, and West Virginia meet by employing a variety of protection tools and taking a multi-media perspective.

Other Programs and Projects: several programs which help to prevent pollution prevention concepts through waste exchanges, recycling information and regulatory integration programs in such industrial applications such as dry cleaning, energy, metal finishing / metal working, paints and coatings, and solvents.

Oklahoma: www.deq.state.ok.us/P2intro.htm

Dianne Wilkins, P2 Program Director
E: dianne.wilkins@deqmail.state.ok.us
T: Customer Assistance Program: (405) 271-1400 or 1-800-869-1400

P2 Clearinghouse: provides information, either industry or process specific, on new technologies, methods and techniques available for pollution prevention and waste management.

Tax Credit: offers a 20% tax credit toward the cost of equipment for the reduction of hazardous waste.

Target 98 Program: voluntary toxics use reduction program.

On-Site Assistance: facility visits to help identify pollution prevention opportunities are available on request. Staff will listen, observe, and respond to needs in a confidential, non-regulatory manner.

Training and Workshop: program which sponsors a variety of training sessions and workshops annually. These sessions focus on pollution prevention and waste management.

Other Programs: includes awards, certification, directory of recyclers, and case studies.

Oregon: www.deq.state.or.us/hub/p2.htm

Oregon Department of Environmental Quality
Pollution Prevention Coordinator
811 S.W. Sixth Avenue
Portland, OR 97204

Industry and the Environment Pollution Prevention Opportunities: provides pollution prevention techniques, strategies, and tips.

TURHWR (Toxics Use Reduction and Hazardous Waste Reduction) Program: provides technical assistance in developing reduction plans, encourages implementation of plans, and encourages continued facility-wide reduction efforts. Some key program elements include plan and progress report reviews, workshops, on-site visits, and an achievement award program.

Pennsylvania: www.dep.state.pa.us/dep/deputate/pollprev/pollution_prevention.html

Stacy A. Richards, Deputy Secretary
E: richards.stacy@dep.state.pa.us
T: (717) 783-0540
F: (717) 783-8926

Governor's Awards for Environmental Excellence: recognizes outstanding business and individuals for their achievements in pollution prevention.

Pennsylvania Environmental Assistance Network: provides small businesses with pollution prevention alternatives.

Global Outreach and Environmental Technology: provides information on environmental technology including grants and financing.

P3ERIE: Prevent Mercury Pollution: voluntary pollution prevention program concentrating on mercury pollution in the Lake Erie area.

Pennsylvania's Environmental Technology Investment through the Ben Franklin Technology Centers: provides details and information on recycled products as pollution prevention alternatives.

Other Programs: other programs include success stories, providing detailed fact sheets, and providing technical assistance concerning zero emissions and the ISO 14000 program.

Rhode Island: www.state.ri.us/dem/org/otca.htm

Richard Enander, Pollution Prevention Supervisor
E: renander@dem.state.ri.us
Room 256
235 Promenade Street
Providence, RI 02908-5767
T: (401) 222-6822 x4411
F: (401) 222-3810

Office of Technical and Customer Assistance: The Office provides pollution prevention assistance to businesses, industry, and governmental agencies to help them prevent and minimize pollution at the source of generation. This outreach function includes: on-site technical assistance; training programs, conferences, and workshops; and both regulatory and economic incentives to prevent pollution and to minimize the generation of pollutant wastes associated with industrial processes. This program works with businesses to develop cost-effective ways to reduce toxic and hazardous material use and waste in the workplace.

South Carolina: www.state.sc.us/dhec/eqchome.htm ("Outreach and Assistance Programs" link)

Robert E. Burgess, Section Manager
E: burgesre@columb30.dhec.state.sc.us
Center for Waste Minimization
2600 Bull Street
Columbia, SC 29201
T: (803) 734-4761
F: (803) 734-9934

Center for Waste Minimization: a non-regulatory technical assistance program established to help business and industry identify waste reduction and recycling opportunities. The center consists of an information clearinghouse and a technical assistance office.

Center for Environmental Policy: provides in-plant waste assessments and other pollution prevention assistance.

South Carolina Environmental Excellence Program: a voluntary program which rewards and recognizes public and private entities that have demonstrated a commitment to move their operations beyond environmental compliance through pollution prevention and resource conservation.

South Carolina Environmental Network: a partnership of 16 organizations including technical colleges, universities, state agencies and small business centers that provides waste assessments and pollution prevention training.

Office of Solid Waste Reduction and Recycling: consists of a recycling awareness program, a used oil recycling program, an information resource center, an energy awards recognition program, and a series of recycling workshops.

Small Business Assistance Program: a free, non-regulatory service which offers technical assistance to small businesses by providing information including technical and compliance information, information on pollution prevention and accidental release prevention and detection, and providing confidential one-on-one consultation through an audit program.

WaterWatch program: a unique effort to involve the public and local communities in water quality protection. Volunteers select a water resource on which to focus and perform one or more activities aimed at protecting water quality.

South Dakota:

Dr. Dennis C. Clarke, Pollution Prevention Coordinator
Department of Environment and Natural Resources
Joe Foss Building
523 East Capitol
Pierre, SD 57501-3181

Municipal Water Pollution Prevention (MWPP) Program: a voluntary and cooperative effort by the EPA, the Department of Environment and Natural Resources (DENR) and South Dakota municipalities to: prevent National

Pollutant Discharge Elimination System (NPDES) permit violations and maintain high compliance rates by publicly owned treatment works (POTWs); maximize the useful lives of POTWs through reduced wastewater flows and loading, and effective operation and maintenance; and ensure effective and timely planning, financing and construction for future needs and growth, before permit violations occur.

Small Business Air Pollution Assistance Program: helps small businesses identify their sources of air pollution and information about recommended air pollution control devices and alternative processes or products that produce less pollution.

Center for Advanced Manufacturing and Production (CAMP) Program: provides assistance to selected manufacturers in the form of source reduction, energy efficiency and waste management recommendations.

EnviroScape: a watershed model used to demonstrate the movement of water and potential pollution that may occur as water passes through a watershed. It is used for environmental training purposes.

Central SD Source Reduction Project: selected towns were analyzed to determine source reduction, recycling and P2 strategies that may succeed in rural areas with low populations spread over a large area.

Wastewater System Infiltration P2 Project: provides technical assistance and training to small towns and rural communities in the use of source reduction strategies.

P2 for the Livestock Confinement Industry Program: an outreach program which assists system owners/operators with permit related issues and the development of P2 plans for operation of the systems.

Project Save: multimedia environmental awareness program for teachers and educators.

On-site Wastewater Treatment (Septic) System Education Project: provides wastewater treatment information to installers and owners of on-site wastewater treatment systems and planning and zoning officials.

Wellhead Protection Area Establishment for Small Communities: assists select areas with the development of low cost, prevention based environmental compliance solutions concerning wellheads.

Home*A*Syst: program which developed a P2 assessment tool for use by homeowners.

Tennessee: www.state.tn.us/environment/p2.htm

E: environment@mail.state.tn.us
Department of Environment and Conservation
Division of Community Assistance
8th Floor, L & C Annex, 401 Church Street
Nashville, TN 37243-1551

Division of Community Assistance: programs which provide technical and financial help to various communities of the state including counties, cities, utilities districts, businesses, industries, schools, associations, and the general public. Among its programs are loan and grant programs, training, testing, and certification programs, assessment and review programs, a household hazardous waste collection program, and an agricultural pesticide waste disposal program.

Pollution Prevention Programs: programs which provide multimedia pollution prevention assistance to industry, commercial establishments, schools, institutions, homes, government, etc., through the preparation and review of pollution prevention plans, on-site visits, market development, general outreach and training.

Recycling Program: provides a recycling program for state employees, coordination of recycling events, and database maintenance and referral for materials exchange, used oil, antifreeze, and battery collection sites and transporters.

Tire Management Program: program established to reduce waste through the routine shredding of scrap tires for disposal, the abatement of unpermitted disposal sites, and assistance to reuse markets.

Texas: www.tnrcc.state.tx.us/exec/oppr/index.html

Andrew Neblett, Director
E: OPPR@tnrcc.state.tx.us
Bldg.E/1 MC 112, TNRCC
P.O. Box 13087
Austin, TX 78711-3087
T: (512) 239-3166
F: (512) 239-3165

Awards and Special Events: Texas has several programs which help to train communities in pollution prevention techniques and recognize those organizations and individuals for their outstanding achievements in the field of pollution prevention.

Buying, Selling, and Trading Recyclables, Compost, and Industrial By-products: programs which achieve pollution reduction through the promotion of purchasing and trading of recycled-content products and compost.

Disposal and Recycling Opportunities: includes programs such as an agricultural waste pesticide collection program, an environmental hotline and website containing community-specific environmental and recycling information, a household hazardous waste collection program, and an annual, one-day event that offers citizens disposal opportunities for used oil, oil filters, antifreeze, and empty, plastic pesticide containers.

Environmental Education: consists of state programs which educate the public from elementary school aged children to adults on pollution prevention techniques.

Recycling and Composting: consists of several programs that help prevent pollution through recycling and composting.

Reducing Pollution, Hazardous and Non-hazardous Waste, Toxics Release Inventory: consists of a border pollution prevention program which helps to reduce the amount of pollution along the Texas-Mexico border, several permanent pollution prevention workshops, site assistance visits, and a database consisting of case studies which are organized by industry, waste, or process.

Regulatory Flexibility and Reinvention: consists of the consolidated reports program which helps to streamline environmental reports and improve the public's access to information and the pollution prevention integration program which integrates prevention strategies into regulatory functions.

Utah: www.eq.state.ut.us/eqoas/p2/p2_home.htm

Sonja Wallace
E: swallace@deq.state.ut.us
168 North 1950 West
P.O. Box 144810
Salt Lake City, UT 84114-4810
T: (801) 536-4480
F: (801) 536-0061

Pollution Prevention Awards: state program which recognizes outstanding businesses and individuals for their achievements in the field of pollution prevention.

Business Assistance Program: program which helps businesses cut down on pollution before it even starts. It includes free information sessions on pollution prevention methods and free voluntary on-site consultations which include pollution prevention recommendations.

Recycling Program: state program which provides information on recyclers and what they recycle, compost information, seasonal information, new recycling technologies, new recycling businesses and manufacturers, and proposed and current recycling contacts.

Used Oil Program: regulates facilities that transport, market, process and re-refine, or burn off-specification used oil. It also promotes used oil recycling including the setup and registration of do-it-yourself used oil collection centers.

Watershed Management: program aimed at improving the protection of Utah's surface and ground water resources. It features a high level of stakeholder involvement, water quality monitoring and information gathering, problem targeting and prioritization, and integrated solutions that make use of multiple agencies and groups.

Vermont:

Paul Van Hollebeke, Pollution Prevention Specialist
E: paulv@wasteman.anr.state.vt.us
Vermont Agency of Natural Resources
Environmental Assistance Division
103 South Main Street
Waterbury, VT 05671-0411
T: (802) 241-3629

REAP (Retired Engineers and Professionals Program): helps small businesses like vehicle service, offset printers and auto body identify cost-cutting pollution prevention or other waste reduction measures through their on-site assistance efforts.

Small Business Compliance Assistance Program: helps businesses in their efforts to comply with the Department of Environmental Conservation's environmental regulations through on-site environmental compliance reviews and a toll-free hotline.

Vermont Business Environmental Partnership: a voluntary, environmental assistance and business recognition program offered by the Environmental Assistance Division and the Vermont Small Business Development Center which aims to achieve greater environmental and economic performance and to promote public recognition of environmental excellence.

Governor's Awards for Pollution Prevention: honors Vermont individuals, organizations, institutions, businesses, and public agencies using innovative approaches that reduce or eliminate the generation of pollutants and wastes at the source.

Compost Center: provides technical assistance for new or developing compost operations, through market development assistance for compost use, and through providing educational materials for backyard to large scale composting.

Plastics Recycling Market Development: focuses on developing a recycled plastics market in Vermont.

Vermont Business Materials Exchange: provides businesses with the opportunity to find outlets for their potentially reusable materials.

Other Programs: pollution prevention programs include waste reduction assessments, research assistance, P2 planning, conferences, workshops, training, and telephone assistance.

Virginia: www.deq.state.va.us/opp/opp.html

Sharon K. Baxter, Program Manager
E: skbaxter@deq.state.va.us
Office of Pollution Prevention
Virginia Department of Environmental Quality
P.O. Box 10009
Richmond, VA 23240-0009
T: (804) 698-4235/4545
F: (804) 698-4500

P2 Grants, Fact Sheets, Case Studies, and Library: Virginia supports several programs which provide businesses and communities with financial assistance, financial incentives, case studies, publications, newsletters, videos, and fact sheets which provide information to businesses, consumers, and local governments which provide information on pollutants linked with several industries, household wastes, and techniques and strategies to reduce the pollution.

Businesses for the Bay: a program consisting of a voluntary team of industries, commercial establishments, and small business within the Chesapeake Bay watershed which are committed to implementing pollution prevention in daily operations and reducing the chemical releases to the Chesapeake Bay.
State Agency Material and Products Listing/Exchange (SAMPLE): an information clearinghouse to help Virginia businesses reuse materials and find alternatives to throwing valuable materials into local landfills.

Washington: www.wa.gov/ecology/pie/98overvu/98aohwtr.html

Greg Sorlie
E: gsor461@ecy.wa.gov
T: (360) 407-6702
Pollution Prevention Planning and Technical Assistance: program in which facilities are encouraged to establish reachable goals for reduction, recycling, and treatment and to report their progress annually. The Department of Ecology also provides technical assistance in preparing plans and progress reports and during implementation. Technical assistance can include on-site visits, phone consultations and workshops.
Toxics Reduction Engineer Exchange (TREE): This project teams the Department of Ecology toxics reduction engineers with businesses for an in-depth look at free or low-cost techniques to reduce waste and save money.
Governor's Awards: honors facilities for their leadership and innovation in pollution prevention.

West Virginia: www.dep.state.wv.us/p2/index.html

Municipal Wastewater assistance
David Byrd
Pollution Prevention Services
617 Broad St.
Charleston, WV 25301-1218
T: (304) 558-0633
F: (304) 558-3778
Pollution Prevention Services: consists of programs which prevent violations of environmental regulations; help industry and government official develop and coordinate P2 programs; maximize the useful life of treatment units; establish early warning indicators for planning process improvement; conduct audits to evaluate facility processes; conduct training programs; work with pretreatment facilities to identify source reduction opportunities; and evaluate sludge/septage disposal processes.

Wisconsin: www.dnr.state.wi.us/org/caer/cea

Lynda Wiese, Director
E: wiesel@dnr.state.wi.us
Wisconsin Department of Natural Resources
Bureau of Cooperative Environmental Assistance
P.O. Box 7921
Madison, WI 53707-7921
T: (608) 267-3125

Awards and Case Studies: Wisconsin supports an awards program which recognizes businesses and individuals for their outstanding achievements in pollution prevention. It also has a database of case studies which provides examples of the success of pollution prevention in industry.

Environmental Cooperation Pilot Program: a pilot program designed to evaluate innovative environmental regulatory methods. The program provides the Department of Natural Resources with the authority to enter into cooperative environmental agreements with persons who own or operate facilities that are covered by licenses or permits under current law.

The Great Printers Project: established to incorporate pollution prevention as the standard business practice for the lithographic printing industry.

Climate Wise: emphasizes the ability of individual companies to set goals for energy and emissions reductions.

Dry Cleaners' Partnership: focuses on reducing the toxicity of wastes used in the dry cleaning process and rewards participants through a five-star recognition program.

Pulp and Paper Pollution Prevention Partnership: aims at creating cost-effective environmental improvement opportunities.

WasteCap Wisconsin: a public/private partnership aimed at businesses helping businesses with waste reduction and recycling.

Wyoming: deq.state.wy.us/outreach1.htm

Office of Outreach & Environmental Assistance
T: (307) 777-6105

Pollution Prevention Program: consists of a pollution prevention library, case studies, newsletters, and other publications.

APPENDIX C

ENVIRONMENTAL ORGANIZATIONS
Contributing Author: Robert Molter

The following is a list of addresses of major environmental organizations grouped in international, U.S. government, and U.S. non-government categories. These organizations can readily provide a wealth of environmental information upon a simple request. For each organization, there is an address to contact followed by a telephone number, fax number, internet address, and/or e-mail address if available. At the end of this section is a list of internet sources which provide environment related links and search engines for quick access to information on a wide variety of topics.

International Organizations

Beauty Without Cruelty
1244 19th Street, NW
Washington, DC, 20037 USA
Tel: (202) 659-9510

Canada-United States
Environmental Council
11 Limehill Road
Tunbridge Wells
Kent YNI 1LJ, England

Caribbean Conservation
Corporation (CCC)
P.O. Box 2866
Gainesville FL 32602 USA
Tel: (904) 373-6441

Clean World International (CWI)
c/o Keep Britain Tidy Group
Bostel House
37 West Street
Brighton BN1 2RE, England
Tel: (44) 273-2358

International Association
on Water Pollution
Research and Control (IAWPRC)
1 Queen Anne's Gate
London SW1 H9BT, England
Tel: +44 (171) 222 3848
Fax: +44 (171) 233 1197

International Board for Soil Res.
and Management (IBSRAM)
P.O. Box 9-109 Bangkhen
Bangkok 10900, Thailand
Tel: (662) 561-1230

The Secretariat for the
Protection of the
Mediterranean Sea
Place Lesseps 1
E-08023 Barcelona, Spain
Tel: (343) 217-1695

The United Nations
Development Programme (UNDP)
1 United Nations Plaza
New York, NY 10017 USA
Tel: (212) 906-5000
http://www.undp.org/index5.html

World Environment Center
605 Third Avenue
New York, NY 10156
Tel: (212) 986-720

WWF International (World Wildlife Fund)
Avenue du Mont-Blanc
CH-1196, Gland
Switzerland
Tel: +41 22 364 91 11
http://www.panda.org/

U.S. Government Organizations

Council on Environmental Quality
722 Jackson Place, NW
Washington, DC 20006
Tel: (202) 395-5750

Department of Agriculture
U.S. Forest Service
Soil Conservation Service
Washington, DC 20250
Tel: (202) 655-4000
http://www.usda.gov

Department of Energy
Washington, DC 20545
Tel: (202) 252-5000
http://198.124.130.244

Department of Interior
Interior Building
C Street, between 18th and 19th, NW
Washington, DC 20240
Tel: (202) 343-1100
http://www.doi.gov

Department of Justice
Land and Natural
Resources Division
10th Street and Pennsylvania
 Avenue, NW
Washington, DC 20530
Tel: (202) 633-2701
http://www.us doj.gov

United States Customs Service
1301 Constitution Avenue, NW
Washington, DC 20229
Tel: (202) 566-5104
http://www.ustreas.gov

National Oceanic and
Atmospheric Administration
14th Street, NW
Washington, DC 20230
Tel: (202) 377-3567
http://www.noaa.gov

The United States Environmental
Protection Agency (EPA)
401 M Street, SW
Washington, DC 20460
Tel: (202) 755-2673
http://www.epa.gov

U.S. Non-Governmental Organizations

Endangered Species Committee
Interior Building
Room 4160
Washington, DC 20240
Tel: (202) 235-2771

Marine Mammal Commission
1625 I Street, NW
Washington, DC 20006
Tel: (202) 653-6237
http://citation.com/hpages/mmc

National Wildlife Federation
1412 16th Street, NW
Washington, DC 20036
Tel: (202) 797-6800
http://www.nwf.org

Sierra Club
530 Bush Street
San Francisco, CA 94108
Tel: (415) 981-8634
http://www.sierraclub.org

World Wildlife Fund
1250 24th St., NW
Washington, DC 20037
Tel: 1-800-225-5993
http://www.worldwildlife.org

Greenpeace, U.S.A., Inc.
2007 R Street, NW
Washington, DC 20009
Tel: (202) 462-1177
http://www.greenpeace.org

National Audubon Society
950 Third Avenue
New York, NY 10022
Tel: (212) 832-3200
http://www.audubon.org

Natural Resources Defense
Council (NRDC)
Suite 300
1350 New York Avenue, NW
Washington, DC 20005
Tel: (202) 727-2700
E-mail: nrdcinfo@aol.com

Union of Concerned Scientists
Two Brattle Square
Cambridge, MA 02238
Tel: 617-547-5552
Fax: 617-864-9405
http://www.ucsusa.org

Internet Resources

EE Link: Environmental Education on the Internet
http://www.eelink.net

EE-Link provides a strong presence for Environmental Education (EE) on the Internet. Consistent with the Key Principles of Environmental Education, EE-Link develops and organizes Internet resources to support, enhance and extend effective environmental education. The EE-Link project is responsible for conducting training workshops, and promoting interaction and exchange among environmental education students, teachers and professionals. It also features the GAIN Directories of Environmental Organizations which consists of over 13,000 records of environmental organizations throughout the continental US and Hawaii.

Ecological Operating Systems: Environmental Quality and Efficiency (EQE) WWW
http://www.igc.apc.org/eco-ops/welcome.html

This WWW site includes an extensive directory of links to sources on environmental quality issues, including environmentally aware business, design, industrial production and consumption practices. It also includes sections on ISO standards, recycling, energy consumption, and general environmental sources.

EcoNet
http://www.igc.org/igc/en

The goal of this site is "to support ecological sustainability and environmental justice." It contains numerous articles, headlines, environmental alerts, as well as numerous links to other organizations.

EnviroLink Library
http://library.envirolink.org

This is one of the most comprehensive resources of environmental information available on the Internet, organized to make information easier to access, interact with and use. Information is grouped under subjects within each main area of the library.

Environmental Links
http://www.pacific.net/~dglaser/ENVIR/LINKS/*links.html

This site contains an extensive list of links to environmental-related information around the world grouped by categories such as alternative energy, forestry, and government.

APPENDIX D

Pollution Prevention Software

Contributing Author: Robert Molter

Due to the widespread use of computers in industry and education, many organizations concerned with pollution prevention are developing software to aid and educate companies. The following list provides information for obtaining such software.

U.S. Environmental Protection Agency and Purdue University
Software for Environmental Awareness

The U.S. Environmental Protection Agency and Purdue University have designed a pollution prevention software program available free to the public. This program provides an overview of pollution prevention concepts and describes, in detail, pollution prevention opportunities in the industry, agriculture, energy, government, and consumer sectors. It features case studies and provides pollution prevention information resources. The programs are available from either EPA Region 5 or Purdue University and can be found on the EPA Public Access Web Server at http://www.epa.gov/grtlakes/seahome. The programs run on IBM compatible systems with high-density disk drives, and are available on diskette and CD-ROM. For further information contact the following:

U.S. EPA Region 5
77 W. Jackson, P-19J
Chicago, IL 60604-3590
Tel: (312) 353-6353

The National Coalition of Advanced Technology Centers (NCATC) and
The Center for Applied Competitive Technologies (CACT)
Pollution Prevention Assessment Tools for Manufacturers

This software was developed to provide environmentally conscious manufacturing assessment tools for small to mid-size manufacturing firms. The software contains one hour of video clips and approximately 500 pages of text. It also includes 23 industry and waste-specific fact sheets, 25 case studies, 147 pages of regulations, 24-page checklist, EPA's Toxic Release Inventory forms, and numerous links to informative sites. Software is available on CD-ROM for only $15 plus shipping. The software is best run on Windows 95 or Macintosh operating systems and requires Apple Quick Time to run. To order the CD-ROM call: (714) 695-1501 (ext. 226).

Pollution Prevention Information Clearinghouse (PPIC)
P2/Finance

P2/Finance, a new spreadsheet software tool that helps companies to collect and analyze data essential to providing a clearer financial evaluation of product/process costs and pollution prevention projects, is now available free of charge for any federal, state, or local government employee from the Pollution Prevention Information Clearinghouse (PPIC). The software offers a valuable starting point for introducing a Total Cost Assessment (TCA) approach to companies. Users input capital and operating costs for a product/process and an alternative product/process, and the program outputs a fifteen-year cash flow analysis and a profitability analysis that calculates three financial indicators:

1. Net Present Value
2. Internal Rate of Return
3. Simple Payback

P2/Finance comes with a user manual and free access to a user hotline. For more information on how to obtain the software call PPIC at: (202) 260-1023.

APPENDIX E

Pollution Prevention Contacts in Education
Contributing Author: Karen L. Counes

The following is a list of educators from several different colleges and universities who teach and conduct research in pollution prevention and related fields. For each entry, an address and method of contact are given (P=Phone, F=Fax, E=Email), as well as the individual's involvement in pollution prevention education and research. Special research opportunities offered at each school are listed, along with the approximate enrollments of Undergraduate (U), Masters (M) and Doctoral (D) students.

CHEMICAL ENGINEERING

DR. DAVID ALLEN
Department of Chemical Engineering
University of Texas at Austin
CPE 2-8-2 26th and Speedway
Austin, TX 78712-1062

P: (512) 471-0049
F: (512) 471-7060
E: allen@che.utexas.edu

Dr. Allen teaches *Design for Environment,* a course offered through the chemical engineering department dealing with methods for incorporating environmental objectives into product and process designs. He is also the co-author of Pollution Prevention for Chemical Processes, a Wiley textbook published in 1997.

The University of Texas at Austin also offers several engineering research opportunities, including such topics as Air Pollution, Hazardous Waste Management, and Water and Wastewater Treatment.

Approximate enrollment: U=3,700; M=900; D=900

DR. ERIC J. BECKMAN
Department of Chemical Engineering
University of Pittsburgh
Pittsburgh, PA 15261

P: (412) 624-9641
F: (412) 624-9639
E: beckman@vms.cis.pitt.edu

Dr. Beckman has established *Pollution Prevention,* a course covering the basic concepts of life cycle analysis, process modification for pollution prevention, and the use of new synthetic pathways for pollution prevention.

Research opportunities include the use of clean-coal technology by-products from coal-fired plants and evaluation of natural gas vehicles (NGVs). The Engineering Center for Environment and Energy is devoted to working with industry, government agencies, and the general public to improve energy and natural resource management.

Approximate enrollment: U=1,250; M=500; D=200

DR. MARTIN BIDE
Rhode Island Center for Pollution Prevention   P: (401) 874-2276
University of Rhode Island   F: (401) 874-2581
Chemical Engineering Department   E: mbide@uriacc.uri.edu
Crawford Hall
Kingston, RI 02881

Dr. Bide teaches both graduate and undergraduate courses in textile wet processing. These courses address pollution prevention techniques in the preparation, dyeing, printing, and finishing of textiles.

The Pollution Prevention Center of Rhode Island offers research opportunities focused on source reduction. Case studies have been conducted for more than 200 companies in a variety of industries, including metal finishers, textiles, and jewelers.

Approximate enrollment: U=650; M=150; D=65

DR. CAROLYN BOLTON
Department of Chemical Engineering   P: (803) 777-7219
School of the Environment   F: (803) 777-8265
University of South Carolina   E: bolton@sun.che.sc.edu
Swearingen Engineering Center
Columbia, SC 29208

Dr. Bolton teaches the new *Environmentally Conscious Manufacturing* course, which deals with several different topics including legal and regulatory frameworks, pollution prevention and waste minimization, total cost assessment, and management of technological change.

Approximate enrollment: Not readily available

DR. YORAM COHEN
Department of Chemical Engineering   P: (213) 825-8766
University of California, Los Angeles   F: Unavailable
School of Engineering   E: yoram@seas.ucla.edu
405 Hilgard Avenue
Los Angeles, CA 90024

Dr. Cohen teaches and conducts research in the fields of pollution prevention evaluation strategies, multimedia transport of pollutants, and multipathway exposure analysis. Dr. Cohen has developed a number of computer models that are currently used in the United States and abroad.

Several research opportunities are available, including topics such as process design with minimal waste by-product formation.

Approximate enrollment: U=2,100; M=450; D=550

DR. ROBERT M. COUNCE
Department of Chemical Engineering
University of Tennessee
Knoxville, TN 37996-2200

P: (423) 974-5318
F: Unavailable
E: counce@utkux1.utk.edu

Dr. Counce teaches courses in process design with a focus on pollution prevention. His research includes designing and modifying industrial processes for pollution prevention. He is involved in the development of environmentally friendly industrial washing and degreasing technology, and recovery and recycling of various process materials.

The Maintenance and Reliability Center is an industry-sponsored research facility whose mission is to improve industrial productivity and safety through advances in maintenance and reliability engineering. Other related research opportunities include air pollution control technologies, alternative fuels, environmental quality and resource conservation, industrial pollution prevention, injury prevention, product evaluation, and nuclear criticality safety.

Approximate enrollment: U=1,700; M=500; D=200

DR. MICHAEL B. CUTLIP
Department of Chemical Engineering
University of Connecticut
Box U-222
Storrs, CT 06269-3222

P: (203) 486-0321
F: (203) 486-2959
E: mcutlip@uconnvm.edu

Dr. Cutlip specializes in the use of numerical methods to solve engineering problems. He is the co-author of POLYMATH, a user-friendly software package that solves complex problems involving Simultaneous Ordinary Differential Equations, Simultaneous Linear and Nonlinear Algebraic Equations and Polynomial, Multiple Linear and Nonlinear Regressions. This program has been incorporated in the chemical engineering curriculum at UCONN and in 120 other chemical engineering departments. Dr. Cutlip's research involves adsorption, photocatalysis, and steady state/transient catalysis.

The Pollution Prevention Research and Development Center, a part of the Environmental Research Institute, is involved in the development and improvement of pollution prevention technologies using research, collaboration with government and industry, and education initiatives.

Approximate enrollment: U=900; M=210; D=260

DR. DIANNE DORLAND
Department of Chemical Engineering
University of Minnesota, Duluth
231 Engineering Building
10 University Drive
Duluth, MN 55812

P: (218) 726-7127
F: (218) 726-6360
E: ddorland@d.umn.edu

This Chemical Engineering department has a new program incorporating the chemical engineering aspects of pollution prevention in junior and senior level courses. Curriculum stresses planning for waste reduction, especially in design, and urges students to reassess the current management of chemical processes.

Approximate enrollment: U=470

DR. SHELDON DUFF
Forest Products Waste Management
Department of Chemical Engineering
University of British Columbia
2216 Main Mall
Vancouver, British Columbia
CANADA V6T 1Z4

P: (604) 822-9485
F: (604) 822-6003
E: sduff@chml.ubc.ca

Dr. Duff teaches a course entitled, *Pollution Prevention and Waste Minimization Engineering for Chemical and Process Industries*, which was added to the curriculum in the fall of 1996. His research involves process steam treatment and recycling, and the integration of pollution prevention and industrial ecology concepts into water and air pollution control courses.

Approximate enrollment: Not readily available

DR. MAHMOUD EL-HALWAGI
Chemical Engineering Department
Auburn University
Auburn, AL 36849

P: (334) 844-2064
F: (334) 844-2063
E: mahmoud@eng.auburn.edu

Dr. El-Halwagi is currently developing integrated methods for cost-effective, energy efficient pollution prevention based on chemical engineering principles.
Research opportunities exist in several areas of process engineering, including textile manufacturing, pulp and paper engineering, and asphalt processes. An opportunity to study the design of hybrid waste recovery processes is also offered.

Approximate enrollment: U=2,700; M=350; D=150

DR. REX T. ELLINGTON
Chemical Engineering Department
University of Oklahoma
Science and Public Policy Program
100 E. Boyd, Room S206
Norman, OK 73019

P: (405) 325-2554
F: (405) 325-7695
E: ellingto@gslan.offsys.uoknor.edu

Dr. Ellington is involved in pollution prevention course and curriculum development for engineering, business, and continuing education students at all levels. A multi-disciplinary view is taken as students study pollution prevention toward sustainable development with economic, environmental, energy use, and product quality issues considered.

The Institute for Gas Utilization Technologies offers research opportunities in gas conversion, gas storage and separations, combustion, and vehicle fuels. The utilization of natural gas as an environmentally friendly fuel source and chemical feed stock is emphasized.

Approximate enrollment: U=110; M=170; D=95

DR. ROBERT M. ENICK
Department of Chemical and Petroleum
  Engineering
University of Pittsburgh
1249 Benedum Hall
Pittsburgh, PA 15261

P: (412) 624-9649
F: (412) 624-9639
E: enick@engrng.pitt.edu

Dr. Enick's research is related to high-pressure phase behavior in petroleum engineering problems. He is also involved in a plastics recycling project involving a separation technique that has wide range industry applications. Dr. Enick is currently developing a pilot-scale apparatus for this technology.

See entry for Dr. Eric J. Beckman in this section for student research opportunities and enrollment data at the University of Pittsburgh.

DR. CHENG-SHEN FANG
Department of Chemical Engineering
University of Southwestern Louisiana
Box 42251
Lafayette, LA 70504

P: (318) 231-5350
F: (318) 231-6688
E: Unavailable

Dr. Fang conducts research and teaches courses in petrochemical waste treatment and minimization and pollution prevention. The primary focus of Dr. Fang's work is end-of-pipe treatment methods.

There are four special research centers for science and engineering that offer research opportunities for students.

Approximate enrollment: Not readily available

DR. JIM FERRELL
Pollution Prevention Research Center
North Carolina State University
Chemical Engineering Program
Raleigh, NC 27695-7905

P: (919) 515-1818
F: (919) 515-3465
E: Unavailable

Dr. Ferrell has taught a course in industrial waste reduction in the past, and pollution prevention has been integrated into the chemical engineering curriculum. Dr. Ferrell's research is related to pollution prevention in petroleum refining and silicon chip manufacturing.

Several research opportunities exist at North Carolina State University in various fields. One facility of particular interest is the Kenan Center for the Utilization of CO_2 in Manufacturing. At this facility, research revolves around the development of engineering and scientific principles required to replace aqueous and organic solvents in manufacturing operations with carbon dioxide, a low-cost, environmentally and chemically benign alternative. Research at the Kenan Center is conducted jointly with UNC-Chapel Hill, and is supported by 16 industrial members, government grants and contracts.

Approximate enrollment: U=4,400; M=820; D=530

DR. MARVIN FLEISCHMAN
Industrial Assessment Center
University of Louisville
Department of Chemical Engineering
Louisville, KY 40292

P: (502) 852-6357
F: (502) 852-6355
E: mofleio1@ulkyvm.louisville.edu

Dr. Fleischman, one of the premier educators in the pollution prevention field, teaches *Pollution Prevention, Waste Treatment, and Disposal*, a course dealing with pollution prevention concepts, applications and issues based on actual assessments at various manufacturing plants. The chemical engineering department also offers courses in industrial waste management, pollution prevention and waste minimization. These courses include guest speakers, field trips, and projects at a local manufacturing facility. The Industrial Assessment Center conducts combined full facility quantitative energy and multi-media waste minimization at manufacturing facilities using students and faculty.

Research opportunities in Waste Minimization and Pollution Prevention are offered through the Industrial Assessment Center.

Approximate enrollment: U=Not readily available; M=500; D=70

DR. WILLIAM JAMES FREDERICK
Chemical Engineering Department
Oregon State University
Gleeson 103
Corvallis, OR 97331

P: (503) 737-2496
F: (503) 737-4600
E: frederiw@ccmail.orst.edu

Dr. Frederick teaches senior level undergraduate and graduate level courses in pollution prevention and waste minimization. He also conducts and directs research in industrial waste minimization in microelectronics and pulp and paper manufacture.

Special research facilities at Oregon State University include the Center for Advanced Materials Research, Nuclear Science and Engineering Institute, Engineering Experiment Station, and Water Resources Research Institute.

Approximate enrollment: U=2,100; M=320; D=180

DR. JEANETTE GARR
Department of Chemical Engineering
Youngstown State University
Youngstown, OH 44555

P: (216) 742-1737
F: (216) 742-1567
E: Unavailable

Dr. Garr teaches *Industrial Pollution Control, Wastewater Treatment,* and *Accident and Emergency Management* with an emphasis on pollution prevention. Her research interests include air pollution from fossil fuel combustion and steel industries, and BUSTR program for fuel storage. Dr. Garr is a fellow of the USDOE/PETC.

Approximate enrollment: Not readily available

DR. RAKESH GOVIND
Department of Chemical Engineering
University of Cincinnati
697 Rhodes Hall (ML 171)
Cincinnati, OH 45221

P: (513) 556-2761
F: (513) 556-3473
E: Unavailable

Dr. Govind teaches pollution prevention and waste minimization concepts through process synthesis and optimization techniques. Dr. Govind's research interests lie in detailed computer analyses of plants for efficiency.

Related research opportunities include topics in Air Pollution Engineering, Environmental Radiological Assessment, Separation Processes, Water Quality Processes, and Hazardous Waste Engineering and Management.

Approximate enrollment: U=1,500; M=700; D=260

DR. CHRISTINE S. GRANT
Department of Chemical Engineering　　P: (919) 515-2317
North Carolina State University　　　　F: (919) 515-3465
Box 7905　　　　　　　　　　　　　　E: grant@eos.ncsu.edu
Raleigh, NC 27695-7905

Dr. Grant teaches a graduate level course for chemical and non-chemical engineers entitled, *Advances in Pollution Prevention: Environmental Management for the Future*. This course focuses on developing strategies for pollution prevention and waste minimization, with an emphasis on the design of industrial processes that minimize or eliminate chemical waste production using ASPEN Model Manager. Included in the course of study are regulatory efforts and legislation, case studies, current research efforts, life cycle analysis, and speakers from the manufacturing and pollution prevention fields.

See entry for Dr. Jim Ferrell in this section for student research opportunities and enrollment data at North Carolina State University.

DR. WILLIAM HECKER
Department of Chemical Engineering　　P: (801) 378-6235
Brigham Young University　　　　　　F: (801) 378-7799
350 CB　　　　　　　　　　　　　　　E: hecker@byu.edu
Provo, UT 84602

Dr. Hecker has cultivated a comprehensive air pollution control course at the undergraduate level that introduces students to pollution prevention concepts and technologies. His research involves the use of catalytic converters as an end-of-pipe treatment method for NOx reduction.

The Advanced Combustion Engineering Research Center (ACERC) offers research opportunities in the development of advanced combustion technology through fundamental engineering research and educational programs aimed at the solution of critical national combustion problems. These programs are designed to enhance the international competitive position of the U.S. in the clean and efficient use of fossil fuels and waste materials, primarily coal and other low-quality fuels.

Approximate enrollment: U=2,000; M=200; D=50

DR. J. R. HOPPER
Department of Chemical Engineering　　P: (409) 880-8784
Lamar University　　　　　　　　　　F: Unavailable
P.O. Box 10053　　　　　　　　　　　E: Unavailable
Beaumont, TX 77710

Dr. Hopper teaches a graduate course called *Waste Minimization*. He has also successfully performed a simulation of the Sohio process for the production of acrylonitrile from the catalytic ammoxidation of propylene using published kinetic and thermodynamic data to illustrate the concepts of pollution prevention by process modification.

Special research facilities at Lamar University include the Gulf Coast Hazardous Substance Research Center, Environmental Chromatography Institute, and the Space Exploration Center.

Approximate enrollment: Not readily available

DR. KRISTINA IISA
Chemical Engineering Department
Oregon State University
Gleeson 103
Corvallis, OR 97331

P: (503) 737-2346
F: (503) 737-4600
E: iisam@ccmail.orst.edu

Dr. Iisa teaches an air pollution control course for graduates and undergraduates emphasizing pollution prevention strategies.

See entry for Dr. William James Frederick in this section for student research opportunities and enrollment data at Oregon State University.

DR. RALPH KUMMLER
Hazardous Waste Management Program
Wayne State University
Department of Chemical Engineering
Detroit, MI 48202

P: (313) 577-3800
F: (313) 577-3810
E: Unavailable

The Hazardous Waste Management Program teaches pollution prevention strategies to engineers through its *Waste Minimization* course. It provides interns to small businesses in Michigan for waste reduction management and provides technical assistance for industry.

Several research opportunities are available in the area of Hazardous Waste Management, such as Waste Containment Systems, Environmental Transport and Management of Hazardous Wastes, Process Design and Synthesis based on Waste Minimization, and Diesel and Alternative Fuels.

Approximate enrollment: U=600; M=600; D=150

DR. GENNARO J. MAFFIA
Department of Chemical Engineering
Widener University
One University Place
Chester, PA 19013-5792

P: (215) 499-4089
F: (215) 499-4059
E: pfgjaffia@cyber.widener.edu

Dr. Maffia teaches pollution prevention concepts in senior level design courses, as well as a pollution prevention seminar for freshman level engineering and non-engineering students. Dr. Maffia has developed True

Basic computer models for the solution of pollution prevention problems that he uses in his classes. He also develops case studies that develop unsteady models for real-world events.

Related research opportunities exist in the areas of Energy Conservation, Environmental Engineering, and Process Design.

Approximate enrollment: U=200; M=150; D=0

DR. VASILIOS MANOUSIOUTHAKIS
Pollution Prevention Fellowship Program   P: (310) 825-9385
Chemical Engineering Department   F: (310) 206-4107
University of California, Los Angeles   E: vasilios@seas.ucla.edu
5531-J Boelter Hall
Los Angeles, CA 90095

Dr. Manousiouthakis teaches a course on pollution prevention as a synthesis activity. He also directs the Ph.D. fellowship and summer internship programs in pollution prevention and is preparing a pollution prevention textbook.

See entry for Dr. Yoram Cohen in this section for student research opportunities and enrollment data at the University of California, Los Angeles.

DR. JIM McCUNE
Chemical Engineering   P: (714) 563-0866
Fullerton College   F: (714) 563-0189
Anaheim Higher Education Center   E: Unavailable
100 Anaheim Avenue
Anaheim, CA 92805

Dr. McCune teaches pollution prevention and chemical engineering courses. He has conducted Pollution Prevention Planning for the plastics and plating industries.

Approximate enrollment: Not readily available

DR. JEFFREY MENSINGER
Department of Chemical Engineering   P: (313) 577-1200
Wayne State University   F: (313) 961-5603
Detroit, MI 48202   E: Unavailable

Dr. Mensinger teaches a course focusing on the overall management requirements for conducting waste minimization and pollution prevention assessments with insights to achieve the proper implementation of proposed programs. Case histories of successfully implemented programs are studied in this course.

See entry for Dr. Ralph Kummler in this section for student research opportunities and enrollment data at Wayne State University.

DR. SUSAN MONTGOMERY
Chemical Engineering Department
University of Michigan
3074 Dow Building
Ann Arbor, MI 48109-2136

P: (313) 936-1890
F: (313) 763-0459
E: smontgom@umich.edu

Dr. Montgomery is involved in the development of multimedia materials for pollution prevention education. She represents the American Society for Engineering Education (ASEE) on the NPPC External Advisory Committee and serves as a Mentor for the NPPC Internship Program.

Special research facilities include the Air Pollution Modeling and Monitoring Laboratory, Environmental and Water Resources Laboratory, and the Great Lakes and Mid-Atlantic Hazardous Substance Research Center.

Approximate enrollment: U=4,300; M=1,500; D=450

DR. JAMES NOBLE
Department of Chemical Engineering
Tufts University
4 Colby Street
Medford, MA 02155

P: (617) 628-5000 x2089
F: (617) 627-3991
E: Unavailable

Dr. Noble has developed a course for chemical and civil engineers called *Hazardous Waste Treatment Technologies* which exposes students to pollution prevention concepts focusing on pollution control and waste management.

Research opportunities are available in the areas of air pollution and several types of process engineering technologies.

Approximate enrollment: U=700; M=300; D=50

DR. S. TED OYAMA
Department of Chemical Engineering
Virginia Polytechnic Institute &
  State University
Blacksburg, VA 24061-0211

P: (540) 231-5309
F: (540) 231-5022
E: oyama@vt.edu

Dr. Oyama teaches an undergraduate course for sophomores entitled *Environmental Issues in Technology*. His research interest is environmental catalysis. His research topics include the cleanup of petroleum feedstocks, utilization of carbon dioxide, selective oxidation with ozone, and conversion of chlorofluorocarbons.

Pollution prevention and waste minimization research opportunities are available at Virginia Polytechnic Institute. Special research facilities include the Center for Energy and the Global Environment, the Center for Power Engineering, the Generic Mineral Technology Center in Mine Safety and Environmental Engineering, the Energy Management Institute, the Virginia Center for Coal and Energy Research, the Virginia Center for Coal and Mineral Processing, and the Waste Policy Institute.

Approximate enrollment: U=5,000; M=1,050; D=500

DR. VITO PUNZI
Department of Chemical Engineering   P: (610) 519-4946
Villanova University   F: (610) 519-7354
800 Lancaster Avenue   E: Unavailable
Villanova, PA 19085

Dr. Punzi teaches an undergraduate elective course dealing with industrial wastewater and hazardous waste handling and minimization for junior and senior level chemical engineering students. The viewpoint of the course is to approach pollution prevention technologies based on bottom-line economic amelioration. Dr. Punzi's research includes treatment and recovery of heavy metals from industrial wastewaters, and environmental applications of reverse osmosis.

Related research opportunities are available through the Environmental Engineering Department.

Approximate enrollment: U=700; M=250

DR. CHRISTIAN ROY
Department of Chemical Engineering   P: (418) 656-7406
Université Laval   F: (418) 656-2091
Ste-Foy, Quebec   E: croy@gch.ulaval.ca
CANADA G1K 7P4

Dr. Roy teaches a pollution prevention course for junior and senior level undergraduate chemical engineering students. The course includes materials from the EPA and other government agencies, scientific literature and the study of the pyrolysis process. Dr. Roy heads a research team that studies the vacuum pyrolysis process.

Seven research centers offer opportunities for students to study subjects such as mineral processing and polymer engineering. Among the seven centers are two Canadian Centers of Excellence and two multifaculty research centers.

Approximate enrollment: U=1,300; M=Not readily available; D=Not readily available

DR. DALE F. RUDD
Chemical Engineering Department
University of Wisconsin, Madison
Madison, WI 53706

P: Unavailable
F: Unavailable
E: rudd@engr.wisc.edu

Dr. Rudd's research interests include pollution prevention in chemical process industries, process design and catalysis industrial development.

The Energy Technology Center, Nuclear Safety Research Center, Solar Energy Center, and the Waste Research and Education Center/Solid and Hazardous Waste Education Center all offer students research opportunities in pollution prevention and energy conservation.

Approximate enrollment: U=3,200; M=520; D=550

DR. HENRY SHAW
Department of Chemical Engineering,
  Chemistry, and Environmental Science
New Jersey Institute of Technology
University Heights
138 Warren Street
Newark, NJ 07102

P: (201) 596-2938
F: (201) 802-1946
E: shaw@admin.njit.edu

Dr. Shaw teaches graduate and undergraduate courses in air pollution control, global environmental problems, catalysis, diffusional systems, and process and plant design. Pollution prevention issues are covered in the plant design course. Dr. Shaw's research includes the incineration of hazardous substances, soot/NO_x control in diesel engines using a fluidized bed reactor, absorption of acid gases, and the scale-up of organic processes in multiphase aqueous systems in order to avoid using polluting solvents as an approach to pollution prevention.

The Center for Environmental Engineering and Science is internationally recognized as the largest industry/university cooperative program. The Center specializes in researching hazardous substance management techniques.

Approximate enrollment: U=1,750; M=630; D=180

DR. DAVID R. SHONNARD
Department of Chemical Engineering
Michigan Technological University
1400 Townsend Drive
Houghton, MI 49931

P: (906) 487-3468
F: (906) 487-3213
E: drshonna@mtu.edu

Dr. Shonnard teaches an undergraduate course for chemical and environmental engineering students on the fundamentals of industrial pollution assessments on the environment and on practiced pollution prevention methods in the chemical industry. Basic chemical engineering principles and software models are applied to solving problems, and economic issues of pollution prevention are also discussed.

The National Center for Clean Industrial and Treatment Technologies offers research opportunities for students in a variety of areas of pollution prevention. The Center is dedicated to creating industrial facilities where waste is minimized by applying economically sound technologies, optimized manufacturing processes, treatment operations, and the reuse of materials.

Approximate enrollment: U=3,500; M=230; D=130

DR. DILIP SINGH
Department of Chemical Engineering   P: (216) 742-1737
Youngstown State University   F: (216) 742-1998
Youngstown, OH 44555   E: Unavailable

Dr. Singh teaches courses on industrial pollution control, wastewater treatment, and accident and emergency management with an emphasis on pollution prevention. Dr. Singh's research includes the application of artificial intelligence and neural network methodologies to process dynamics and control.

Approximate enrollment: Not readily available

MR. WIBOWO M. H. SURJOWIDJOJO
Microbiology and Bioprocess   P: (62) 22 250 6454
 Engineering Lab   (62) 22 250 0989 x331
Department of Chemical Engineering   F: (62) 22 250 1438
Institut Teknology Bandung   E: wibowo@bandung.
Jalan Ganesha 10   wasantara.net.id
Bandung INDONESIA 40132

Dr. Surjowidjojo is currently working on incorporating general concepts of pollution prevention into teaching and research and establishing courses on pollution prevention and environmental management.

Approximate enrollment: Not readily available

DR. LOUIS THEODORE
Department of Chemical Engineering   P: (718) 862-7185
Manhattan College   F: (718) 862-7189
Bronx, NY 10471   E: marseykt@aol.com

Dr. Theodore teaches a graduate engineering course entitled *Pollution Prevention*. The course introduces students to equipment and process calculations and devotes a considerable amount of time to the overall philosophy and the general economic issues of pollution prevention. Dr. Theodore has developed an USEPA training course (with slides) titled *Pollution Prevention*. Dr. Theodore has co-authored a 1992 Van Nostrand Reinhold graduate level textbook called Pollution Prevention, which served as a basis for this textbook. He has also published a tutorial entitled *Pollution Prevention*, with sixty problems dealing with topics from energy conservation to domestic issues (ETS Theodore Tutorials', Roanoke, VA, 1994, (800) 424-7184), a *P2 Problems and Solutions* text in 1994 (Gordon and Breach, Newark, NJ), and a non-technical text keying on pollution prevention titled, Fifty Major Environmental Issues Facing the 21st Century (Prentice-Hall, 1996).

The Environmental Engineering Department offers related research opportunities in fields such as water quality, hazardous material containment from spills, and toxic chemical modeling in natural water systems.

Approximate enrollment: U=510; M=200

DR. EDWARD M. TRUJILLO
Department of Chemical and Fuels
  Engineering
University of Utah
3290 Merrill Engineering Building
Salt Lake City, UT 84112

P: (801) 581-4460
F: (801) 581-8692
E: edward.trujillo@m.
    cc.utah.edu

Dr. Trujillo is currently developing a pollution prevention course for the Department of Chemical and Fuels Engineering, and developing mathematical models of acid mine drainage. He conducts research on minimizing pollution effects from mining operations.

Related research opportunities are offered through the Chemical and Fuels Engineering and Mining Engineering in the areas of catalysis, fossil-fuels conversion, hazardous waste management, innovative mining systems, and mine reclamation.

Approximate enrollment: U=1,400; M=330; D=300

DR. DEAN ULRICHSON
Department of Chemical Engineering
Iowa State University
Sweeney Hall
Ames, IA 50011

P: (515) 294-6944
F: (515) 294-2689
E: dlulrich@iastate.edu

Dr. Ulrichson teaches a senior level undergraduate course that introduces students to pollution prevention concepts. He also teaches courses in process simulation and design that include safety, health and environmental considerations. Dr. Ulrichson is also coordinating the development of an environmental engineering curriculum.

Iowa State University has several research facilities that offer research opportunities for students. The Ames Research Laboratory of the U.S. Department of Energy, the Center for Building Energy Research, the Center for Coal and the Environment, the Industrial Assessment Center, and the Iowa Energy Center are all devoted to using cleaner energy sources and minimizing production wastes.

Approximate enrollment: U=4,100; M=500; D=300

DR. MARGRIT VON BRAUN
Department of Chemical Engineering
University of Idaho
Buchanan Engineering Lab 315
Moscow, ID 83844-1025

P: (208) 885-6113
F: (208) 885-7462
E: cdixon@uidaho.edu

Dr. Von Braun teaches courses in pollution prevention, environmental audits and hazardous waste management for junior and senior undergraduate and graduate students. Dr. Von Braun has also incorporated pollution prevention concepts into two chemical engineering courses entitled *Advanced Plant Design* and *Hazardous Chemical Waste*, which are open to undergraduate seniors and graduate students.

The Idaho Water Resources Research Institute and the Institute for Materials and Advanced Processing offer students the opportunity to work on research projects in energy conservation and pollution prevention.

Approximate enrollment: U=1,000; M=200; D=70

DR. JOHN W. WALKINSHAW
Chemical Engineering
University of Massachusetts, Lowell
One University Avenue
Lowell, MA 01854

P: (508) 934-3159
F: (508) 934-3047
E: Unavailable

Dr. Walkinshaw researches problems in recycled paper and paper products for the remanufacture of paper goods. He has worked with first year students on the development of a design module for the manufacture of paper and cleaning of printed recycled pulps.

The Center for Environmentally Appropriate Materials, the Sustainable Energy Center, and the Toxics Use Reduction Institute all offer research assignments in areas related to pollution prevention.

Approximate enrollment: U=650; M=530; D=110

DR. GREGORY YAWSON
Department of Chemical and Metallurgical
  Engineering   P: (313) 577-3848
Wayne State University   F: (313) 577-3810
Detroit, MI 48202   E: Unavailable

Dr. Yawson currently teaches pollution prevention as a part of a graduate level course and as a pre-college program. His research includes industrial and agricultural waste recycling, recovery and reuse.

See entry for Dr. Ralph Kummler in this section for student research opportunities and enrollment data at Wayne State University.

CIVIL AND ENVIRONMENTAL ENGINEERING

DR. PAUL ANDERSON
Department of Chemical and Environmental
  Engineering   P: (312) 567-3531
Illinois Institute of Technology   F: (312) 567-8874
10 West 33rd Street   E: enveanderson@minna.iit.edu
Chicago, IL 60616

Dr. Anderson has developed a one-semester course designed to expose engineering students to quantitative aspects of pollution prevention.

Several research opportunities are available through the chemical and environmental engineering department. The IIT Energy and Power Center is devoted to sustainable global energy development. Research projects at the Center include topics in enhanced oil and gas recovery, process monitoring and control, and pollution prevention.

Approximate enrollment: U=940; M=640; D=220

DR. C. ROBERT BAILLOD
Civil and Environmental Engineering   P: (906) 487-2520
Michigan Technological University   F: Unavailable
1400 Townsend Drive   E: baillod@mtu.edu
Houghton, MI 49931

Dr. Baillod is currently developing engineering design projects for use in senior and graduate level courses. The EPA sponsored CenCITT research program provides information on clean technologies and pollution prevention for use in graduate and senior level engineering courses.

See entry for Dr. David R. Shonnard in the chemical engineering section for student research opportunities and enrollment data at Michigan Technological University.

DR. PAUL L. BISHOP
Department of Civil and Environmental
  Engineering
University of Cincinnati
P.O. Box 210071
Cincinnati, OH 45221-0071

P: (513) 556-3675
F: (513) 556-2599
E: pbishop@boss.cee.uc.edu

Dr. Bishop teaches a course for senior and graduate level engineering students called *Environmentally Conscious Engineering*. Dr. Bishop is University Coordinator for Interdisciplinary Environmental Affairs and Director of the Center for Hazardous Waste Research and Education. He has authored a textbook and over 160 research publications. Among Dr. Bishop's research interests are the stabilization and solidification of hazardous wastes and biofilm-based water treatment processes.

See entry for Dr. Rakesh Govind in the chemical engineering section for student research opportunities and enrollment data at the University of Cincinnati.

DR. CURTIS BRYANT
Department of Civil Engineering
University of Arizona
Tucson, AZ 85721

P: (602) 621-2266
F: (602) 621-2550
E: Unavailable

Dr. Bryant is interested in developing pollution prevention education programs from a human and economic perspective. He has worked with anthropologist Dr. Rathje on the psychology of garbage production and opportunities for reuse.

Many research opportunities are offered through the Chemical and Environmental Engineering Department, including aerosol technology, hazardous and toxic wastes, and industrial wastewater treatment.

Approximate enrollment: U=2,200; M=430; D=350

MR. EDWARD CHIAN
School of Civil and Environmental
  Engineering
Georgia Institute of Technology
Atlanta, GA 30332-0512

P: (404) 894-7694
F: (404) 894-8266
E: edwards.chian@ce.gatech.edu

Mr. Chian teaches classes in solid and hazardous waste management, industrial waste treatment, and sustainable development and technology. His research interests include developing sustainable technology, pollution prevention in metal finishing industries, and dissolved air flotation membrane processes.

Research opportunities for students are available in several engineering departments. Research topics include catalysis, process control, hazard mitigation and safety, and waste management.

Approximate enrollment: U=5,500; M=1,000; D=1,100

DR. ANTHONY COLLINS
Department of Civil and Environmental
  Engineering
Clarkson University
Rowley Laboratories
Potsdam, NY 13699-5715

P: (315) 268-6490
F: (315) 268-7636
E: adminnyjb@clvm.clarkson.edu

Dr. Collins teaches courses on the design of water distribution and wastewater collection systems, and water and wastewater treatment processes. His research focuses on physical-chemical and biological treatment processes, application of expert systems, hazardous waste management, and the disposal of treatment residuals.

The Hazardous Waste and Toxic Substances Research and Management Center provides students with opportunities to conduct research in waste reduction and minimization, multimedia-exposure assessment of hazardous wastes and toxic substances, and waste treatment, remediation and disposal technologies.

Approximate enrollment: U=1,300; M=100; D=60

DR. MOHAMED DAHAB
Department of Civil Engineering
University of Nebraska-Lincoln
W348 Nebraska Hall
Lincoln, NE 68588-0531

P: (402) 472-5020
F: (402) 472-8934
E: mdahab@unl.edu

Dr. Dahab teaches courses titled *Solid Waste Management Engineering* and *Hazardous Waste Management Engineering,* which incorporate pollution prevention and waste reduction strategies. He also teaches *Pollution Prevention*, a new course that deals with issues in both government and industry. Dr. Dahab's research interests lie in waste minimization and pollution prevention.

The Civil Engineering Department offers students research projects in hazardous waste management and remediation, and the Center for Nontraditional Manufacturing Research offers students research opportunities in areas such as exploring alternative fuels for new manufacturing methods and industrial processes.

Approximate enrollment: U=1,200; M=380; D=120

DR. CAROL DIGGELMAN
Department of Physics and Chemistry
Milwaukee School of Engineering
P.O. Box 644
Milwaukee, WI 53201-0644

P: (414) 277-7320
F: (414) 277-7470
E: diggelman@warp.msoe.edu

Dr. Diggelman developed and teaches *Introduction to Hazardous and Solid Waste Management,* a course based on pollution prevention strategies for waste management. Students complete a term project under the guidelines of RCRA Subtitle C Management of a waste system to become familiar with common industry situations and practices.

Students are offered research opportunities in many fields, including energy management and hazardous waste treatment.

Approximate enrollment: U=Not readily available; M=Not readily available

DR. HADI DOWLATABADI
Engineering and Public Policy
Carnegie Mellon University
Pittsburgh, PA 15213

P: (412) 268-3031
F: (412) 268-3757
E: hd01@andrew.cmu.edu

Dr. Dowlatabadi's research interests involve the environmental impacts of energy use.

The Environmental Institute, Green Design, and the Spray Systems Technology Center are three research facilities that offer students the opportunity to conduct research in several areas related to pollution prevention and waste minimization.

Approximate enrollment: U=1,350; M=220; D=350

DR. RYAN DUPONT
Utah Water Research Laboratory
Civil and Environmental Engineering
Utah State University
UMC-8200
Logan, UT 84322-8200

P: (435) 797-3227
F: (435) 797-3663
E: rdupo@pub.uwrl.usu.edu

Dr. Dupont teaches a senior-level undergraduate environmental engineering elective course called *Air Toxics and Pollution Prevention.* This course discusses pollution prevention concepts in industry and the private sector, and emphasizes air emissions and risk factors associated with the 1990 Clean Air Act amendments. Dr. Dupont's research focuses on treatment methods for in situ bioremediation of contaminated soils and groundwater.

Related research opportunities are available in the fields of Toxic Waste Management and Treatment, and Waste Recycling.

Approximate enrollment: U=1,000; M=180; D=70

DR. HECTOR FUENTES
Department of Civil and Environmental
 Engineering
Florida International University
University Park, VH-160
Miami, FL 33199

P: (305) 348-2837
F: (305) 348-2802
E: fuentes@eng.fiu.edu

Dr. Fuentes teaches a graduate environmental engineering course in pollution prevention. His research involves pollution prevention initiatives associated with NAFTA and Latin America. His graduate research assistants test computer models of industrial pollution in Dade County, Florida.

Approximate enrollment: U=Not readily available; M=Not readily available

DR. KUMAR GANESAN
Department of Environmental
 Engineering
University of Montana
West Park Street
Butte, MT 59701

P: (406) 496-4239
F: (406) 496-4133
E: kganesan@mtvms2.mtech.edu

Dr. Ganesan teaches courses on pollution prevention, air pollution control, industrial ventilation, and particle technology.

Approximate enrollment: U=Not readily available; M=Not readily available; D=Not readily available

DR. BERNICE GOLDSMITH
Social Aspects of Engineering
Concordia University
1455 de Maisonneuve Boulevard West
Montreal, Quebec
CANADA H3G 1M8

P: (514) 848-3071
F: (514) 848-4509
E: bernice@VAX2.concordia.ca

Dr. Goldsmith teaches a graduate engineering course called *Environmental Life Cycle Assessment*, which investigates the environmental and social profiles of products, processes, and services. She also teaches a course called *Engineering, Resources and Environment*.

Approximate enrollment: U=Not readily available; M=Not readily available; D=Not readily available

DR. NANCY J. HAYDEN
Department of Civil and Environmental
  Engineering
University of Vermont
Burlington, VT 05405

P: (802) 656-1924
F: (802) 656-8446
E: nhayden@emba.uvm.edu

Dr. Hayden teaches *Hazardous Waste Management Engineering*, a course where students explore source reduction alternatives for managing hazardous wastes through an interactive term project with a local business or industry.

Related research opportunities are available in air pollution, hazardous waste management, and manufacturing processes.

Approximate enrollment: U=450; M=110; D=40

DR. ISABEL HEATHCOTE
Environmental Engineering and
  Environmental Sciences
University of Guelph
226 Thornbrough Building
Guelph, Ontario
CANADA N1G 2W1

P: (519) 824-4120 x3072
F: (519) 836-0227
E: heathcot@net2.eos.uoguelph.ca

Dr. Heathcote co-teaches a course on pollution control planning based on Ontario legislation regarding pollution control standards for surface water and non-point source pollution.

Approximate enrollment: U=Not readily available; M=Not readily available; D=Not readily available

DR. SUNIL HERAT
Environmental Engineering
Griffith University
Faculty of Environmental Sciences
Queensland, AUSTRALIA 4111

P: +61 7 3875 5288
F: Unavailable
E: s.herat@ens.gu.edu.au

Dr. Herat teaches courses in graduate and undergraduate environmental engineering, including a pollution prevention course that includes site visits to plants.

Approximate enrollment: U=Not readily available; M=Not readily available; D=Not readily available

DR. WILLIAM JAMES
School of Engineering
University of Guelph
Guelph, Ontario
CANADA N1G 2W1

P: (519) 824-4120 x2433
F: (519) 767-2770
E: wjames@net2.eos.uoguelph.ca

Dr. James co-teaches *Pollution Prevention Planning,* a course based on Ontario legislation for pollution prevention in surface water and non-point source pollution. His research includes modeling the long-term impacts of surface water pollution and flows resulting from development.

Approximate enrollment: U=Not readily available; M=Not readily available; D=Not readily available

DR. DAVID KIBLER
Civil Engineering Department
Virginia Polytechnic Institute and State University
200 Patton Hall
Blacksburg, VA 24061-0105

P: (540) 231-8309
F: (540) 231-7532
E: kiblerdf@vt.edu

Dr. Kibler's research and teaching interests involve urban hydrology and flood control, stormwater management, and management practices for reducing non-point source pollutant entry into receiving waters.

See entry for Dr. S. Ted Oyama in the chemical engineering section for student research opportunities and enrollment data at Virginia Polytechnic Institute and State University.

DR. EDWARD KLEVANS
Nuclear Engineering
The Pennsylvania State University
Sackett Building
University Park, PA 16802

P: (814) 865-1341
F: (814) 865-8499
E: Unavailable

Dr. Klevans teaches both an undergraduate and a graduate course in radioactive waste management. His research interests include modeling cement behavior for long term low-level waste storage, thermal hydraulic safety, and fuel management.

The Metal Casting Research Center is one of many research facilities at Penn State. The Center is dedicated to finding ways to adapt to regulatory and environmental mandates, and find cost-effective solutions for reclamation and material reuse in foundries.

Approximate enrollment: U=7,300; M=970; D=870

DR. REID LEA
Department of Civil Engineering
University of New Orleans
823 Engineering Building
New Orleans, LA 70148

P: (504) 286-7089
F: (504) 286-5586
E: wrlce@basin.crc.uno.edu

Dr. Lea teaches *Pollution Prevention Plans*, a course designed to develop pollution prevention agendas for industry as required by law. Dr. Lea's research involves waste management and environmental engineering.

The Urban Waste Management and Research Center coordinates studies of the environmental interactions of air, water and land as they relate to preventative and corrective measures used to eliminate sources of pollution in urban areas.

Approximate enrollment: U=980; M=140; D=20

LT. COL. STEVEN L. LOFGREN
Department of Engineering and
  Environmental Management
Air Force Institute of Technology
AFIT/ENV
2950 P Street, Building 640
Wright-Patterson AFB, OH 45433-7765

P: (513) 255-6565 x4314
F: (513) 476-4699
E: slofgren@afit.af.mil

Lt. Col. Lofgren teaches and leads graduate student research groups in the field of environmental management. The Master's degree program stresses the importance of implementing strategic operational decisions considering their effects on interrelated systems.

Several industry and federally funded research projects in many fields are available to students. Some topics include air pollution transport modeling, environmental management in organizations, hazardous waste minimization, risk analysis, and solid and hazardous waste control.

Approximate enrollment: M=310; D=80

DR. KRISHNANAND MAILLACHERUVU
Department of Civil and Environmental
  Engineering
Polytechnic University
Six Metrotech Center
Brooklyn, NY 11201

P: (718) 260-3260
F: (718) 260-3433
E: kmaillac@duke.poly.edu

Dr. Maillacheruvu teaches pollution prevention concepts in undergraduate and graduate engineering courses. Some of Dr. Maillacheruvu's research work includes pollution prevention source reduction measures, landfill reclamation strategies and decision support system development, and bioremediation enhancement in subsurface environments. Several research opportunities are available for students. Related topics include energy and power, and environmental issues and government regulations.

Approximate enrollment: U=1,600; M=850; D=140

DR. JOSEPH M. MARCHELLO
Old Dominion University
Kaufman-Duckworth Room 35
Norfolk, VA 23529-0241

P: (804) 683-3753
F: (804) 683-5354
E: jmm100u@triton.kdh.odu.edu

Dr. Marchello teaches graduate civil and environmental engineering courses in pollution prevention, air quality, and solid and hazardous wastes. Some of Dr. Marchello's recent research projects include refuse-driven fuel and bioremediation, and the control of diesel engine air emissions.

The Industrial Assessment Center offers students research opportunities in discovering ways to reduce energy expenditures. The Center offers free student and faculty assessments to companies wishing to reduce their energy costs.

Approximate enrollment: U=580; M=250; D=110

DR. DAVID H. MARKS
Department of Civil and Environmental
  Engineering
Massachusetts Institute of Technology
Room 1-123
Cambridge, MA 02139

P: (617) 253-1992
F: (617) 258-6099
E: dhmarks@mit.edu

Dr. Marks serves as the Director of the MIT Program for Environmental Engineering Education and Research. His research interests include environmental systems, water resource systems, environmental management, industrial ecology, and environmental remediation.

The Center for Technology, Policy and Industrial Development, the Industrial Research Lab, and the Leaders for Manufacturing Program all sponsor student research projects in the fields of pollution prevention and waste minimization in industry.

Approximate enrollment: U=2,100; M=1,300; D=1,100

DR. MALCOLM J. McPHERSON
Department of Mining and Minerals Engineering
Virginia Polytechnic and State University
Blacksburg, VA 24061-0239

P: (540) 231-8109
F: (540) 231-4070
E: mmcphrsn@vt.edu

Dr. McPherson is the author of a book entitled <u>Subsurface Ventilation and Environmental Engineering</u> (Chapman & Hall, 1993). His research interests include air pollution resulting from emissions of liquids, gases, and particulates from mining activities, land pollution resulting from land disturbance from mining, and the control of ventilation and air quality in subsurface environments.

See entry for Dr. S. Ted Oyama in the chemical engineering section for student research opportunities and enrollment data at Virginia Polytechnic Institute and State University.

DR. CAROL J. MILLER
Department of Civil and Environmental
  Engineering
Wayne State University
Detroit, MI 48202

P: (313) 577-3876
F: (313) 577-3881
E: cmiller@eng.wayne.edu

Dr. Miller teaches courses in *Groundwater Modeling* and *Groundwater/Hydraulics*. Her research includes the design and construction of landfills and computer modeling of subsurface environmental problems.

See entry for Dr. Ralph Kummler in the chemical engineering section for student research opportunities and enrollment data at Wayne State University.

DR. DONALD MODESITT
Environmental Engineering Program
University of Missouri
Department of Civil Engineering
Rolla, MO 65401

P: (314) 341-4452
F: (314) 341-4729
E: Unavailable

Dr. Modesitt teaches undergraduate environmental engineering courses that introduce concepts such as process modification to prevent industrial pollution. His research interests include water quality and treatment, hazardous waste treatment, public health concerns, and aquiculture.

The Environmental Research Center, the Center for Environmental Science and Technology, the Center for Technology Transfer, and the Cloud and Aerosol Science Lab are some of the many facilities that offer students the opportunity to conduct research in environmental processes and pollution prevention control technologies.

Approximate enrollment: U=2,900; M=480; D=160

DR. DAVID MOY
Griffith University
School of Environmental Engineering
Queensland, AUSTRALIA 4111

P: 61 7 3875 5506
F: 61 7 3875 5288
E: d.moy@ens.gu.edu.au

Dr. Moy teaches environmental engineering undergraduate, graduate diploma, and Master's degree courses.

Approximate enrollment: U=Not readily available; M=Not readily available; D=Not readily available

MR. EL KHOBAR M. NAZECH
Department of Civil Engineering
University of Indonesia
Campus Ul Depok
Jakarta, INDONESIA 16424

P: (62) 21 727 0029
F: (62) 21 727 0028
E: khobar@makara.cso.ui.ac.id

Mr. Nazech designed the undergraduate and graduate engineering courses in *Pollution Prevention* and *Clean Industrial Production.* The courses are taught using Indonesian case histories. Mr. Nazech also assisted in the implementation and planning of US AID's waste reduction assessment studies on specific Indonesian industries with US experts.

Approximate enrollment: U=Not readily available; M=Not readily available; D=Not readily available

DR. FREDERICK G. POHLAND
University of Pittsburgh
Department of Civil and Environmental
  Engineering
1141 Benedum Hall
Pittsburgh, PA 15261-2240

P: (412) 624-1880
F: (412) 624-0135
E: pohland@civ.pitt.edu

Dr. Pohland teaches graduate environmental engineering courses focused on pollution prevention and waste minimization techniques. His research includes environmental engineering operations and processes, industrial and hazardous waste management, and environmental impact assessment.

See entry for Dr. Eric J. Beckman in the chemical engineering section for student research opportunities and enrollment data at the University of Pittsburgh.

DR. ROBERT B. POJASEK
Senior Program Director, Cambridge
  Environmental Inc.
58 Charles Street
Cambridge, MA 02141

P: (617) 225-0812
F: (617) 225-0813
E: camenv58@aol.com

Dr. Pojasek is a lecturer at Tufts University and an adjunct professor at Harvard University. He teaches graduate level engineering courses on pollution prevention with an emphasis on waste minimization rather than the design of new treatment facilities.

DR. ANGELOS PROTOPAPAS
Department of Civil and Environmental
 Engineering
Polytechnic University
Six Metrotech Center
Brooklyn, NY 11201

P: (718) 260-3632
F: (718) 260-3136
E: Unavailable

Dr. Protopapas teaches two groundwater hydrology courses that introduce pollution prevention concepts. These courses deal with contaminant transport, treatment technologies, and pollution prevention.

See entry for Dr. Krishnanand Maillacheruvu in this section for student research opportunities and enrollment data at Polytechnic University.

DR. SHELDON J. REAVEN
State University of New York
College of Engineering and Applied
 Sciences
210 Old Engineering
Stony Brook, NY 11794-2250

P: (516) 632-8765
F: (516) 632-7809
E: sreaven@ccmail.sunysb.edu

Dr. Reaven teaches a graduate course called *Diagnosis of Disputes in Pollution Prevention*. He serves as the Executive Editor of *Journal of Environmental Systems*, which publishes articles on pollution prevention. Dr. Reaven has devised pollution prevention and waste minimization plans for several different industries, including electronics and petrochemicals.

Approximate enrollment: U=Not readily available; M=Not readily available; D=Not readily available

DR. LISA RIEDLE
Department of Civil Engineering
University of Wisconsin, Platteville
1 University Plaza
Platteville, WI 53818

P: (608) 342-1539
F: (608) 342-1566
E: Unavailable

Dr. Riedle is developing ways to incorporate pollution prevention and source reduction concepts into the civil engineering curriculum. She teaches substitution and avoidance as sound environmental practice with respect to the use of many consumer products.

Approximate enrollment: U=1,400

DR. DIPAK ROY
Department of Civil and Environmental
  Engineering
Polytechnic University
Six Metrotech Center
Brooklyn, NY 11201

P: (718) 260-3768
F: (718) 260-3433
E: droy@duke.poly.edu

Dr. Roy teaches courses in biological and chemical unit processes for water and wastewater treatment, hazardous waste management, hazardous waste site remediation, and air pollution. Dr. Roy's recent research activities included hazardous waste management using in situ remediation techniques like bioremediation and surfactant soil flushing, and soil washing using microbubble gas suspensions known as Colloidal Gas Aphrons.

See entry for Dr. Krishnanand Maillacheruvu in this section for student research opportunities and enrollment data at Polytechnic University.

DR. MICHAEL L. SHELLEY
Department of Engineering and
  Environmental Management
Air Force Institute of Technology
AFIT/ENV
2950 P Street, Building 640
Wright-Patterson AFB, OH 45433-7765

P: (513) 255-2998
F: (513) 476-4699
E: mshelley@afit.af.mil

Dr. Shelley teaches graduate engineering courses in pollution prevention, environmental management and policy, economic and decision analysis, environmental systems engineering, and quality control and management.

See entry for Lt. Col. Steven L. Lofgren in this section for student research opportunities and enrollment data at the Air Force Institute of Technology.

DR. SAMPAT SRIDHAR
Carleton University
Department of Civil and Environmental
  Engineering
1125 Colonel Bay Drive
Ottawa, Ontario
CANADA K1S 5B6

P: (613) 520-2600 x8280
F: (613) 520-3951
E: ssridhar@ccs.carleton.ca

Some of Dr. Sridhar's teaching and research topics include solid, liquid and gaseous pollution abatement, hazardous and radioactive waste management, industrial wastewater treatment, and the use of UV technologies for pollution abatement.

Approximate enrollment: U=Not readily available; M=Not readily available; D=Not readily available

DR. MICHAEL K. STENSTROM
Civil and Environmental Engineering
 Department
University of California-Los Angeles
4173 Engineering I
Los Angeles, CA 90095

P: (310) 825-1408
F: (310) 206-5476
E: stenstro@seas.ucla.edu

Dr. Stenstrom teaches courses in treatment and prevention of water pollution. His research interests focus on process development for water and wastewater treatment systems, including mathematical modeling and optimization.

See entry for Dr. Yoram Cohen in the chemical engineering section for student research opportunities and enrollment data at the University of California, Los Angeles.

MR. ENDRO SUSWANTORO
Department of Environmental Engineering
University of Trisakti
JL-Kiyai Tapa No 1
Jakarta, INDONESIA 11440

P: (62) 21 566 3232 x767
F: (62) 21 560 2575
E: Unavailable

Mr. Suswantoro is planning to develop courses in pollution prevention. He conducts research in environmental impact assessment and waste reduction, concentrating on industry.

Approximate enrollment: U=Not readily available; M=Not readily available; D=Not readily available

DR. KEN WILLIAMSON
Department of Civil Engineering
Oregon State University
Apperson Hall 202
Corvallis, OR 97331-2301

P: (503) 737-6836
F: (503) 737-3052
E: williamk@ccmail.orst.edu

Dr. Williamson teaches graduate and undergraduate courses in pollution prevention.

See entry for Dr. William James Frederick in the chemical engineering section for student research opportunities and enrollment data at Oregon State University.

DR. SANDRA WOODS
Department of Civil Engineering
Oregon State University
Apperson Hall
Corvallis, OR 97331-2301

P: (503) 737-6837
F: (503) 737-3099
E: Unavailable

Dr. Woods is working to incorporate pollution prevention into the engineering curriculum at Oregon State.

She has worked with Dr. Williamson in developing Waste Reduction seminars that were used to introduce pollution prevention at the graduate level.

See entry for Dr. William James Frederick in the chemical engineering section for student research opportunities and enrollment data at Oregon State University.

DR. NAZLI YESILLER
Department of Civil and Environmental
 Engineering
Wayne State University
Detroit, MI 48202

P: (313) 577-3766
F: (313) 577-3881
E: yesiller@eng.wayne.edu

Dr. Yesiller teaches *Land Disposal of Hazardous Waste*, *Landfill Design*, and *Remediation Geotechnics*. Dr. Yesiller's research involves landfill design and construction, and computer modeling of subsurface environmental problems.

See entry for Dr. Ralph Kummler in the chemical engineering section for student research opportunities and enrollment data at Wayne State University.

DR. THOMAS YOUNG
Civil and Environmental Engineering
Clarkson University
Rowley Laboratories
Potsdam, NY 13699-5715

P: (315) 268-4430
F: (315) 268-7636
E: adminnyjb@clvm.clarkson.edu

Dr. Young teaches several undergraduate and graduate courses in water quality simulation modeling for surface and subsurface aquatic systems, water renovation processes, and wet chemical and instrumental environmental analysis. His research interests include contaminant fate and transport modeling in aquatic systems, fluvial load estimation, and applications to industrial and hazardous waste management.

See entry for Dr. Anthony Collins in this section for student research opportunities and enrollment data at Clarkson University.

DR. AMY ZANDER
Civil and Environmental Engineering
Clarkson University
Rowley Laboratories
Potsdam, NY 13699-5715

P: (315) 268-6532
F: (315) 268-7636
E: adminnyjb@clvm.clarkson.edu

Dr. Zander teaches courses in water and wastewater quality and treatment. Her research includes physical, chemical and biological processes in wastewater treatment, and membrane phase contact processes and pressure-driven membrane processes used in wastewater treatment.

See entry for Dr. Anthony Collins in this section for student research opportunities and enrollment data at Clarkson University.

INDUSTRIAL ENGINEERING

DR. DAVID CRESS
Petroleum Engineering Department
Marietta College
215 Fifth Street
Marietta, OH 45750

P: (614) 376-4780
F: (614) 376-4777
E: cressd@mcnet.marietta.edu

Dr. Cress teaches an introductory course in industrial engineering that focuses on life-cycle analysis and design for the environment. He is currently developing pollution prevention sections for courses in air pollution control and environmental science.

Approximate enrollment: U=Not readily available; M=Not readily available; D=Not readily available

DR. MIRIAM HELLER
Department of Industrial Engineering
University of Houston
4800 Calhoun Road
Houston, TX 77204-4812

P: (713) 743-4193
F: (713) 743-4190
E: heller@jetson.uh.edu

Dr. Heller developed and introduced the *Industrial Ecology* course into the Industrial Engineering curriculum. Her research includes expert design and selection of pollution prevention technology in the metal finishing and electroplating industries.

Related research subjects include Manufacturing Systems, Environmental Systems Modeling, and Facility Layout. One example of a student project is the Remediation and Disposal of High Efficiency Particulate Air (HEPA) Filters.

Approximate enrollment: U=1,100; M=420; D=220

DR. K. B. RUNDMAN
Metallurgical and Materials Engineering
Michigan Technological University
1400 Townsend Drive
Houghton, MI 49931

P: (906) 487-2632
F: (906) 487-2934
E: krundman@mtu.edu

Dr. Rundman is developing a senior and graduate level course on material and energy flow in an industrial society, and has co-developed a sophomore level course called *Engineering for the Environment*. Both courses address pollution prevention concepts. Dr. Rundman is also developing a course on specific pollution prevention problems in the foundry industry, such as recycling, air quality, and solid waste management.

See entry for Dr. David R. Shonnard in the chemical engineering section for student research opportunities and enrollment data at Michigan Technological University.

DR. JULIE ANN STUART
Department of Industrial, Welding, and
  Systems Engineering
The Ohio State University
1971 Neil Avenue
Columbus, OH 43210-1271

P: (614) 292-6239
F: (614) 844-7852
E: jastuart@cadcam.eng.ohio-state.edu

Dr. Stuart teaches courses in industrial and systems engineering with pollution prevention concepts and environmental issues included in the coursework.

The Center for Industrial Sensors and Measurements offers students the opportunity to conduct research in the improvement of manufacturing processes and the Water Resources Center offers students the chance to discover new beneficial uses for waste flue gas and develop treatment technologies for wastewater.

Approximate enrollment: U=4,300; M=600; D=800

MECHANICAL ENGINEERING

DR. JAMES W. BLACKBURN
Mechanical Engineering
Southern Illinois University at Carbondale
Carbondale, IL 62901

P: (618) 453-7008
F: (618) 453-7455
E: blackburn@engr.siu.edu

Dr. Blackburn is currently teaching graduate and undergraduate courses and implementing a research program in bioremediation and pollution prevention. One of Dr. Blackburn's research interests is pollution prevention through tuning complex chemical processes and bioprocesses for higher yields and lower by-product and waste generation.

Related research opportunities are available in the areas of coal combustion residues management, engine pollution control, recycling and utilization of industrial wastes for construction applications, and toxic waste treatment.

Approximate enrollment: U=800; M=100; D=0

DR. BARNEY L. CAPEHART
Industrial and Systems Engineering
University of Florida
303 Weil Hall
Box 116595
Gainesville, FL 32611

P: (352) 392-3180
F: (352) 392-3537
E: capehart@ise.ufl.edu

Dr. Capehart teaches a course called *Energy Management*. He is the author of a textbook entitled Guide to Energy Management (Atlanta: Fairmont Press, 1994).

Student research opportunities are available in air quality, alternative energy techniques, environmental resource management, and nuclear waste technology.

Approximate enrollment: U=3,700; M=1,100; D=350

DR. SHIRLEY FLEISCHMANN
Seymour and Esther Padnos School
  of Engineering
Grand Valley State University
301 West Fulton, Suite 618
Grand Rapids, MI 49504

P: (616) 771-6762
F: (616) 771-6642
E: Unavailable

The Padnos School of Engineering has completed a two year curriculum development project that incorporates design principles and projects with environmental considerations.

The focus of the engineering program is environmentally responsible design. All students complete a year of integrated cooperative education and two-thirds of the coursework includes laboratory and practice-oriented projects.

Approximate enrollment: U=300

DR. MAHENDRA S. HUNDAL
Mechanical Engineering
The University of Vermont
Votey Building
Burlington, VT 05405-0156

P: (802) 656-1930
F: (802) 656-1929
E: hundal@emba.uvm.edu

Dr. Hundal teaches a course called *Design for Environment*. Dr. Hundal's past work has included designing for manufacturability, designing for cost, product development, and noise and vibrations.

See entry for Dr. Nancy J. Hayden in the civil and environmental engineering section for student research opportunities and enrollment data at the University of Vermont.

DR. EDWARD S. RUBIN
Center for Energy and Environmental
  Studies
Carnegie Mellon University
128 A Baker Hall
Pittsburgh, PA 15213

P: (412) 268-5897
F: (412) 268-3757
E: rubin+@cmu.edu

Dr. Rubin is working on an NSF project to enhance the environmental content of the freshman and sophomore curricula. Dr. Rubin's research involves the green design of electric power systems and chemical processes.

See entry for Dr. Hadi Dowlatabadi in the civil and environmental engineering section for student research opportunities and enrollment data at Carnegie Mellon University.

DR. J. K. SPELT
Department of Mechanical Engineering
University of Toronto
5 King's College Road
Toronto, Ontario
CANADA M5S 1A4

P: (416) 978-5435
F: (416) 978-7753
E: spelt@me.utoronto.ca

Dr. Spelt teaches a course called *Environmental Engineering*, which deals with pollution prevention, applied ecology, regulatory theory, the causes of environmental disturbances, and various aspects of energy conservation.

Approximate enrollment: U=Not readily available; M=Not readily available; D=Not readily available

NAVAL ENGINEERING

DR. ANASTASSIOS PERAKIS
Naval Architecture and Marine
  Engineering
University of Michigan
College of Engineering
218 NAME Building
Ann Arbor, MI 48109

P: (313) 764-3723
F: (313) 936-8820
E: tassos@engin.umich.edu

Dr. Perakis teaches a graduate Marine Systems course called *Reliability and Safety*, and an advanced seminar on marine systems safety. Dr. Perakis's research focuses on probabilistic modeling and optimization of marine systems.

See entry for Dr. Susan Montgomery in the chemical engineering section for student research opportunities and enrollment data at the University of Michigan.

INDEX

Absorbers, 92-93
Absorption, 46, 62
Accounting:
   environmental, 151-160, 177-178
   financial, 152
   national income, 151-152
   management, 152
Accident and emergency
   management, 227-229
Activated sludge, 106
Adiabatic, 19-20
Adsorbers, 93-94
Adsorption, 49, 50, 61-62, 93-94, 96, 100, 302
Air classification, 110-111
Air stripping, 98-99
Anaerobic digestion, 108
Ancillary processes, 69-86
Arrhenius equation, 36
Autoignition temperature, 17-18
Avegadro's number, 13

Baghouse, 91
Batch reactors, 56
Biochemical oxygen demand (BOD), 23
Biological treatment, 105-109
Block diagrams, 130-131
Blowers, 74
Boilers, 114. 222
Boiling point, 15
Boyle's Law, 29-30
Budgeting, capital, 159-160

Calcination, 101-102
Capital recovery factor (CRF), 162, 164, 176
Centrifugation, 95, 98, 302-303
Charle's Law, 29-30
Chemical kinetics:
   velocity constant, 35-36
   collision frequency factor, 36
Chemical Manufacturing Association (CMA), 201-204
Chemical Oxygen Demand (COD), 20
Chemical properties, 16-20

Chemical reaction equilibrium:
   equilibrium constant, 34-35

Chemical reactors:
   reactor types, 56-58
   batch, 56
   stirred tank, 56, 57
   tubular, 56, 57
   packed bed, 56, 57, 93
   fluidized bed, 56, 57-58, 116-117
Chemical synthesis, 361-363, 364-365
Chemical treatment, water, 101-105
Chillers, 23
Clean Air Act (CAA), 190
Clean Water Act (CWS), 190
Climate Wise, 218-219
Closed-loop recycling, 1, 3
Coagulation, 97
Compaction, 111
Composting, 108-109
Compressors, 75, 78-79
Conditionally exempt small quantity generators, 245-246
Conduction, 58
Conservation law:
   of mass, 27-28, 37
   of energy, 8-29
Convection, 58
Conveyance systems, 69-75
Conveyors, 79-80
Cooling towers, 223
Cooling water, 76-77
Costs:
   capital, 149, 160-162, 178
   contingent, 155
   conventional, 153-155
   environmental, 152-159, 178
   equipment, 160-162, 178
   image and relationship, 155-156
   life cycle, 150, 153, 160
   operating, 149, 153, 162-163, 178-179
   overhead, 157
   potentially hidden, 154-155
Couplings, 71-72
Crystallization, 46, 50
CSTR, 56
Cyclone separator, 90

Dampers, 71
Decantation, 304
Definitions:
   biological properties, 20-21

494 Index

Definitions (continued):
  chemical properties, 16-20, 21-22
  fundamental, 11-13
  physical properties, 13-16, 22
Dehumidification, 47-48
Demineralized Water, 77
Density, 13-14
Design for the Environment (DfE), 174, 179, 194, 240-242
Desorption, 48-49
Dialysis, Fractional, 50
Diffusion:
  coefficient or diffusivity, 15
  gaseous, 50
  thermal, 51
  sweep, 51
Dimensional Analysis, 5
Discounted cash flow method, 164-166, 176-177
Dissolved Air Flotation, 97
Distillation, 46, 47, 60-61, 96, 300-301
Domestic activities, 247-248
Drug Manufacturing and Processing
  case studies, 370
  good housekeeping, 369-370
  pollution prevention options, 365-369
  process description, 361-364
  process modifications, 365-368
  recycling and resource recovery, 368-369
  waste description, 364-365
Dryers:
  continuous tunnel, 63
  rotary, 63
  indirect, 63
  spray, 63
Drying, 46, 48-49, 63
Ducts, 70-71

Economic considerations, 149-181
Economizers, 71
Effusion, 50
Electrodialysis, 50, 100-101, 344-345
Electrolysis, 102, 344-345
Electronics industry:
  case studies, 356-358
  good housekeeping, 356
  pollution prevention options, 355-356
  process description, 351-354
  process modifications, 355-356
  recycling and resource recovery, 356

Electronics industry (continued):
  waste description, 354-355
Electrostatic precipitator, 90
Emergency Planning and Community Right to Know Act (EPCRA), 190-191
Endothermic process, 33
Energy conservation, 217-224
English engineering units, 5, 416-419
Enthalpy, 17-19, 28-29, 33
Environmental accounting, 151-160
Environmental justice, 249-251
Environmental Justice through Pollution Prevention (EJP2), 207
Environmental management system (EMS), 243-245
Environmental organizations, 451-454
Enzyme treatment, 109
EPA:
  pollution prevention policy statement, 195-196
  pollution prevention strategy, 192-195
Equilibrium distribution coefficients, 32-32
Equipment cost index, 161
Ethical considerations, 248-249
Evaporation, 46, 62-63, 301-302, 344
Event tree analysis, 140
Exothermic process, 33
Extraction, 63
Extremely hazardous substances (EHSs), 190-191

Fans, 74
Fault tree analysis, 139-140
Federal regulations, 188-195
Feedback loops, 80-82
Fermentation, 361, 362, 364
Filtration, 98, 302
First law of thermodynamics, 28-29
Fittings, 71-72
Flammability limit, 16-17
Flash point, 17
Flocculation, 96, 97
Flow sheets, 129-140
Fluid flow, 45
Fluidized bed incinerator, 116-117
Formulation, drug:
  direct compression, 364
  slugging, 364
  wet granulation, 364

Fractional sublimation, 48
Free energy, 34
Freeze crystallization, 96
Freezing point, 15

Gaseous control devices, 92-95
Good housekeeping practices:
  in the drug manufacturing
    industry, 369-370
  in the electronics industry, 356
  in the metal finishing industry, 345
  in the paint manufacturing industry,
    377, 383-384
  in the pesticide formulation industry,
    392, 393-394
  in the printing industry, 329, 333-334
  in the pulp and paper industry, 404-405
Graphic flow diagrams, 131
Green Lights program, 196, 213, 224
Gross heating value *see Higher heating value*

Hazardous Air Pollutants (HAPs), 190
Hazard risk assessment, 227-229
Health risk assessment, 229-231
Heat capacity, 14
Heat exchangers:
  direct contact, 60
  double pipe, 9
  evaporators, 62-63
  shell and tube, 59-60
  types, 59
Heat of combustion, 18-19
Heat of reaction, 17, 33
Heat transfer, 45
Henry's Law, 30-31
Heterogeneous reaction, 56-57
Higher heating value (HHV), 19
High pressure sprayer, 376, 391
Homogeneous reaction, 56
Humidification, 47-48
Humidity, 21-22
Hydrolysis, 102-103

Illustrative examples:
  ancillary processes and equipment,
    83-85
  conservation laws, 36-38
  definitions, 22-24
  economics, 174-177

Illustrative examples (continued):
  miscellaneous pollution prevention
    topics, 251-257
  plant equipment, 63-65
  pollution control to pollution
    prevention, 210-212
  pollution prevention opportunity
    assessment, 274-277
  process diagrams, 141-144
  recycling, 307-310
  source reduction, 287-290
  unit operations, 52-53
  units, 8, 416-419
  waste treatment processes, 118-123
Internal energy, 28-29
Ideal gas law, 29-30, 36-37
Ideal gas constant, values, 31
Instrumentation & controls, 80-82
Interlocks, 82
Ion exchange, 96, 305, 344-345
Incineration:
  principles, 112-114
  heat recovery, 114
  incinerator types, 114-118
Integrated waste management, 187-188
Insulation, 221-222
ISO 14000, 242-245
  certification, 243-244

Labor, 169
Lagoons, 107
Land disposal restrictions, 189-190
Leaching, 46, 49, 63
LEL or LFL, 16-17
Liability reduction, 271
Life cycle analysis, 150, 153, 160, 177,
    234-240
Life cycle assessment worksheet, 237-238, 239
Life cycle checklist, 237, 238
Lighting, 224-225
Liquid-liquid extraction, 46, 49, 63, 303

Mass transfer, 44, 46-52, 60-63
  equipment, 60-63
Material storage, 78-80
Material substitution, 366
Material transportation, 78-80
Metal finishing industry:
  case study, 347
  good housekeeping, 345-346

Metal finishing industry (continued):
  pollution prevention options, 337-347
  process description, 337, 338-339
  process modifications, 340-343
  recycling and resource recovery, 343-345, 346
  waste description, 337, 340
Mole fraction, 22, 23, 30-32
Molecular weight, 13
Moles, 13, 32-33
Motors, 225-226
Multimedia approach, 1, 231-234
Multiple hearth incinerator, 117-118

National Emission Standards for Hazardous Air Pollutants (NESHAPs), 190, 287
National Pollution Prevention Center for Higher Education (NPPC), 194-195, 208
Natural extraction, 363, 365
Net heating value (NHV), 19
Neutralization, 103

Osmosis, 51
Ovens and furnaces, 220-221
Oxidation, 104

Paint manufacturing:
  case studies, 377-384
  good housekeeping, 377
  pollution prevention options, 375-377
  process description, 373-374
  process modifications, 375-376
  recycling and resource recovery, 377
  waste description, 375
Partial condenser, 60
Partial pressure, 21-22
Particulate control devices, air, 89-92
Perturbation study, 166-167
Pesticide formulating industry:
  case studies, 392-394
  good housekeeping, 392
  pollution prevention options, 391-392
  process description, 387-390
  process modifications, 391-392
  recycling and resource recovery, 392
  waste description, 390-391
Pfaudler reactor, 362
pH, 21
Phase equilibrium, 30-32
Phase equilibrium coefficients, 31-32

Photolysis, 105
Physical properties, 13-16
Physical separation, 46, 51-52, 110-111
Pigs, 376, 391
Pipes and piping, 69-70
Piping and instrumentation diagrams (P&ID), 132-139
Plant and process design, 173-174
Plant equipment, 55-67
Plant siting and layout, 167-173
Pollution prevention:
  contacts in Higher Education, 457-492
  definition of, 3
  EPA's strategy, 192-195
  future of, 209
  impediments to, 272-274
  incentives for states (PPIS), 204-206
  incentives, 270-272
  industries 2000 project, 194
  information clearinghouse (PPIS), 195
  options for domestic activities, 247-248
  principles, 183-311
  promotion in educational institutions, 207-208
  promotion in federal agencies, 197-198
  promotion in industry, 198-204
  promotion in states and tribes, 204-207
  promotion through community organizations, 208
  resource exchange (P2Rx), 206
  software, 455-456
  state programs, 421-450
  33/50 program, 192-194, 199
Pollution Prevention Act of 1990, 1, 186-187, 191-192
Pollution Prevention industrial applications:
  commercial printing, 315-336
  drug manufacturing and processing industry, 361-372
  electronics industry, 351-359
  metal finishing industries, 337-349
  paint manufacturing industry, 373-386
  pesticide formulating industry, 387-395
Pollution prevention opportunity assessment, 261-279
  assessment, 264-267
  feasibility analysis, 267-268